PIERRE BRICHARD'S BOOK OF
CICHLIDS
AND ALL THE OTHER FISHES
OF LAKE TANGANYIKA

PIERRE BRICHARD'S BOOK OF
CICHLIDS
AND ALL THE OTHER FISHES
OF LAKE TANGANYIKA

Spathodus erythrodon

t.f.h.

Distributed in the UNITED STATES by T.F.H. Publications, Inc., One T.F.H. Plaza, Neptune City, NJ 07753; in CANADA to the Pet Trade by H & L Pet Supplies Inc., 27 Kingston Crescent, Kitchener, Ontario N2B 2T6; Rolf C. Hagen Ltd., 3225 Sartelon Street, Montreal 382 Quebec; in CANADA to the Book Trade by Macmillan of Canada (A Division of Canada Publishing Corporation), 164 Commander Boulevard, Agincourt, Ontario M1S 3C7; in ENGLAND by T.F.H. Publications Limited, Cliveden House/Priors Way/Bray, Maidenhead, Berkshire SL6 2HP, England; in AUSTRALIA AND THE SOUTH PACIFIC by T.F.H. (Australia) Pty. Ltd., Box 149, Brookvale 2100 N.S.W., Australia; in NEW ZEALAND by Ross Haines & Son, Ltd., 82 D Elizabeth Knox Place, Panmure, Auckland, New Zealand; in the PHILIPPINES by Bio-Research, 5 Lippay Street, San Lorenzo Village, Makati Rizal; in SOUTH AFRICA by Multipet Pty. Ltd., Box 235 New Germany, South Africa 3620. Published by T.F.H. Publications, Inc. Manufactured in the United States of America by T.F.H. Publications, Inc.

CONTENTS

To Mireille, my daughter and a keen biologist, who helped me since she was 12, to Jacky, my son-in-law, to Thierry, my son, who developed the photographic techniques so that fishes could be photographed 50 meters down as if they had been in a tank, to all who helped me, and especially Prof. M. Poll and Dr. D. Thys van den Audenaerde, this book is dedicated.

Pierre Brichard

FOREWORD

Lake Tanganyika is by no means just another African Great Lake or just another inland sea. Nowhere else in Africa, and as far as I know in the world, can we find as large and as deep a lake whose lifespan encompasses so many millions of years of uninterrupted and gradual evolution. For an immensely long period of time, during which other lakes dried out or became covered with ice, fishes and many other life-forms were able to live and develop adaptations to the ecological niches that were appearing in the various habitats of the lake. Sudden and dramatic changes of the environment were buffered by the sheer size, location, and isolation of the lake. Because the ancestral forms were not wiped out by dramatic changes in their environment, they could develop, under stable ecological conditions, a succession of adaptations leading to the present fish-flocks. This led to increasingly specialized forms.

The effects of this long-term evolution were compounded by several features:

1. The layout of the coastlines, i.e., alternating rock, sand, and mud, changed many times during the lake's long climb toward its present level. As a result, coastal species of fishes went through a succession of periods of isolation on their local grounds followed by periods during which their biotope expanded and they were able to mix with other populations of the same species or with closely related species, eventually to become isolated again later on. Genetic drift on a very large scale could thus be at work on the various fish stems much more than if the coastlines had been more permanent.

2. The fishes that developed in the lake stemmed from three main groups: (a) mouthbrooding cichlids, (b) nestbreeding cichlids, and (c) non-cichlids, among which the foremost are the Ostariophysi, especially the catfishes.

These three main groups developed in conflict and competition with each other. There are now a little bit less than 40% non-cichlid species, 20% nestbreeding species of cichlids, and about 40% mouthbrooding cichlids among the nearly 300 species

Pierre Brichard setting up a specially built tank in preparation for fish photography. Photo by Dr. Herbert R. Axelrod.

currently described as found in the lake basin. This is a unique situation among the African Rift Lakes—none of the other lakes harbors sizable populations (if any at all) of nestbreeding cichlids and non-cichlids when compared with mouthbrooding cichlids. Nowhere else can the adaptations of each group be as well studied as in Lake Tanganyika.

Lake Tanganyika is often given as an example of endemism, as more than 95% of the lake cichlids are not found anywhere else. What is less known is the fact that many coastal species are endemic to one half of the lake and are to be found either only in the southern part or only in the northern part. The list of species as per June, 1983, has reached a total of 277 species, some of which were discovered as late as March, 1983, with more awaiting early publication by the author. About a dozen more will be described.

Extensive research and new exploration in quest of the why's and how's of the fauna of this lake explain why it has taken more than six years to present the second edition of my book. The goal of the recent explorations and those that are already planned for the future is not so much the discovery of new species and local varieties, but to elucidate points in the vertical distribution of coastal species and their range along the coastlines.

The last three years have been devoted to exploring the northern basin and its approaches, namely the Ubwari peninsula, jutting out from the western coast of the lake for 40km toward the

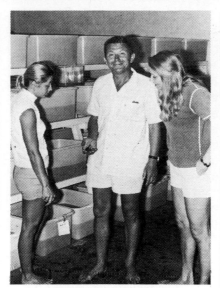

Pierre Brichard showing off a rare fish to his daughter Mireille and Mrs. Evelyn Axelrod.

eastern coast and leaving a gap 400m deep and 20km wide. The peninsula and this gap are very important features of the lake because they have been in the past (probably for several million years) the northernmost limit of the lake, one along which rock-dwelling species from the western coast and eastern coast could migrate and mix. The present distribution of rock-dwelling species north of the mouths of the Malagarazi and Lukuga Rivers cannot be understood if one does not keep in mind the significance of this fact.

It is along the Ubwari, Ngoma, and Ngombe escarpments that clues about the early distribution of several rock-dwelling species as well as in the northern basin about the chronology of the build-up of ecological barriers shall be found.

The first edition of this book has been widely used by biologists in their field studies. Thus special care has been devoted to updating the identification keys and making them more reliable and easy to use. They do not, however, replace the original descriptions to which one will often have to refer when in doubt about an identification.

In a lake where local endemism is rampant and local varieties so many, proper identification of collecting sites is essential. Several maps as well as an index of localities have therefore been added.

During more than 1,000 hours personally spent in the lake making scuba dives, and that not counting the many additional hours spent by our team, countless data have been recorded on the distribution of fishes with respect to the layout of the underwater slope, as well as their breeding, feeding, and fear behaviors. These are dealt with in special chapters.

Hobbyists also have welcomed the first edition, and information on the habits of their pets in the lake, as well as on the advisability of trying to secure a particular species, is spread throughout this book. A list of compatible species suited to various levels of aquaristic experience has been compiled, as well as information about how the fishes are collected.

People often ask: "Are there many more species remaining to be discovered in the lake?" Disregarding local races and geographical varieties I shall try to give a conservative estimate. One might say that as yet less than 500km of coastline out of 2,500km at least have been explored from the surface down to 15 meters. A thousand kilometers, mainly on the eastern coast, have been spot-checked here and there. So have the wide-open sandy floors, a few major affluent rivers, and the first third of the Lukuga outlet. Thus 1,000km of coastline, essentially on the western coast, the deep slopes everywhere, and many affluents remain virgin territory. Given the vertical stratification of many species and the narrow range of many others, I wouldn't be surprised if in the long run the lake basin were to contain more than 500 species, of which over 400 will be found within the lake itself. Not a few will be sand-dwelling fishes, especially mouthbrooding cichlids, but there will also be nestbreeders. The major harvest, however, will come from the rocky areas and below 20m, with nestbreeders predominating.

How deep will they be found and where around the lake? If pelagic species are capable of ranging from the surface down to 200m and can get used to rapid changes in oxygen levels, then sedentary species living like a *Julidochromis* permanently in a rock anfractuosity could progressively adapt to the lowest bearable oxygen level. This would mean that, provided shelters are available and not silted over, we should find some species living deeper than 100m.

The steepest bottoms are to be found along the Marungu, Ngoma, Ubwari, and Kungwe mountain ranges. There the slopes are very steep indeed, but aeration by strong currents is good and silt has not accumulated.

Out of the 400 species that I expect to be one day registered from the lake, probably less than 100 will be non-cichlids. Cichlids should thus involve more than 300 species, of which at least 100 will be nestbreeders. Because ubiquitous fishes are the most commonly caught and the least specialized, most have already been discovered. The bulk of the new discoveries will probably consist of highly specialized species with unusual adaptations.

Our knowledge of the fishes, of their ecology and behavior for example, will depend on the use of much more sophisticated gear than the type we are using today. One day perhaps scientists will use deep-diving vehicles to explore the dark recesses over 300 meters down. Until then this book will have filled its purpose if scientist and hobbyist alike have shared the thrill that was mine as I stood looking at the fishes around me in their home waters.

The major harvest of new species will undoubtedly come from the rocky shores, especially in deeper water. Collecting in terrain such as this is very difficult. Photo by Dr. Herbert R. Axelrod.

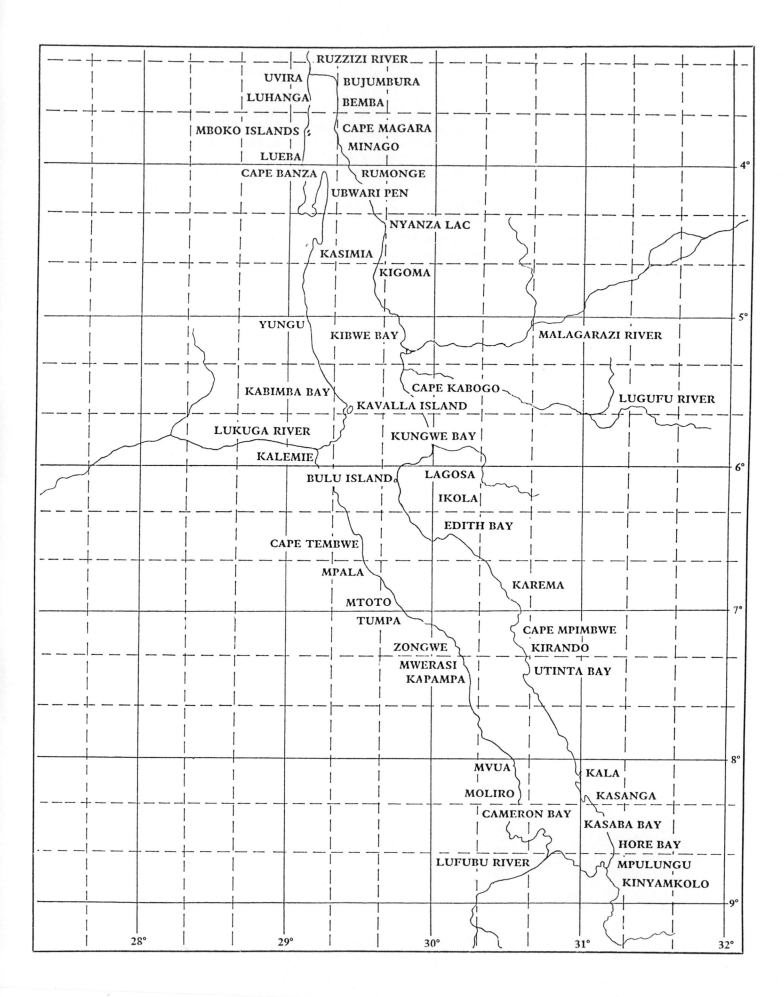

RUZZIZI RIVER

UVIRA   BUJUMBURA

LUHANGA   BEMBA

MBOKO ISLANDS   CAPE MAGARA

MINAGO

LUEBA

CAPE BANZA   RUMONGE

UBWARI PEN

NYANZA LAC

KASIMIA

KIGOMA

YUNGU   KIBWE BAY   MALAGARAZI RIVER

CAPE KABOGO   LUGUFU RIVER

KABIMBA BAY   KAVALLA ISLAND

LUKUGA RIVER   KUNGWE BAY

KALEMIE

BULU ISLAND   LAGOSA

IKOLA

EDITH BAY

CAPE TEMBWE

MPALA   KAREMA

MTOTO

TUMPA

CAPE MPIMBWE

KIRANDO

ZONGWE

MWERASI   UTINTA BAY

KAPAMPA

MVUA   KALA

MOLIRO   KASANGA

CAMERON BAY   KASABA BAY

HORE BAY

LUFUBU RIVER   MPULUNGU

KINYAMKOLO

4°

5°

6°

7°

8°

9°

28°   29°   30°   31°   32°

12

CHAPTER I

LAKE TANGANYIKA: GEOGRAPHY, HISTORY, CLIMATE, PROPERTIES OF THE WATER

Lake Tanganyika spreads its blue waters between 3°20'S and 8°45'S latitude and between 29°E and 31°E longitude. It is thus elongate, the longest dimension on an overall north-south axis. It fills a deep crevice in the continental crust in the middle of the African Rift Valley which extends from the Danakil depression on the Red Sea coastline to Mozambique, thus well over 5,000km.

The lake canyon extends for more than 750km, of which 50 to 60km in the northern portion have been filled in recently by the alluvial deposits of the Lake Kivu outlet, the Ruzzizi River, emptying into Lake Tanganyika.

The lake averages a bit more than 40km in width, the northern basin being the narrowest with an average width of about 30km, the central and southern areas reaching a maximum width of a little more than 80km.

To illustrate these rather abstract figures let us say that, as the crow flies, the lake extends from Paris to Marseilles, or from Brussels to Berlin, or again from New York to Cleveland.

SHORELINE LAYOUT

The northern coast of Lake Tanganyika is made up entirely by the frontage of the Ruzzizi River delta, a low and level plain extending as swamps and sandy beaches for about 35km. The southern coast is lined by a massive escarpment, about 500 meters high, that separates the lake from the swampy Bangweolo, Moero, and Moero Wantipa Lakes, all of which belong to the Congo River basin. The western and eastern coastlines present a succession of mountain ridges rising to 2,500-3,000 meters in the northern part of the lake and to 1,500-2,000 meters in the southern part, the western mountain wall being the highest. Both coasts have shallower grounds around the central part of the lake. The western gap is narrow and provides the lake with an outlet, the Lukuga River, through which the lake overflow is evacuated toward the upper Congo River, a little more than 300km of winding gorges and falls. The eastern central plain is occupied by the Malagarazi River.

South of these plains the ground rises again, on the western coast toward the Marungu mountain range, on the eastern coast toward the Kungwe range, both in excess of 2,100 meters. The lake thus lies in a deep trench mostly lined with abrupt slopes. Given the narrowness of the lake, the underwater profile of its slopes and floors is thus a broad "U", and the gradient of the slopes is usually very steep.

There are three main basins, two of which are very deep. The southern basin, south of the Kungwe mountains, is 1,470 meters deep, while the central basin is 1,300 m deep. The northernmost basin, lying north of the Ubwari peninsula, is only 250 meters deep. The first two basins fringe the western coast, and depths of 1,000 meters have been recorded only 3km from the shore. On the eastern coast, as a result, the underwater slopes are more gentle and extensive shallows prevail south of the Kungwe range. The two deepest basins are separated by a mound whose

The Ruzizi River estuary and alluvial plain viewed from the hills of Burundi. Photo by Pierre Brichard.

archipelago on the western coast is the one located at approximately 5°30′S latitude. There is another archipelago south of the Kungwe on the eastern shore, as well as one in front of Mpulungu in the southeastern corner of the lake. Historical reports mention also that the Ubwari Peninsula was an island, about 60km long, perhaps as recently as 200 or 300 hundred years ago, as was the Katenga promontory just north of the Kavalla Archipelago (5°30′S).

Lake Tanganyika is now 34,000 sq km in area, or a bit more than the surface area of Belgium. The volume of water is 35,000 km³, or half the volume of the North Sea, while the surface of the lake basin covers 250,000 sq. km, or as much as the whole of Great Britain.

top culminates only 50 meters under the surface of the lake, although the mound does not extend across the full width of the lake.

The northern rim of the central basin is lined by a 1,800 meter high wall whose crest, 500 meters above the lake level, is now the Ubwari Peninsula, which extends from the western coast toward the eastern coast, leaving a gap 20km wide between its tip (Cape Banza) and the Rumonge area.

Clues that at one time, not so long ago, this gap was forded by typical rock-dwelling cichlids such as *Tropheus* and *Petrochromis* suggest that by then the northernmost boundary of the lake did not include the basin lying north of the peninsula. This basin is thus the last extension of the lake.

Orientation of the mountain ranges around the lake, the depth of its basin, and the gradient of the slopes help explain why there are few islands. The only small

A view of the northern beaches close to the Ruzizi River delta. Photo by Pierre Brichard.

GEOLOGICAL HISTORY

According to the latest data available (1981) Lake Tanganyika is the oldest lake in Africa and perhaps in the whole world, being formed during the Miocene about 20 million years ago. The rift in which the lake formed was not split open in a spectacular and sudden cataclysm, but developed by a very progressive sinking of the underlying ground. Counterpressures on the crevice edges resulted in mountain-building activity by which the ground around the lake started to rise and the present mountain ranges started to form. This activity was especially strong in the northern sector where the rift had cut through very ancient rock, as much as 2.5 billion years old. Volcanic activity is still manifest in several submerged thermal springs that our team discovered at Bemba and Cape Banza, and might, according to very recent discoveries, be presently on the rise.

In the southern half of the lake rocks are much younger,

The old-fashioned sail-rigged dugouts are seldom ever seen anymore. This native is maneuvering his boat over the rock-strewn shallows. Photo by Pierre Brichard.

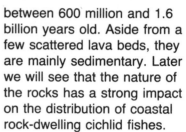

A beach composed of many small pebbles and some larger rocks separates the water's edge from the luxuriant bushes. *Eretmodus* can usually be found in shallow inshore waters such as this. Photo by Pierre Brichard.

between 600 million and 1.6 billion years old. Aside from a few scattered lava beds, they are mainly sedimentary. Later we will see that the nature of the rocks has a strong impact on the distribution of coastal rock-dwelling cichlid fishes.

The rift cut across the bed of ancient rivers, including the Malagarazi, that at that time had been flowing toward the upper Congo River. The rivers started to flow into the depression which they began to fill. Ever since that time the Malagarazi has been building its huge alluvial cone on the lake floor. This cone now reaches halfway toward the western slopes and has in all probability been a major barrier, with its silt and mud, to the passage of rock-dwelling fishes from the southeastern coast to the north.

If the central mound separating the two deep basins already existed, the two basins developed into separate entities. Their levels rose, stabilized, or sank according to the climate

prevailing in the area. The salinity of their respective waters was boosted by the input of salts brought in by rivers and by high levels of evaporation. The two lakes merged into one when the central mound was flanked by the rising waters. Perhaps at times of drought the level fell back and the two basins were separated again. Nobody can tell until drillings in the mile-thick sediments covering the floors of the two basins provide clues as to what happened.

Anyhow, the small size of the lake relative to the size of the basin—which was probably much the same size as it is today—increased the effects of the prevailing climates. The water level could probably rise or fall much more quickly than it can today, and by several meters a year.

Nobody knows when the lake reached its present level of 774-775 meters above sea level. Soundings near the mouths of mountain torrents revealed that for some time the lake stabilized at a level 550 meters below today's level. Deep submerged gullies (beds of ancient mountain torrents) cut into the underwater slopes and then peter out in the depths around 550m below the surface.

By that time the lake's northernmost boundary had risen halfway up the Ubwari wall. Afterwards, the lake invaded the northernmost basin, spilling over the rim of the Ubwari gap and extending northward for another 130 to 150 km until it reached another wall, the one that dammed the Lake Kivu area. Whether or not there had already been a small lake in the northernmost basin before Lake Tanganyika flooded the area is as yet

undocumented, but not impossible.

Afterwards, the Ruzzizi

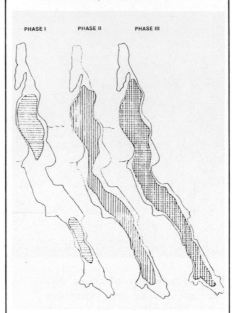

PHASE I PHASE II PHASE III

Three phases in the probable evolution of Lake Tanganyika. **Phase I:** Two separate lakes with a central mound 500 to 600m high. **Phase II:** The two lakes fuse as the lake depth increases to 700m. **Phase III:** The maximum lake depth increases to 900m (but still 550m below present level).

alluvial plain started to fill in the northern basin, a process that is even now under way. As a result, the northern basin probably lost between 50 and 60km of its length.

That at one time the Ruzzizi plain and swamps did not exist and prevent the migrations of rock-dwelling fishes, is born out by the fact that the eastern and western coastlines of the lake north of the Ubwari have very similar populations, and they are quite different from the ones found south of the Ubwari.

Some say, and it is a serious mistake, that the Lukuga outlet developed quite recently, in fact in 1878. They misread the original reports from the first

European explorers who went into the area. Stanley, for example, circumnavigated the lake in two successive explorations (1871-1876). He noted during the last one that the lake was on the rise, went to the Lukuga, and found it clogged by papyrus and floating islands of reed (which had drifted in from the flooded swamps around the lake). Beyond the obstruction he found the dried-out bed of the Lukuga, which shows that the river had drained the lake before. In 1878 the natural dam yielded to the tremendous pressure built up by the rising lake and the Lukuga resumed flowing. A similar obstruction occured in 1962-63. Again the lake rose to 778 meters before the obstruction was cleared by engineers. It took the lake several years to drop back to its present level.

No traces can be found on the hills around the lake indicating that the level ever crept higher than it is today by more than a few meters. On the other hand, nothing could prevent the lake level from falling lower than the outlet were a dry climatic period to prevail in the lake area, in which case evaporation would increase and would not be compensated for by rain. The lake would then again be an inland sea without any connection with another hydrographical basin.

The Lukuga drains off only a very small part of the 150cm of rain falling on the lake, to which rainfall over the lake basin should be added. The lake thus probably receives more than 200cm of rainfall a year. More than 95% of this is lost by evaporation. Thus in very wet years, when the input is high and evaporation is

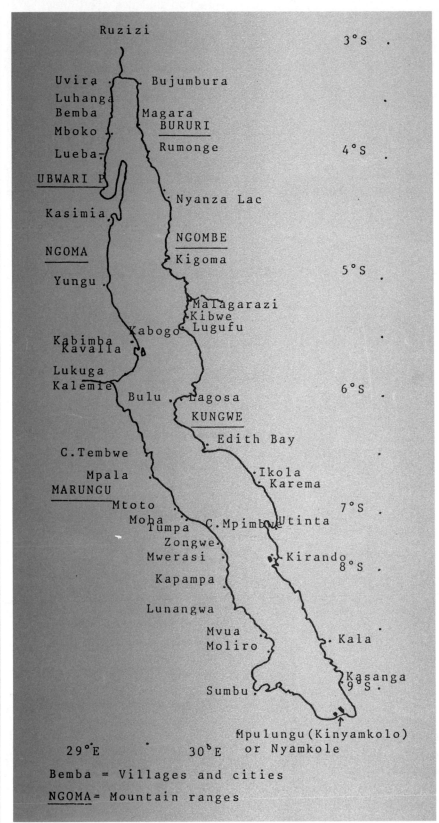

Ruzizi

Uvira .
Luhanga   . Bujumbura
Bemba   . Magara
Mboko .   BURURI
Lueba ̄   Rumonge

UBWARI P

          . Nyanza Lac
Kasimia .

NGOMA   NGOMBE
              . Kigoma

Yungu .

              . Malagarazi
              . Kibwe
Kabogo . Lugufu
Kabimba .
Kavalla .

Lukuga
Kalemie
    Bulu . . Lagosa
          KUNGWE
          . Edith Bay

C. Tembwe
    Mpala .   . Ikola
MARUNGU   . Karema
          Mtoto .
    Moha   C. Mpimbwe . Utinta
    Tumpa
    Zongwe .
Mwerasi .   .. Kirando
Kapampa .

Lunangwa

    Mvua
    Moliro .   . Kala

                   . Kasanga
Sumbu .

              Mpulungu (Kinyamkolo)
29° E   30° E   or Nyamkole

Bemba = Villages and cities
NGOMA = Mountain ranges

3° S .

4° S .

5° S .

6° S .

7° S .

8° S .

9° S .

A map of Lake Tanganyika showing the localities discussed in this book, as well as the major rivers and mountain ranges.

hampered by heavy cloud cover, the lake level is bound to rise. At the end of the rainy season the lake has risen by 50 to 70cm, but then it recedes again until September. As far as we can see the lake is now stabilized at around 774-775 meters.

To base a study of the distribution of the many endemic coastal species of fishes on the present layout of the various coastal habitats would be misleading because of the fluctuations of the lake levels in the past. Sand beaches, swamps, and rocky coasts alternate. In the past they shifted about, expanded and/or contracted according to the profile of the coastline at the height the level had reached during the lake buildup. Fishes that had adapted to the various biotopes would see their habitat expand, contract, or even disappear altogether.

To reconstruct how the forces at work acted on the shorelines is no idle speculation, as it throws light on the distribution of present day fishes.

The 1962-63 floods, which increased the lake level by a mere 4 meters, provided us with clues on long-past fluctuations and their effects on the littoral landscape. From my observations at the time I noted that as the lake flooded, low-lying land, beaches, and swamps disappeared under the surface of the expanding lake. Aquatic plants and reeds in the swamps and vegetation in the plains were uprooted or pieces were ripped off and drifted away and ground features were leveled by the surf as the lake crept far inland over low-lying ground. As soon as the water reached rising

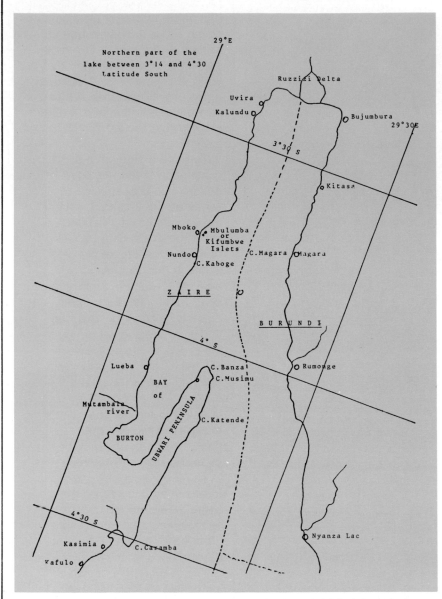

The northern part of Lake Tanganyika between 3°14' and 4°30' S latitude.

reached shallows where they settled, forming sandy bottoms. Were the lake to enter a valley, low-lying ground would disappear, but as soon as the lake reached the foot of the hills, erosion would again nibble the topsoil away to reveal the rock underneath. *Thus, when the lake was on the rise, the habitats of sand and mud lining beaches, coastal plains, and river estuaries would contract or disappear. Rocky coasts would expand and merge.*

As the lake level stabilized or dropped, flat plains, river estuaries, and sandy beaches lining the lowlands expanded. As dominant winds blew from the south, the particles drifting along the coastlines headed north and settled on the leeward side of obstacles. They eventually linked crescent-shaped sand bars with the coast lying north of river estuaries or promontories (Mutumba and Katibili) to form lagoons.

In the past, during the periods when the lake was on the rise, rocky coasts probably linked most of the lake perimeter in what we might call megarock biotopes. *When the water level remained stable, the materials brought in by erosion and alluvial material carried by mountain torrents expanded the sand beaches or built river deltas* with their sand and mud whenever the slope gradient and the depth of the bottom so permitted within the rocky biotopes. When the underwater slopes were too steep and led to deep floors, this process of fragmentation of the rock habitats was inhibited. The impact of these phenomena over such a long period of time was tremendous on the fish's capacity to move

ground, waves battered the shore, washed away the topsoil, and laid bare the hard rocky core of the hills. Rocks eroded or fragmented more or less according to their nature. Solid rock hardly split at all leaving huge cyclopean boulders lining the shore. Where softer rocks were more prevalent, they broke down and collapsed into rubble. The rubble disintegrated further crumbling into smaller rocks, stones, pebbles, gravel, and finally coarse sand. Sandstones fractured into slabs in the pounding surf that were returned to sand. Where the shore lined precipitous slopes, heavy material sank to the deep regions, but light particles and dust drifted with coastal currents until they

about and colonize new territories.

When the lake was on the rise and megarock biotopes prevailed coastal fishes that had become specialized as rock-dwellers saw their habitat expand and perhaps cover most of the lake perimeter except for a few major ecological barriers. Local populations could thus come in contact with each other and interbreeding would eventually occur. *But when the level remained stable, the fragmentation of the rock-biotopes increased with time* until rock-dwelling fishes became isolated on small stretches of coastline and could not move about over the sandy or muddy bottoms, and inbreeding was then the rule. Local races and species evolved during those times.

As the fluctuations of the water level probably occured many times in the past, fords between similar biotopes alternately opened and closed. Fishes that had developed local forms, races, and subspecies could come into contact again. Some could not interbreed anymore, so that species still very close genetically, but distinct, overlapped and came to live together in the same biotope on the same slope.

Thus the present distribution of many fishes in the lake, especially the rock-dwelling cichlids, arose from their specific mobility, but also, we might say, from the mobility of the boundaries of their habitat during the lake's geological history.

CLIMATE

Along the Equator rains are very heavy. In the northern

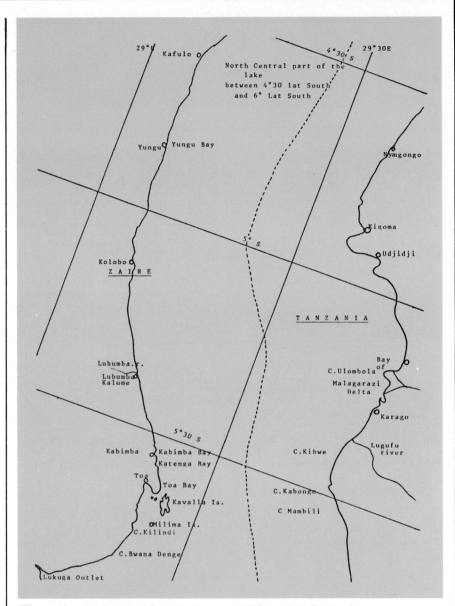

The north-central part of the lake between 4°30' and 6° S latitude.

region of the lake, which lies just south of the Equator, the rainy season starts early in October and lasts until the end of December. There is a lull from 4 to 8 weeks long in January-February, and they resume in earnest in March and last until the end of May. The dry season then settles in. As a result, rainfall in the northern part of the lake is spread over about eight months.

The southern coast (at about 9° S lat.) feels the impact of the Zambezi tropical climate. As a result, rains start early in December in earnest and wind up at the end of March. Rains are thus much more concentrated in the south, lasting only four months, but with a quantity of rainfall about the same as that up north.

There is an important lag in the start of rains, between 6 and 8 weeks, between the two

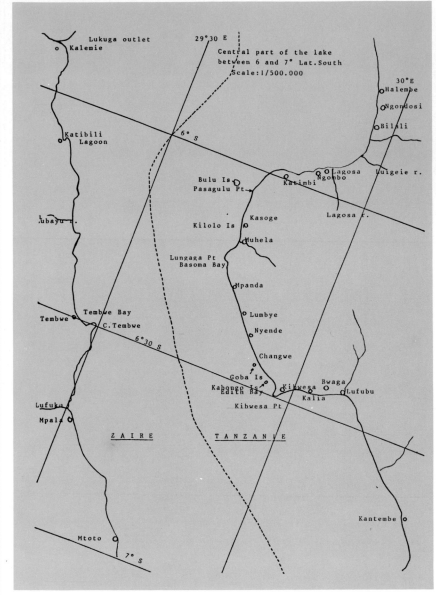

The central part of the lake between 6° and 7° S latitude.

ends of the lake. It takes that long for the sun, followed by its retinue of cloud formations, to cover the distance.

Rains swell mountain torrents which, as they cascade down the hill slopes, carry with them sediments stripped from their valleys and gullies. Mineral salts carried by the affluents act as fertilizers on the lake water for the phytoplankton bloom to follow.

The phytoplankton is fed upon mainly by crustaceans which make up the zooplankton. They, in turn, are the staple food of many fishes, pelagic and coastal alike.

The fact that the mineral salt input lasts from October to May in the northern part of the lake and only from December to March in the southern part, has had far-reaching results on the fishes. By mid-October the

phytoplankton bloom is in full swing in the northern part of the lake and as the affluent rivers in the central parts and then the southern parts of the lake start to swell and carry their alluvial sediments to the lake, the phytoplankton bloom spreads from north to south, followed by the zooplankton bloom and the fishes.

Pelagic fishes feeding on the plankton and their predators roam the lake in search of food, and at some periods of the year it has been found that one half of the lake was, for fishing purposes, practically devoid of schools of pelagic fishes. There are, then, seasonal migrations of pelagic fishes.

Coastal species feeding on drifting plankton need not migrate in search of their food. At all times, as might be expected with mineral salts fertilizing the coastal waters, plankton is more abundant along the coastal waters than in the open waters.

At the end of the dry season, plankton density drops as low as 3,000 per m^3 but reaches 350,000 per m^3 at its peak season. The northern basin has the highest density at all times for several reasons:

1. The northern basin is rather shallow and the salts, having sunk toward the deep regions, can be brought back toward the surface by a turnover of the water layers.

2. Regular rainfall feeds a very high number of springs, brooks, torrents, and more substantial rivers in the basin. There are more than 100 altogether, with 47 affluents reaching the lake in the Bay of Burton area, of which 13 are on the western coast of the Ubwari Peninsula coast alone.

Because of the plankton blooms, northern waters have less average transparency than the southern. Lateral visibility, often 15 meters in the central and southern areas with a minimum of five or six, seldom reaches 10 meters in the north, and often is curtailed to 1 or 2. As water transparency is essential for the sun's rays to penetrate the water layers, plant growth is inhibited in the northern sector and the vegetal carpet, known as biocover, grows in a shallower layer. Many fishes in the northern sector that depend on the biocover for their diet are thus restricted to a shallower vertical range (ex. *Tropheus, Petrochromis*). Thus, one can see that the differences in plankton densities between the northern and the southern ends of the lake can have an impact on the density ratios, respectively, of plankton-picking fishes and rock-grazers.

Rains, as they fall on the lake, are not distributed evenly. As they near the lake coastline, clouds are deflected sideways along the coastline by the huge thermal barrier over the lake. Rains are therefore three times heavier on the coasts than over the open waters of the lake.

Very often the clouds are funneled during their travels over the lake hinterland by ground features, such as valleys or mountain ridges. They very often appear on a very short stretch of coastline, which then receives much more rainfall than adjoining areas. There are such rainy areas south of Cape Chipimbi in the southern part and 30km south of Bujumbura in the northern part of the lake.

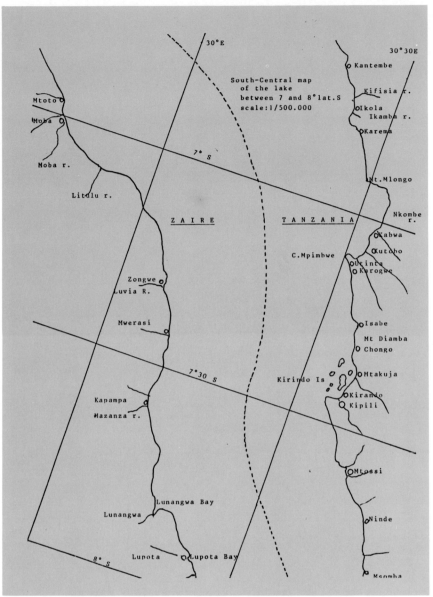

The south-central part of the lake between 7° and 8° S latitude.

PREVAILING WINDS

Dominant winds blowing from the southeast push the moisture born from the lake evaporation against the western ridges. As the moist air rises along the slopes it cools down and condenses as clouds, eventually to fall back again as rain and feed the countless small affluents. In the northern half of the lake rainfall is higher on the western coast than on the eastern.

Aside from their effect on rainfall and the one we have already discussed on the formation of coastal lagoons, the prevailing winds from the southeast have had a major impact on the distribution of the three types of coastal habitats because of the N-S bearing of the lake axis and the localization of the deep

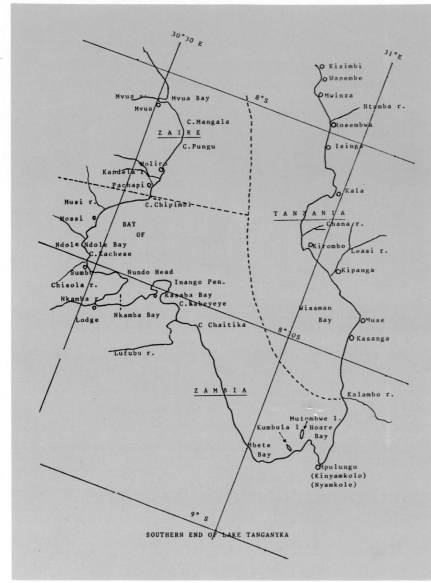

The southern end of Lake Tanganyika between 8° and 8°45′ S latitude.

can be said to be mainly of eolian origin.

North of Kalemie and the Lukuga outlet the mountain ridges of the Ngoma line the central deep basin on the same N-S axis and again prevailing winds erode the shoreline. Under the battering of waves fine silt is carried northward until it settles on top of the isthmus separating the Ubwari peninsula from the western mainland. Whether or not the precipitous coastline of Ngoma extended at one time toward the northwestern corner of the lake to Uvira and the Ruzzizi delta remains to be seen. It is entirely possible that at one time the isthmus of the Ubwari, now a very shallow swampy plain, did not exist. The bottom of Kasimia Bay, as far as I could check, is rock, silted over with sand.

The impact of winds and waves can also be assessed on a much smaller scale at the Inango peninsula between Kasaba Bay and Nkamba Bay in the south. East of Kasaba Bay the Lufubu River drops from the escarpment. The sediments carried along have filled the lower course of the river and as soon as they reach the lake are deflected westward by prevailing winds. They settle on the shallow floor of Kasaba Bay and have built the large beach on the eastern side of the peninsula. Very probably not so long ago this peninsula was an island and the two bays were merged into a single entity.

A clue to the past history of the area is provided by the partition of *Tropheus* races in the two bays. To the east we find the rainbow *Tropheus*, which in Kasaba Bay is represented by the

basins close to the western shores.

The western coastline has three major coastal biotopes: The southwestern biotope extending from Cameron Bay to a little north of Moba more than 300 km away, is entirely made of rock, uninterrupted by river deltas or sandy shallows and beaches. The coastline, battered daily by waves, has had its topsoil washed away,

with the heavy particles sinking toward the deep regions of the lake. Next, the lighter particles were pushed northward by prevailing currents and settled in the lee of the coastline around Cape Tembwe.

The coastline from Tembwe north, for more than 150 km, was heavily silted over by these sediments and the expansive shallow sand bottoms in front of the Lukuga

Some stretches of coastline are lined with reedy growth, providing shelter for certain species and/or their young. Photo by Hans-Joachim Richter.

The point at Bemba

Bemba, looking toward Mako

Mbulumba à Mboko islet.

Beach at Mboko

Highest peak on the northwestern coast, 3000 m (9,900 ft.)

Cape Muzimu, Ubwari peninsula.

Virgin forest at the edge of Kavalla.

Cape Baiza and Muzimu, Ubwari. Photos by Pierre Brichard.

Bay of Kasimia at Ubwari.

Ngoma, west coast, 5°30′ S.

Kabimba islet, 5°40′ S.

Bay of Kabimba.

Swamp in virgin forest on Kavalla Island.

Bay of Moba, Marungu.

Estuary forming an ecological barrier.

Fleet fishing at Kitasa for herring. Photos by Pierre Brichard.

The shoreline of Lake Tanganyika near the town of Kigoma (Tanzania) as seen from the window of an airplane. Photo by Glen S. Axelrod.

Flying over Kigoma. The greenish inshore water indicates a high photosynthetic area. Photo by Glen S. Axelrod.

"Kabeyeye" color morph. At one time they could travel freely over rocky areas to colonize Inango Island and part of Kamba Bay. There they encountered another form of *Tropheus* coming from the West, the form "with the double V on the nose," and the two races overlapped in the central part of the Bay. Afterward the "Kabeyeye" population of *Tropheus* was cut off from the Kasaba Bay and Inango main population by the silting of the area between the island and the mainland.

Thus, prevailing winds have had far-reaching effects on the coastlines, especially on the western shores by alternately shaping megarock biotopes or, to the contrary, building intercalate ecological barriers of sand and mud.

WATER ISOTHERMY

Lake Tanganyika is one of the few deep tropical lakes in which water is not stratified into very distinct temperature layers. In open waters temperatures close to the surface are on average about 26.5°C. There is then a sudden drop between 40 meters (in the north) and 100 meters depth (in the center and the south) to 24°C. Deeper yet the water is still around 23.5°C until by 1,000 meters depth we record a temperature of 23.3°C and on the deepest floors even a bit more. So the temperature drop through the entire depth of more than 1,300 meters is only about 3.2°C, which is very little indeed.

It has been said that slow convection currents might bring cooler water toward the top layer, and that these currents might have arisen from the constant motion given to surface waters by prevailing winds. Such currents, sometimes powerful, have been recorded by our team at the tip of promontories and they were always headed north.

One has to remember as well that the lake grew over a deep rift in the Earth's crust and that the central magma should not be far off. In fact, hot spots have been found recently under the northern floor. Volcanic activity has not stopped and heat could very well radiate through the mile-thick sediments lying on the lake floor to heat the deepest water layers.

If, as a rule, the lake is isothermic it doesn't mean that local temperature variations cannot be found. In shallow and protected coves temperatures can reach 29.5° C, and underwater thermal springs can locally raise the temperature by several degrees. At Cape Banza (tip of the Ubwari) the water gushing out from vents 5-6cm across is at the boiling point.

The dry season in the southern part of the lake is a time of cold weather. July is especially cold on the high plateaus surrounding the southern shores and temperatures can drop to the freezing point at night. As a result shallow coastal waters cool down by one or two degrees celsius, but not as much as in Lake Malawi where, as a result, rock-dwelling cichlids tend to live a bit deeper during the austral winter months (personal communication from Mr. Eric Fleet, a former fish collector in the area).

The 24°C thermocline between 40 and 100 meters deep is a very important ecological parameter for the

Even though this is a lake, there is a surf zone created by prevailing winds. Some species prefer this turbulent water that provides them with high oxygen levels and some protection. Photo by Glen S. Axelrod.

The great cliffs called "Mgu wa Tembo," which in Swahili means "Elephant's Foot," at the entrance to Kigoma Bay. Photo by Glen S. Axelrod.

fishes. Many coastal species are territorial and don't move about much. Many spend their lives in an area barely a few meters across. Others, like rock-grazing cichlids, because of their feeding habits remain in the top layers. As a result, those fishes living above the thermocline have adapted to a temperature at or above 26°C; those living below it are adapted to 24°C. None has ever experienced, except for a short period of time during which rainfall sinks toward the deeper areas, any cold spell.

This is why the lake cichlids are the most difficult to ship abroad—because they are not accustomed to dealing with drastic temperature changes.

The thermocline has other effects that we were able to ascertain during deep dives. From the surface to about 40 meters (in the northern part of the lake) plankton can be very dense and visibility curtailed to a few yards. Within a few meters and quite suddenly plankton disappears, the water becomes crystal-clear, and one can actually feel it cool down.

A rocky beach can be seen along this coastline, which is about a mile from the southern escarpment. Photo by Pierre Brichard.

The mountain slopes seen along this shore continue beneath the lake's surface at about the same angle. Photo by Pierre Brichard.

Several other features make the layer beneath the thermocline a different habitat from the one we investigated in the upper layer. First and foremost is the drop in light intensity. By 40 meters (a bit deeper in the south) the light becomes very subdued, contrasts are muted, and everything looks leaden gray.

In this twilight fishes move about as off-color shapes. To survive they had to develop adaptations entirely different from the ones developed by fishes living on the brightly-lit slopes near the surface. No wonder our dives in the area close to or below the thermocline increasingly bring about the discovery of species that had not been identified before.

PHYSICAL AND CHEMICAL PROPERTIES OF THE WATER
The lake water is very mineralized. Among the major Rift Valley lakes it has the highest concentration of mineral salts, double that found in Lake Malawi, triple that of Lake Victoria. Three

quarters of the total concentration is made up of carbonates—sodium, calcium, and magnesium (appr. 300mg/L out of 400mg), the pH is about 9.5, with slight local variations, the German hardness is about 12-14, and the conductivity is about 500-600 mS. As such the lake is very different from a tropical African river habitat—usually acid, very soft, and with a very low conductivity. This helps explain why some African fish families didn't colonize the lake when they had the opportunity. On the other hand, it helps explain why the lake fishes are restricted to the lacustrine biotopes and didn't migrate toward the Congo River once the Lukuga outlet opened up.

The high concentrations of carbonates had a very unusual impact on the coastal rock biotopes. They become concentrated in a kind of calcite cake on the rocks from 6 to 8 meters down to a depth of approximately 25 meters. Acting on the rubble as a cement, this material welds all rocks, small and large, together, prevents rock slides

A virgin forest lines this boulder and rubble shore. This area teems with wildlife, with lions, leopards, elephants, and antelopes inhabiting the forest and rock-grazing fishes and crocodiles in the water. Photo by Pierre Brichard.

on steep bottoms, and prevents holes and cavities from caving in or from being filled with silt and sand. Labyrinths and caves, overhanging roofs, and pinnacles of rock in which rock-dwelling fishes from the giant catfishes to the tiniest species or fry from other species can find shelter are thus multiplied

Without the calcite cakes many ecological niches would be missing from the underwater scenery, and one might say it is one of the most important components of a rocky habitat in the lake, one that is missing from the other African Lakes.

Sometimes the shoreline is composed of larger rocks with no area that could be considered a beach. Photo by Glen S. Axelrod.

Calcite formations have patterns to their shapes. Rocky walls facing upward, toward the surface, have a thin, smooth crust of calcite, while vertical walls are often strongly ribbed or display sharp-pointed concretions sometimes as much as 15cm high, not unlike those of branched coral. Sometimes tiny knobby pinnacles 10mm high grow on flat surfaces. Whether or not calcite formations are built by encrusting algae, why there are patterns and a set of shapes, why the growth is not at random, and whether or not the calcite is formed only by chemical reaction, have not been investigated since we discovered the phenomenon back in 1971.

This shoreline appears to be less rocky, but appearances may be deceiving: the scenery may change drastically as one looks below the surface. Photo by Pierre Brichard.

Below 25 meters calcite appears to be eroded and to "melt" away—again we are at a loss to understand why.

CHEMICAL ANALYSIS OF THE WATER

CHEMICAL ANALYSIS OF LAKE TANGANYIKA WATER
(Through the courtesy of Dr. Kuferath, member of the 1946-1947 Belgian Hydrobiological Mission.)

Salt	*Mg/l*
Na_2CO_3, anhydrous	125
KCl	59
KNO_3	0.5
Li_2CO_3	4
$CaCO_3$	30
$MgCO_3$	144
$Al_2(SO_4)_3 \cdot 18H_2O$	5
K_2SO_4	4
Na_2SO_4	1
$FeCl_3 \cdot 6H_2O$	0.5
$Na_3PO_4 \cdot 12H_2O$	0.4
Na_2SiO_3	13.5

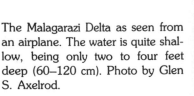

The Malagarazi Delta as seen from an airplane. The water is quite shallow, being only two to four feet deep (60–120 cm). Photo by Glen S. Axelrod.

A pebble beach leading to a patch of reed growth.

A very steep shoreline with large boulders.

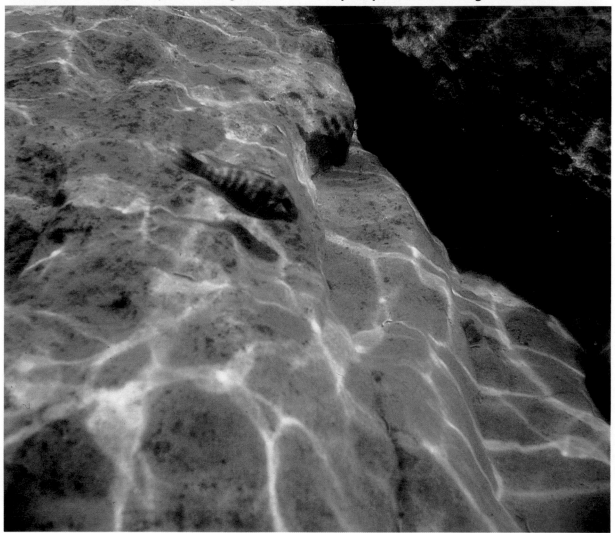

Below the surface a cichlid makes a bee-line to shelter as the diver approaches. Photos on this page by Hans-Joachim Richter.

Silica is mainly brought into the lake by affluents in the northern half of the lake and has a strong impact on the ecological barriers made up of sand beaches, because it combines with quartz to build extensive sandstone slabs along the beaches. The slabs, often fractured by pounding

constitutes a major problem with African river fishes, are unheard of in the lake. Fishes brought from rivers and which had been infected with Ich were cleared of the disease in a couple of days when they were put in a tank filled with lake water and without any additive.

The Zaire mountains can be seen across Lake Tanganyika. Photo by Dr. Herbert R. Axelrod from the Bujumbura, Burundi, shore.

surf, play host to a variety of rock-dwelling fishes during their migrations across sand barriers from one rock biotope to the next.

The combination of the various salts in the lake water has some unexpected results with respect to the treatment and packing of the fishes intended for the aquarium trade. It was found to inhibit the growth of the green algae that occurs in tanks when the lighting is too strong. It buffers the effects of tranquilizers used when shipping fishes.

Doses that would have killed a river cichlid have no effect whatsoever on lake fishes packed in lake water. Copper sulfate saturated solutions can also be used at a concentration of several cc's per 50 liters of water without any harmful effects on the fishes.

On the other hand, several drugs used commonly in the trade to cure fishes from wounds and abrasions were found to be useless.

Parasitic diseases such as the deadly Ich, which

OXYGEN LEVELS

If oxygen levels close to the surface of the lake are saturated, they still remain quite high down to 40 meters in the northern portion and 100 meters in the southern portion of the lake. They then start to drop drastically so that below 250 meters at most there is not enough oxygen left to support life. Thus, from 250 meters down the lake is devoid of any life except for anaerobic bacteria. These deep layers are for fishes a major horizontal barrier.

A contrast of habitats from different parts of Africa. This is a scene in West Africa where *Hemichromis* occurs. All photos on this page by Hans-Joachim Richter.

An area of rapids in West Africa where *Pelvicachromis* may be found.

The typical rocky shore of Lake Tanganyika.

A scenic view taken from the top of the Congo-Nile divide, about 8,000 feet (2,400 m) high. Photo by Pierre Brichard.

In the past as in present days this barrier prevented fishes from using rock formation as stepping stones and fords to cross sand and mud bottoms and slopes during the lake rise, once they had sunk beneath the oxygen-rich layers.

As the lake crept along the hills it lost old biotopes at the same pace it gained new footholds.

The oxygen-bearing layer, although shallow by the lake standard, is anyhow much deeper than any African river and has provided many fishes with an opportunity to become semibenthic in their habits.

WATER TRANSPARENCY

We have seen the impact of crystal-clear water quality on the growth of biocover. Though it is quite exceptional overall, transparency can be curtailed by a number of factors as in the northern basin. Rivers and mountain torrents can have a strong impact when they reach the lake, not so much because the quality of their water is very different from that of the lake, but because they act on the local ecology through the silt they carry, particularly when they are swollen by seasonal rains.

Usually river water cascading down from the hill crests is very cool. As it reaches the lake the river water doesn't mix with it but heads straight toward deep water, following the contours of the submerged slope. In front of a brook laden with clay, especially kaolin, a diver will see the bottom features veiled in a thick fog only 20 to 25 cm high gliding down toward the abyss.

After a storm, the rocks close to a river mouth are covered with sediments and typical rock-dwelling cichlids, mainly biocover-grazers, disappear for a while. They are replaced by typical sand-dwellers, such as *Callochromis* and *Xenotilapia*, until a few days later, when, under the pounding of the surf, the biocover is laid bare again and the silt settles in sandy patches on the bottom.

Large rivers meandering through low-lying lands before reaching the lake have time to warm up and their waters, instead of sinking toward the deeps when they enter the lake, glide along the surface for miles. The murky waters of the Malagarazi River, for example, still float 10km offshore in front of its delta. The brown waters of the Ruzzizi extend for miles around its estuary, but 30 meters down the lake water is crystal-clear.

CONNECTION WITH THE CONGO RIVER

Very misleading theories have been brought into the open about the relationship of the lake fauna to the neighboring basins. It has been said that the lake during most of its history was linked with the Nile basin and that only recently (as we have seen when dealing with the Lukuga outlet) was the lake captured by the Congo River basin.

No evidence of this theory can be found as yet among the lake fish flocks, except for a few species of cichlids living along the northernmost swamps (like *Tilapia nilotica*) and which had probably entered the lake when the area around Lake Kivu was captured by Lake Tanganyika a few thousand years back.

Strong evidence can be found that the lake had been linked with the Congo basin a long time ago. That the Congo River affluents and their fish fauna played a very important part in seeding the lake with the ancestors of existing species is borne out by the number of the latter that are directly related to or identical with typical Congolese species.

Among primitive fishes we find two *Polypterus* (*P. ornatipinnis* and *P. endlicheri congicum*). We also find more than 50 known species (with many more left to be discovered) of *Lamprologus*-related cichlids. They are so well diversified that their evolution couldn't possibly have been brought about by the recent invasion of the lake by Congolese forebearers.

One might object on the

The 10,000 foot-high (3,000 m) ridge of the western wall rises across the lake in Zaire 25 miles (15 km) distant. Photo by Pierre Brichard from the Congo-Nile divide.

basis that many more lake cichlids belong to *Tilapia* and *Haplochromis* stems that abound in the upper Nile basin, and that they, too, became diversified in such a way that their colonization of the lake waters must be very old. These two lineages are not typical of the Nile basin. There are, however, many *Tilapia* and *Haplochromis* species in the Congo and Zambezi River basins and the lake could well have been seeded with species of *Lamprologus, Tilapia*, and *Haplochromis* living in the Congo River affluents.

More than 100 species of fishes belonging to families other than cichlids have also colonized the lake basin, of which more than 50 live in the lake proper. Some of them belong to species that are found all over central Africa, and therefore not providing clues about their origin. A few are Congolese, and given the lake size it is difficult to see how they could possibly have settled all around the lake had they only been able to reach the lake a little more than 100 years ago.

In sharp contrast to the other African Rift lakes, Lake Tanganyika could well have been seeded at one time or another with fishes living in three main basins, the Nile, the Congo and the Zambezi, but undoubtedly it is the Congo River that had the main impact; this happened a long time ago, not just in modern times.

CONCLUSIONS

To summarize this chapter we have to keep in mind:
1. The lake is very large, very deep, very old, and has been a stable entity for most of its lifespan.
2. The layout and position of the lake just south of the Equator had considerable influence on the climate, the development of shorelines, and by the progressive seasonal spread of the plankton bloom on the distribution of the main fish flocks.
3. Present coastlines do not reflect the fluctuations and localization of the various types of habitats in the past.
4. Only the top 250 meters of the water column is capable of sustaining life. Dead

A contrasting view of the rocky coast of Lake Malawi. The rock-dwelling fishes in Malawi are called mbuna by the natives. Photo by Dr. Herbert R. Axelrod.

An aerial view of Cape Maclear, Lake Malawi, about 16 kilometers from the famous Monkey Bay. Photo by Dr. Herbert R. Axelrod.

water below this acts as a major ecological barrier preventing coastal fishes with strong adaptations to a given type of biotope from moving around the lake freely.

5. The salinity of the lake water is high. Because of this feature it is difficult for a river fish to settle in the lake and for a lake species to migrate through the Lukuga to the Congo basin rivers. The fish species that developed in the lake lived within an isolated basin.

6. When the lake overflowed in the Lukuga area is not known, but it certainly was not in very recent times.

7. The ancestors from which the lake species evolved probably came from the Congo basin.

8. Two sand or mud ecological barriers, on the eastern and western coasts, separate the rock biotopes lying north and south of the barriers.

With the fluctuations of the depth of this lake, these rocks might be future rocky habitats for fishes or may have been such in the past. Photo by Pierre Brichard.

CHAPTER II

LIFE IN THE LAKE OTHER THAN FISHES

Lake Tanganyika is an incredibly rich reservoir of plant and animal life, mostly endemic, and unequaled in the world save for perhaps Lake Baikal in Siberia. We have already seen why—an unbroken evolution spanning countless millions of years during which life could develop and adapt and fit into the many ecological niches.

The impact of these lifeforms on fishes was enormous and it is because the living organisms were so many and so varied that the fishes could occupy the lake biotopes as they did. Just to give one example: dwarf species of nestbreeding cichlids couldn't have colonized the barren lake floor had it not been for the empty snail shells of *Neothauma* in which they could find shelter and a breeding place. Did these fish perhaps become miniaturized because of the shells?

Because of this impact, and before looking at the fishes themselves, we must enter their world, dwell there at some length, and gain understanding about the lifeforms with which they share their habitat, much moreso than we did in my other book.

PLANKTON

Minerals flushed into the lake during the rainy season, brought up from the deep layers by currents or seeping in from the lake basin through rock and the water table, act as fertilizers for the algal bloom. Phytoplankton provides food for the invertebrates making up the zooplankton which in turn provide food for

Empty *Neothauma* shells are utilized by some of the smaller fishes as a refuge in sandy areas. Photo by Hans-Joachim Richter.

many of the fishes and other aquatic animals.

Bacteria feed on the organic wastes sinking slowly to the lake bottom. So do deep-living crustaceans which also provide food for benthic fishes.

According to Prof. Van Meel, the plankton specialist who participated in the 1946-1947 Belgian Mission, the density of plankton shows seasonal variations from a low of 3,000 organisms per ton of water during the dry season to a high of 350,000 organisms a few weeks after the start of the first rains in the area. The highest densities usually occur closest to the shore where the mineral input is the highest. This means at least a 120 fold increase occurs over only a few weeks time. There are

Occasionally a red plankton tide turns the normally blue lake waters a deep rusty red. This particular event lasted several days and covered a wide area.

even higher concentrations yet in the northernmost basin where the narrowness and shallowness of the basin compound the effects of the high river input and the effects of prevailing winds increase the drifts toward the basin.

By 1954 Prof. Van Meel had identified more than 240 species of phytoplankton alone (Lake Malawi has about 340 species and Lake Victoria about 370). Chlorophyta (green algae) accounted for about 90 species (Lake Malawi has 122, Lake Victoria 234), diatoms about 103 (Lake Malawi has 178, Lake Victoria 85), and Cyanophyta (blue-green algae) about 54 (Lake Malawi 29, Lake Victoria 39). One might conclude from these figures that Lakes Tanganyika and Malawi have half the variety of green algae as Lake Victoria, which might explain the green color and lack of visibility of the water in Lake Victoria. The ratio between the three types of

phytoplankton is more balanced in Lake Tanganyika (4:4:2) than in the two other lakes (5:7:1 and 10:3:2 respectively).

Zooplankton involves about 100 species of invertebrates, among which are included copepods, shrimp, etc. There are about a dozen species of daphnia, 34 species of copepods, and about a dozen shrimp.

Pelagic fishes feed on clouds of zooplankton drifting in the open waters. If, in 1972, 12,000 tons of pelagic fishes, most of them clupeids that had been fished in the eastern half of the northern basin, were brought to the central market in Bujumbura, it tells much about the abundance of plankton available to feed these fishes.

In 1950, Prof. Max Poll said that 100,000 tons of pelagic fishes could be caught each year from the lake without depleting the natural stocks. Since then studies have put

the total even higher. They set the weight of pelagic fishes at all times in the lake at 2.5 million tons with an average of 30 kg (or perhaps even more) per hectare (2.4 acres). This is not especially high for a lake. Lake Edward, for example, fertilized almost entirely by hippo dung (there are 20,000 hippos in the lake!) is said to harbor 100kg of fishes per hectare. But given the size of Lake Tanganyika, this is enough to provide several million people with much needed animal proteins.

Coastal waters are bathed by the drifting pelagic plankton clouds as well. They bring in a constant, and in many places overabundant, supply of food to coastal species of fishes. This supply is complemented by the local production of invertebrates feeding on organic matter, which is more abundant in coastal waters as it is contributed to the shore biotopes by affluent rivers and the local algal growth on rocks. The amount of insects falling into the water and their aquatic larvae is also not at all negligible in coastal waters.

We have seen that plankton blooms explosively after the first rains. This does not mean that all the varieties of phytoplankton and invertebrates start multiplying at the same time. The green *flos aquae* appears very quickly after the start of the rainy season, but its bloom is followed during the rainy season by the bloom (much less spectacular) of other species. It might even be that some blooms, among them the red bloom which I saw only in the southern part of the lake at the end of the season, are local phenomena.

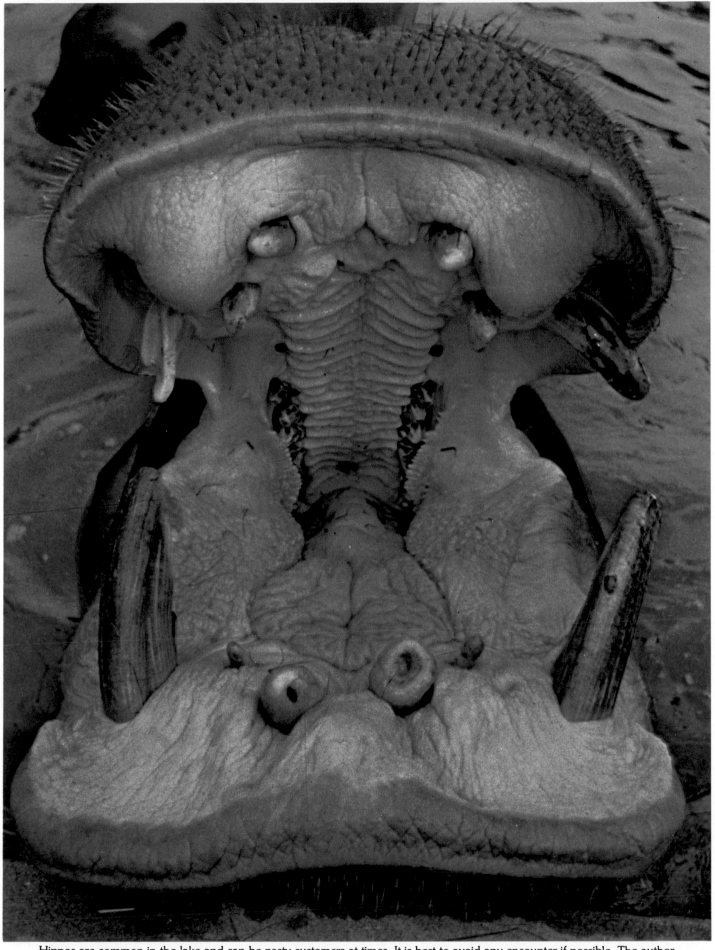

Hippos are common in the lake and can be nasty customers at times. It is best to avoid any encounter if possible. The author recommends diving to deep water if one has on SCUBA equipment.

Most of the time hippos can be seen wallowing in the lake waters or along its shore. They are excellent swimmers and can remain underwater for long periods of time but usually are seen with just their heads protruding.

With some varieties of phytoplankton and zooplankton starting to develop at a later time, it is not at all impossible that the clouds of plankton drifting in the lake do not show the same associations of vegetable and animal life throughout the whole year. Specific associations between phyto- and zooplankton are probable, but to my knowledge they have never been investigated.

Each species of planktonic animal has its own ecological niche. For example, there are light-shy shrimp hidden deep within the rock recesses or deep down on the bottom, while others are attracted by light. So are the copepods, "hopping" about in the clouds rising in the water column at night during their nyctemeral cycle of activity. They, in turn, are followed by benthic or pelagic fishes. As each organism has its specific requirements as to the amount of light, one might guess that the planktonic clouds are not all exactly the same, not only during the course of the seasons but also with respect to the depth at which they occur.

BIOCOVER

Much has been said and written about biocover and the fishes feeding on it, the typical rock-grazing cichlids foremost among them. It's funny, apparently nobody has gone to the trouble of determining exactly what grows and lives on the slippery mantle covering the rocks (or any other substrate for that matter). People say it is an assemblage of algae, bacteria, and invertebrates, the main feature being the algal carpet providing the fish with a field to

An underwater macrophotograph showing a biocover consisting entirely of a thin growth of brown algae. Photo by Pierre Brichard.

graze on, and leave it at that. It is not all that simple, because there are many algae involved, each with its own set of requirements concerning light and the substrate it can grow on, as well as different invertebrates, also with their own preferences.

Some rocks are totally devoid of any algae while rocks nearby are covered with a lush biocover. The nature of the rock, just like the nature of the soil for terrestrial plants, will boost or hinder the growth of algae in general or of some particular algae only.

The amount of light penetrating the water at a given depth is of course essential, again as it is with terrestrial plants. Some algae, such as the green filamentous algae, need strong sunlight, while others will thrive in more subdued light conditions, which usually means at deeper levels. This all means that we find a layering of the various species of algae and wide local variations in the thickness

and composition of biocover from one coastline to the next and sometimes between two stretches of slope only a few meters apart.

As we did with plankton, we should divide the biocover between its two major components, the plant (phyto-) and animal (zoo-) biocovers. In the top layer, close to the shore line, we will find many insect larvae (such as those of Ephemera—may flies) grazing on the biocover of green filamentous algae. These don't penetrate very deep, however, and a few meters down the assemblage of filamentous algae and insect larvae is replaced by encrusting algae and small crustaceans.

The magmatic or volcanic rocks of the northern part of the lake are mostly replaced in the deep southern part by sedimentary rocks and one would expect the composition of algae to be very different as they are successful or not successful in finding a foothold.

41

Many of the lake fishes are biocover feeders, grazing on the algae or selecting the smaller animals living in or on the algae. Photo by Glen S. Axelrod.

I have already discussed the calcite that covers the rocks as soon as we get a few meters deep. This has been colonized—if not built—by algae with low light requirements. Anyhow, the vegetal carpet is much thinner there than higher up on the slope, the algae finding a hold and sheltering in the tiny crannies between the minute calcite spires. The broad mouth of most rock-grazers cannot squeeze into the small recesses—no wonder they prefer the lush gardens of the top of the slope. But *Tropheus duboisi*, though a rock-grazer, does not like the topmost slope and prefers to live 10 meters deep. Could it be that this fish developed a preference for the type of biocover it finds at this depth and is able to utilize because its narrow mouth fits into the crannies?

Not only rock-grazing cichlids live on the biocover. Although the vegetal carpet has all but disappeared from the rocks, many rock-pecking species live on the tiny animals creeping or hidden on the rock surface.

On a coastline herbivorous species will be seen only in the top 15 meters, while microcarnivorous fishes will thrive much deeper. And, as the invertebrates are stratified and other ecological conditions change we will find a layering of fishes living on the zoo-biocover as well.

To grow in a rich biocover, algae need good anchorage. Rocks that roll about a lot, like on pebble beaches in the surf, do not permit a wide variety of algae to grow. Neither do areas where rocks are heavily silted over by sediments and dust. They are deserted by

most biocover feeding fishes, be they herbivores or carnivores.

The variations of biocover density help explain the variations in the density of the fishes living on it. Along with the variable input of drifting plankton along a coast, this explains why some places around the shores can harbor as little as 1 fish per 100 square meters of bottom area, or as much as 10 per square meter or even more, which means a variability of 1 to 1,000 in habitat occupation.

This is why travelers to the lake should avoid generalizing about local observations and reporting about the incredible density of *Tropheus*, for example, which they found in selected spots. They might be correct locally but misleading when extrapolating over a whole coastline and even moreso of course when dealing with the entire perimeter of the lake.

AQUATIC PLANTS

The average depth of the lake, the sharp U-shaped profile of its basin, and the predominance of rock habitats prevent extensive plant beds. Plants thrive in shallow and protected coves and of course river estuaries and coastal lagoons. Thus they have only a local impact on fishes.

Specialized plant eaters, such as *Sarotherodon*, *Tilapia*, and related species, have selected the planted areas as their habitat. As swamps are especially associated with rivers, no wonder the fishes living in the few lush gardens of the lake are less endemic, less typical of the lake fish fauna, and more of a riverine type of fish. The endemic fishes by and large don't feed on aquatic plants because they have had less opportunity to develop a large flock of plant-eating species and more opportunities to feed on flesh. Those that were vegetarian when the lake started turned to the biocover for food and became specialized as microherbivorous species. This is exemplified by the main lake tilapiine cichlid *Sarotherodon tanganicae*, which has a biocover-grazing type of teeth, whereas riverine *Tilapia* have teeth with cutting edges, the better to cut higher plants.

The plant inventory of the lake includes several riverine plants, such as *Myriophyllum*, *Potamogeton*, *Vallisneria*, *Ceratophyllum*, etc., and lacustrine species, such as *Azolla*, *Najas*, and the funny looking, spiny, and very brittle *Chara*.

Encrusting algae have covered the vertical sides of the rocks with coral-like calcite formations. This photo was taken by Pierre Brichard at a depth of about 12 m.

The plant inventory of the lake includes *Myriophyllum*, a riverine plant. Photo by Ruda Zukal.

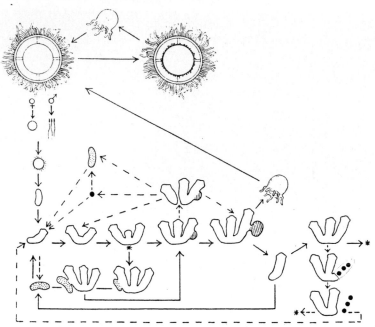

A schematic diagram of the life cycle of *Limnocnida*. Note the alternation of generations. (from Bouillon, 1959, *Ann. Mus. Roy. Congo B., 71*).

JELLYFISH

I had the privilege back in the fifties to discover a tiny freshwater jellyfish, *Limnocnida congoensis*, in the Kinshasa rapids of the Congo River. Lake Tanganyika harbors another—*L. tanganicae*. If the riverine species is rather rare and seasonal, the lacustrine species, at most 25mm across, often much less, and translucent white, rhythmically pulsates in dense clouds, especially in the northern basin. In all likelihood it feeds on organisms in the clouds of drifting plankton in which it gets mixed.

Entering a cloud of these eerie umbrellas, the diver feels as if engulfed in a snow storm. They are harmless to human beings and the only feeling I had when they stung my lips was a very slight tickle.

For fishes, however, they can be a deadly menace when their clouds, sometimes so dense that visibility is cut to as little as a foot, reach coastal waters or the bottom. When they reach a rocky slope, usually the scene of much activity, everything looks dead. Every single fish disappears from view, either hiding in the recesses or fleeing the area. As the jellyfish enter the dark cavities in which so many fry develop, whole spawns are decimated from the stings. Even larger fishes, like fully grown *Tropheus* or *Julidochromis*, died in our nets as they came into contact with the hundreds of jellyfish that we couldn't avoid sweeping into the net when catching the fish. We soon learned to stop fishing, as do the local commerical fishermen with their trawlers, when the jellyfish clouds are sighted. They know it is useless to try— the fishes are gone!

We feel that the clouds of jellyfish are potentially one of the main causes of mass mortality of juvenile fishes, at least in the northern part of the lake. Whole populations of sedentary species might at times be entirely wiped out.

The normal diet of jellyfish is not known, and it is not known if they have natural enemies, which they should have. But as yet no organism is known to prey on the ghost-white umbrella.

Jellyfish don't like too much sunlight and thus have a nyctemeral cycle, which means that they are seldom seen close to the surface in large numbers by daytime and are much more common there at night. Local fishermen call them the "eyes" of the lake and say, "At night Tembwe the

Myriophyllum is commonly used as plant cover for freshwater aquarium fishes. At least one species of this plant can normally be found in any well-stocked pet shop. Photo by R. Zukal.

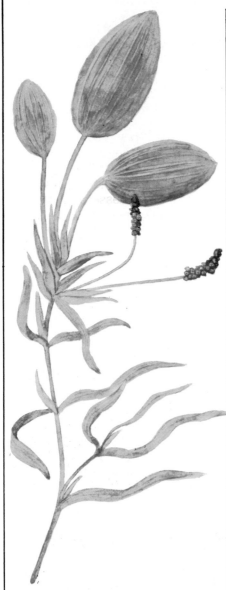

Potamogeton is another of the normally riverine plants of Lake Tanganyika and its environs. It will also be commonly available to freshwater aquarists.

lake opens his eyes and looks at the stars, and then when dawn is near and the stars disappear from the sky, the lake also closes its eyes."

One would gather from what I have said that *Limnocnida* would be an unwelcome aquarium addition. In fact this is not entirely the case. The jellyfish is too small and too slow to hurt any fish over a few cm long, or even the young fry because nobody would put a small cloud of jellyfish into his tank. A few or even a single jellyfish would bring all the excitement that one would want. It could be fed with daphnia and cyclops and perhaps live artemia larvae.

SPONGES

Sponges grow everywhere in the lake but they are much more abundant in the northern basin, probably because the plankton is so dense in this area. Seldom did I find rocks as heavily festooned and colored by broad and ubiquitous patches of multi-hued sponges in the southern part as I did south of Bujumbura.

There are nine species known from the lake so far, of which I was lucky enough to

The medusae of the jellyfish may occur in dense clouds and actually be a hazard to some of the fishes in the lake. Photo by Arnost Pustka.

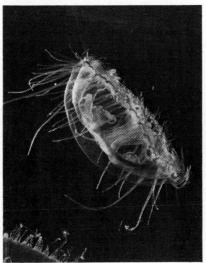

find two. They are divided into two groups, the first including soft-tissued sponges growing close to the surface in well-lit areas. When plankton drifts are plentiful they can grow spires and arches several cm high. They might also spread onto driftwood and aquatic plants.

There is a symbiotic relationship between sponges and green algae. All sponges are white when they first appear, and then progressively turn green. In a given area one can thus tell the oldest sponges by their deep green color, while the youngest patches are white—with all the shades of green in-between. This holds true with the hard spiny sponges that live deeper and even in dark recesses. On the other hand, I have seen sponges that were a dark brown when old and a bright vermilion red when young under large overhanging rocks, deep in the recesses. They are still unidentified.

I have the feeling that in areas where sponges cover most of the rock surfaces fishes are less common, but I must admit that the other members of the team do not concur with this opinion. It is possible that sponges exude a kind of toxic repellent. Unusually high mortalities of fishes carried along with sponges in a pail might perhaps be attributed to the sponges. On the other hand, this phenomenon might have been seasonal and related to sexual activity by the sponges. Anyhow, it is a fact that I have never seen a fish eat a sponge, nor do sponges display traces that they have been preyed upon.

If sponges could be compatible with life in a tank

45

Two different sponges growing on the same rock slab. One will probably become dominant and kill the other. Photo by Pierre Brichard.

MOLLUSKS

The lake has been endowed with one of the largest collections of freshwater mollusks in the world. Many are bivalves, many are snails in a flabbergasting array of sizes and shapes. They reflect the many biotopes they live in, from river swamps and lagoons to the deep floors, and even the dark recesses where I have discovered small snails that are still not identified. Some of them look so much like marine species that at one time they buttressed the theory

and one could remove them from the rocks without hurting them, they could well become one of the most valuable additions to a Rift Lake cichlid tank. But how would we feed them?

One might try to raise them in the lake. Put pebbles close to sponges and hope they will seed the pebbles. One would have then only to harvest the pebbles in due time. There are so many other things to do...

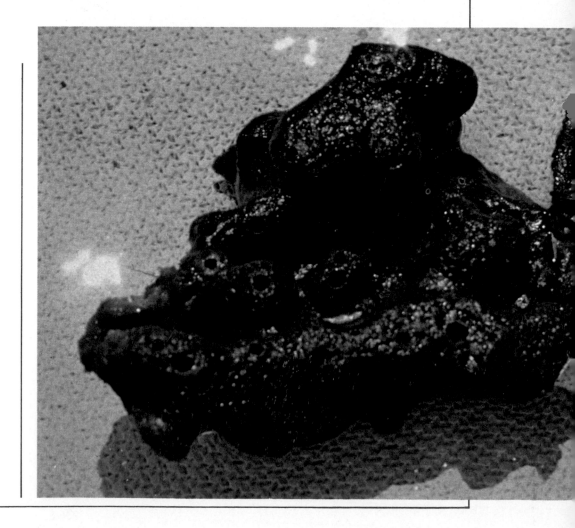

This as yet unidentified sponge was collected and photographed by Dr. Herbert R. Axelrod in Lake Tanganyika. The oscula are clearly evident in this photo.

that the lake at one time had been linked to the Indian Ocean, which ultimately proved to be a fallacy.

The lake shells show a remarkable adaptation to various ecological conditions. Snails living on wave-battered shores are heavy, those creeping on mud-floors have lightweight shells with a broad surface, long spikes, and ridges so that they can remain on top of the soft mud. Two of the most common shells are *Iridina*, a giant mussel 15cm or more long with a very gorgeous mother-of-pearl

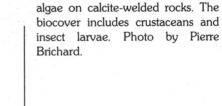

An underwater photo of encrusted algae on calcite-welded rocks. The biocover includes crustaceans and insect larvae. Photo by Pierre Brichard.

iridescent lining (hence its name), and *Neothauma*, a 3-4cm diameter snail that is the favorite diet of the very beautiful *Synodontis multipunctatus*, of other *Synodontis*, and of the giant catfishes.

This snail, the only feature of the barren sandy floors when they are dead, offers the shelter of its shell to an astounding variety of fishes. Best known among them and often cited in magazines are the dwarf nestbreeding *Lamprologus* species, such as *L. brevis*, *L. ocellatus*, or *L. meeli*. But aside from the six species of *Lamprologus* known to breed in the snail shells (*brevis, ocellatus, meeli, ornatipinnis, multifasciatus,* and *signatus*) there are many more, either nestbreeding cichlids, like *L. callipterus* or probably *L. leloupi, L. caudopunctatus, L. wauthioni, L. kungweensis, L. hecqui,* and

the dwarf limnochromine *Tangachromis dhanisi,* of which nobody could hope to succeed in spawning it were it not for the shelter offered by the shells.

The list of species making good use of the shells is not yet exhausted and we should add to it the baby *Mastacembelus ophidium* and *Chrysichthys*. Not that all of them, by their size, can live in the shells but they can release their eggs into a shell pile and raise the young there until they are large enough to venture out into the open.

This type of ecological niche (shell-dwelling) is perhaps most widespread in Lake Tanganyika, and it is a very useful one if we take into account the number of species that make use of the shells. At first glance the niche appears to be rather simple. But there is more to this phenomenon than meets the eye.

First, the shells are not mostly found alone, but assembled in piles. What brought them together? Like elephants in their mythical graveyard... Then, only a few shells are not broken as if they had been crushed. This might stem from a natural cause, such as in the acid waters of a river where they dissolve away in due time. In the lake the carbonates in the water settle on the shells, so that they remain and even get stronger with time. The second question is: what animal is crushing and eating the snails? We know that catfishes, especially *Synodontis*, rip off the snail's operculum and gobble up the animals, but it is doubtful that they can crush the shell. Perhaps large bagrids (ex. *Chrysichthys*) pick up the snails as they forage and bring them to a particular spot, perhaps their nest, where they

Shell-dwelling perhaps is widespread in Lake Tanganyika. The current fad in African lake cichlids is the keeping of the small species of *Lamprologus* that use these shells as their home and nesting site. Photo by Hans-Joachim Richter.

are able to gobble them at ease. The piles are not found at depths where wave action can push them together (in which case the piles would normally be in the shape of long ribbons and not circular).

Bivalve mollusks are eaten by species of *Lamprologus*, like *L. tretocephalus*, and also by the large cichlid *Lobochilotes*. They are usually small, perhaps on average 15mm across. I have seen piles of the broken bivalve shells on a large stone. Obviously they had been brought there on several trips as there were quite a number of them. Also, the small bivalves live embedded in the sand, not in the open nor on rocks.

CRUSTACEANS, LEECHES, AND PARASITES

The multitude of copepods and other tiny invertebrates in the plankton are not the sole representatives of crustaceans in the lake. Many endemic shrimp and crabs take shelter in the sand or in the rock anfractuosities. In the dark recesses, for example, one finds a tiny bright red shrimp 5mm long that is never seen in the open by daytime and that is one of the staple foods found in rock habitats by the local fishes. At various times of the year these shrimp are said to be caught by the millions by local fishermen, but I never had the opportunity of witnessing the phenomenom, although during a night dive with spotlights one is really bothered by the shrimp swarming around.

Six of the seven species of crabs found in Lake Tanganyika belong to the genus *Platythelphusa*. The other is a *Potamon*. Shown here is a species of *Platythelphusa* in dorsal and ventral views. The expanded abdominal plates indicate that this specimen is a female. Photos by Dr. Herbert R. Axelrod.

49

Crabs are scavengers and fish-eaters and in their turn are eaten by fishes (such as large *Chrysichthys).* They carry their eggs under the female's abdomen, where they are protected from predation.

Crabs live mainly among rocks in which they hide during the day. They are out feeding at night. Sometimes they assemble in large numbers. Photo by Pierre Brichard.

Seven species of crabs have been reported from the lake, all of them endemic. Six of them belong to the genus *Platythelphusa,* the other one being a *Potamon.* They live mainly over rocks, in which they hide during the daytime. They are scavengers and fish-eaters. In isolated rocky outcrops in the middle of huge sand bottoms they assemble in such quantities that every nook and cranny shelters several specimens. In the breeding season they carry their eggs in the protection of the female's enlarged abdomen, where they incubate in relative safety. This egg protection may be a reason for the crabs' abundance. This does not mean that the crabs themselves are free from predation by the fishes. They are sucked out of their shelters and crushed by the tremendous jaws of the giant catfishes, like *Chrysichthys.*

Parasitic crustaceans, such as the horrible *Lernaea,* have been reported from the lake, but I have only once seen a fish parasitized by this monster which can reach 30cm in length. Endoparasites in the shape of worms are occasionally found, but as yet never in fishes in the northern basin. In the southern part of the lake it appears there exists a specific association between rock-grazing cichlids and a worm that infiltrates fish muscles. From my notes taken during the three years I spent in the southern part of the lake as a fish collector it appears that about 50% of the *Tropheus* collected were infected with this parasite. Many were already in the terminal stages of emaciation when they were collected. Nothing could be done to save them. *Petrochromis* are also infected by the parasitic worm, but to a lesser extent.

Leeches are found everywhere in the lake, as they are also found everywhere in central African rivers and swamps. There are several varieties. The common black leech is found in river swamps and deltas, and several very small pale purple leeches are found in the mouths of hippos and crocodiles. These do not exceed 2cm in length.

By far the most common and spectacular leech in the lake is the beige banded leech, commonly reaching 15cm in extended length. One specimen, a female that apparently had some trouble releasing her eggs, was captured and preserved by me at Kasaba Bay. It was more than 30cm long when extended and thicker than my thumb when contracted. Leeches are a serious nuisance in the southern part of the lake, where they are especially abundant. I remember one day at a place between Ndole Bay and Mossi, we had to abandon fishing after 30 minutes as we were the obvious targets of a colony of leeches. We had already removed 22 of them between the two of us when we decided enough was enough.

As a rule, leeches are not dangerous to one's health. This I made absolutely sure of. They don't transmit parasitic or other diseases when they inflict their bite and inject a drug combining anesthetic and anticoagulant properties. Once their hold is secured it is difficult to get rid of them and pry them loose, especially when under water. It happened one day when I was diving at Cape Chipimbi that one leech

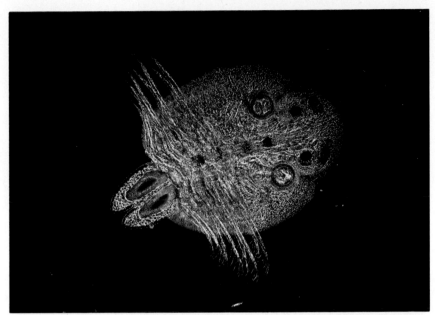

Argulus is a parasitic crustacean that in Lake Tanganyika often parasitizes catfishes, even big *Bagrus, Chrysichthys, Heterobranchus,* and *Auchenoglanis* (but usually not *Malapterurus*). Photo by D. Untergasser.

Argulus also parasitize *Lates*, but most fishes in the lake seldom have more than a few attached. Perhaps the scales protect them. Photo by Dr. E. Elkan.

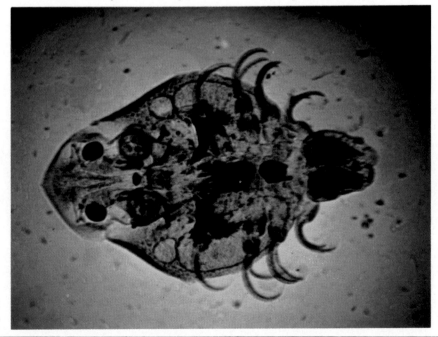

Adult female.

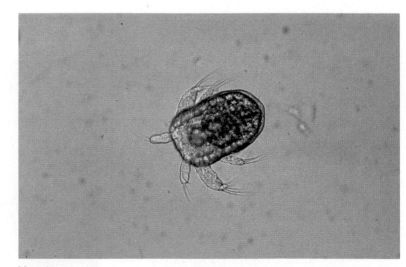

Nauplius.

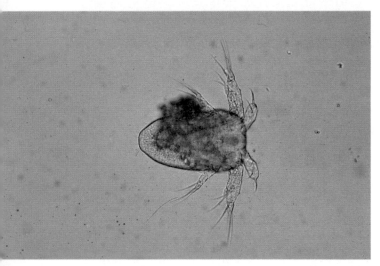

Metanauplius.

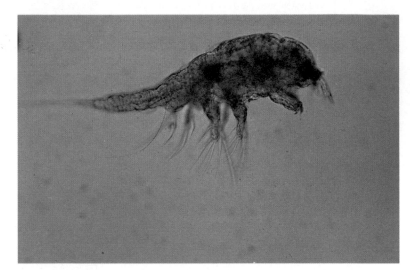

First copepodid.

Sixth copepodid female.

Sixth copepodid female with two spermatophores attached.

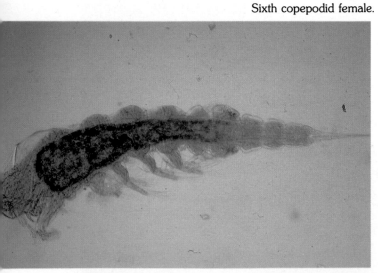

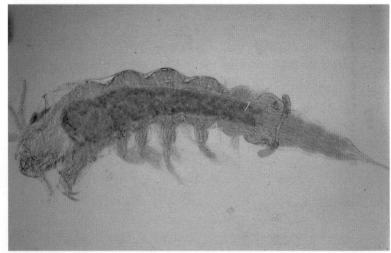

The life cycle of *Lernaea*. Each copepodid stage adds thoracic segments and appendages until 5 segments and 4 pairs of swimming legs are present (third copepodid). The male deposits his spermatophores on the genital segment of the female and then dies. The female attaches to a host and begins producing eggs. Photos by Dr. George Post.

entered my ear, a fact I realized only when it started chewing on my eardrum. As the doctor told me afterward, I am probably the only man on earth who heard a leech bite. I lost most of my hearing from this attack permanently.

Argulus, a ray-shaped crustacean, often parasitizes catfishes, which host them by the hundreds. Given their numbers, the bloodsuckers can seriously weaken a fish even as large as adult *Bagrus, Chrysichthys, Heterobranchus*, and *Auchenoglanis*. But the electric catfish, *Malapterurus*, is remarkably free from *Argulus*, and it is not at all impossible that they can shake them off with a jolt.

Lates often have many individuals of *Argulus* stuck to their gills, but most fishes, except the large cats, seldom have more than a few. Given the protection provided by their scales cichlids are remarkably free from the parasites.

A serious study of the lake parasites and their impact on fishes must still be undertaken. Whenever the subject is mentioned in a book, the author is satisfied with a list of the parasites that have at one time or another been found in the lake, so that one gains the feeling that the parasites are much more prevalent than they really are. In fact, no fishes are infested on a large scale, save for those mentioned in the example given regarding the southern rock-grazers.

Fishes coming from the lake destined for aquarist's tanks seldom play host to parasites or diseases. It is standard practice for experienced fish collectors to treat their fishes properly for parasites, etc.

REPTILES

Water Snakes

Climbing the ladder of aquatic life from the simplest forms to the more sophisticated, we unaviodably reach a point where the animals we deal with are getting a bit more worrisome as an occupational hazard for the fish collector or a scientist doing field work. Reptiles, mainly water snakes and crocodiles, not only play upon one's imagination, but are all too real a danger when improperly dealt with. With the number of unprepared people coming to Lake Tanganyika to spend a few weeks exploring increasing, no account of the lake lifeforms can overlook the danger of these reptiles.

The most common encounters (if, by principle, crocodile infested waters are avoided) are with snakes. Their archetype in the lake waters is *Boulengerina annulata* var *stormsi*. We have met all sizes of this very deadly snake (a cobra), and very often indeed.

When one of the fishing grounds harbors a *Boulengerina* it is bound to happen often that one or all of the divers will meet it during a dive and from one dive to the

One of the bush vipers (genus *Atheris*) is commonly found close to reeds and papyrus stems. It is a venomous snake and should be avoided. Photo by S. Kochetov.

This puff adder (a member of the genus *Bitis*) is extremely venomous. The colorful individual seen here is just a juvenile but is still deadly. Photo by Dr. Guido Dingerkus.

A "rhinoceros"-type of gaboon viper, also a member of the genus *Bitis*. Note the so-called "horns" on the snout. This is another extremely venomous snake. Photo by S. Kochetov.

next. Most of the time one spots the slithering body gliding slowly through the water or on the bottom between the rocks. It will enter anfractuosities in search of rock-dwelling fishes, remain hidden for quite a while, then poke its head through an opening in the rubble quite a few yards from the place it entered.

It could happen that if the snake entered the rubble before you appeared on the scene, as you are investigating a small cave, all of a sudden you are face to face with the cold eyes of the snake—and just a few inches away. This happened once to my servant in a place from where he could not beat a retreat as he was squeezed between two rock walls. Slowly reversing the beat of his fins, ever so slightly he managed to pull back a few inches, enough to have the snake resume its exit from the hole, slither past his shoulder taking what seemed like an eternity, and then enter another cave.

Cobras are very shy in water and I have never seen one showing any intention to bite. More than once they spread their neck in typical threatening cobra fashion, standing poised to strike, but as we remained quiet they just went away. At close quarters, such as they were, on land they would have surely struck. Apparently in the water, which is to them as foreign as it is to man, they by all means will avoid a fight. Which is as well because their venom is exceedingly powerful and bound to leave one a cripple at best.

Cobras are territorial and stay in the same area for several years. They find a den close to the water's edge,

Almost everyone has heard stories about the mambas. This extremely venomous species is the green mamba, a member of the genus *Dendroaspis*.

Mambas are predators. This one has attacked a chameleon that will probably succumb very shortly.

often with one opening underwater so that they can get out to forage and back without being seen in the open. I have seen a youngster bury itself headlong in moist sand on a beach, but I don't know if this is common practice. They reach a length of up to 12 ft. (3.6 meters), but this is exceptional. A large *Boulengerina* averages 8 ft. (2.5 meters) and is less than one's wrist across in thickness. Like all cobras they are dirty yellow or pale green in color and banded on the chest when adult. But some specimens are chestnut, so it might happen, as it once did with me, that as you discover only one part of a coil between the rocks, you mistake this snake for a *Mastacembelus* eel and wish to grab it. Better not!

These snakes can spend more than 15 minutes under water each time they dive, and I have seen several come up from deeper regions when I was 12 meters down. It is said that they can dive to 40 m depth.

Another snake, one that is harmless, can be found in very large numbers. *Glypholycus*, reaching a length of about 50-60cm, specializes in hunting clupeids as they venture inshore. I once met a herpetologist who told me that he had tagged 1,500 of these snakes on a 10km long beach in the southern part of the lake. When he later tried to catch more to figure out what their population was, only one snake out of 100 he caught carried the tag. Which could mean that there were 150,000 *Glypholycus* on this beach. Even if many of the snakes might have lost their tag or died, it still indicates how many

of these snakes are concentrated in the area. They are pale gray in color.

If one wishes to find snakes, just turn over the rubble along the coast (on dry land I mean); you are bound to find quite a few. But beware! There are quite a few deadly ones, including spitting cobras, mambas, *Naja melanoleuca*, Gaboon viper, puff adder, and most abundant, especially close to reeds and papyrus stems, the water viper and *Atheris*. The only snake that is really dangerous is the one you haven't spotted, so be careful where you put your foot down.

Monitor Lizards

Varanus niloticus is very abundant around the lake and preys on coastal fishes as much if not more than it does on birds, frogs, mice, and insects. It is preyed upon when young by birds of prey like eagles and falcons, but as it grows and reaches adulthood it becomes more than a match, with its 20kg of weight and more than 1.5 m in size, for any bird. Monitor lizards are the main enemies of young crocodiles and their eggs. They locate the nest, dig until they discover the cluster of eggs, and start gobbling them up. Female crocodiles remain by their nest while the eggs incubate in the warm soil. They rush to attack any monitor lizard that gets too close, leaving the nest unprotected as they follow the lizard in hot pursuit. The mate of the lizard thus has plenty of time to get at the eggs. A female crocodile, once the nest has been opened, will not try to repair the damage and abandons her brood. Thus

A mamba with the head held rigidly as the mouth is opened to expose the deadly fangs. The drops near the tips of the fangs are actual venom. Many venomous snakes are "milked" for this venom, which is used for varied purposes.

monitor lizards limit the size of a crocodile population. But the largest of the lizards are no match for adult crocodiles.

stems and fail to reach the safety of the shoreline before the sun is up, so that mortality by dehydration can be very

scrambled to come closer. So I started to wonder. If baby turtles can learn so fast, what about crocodiles?

A half-grown monitor lizard (*Varanus niloticus*). Monitor lizards are common around the lake and prey upon (among other animals) the coastal fishes. Photo by Dr. Guido Dingerkus.

Turtles

There are several aquatic turtles found in quiet coves and swamps. They, too, prey upon the fishes, frogs, and eventually birds. They grow quite large, and I have seen some weighing about 20kg. They are preyed upon when young by fishing eagles, crocodiles, and monitor lizards. They breed on the ground, in the sand, and as soon as the eggs hatch the baby turtles head for the water. As they do so, if they have to cross a patch of grass they can become entangled in the plant

high. When in Kasaba Bay, I used to scout the grass lining the beach in search of the poor devils, put most into the lake, but kept quite a few for study in a small pail, feeding them flake food that I had on hand to feed the fishes.

I learned a lot about the adaptability of reptiles. If at first they were scared witless when I came to feed them—probably mistaking me for the looming shape of a bird, something that they were programmed to be wary of—in not more than two days time they realized I was the source of their food and

Crocodiles

Crocodiles in tropical Africa are a major hazard for the fish collector, as they are found in every river and swamp. They have often been wiped out near cities by hunting for commercial purposes or just to get rid of a nuisance.

In my younger days I have killed quite a few and, as they became an occupational hazard when I started to collect fishes, I have tried to accumulate as many data as I could from personal observations and reliable reports on these big saurians.

This very young Nile crocodile (*Crocodylus niloticus*) may start out feeding on small fishes, frogs, and even insects. Photo by Dr. Guido Dingerkus.

The only dangerous one is *Crocodylus niloticus*, which basically is a swamp crocodile. This means that it is much rarer in the Congo River virgin forest affluents, where it is replaced by the African mugger, *C. cataphractus*, a fish eater and not reported to attack man. *C. cataphractus* has been reported from the lake, but only on rare occasions and I have never met up with one.

C. niloticus is found everywhere in the lake but in varying numbers. They are especially common in the southern parts, mainly in Nkamba Bay (which is included in a natural park) but also in Kasaba Bay and along the shores of Cameron Bay. Many are also found around river deltas. The Ruzzizi has a very large population and as

the area is heavily occupied by pastoral tribes as well, autopsies of the saurians killed by hunters have revealed that half of the crocodiles had human artifacts, such as bracelets, necklaces, etc., in their stomachs. This puts to rest recent reports that crocodiles are not that dangerous. In some areas of the lake fishermen say they attack often, in others local people say they don't. Why then the discrepancy between the reports? Cousteau investigated the crocodiles in the lake and came back saying they were scared by the divers and fled. All in all he thought they were less dangerous than hippos, which attacked his divers. I was in the same area, and although I gave a wide berth to a hippo herd, I was much more wary of crocodiles,

and with good reason as it turned out.

The main feature of a crocodile is that it is a very cautious beast and wary of man where it has been hunted down. The reflex move of a crocodile upon being made aware of the approach of man is to dive into the water, and if already in the water to submerge and sink to the bottom. The second feature is its curiousity. A crocodile, whatever its size, will come back to investigate. I have seen crocodiles slither into the water from a sand bar when I was still 50 meters away with my runabout, disappear from sight, and as I stood looking around for a chance to shoot it, I was quite upset to discover the beast alongside my boat with only its eyes ever so slightly protruding from the

An old adult Nile crocodile. These old veterans are truly dangerous to man. Brichard's stories about crocodiles that he has encountered should make potential fish collectors wary. Photo by Dr. Guido Dingerkus.

surface and watching me.

Met with the looming and unusual shape of a diver noisily breathing through his mouthpiece, the first instinctive reaction of a croc will be to flee. All reports I have received tell the same story. But if a crocodile, after several such encounters, feels it has suffered no harm from the meetings, it might decide that more investigation is in order. Cautiously it will remain nearby trying to decide whether to attack or not. If it does and after having entered the water cautiously and had a good look around, you are busy poking your nose into rock cavities for a look at the fishes and you forget to look behind you, what do you think will happen?

This happens more frequently if there are several crocodiles around and they start competing for prey, as I could see so often in Nkamba Bay. Once 11 crocodiles lined up along a small stretch of beach on which aquatic birds were foraging. Throwing caution to the wind they attacked in a frenzy by threes and fours—fighting each other for a snap at one of the birds.

Young crocodiles go after frogs, fishes, and even insects, but as they grow older and become more powerful and wise, they start to go after warm-blooded animals. Aquatic birds are usually first, then small mammals like otters, then hoofed animals like antelopes coming for a drink in late afternoon. When big enough they will start attacking man, mostly children as they splash in the water, but also adults as they wade knee-deep along the shore. Game wardens and fishermen have confirmed what I felt—

crocodiles prefer warm-blooded prey.

This is why hippo herds attract crocodiles. It is seldom that you don't find one or several crocs near the huge hippos. Not that they attack the adults; a very large hippo can chew and puncture any crocodile's thick hide as if it were a sheet of paper. But although a herd will try to protect the youngsters they cannot prevent crocodiles from seizing them. Crocodile predation is the one natural feature limiting the number of hippos. This is born out from surveys made in lakes where crocodiles do not exist, such as in Lake Edward. In Bujumbura, where at night a hippo herd often wanders along the streets lining the lake, there is rarely a very young hippo. They are snatched away by crocodiles.

I once had a very close brush with a crocodile in Cameron Bay. We were fishing in Kachese and one of my helpers not familiar with the use of swim-fins was splashing about wildly as he tried to catch *Tropheus*. A crocodile about 3½ meters long happened to pass in front of the small cove perhaps 100 yards away and heard the splashing. It couldn't see its origin as a crocodile lying very low in the water cannot see anything lying flat on the water more than 15 or 20 meters away. Probably assuming here was a bird or a fish in big trouble, the croc rushed toward the source of the commotion. To better locate it the saurian raised its head as much as it could above water and came in with a ring of foaming water around its neck. My African helper discovered the croc as it was steaming along barely 15 meters away, straightened up, and started floundering toward the pebble beach. The croc stopped short...here was an unexpected development. It had found a man instead of a harmless fish or bird. As it assessed the situation, its eyes fell upon my companion busily retrieving a few *Tropheus* from her handnet and kneeling neck deep next to the trap. The reptile resumed its charge but toward her. She in turn rose full height and waded toward the shore. Another stop, and thinking the effects of surprise had been lost the croc headed for open water. As it did, I poked my head out of the water and appeared on the scene. Again another stop...and this time, probably feeling very frustrated, the beast snapped its mouth open and shut several times with a loud noise and made off. Half an hour later it was back, but we were expecting it and we let it alone and departed. Perhaps fifteen days to a month later this croc, which was the only one in the area, grabbed an African.

If I have dwelt at some length about the psychology of crocodiles it is because unwary aquarists can recklessly get into trouble with the big man-eaters and I will

The Nile crocodile is basically a swamp dweller found everywhere in the lake, but especially in the southern parts. It really is dangerous, as was discovered when half of these animals killed by hunters were found to have human artifacts (jewelry, etc.) in their stomachs. Photo by Dr. Guido Dingerkus.

thus give them a few hints.

—Never go into the water without having asked if there are usually crocodiles around. If the answer is no, still be cautious. Crocodiles should be expected to live near swamps, river estuaries (even small ones), or in quiet coves.

—Although they do not like rocky coasts, especially if they are on the windward side, large crocodiles wander quite a lot daily and one should never go into the water without a good look around the shore and at the water's surface.

—Avoid standing for a long time or wading knee-deep along the shore; you are more visible to a croc standing up than lying prone.

—Avoid splashing and making noises as crocodiles have a keen sense of hearing and sound transmission is excellent in water.

—If you decide to have a look at the fishes in crocodile infested waters, enter the water slowly after having a good look around at the shoreline and at the water's surface. Don't stay too long in the water; the longer you stay the more time you give the reptiles to come. On the Inango Peninsula in the southern part of the lake there were many crocodiles. They did not prevent me from fishing, but I tried not to stay more than 30 minutes in the same spot and did not come regularly to the same place.

—If you are diving with scuba gear and you spot a crocodile try to blow bubbles into the water, grunt loudly, make any kind of loud noises, *and don't flee!* Charge! . . . if you can.

HIPPOS

Hippos are fascinating animals, first because they are often very huge (I have seen big females 2 meters high at the rump) and secondly because they are so unusual.

The best thing to do when one gets in the way of a rogue hippo is to dive and head for deep water. Its no use trying to hide by diving on the spot when the water is clear. Hippos walk on the bottom and, unbelievably, can even run and charge at a very fast pace given their bulk, much faster than a diver could swim. But they don't go deep, and I would think that 4-5 meters down is about the deepest they will go.

The best thing to do is to leave them alone, especially a family with a baby, and never get between them and the

This may be a particularly domestic scene, but hippos are especially dangerous when they have young with them as this one does. Also, never get between them and the water — they will probably charge.

The shores of Lake Tanganyika and its adjacent streams and marshes serve as wintering grounds and/or breeding grounds for a tremendous number of aquatic birds. Flamingos and pelicans are just two of the birds that spend most of the year here, along with masses of ducks and geese of many species. Although some of these birds are found here all year, others migrate north as far as Europe and the Middle East after breeding. The noise on an African lake at sunrise can be deafening, but the colors and activity of the birds make up for any temporary inconvenience.

water when they come on land to graze; they will get angry and charge.

AQUATIC BIRDS

The African fishing eagle is a common sight around the lake, especially along hillsides, which they seem to prefer. Seagulls are common, too.

How they reached the lake, which is more than 1,000km from the Indian Ocean, is a mystery.

Terns are especially common in the southern parts and their staple prey is the clupeid. As soon as they spot a school close to the surface they assemble in large flocks

and dive, so that the fish schools are in turn detected by local fishermen who also try their luck on the hapless Ndakala.

As one can realize from this all too fast survey of aquatic life in the lake, the lifeforms that share the habitats with the

fishes have a very strong impact on the various fish flocks. Much time will be needed before we gain a better insight into the intricate relationships that have developed between the fishes and their neighbors. Let us hope that the pristine beauty of the lake as yet unspoiled by industrial or human pollution will remain as it is for future generations to enjoy.

Sea gulls are usually associated with seashores, but many breed on inland lakes and migrate in large numbers over all the continents. This is why they appear in this lake some 1,000 km from the ocean.

The African fish eagle is a common sight around the lake, especially along the hillsides.

Boulenger identified and described many species of fishes based on collections made before WWI by Prof. J. E. S. Moore. This photo is a portrait of one of the fishes named for Moore, *Tropheus moorii*. Photo by Hans-Joachim Richter.

CHAPTER III

ECOLOGY AND DISTRIBUTION OF FISHES

Four main investigations of the various fish flocks in the lake have been conducted so far. They have successfully brought to light what we know today about their variety and distribution. But after nearly 100 years of collecting there still might remain hundreds of species waiting to be discovered. There are thus broad gaps in our understanding of the distribution of the fishes around the lake and even much more to be learned about their respective ecologies.

The first investigation, by Boulenger, was based upon several collections of fishes made before World War I, especially those made by Moore at the turn of the last century. Boulenger reported again on his finding in his monumental *Catalogue of the Fresh-Water Fishes of Africa* in which he identified more than 100 species from a lake which, according to the first surveys, had been said to be *practically devoid of any fish*!

The second major exploration was made by Professor Max Poll, participating in the 1946-1947 Belgian Hydrobiological Mission to Lake Tanganyika, a model of concerted and sustained effort in all relevant disciplines. Prof. Poll added several dozens of species to the list made by Boulenger, and for the first time called attention to the segregation and specialization of species according to the type of biotope they were living on, namely open, deep, and littoral waters. He split the fishes of the last-mentioned category into mud, sand, or rock-dwelling species flocks, each with its morphological adaptations to a given habitat.

The third investigation was headed for a period of more than 10 years by a team of Belgian scientists headed by Professor Marlier of the IRSAC (Institut de Recherches Scientifiques en Afrique Centrale) at Uvira. They concentrated their studies on the northwestern coast of the northern basin and discovered the effects of isolation and speciation among rock-dwelling species. Professor Marlier's work has become the basic reference for studies of these phenomena in other African lakes and around the world. This work also led to the discovery by Matthes of several new species, especially of *Ophthalmotilapia, Mastacembelus*, and *Synodontis.*

The fourth investigation was started by our team of tropical fish collectors in 1971 which, after more than 2,000 individual dives, is still going on.

For all practical purposes, since 1947 most investigations have concentrated on the northernmost basin, north of the Ubwari Peninsula, although for a few years there was also some work accomplished in the southernmost area. Systematic research has left the main coastlines stretching between the 4°S and 8°45' S latitudes practically untouched.

As we stand now, what do we know about the fish distribution and the biotopes in the lake?

Prof. Poll, as we mentioned, divided the fishes into three main groups:

1. The open water or *pelagic* species, living far from the

shores and more or less independently from them (except for seasonal spawning activities).

2. The deep bottom or *benthic* species, living down to the deepest available reaches (to 250 meters), often under open water and far from the shoreline, but mostly in contact with the bottoms and slopes leading to the shore.

3. The *littoral* species, living near the water's edge and on the slopes leading to the shore on one side and to open water on the other.

Prof. Poll then subdivided the coastal fish flocks into three groups according to the type of coastal habitat they were living on:

1. *Swamp-dwellers*, living in river deltas, coastal lagoons, and mud flats. Although many are endemic to the lake basin, they have retained the morphology and behaviors of riverine fishes and have not colonized the lake far from a river mouth. They are thus not among the typical lacustrine species.

2. *Sand-dwellers*, living almost exclusively in the lake over soft bottoms (sand or mud) and eventually found in front of river estuaries. They show a high degree of adaptation to their biotope and are endemic. They breed in the lake.

3. *Rock-dwellers*, most of which are endemic to the lake. All of the rock-dwelling cichlids are endemic. They live on rock-strewn slopes to which they have strong bonds. Most breed in the lake (except for a few non-cichlids).

Thirty-five years after Prof. Poll's work, his classification of the fish flocks into these groups still holds even the light of new discoveries. In the context of the survey equipment then available for collecting (seines, dragnets, etc.) and without any direct means for observation, such as scuba gear, his findings still hold true.

Our team devoted much time to identifying ecological barriers leading to the isolation of local populations in coastal biotopes, to the study of littoral distributions of species·along the coastlines, and to the systematic explorations of rock biotopes leading to inventories of the fish flocks between the surface and 15 meters at first, but now down to 40 meters. At least 24 new species have been discovered since our team started its work 12 years ago, some of which are still awaiting publication of their description as the present book is going to press.

Most of the research contributing to our present knowledge of the lake fishes and their biotopes has been devoted to pelagic and rock-dwelling species. The first because they are economically important to the well-being of the people living around the lake shores and in the hinterland, the latter because they are very typical and the easiest to observe in their habitat.

The deep-living species live in a world where direct observation is as yet impossible to conduct for any great length of time. Swamp-dwellers live in a world where murky water makes direct observation with scuba gear well-nigh impossible and crocodiles or hippos, as well

as the dreaded disease schistosomiasis (bilharzia), add to the hazards. Swamps around the lake are usually given a wide berth by divers.

Sand-dwelling cichlids, distributed over the wide-open expanses but in local concentrations, are difficult to find and are best observed whenever they appear on sand patches near rocky outcrops and rubble-strewn slopes. Contrary to most rock-dwelling fishes, they move about a lot and wander over the sunken plains. They are hard to follow for any length of time. Moreover, as they are usually more delicate to handle than the hardy rock-dwellers, their study in captivity (in the confines of an aquarium) lags well behind that of rock-dwelling species. This is true also because many species living over sand floors live in mid-water, off the bottom, and are fast-moving species like the pelagic fishes.

The frontiers between the various biotopes are of course schematic and not clear-cut. The boundaries are crossed more often than not by many species. For example, pelagic species often roam in very deep waters. On the other hand, benthic species ascend toward the surface at night in open waters when the bottoms they rise from lie far offshore, or they move to the coastal area when they follow the slope leading to the shore. Sand-dwellers can wander toward the deeper reaches or come into gradual contact with the murky waters in front of a river delta or with rocks.

To decide that a species is bentho-pelagic or bentho-littoral because it has been found occasionally here and there is a decision fraught with

Species of the genus *Triglachromis*, like this *T. otostigma*, live mainly over mud bottoms but can also be found over sand. Photo by Thierry Brichard.

peril so long as systematic observations have not provided us with enough data. Let us admit that for the time being we can only speculate about the ecology of fishes living on slopes deeper than 15 to 20 meters.

Fishes living beyond our present reach, however, can provide us with clues about the broad features of their ecology by their morphological and anatomical adaptations. Mud-dwellers like *Triglachromis*, whose pectoral fins have independent first rays (perhaps acting as feelers), probably live mainly over mud bottoms, but are also found over sand. Fishes with unusual sensory organs on the head, like species of *Trematocara* and *Lepidiochromis*, need them in

the deep, dark reaches of the lake to find their prey. Sand-dwellers commonly have elongated pale beige or pastel-colored bodies and foreward slanted teeth with which they scoop mouthfuls of sand from which they sift crustaceans. Rock-dwellers more often have short, stocky bodies (cichlids), or very elongated shapes with which to slither in and out of recesses (many non-cichlids). Slow-moving fishes often sport long threadlike filaments on their finnage that would slow down roaming pelagic species. The list of these adaptations is endless, but each provides us with a clue about fish ecology and behavior.

Some fishes do not seem to have become specialized at all toward any one of the

A view of the lake in the vicinity of Burundi. Pelagic schools of fishes, mostly clupeids, roam the lake and form the basis for a fishery of sorts. Photo by Dr. Herbert R. Axelrod.

The pelagic fishes are free to move about the lake. Local races are therefore not usually found, which is in direct contrast to the situation in rock-dwellers. Photo by Dr. Herbert R. Axelrod.

habitats—especially in the coastal areas. As a result they are often found everywhere around the lake because whatever adaptations they accumulated did not prevent them from surviving when they came into contact with another type of biotope. There are many such ubiquitous species in coastal waters. Coastal habitats are thus not always very well delimited and the various fish flocks, more often than not, do not lead their respective lives in total segregation.

PELAGIC WATERS

Except for the coastal fringe of slopes and shallows, which is most commonly very narrow, the entire body of the lake, which means 34,000 sq. km, is composed of pelagic waters in which fishes move about in very dense schools. There are few species making up the schools but their economic impact is tremendous. The density of the pelagic fish school has been estimated at between 2.8 to 4

million tons at all times. Their density is said to be between 25 and 30 kg per hectare (2.5 acres) and the annual catch could reach 200,000 tons, it is believed, without depleting the stocks. Many more than 10,000 fishermen around the lake live only from the night catch of the Ndakala (or Ndagaa) as the clupeids are called.

Pelagic species have gone through similar adaptions to live in their particular biotope and there are no unspecialized species among them. Local geographical races are seldom found because the vast expanses they roam in do not present ecological barriers by which populations of the fish might become isolated and start inbreeding.

BENTHIC WATERS

Living in the deeper reaches of the oxygen-bearing layer, much deeper than a fish would ever live in a river, has demanded from the fishes major adaptations to low

oxygen levels and a permanent twilight at best, and more often total darkness. But some of the riverine fishes, mainly catfishes, found in swamps where oxygen levels are low and light very poor have found a habitat to their liking on the lake bottom.

Cichlids, on the other hand, had to develop means by which to live so deep and in the dark but otherwise crystal clear waters in which sound transmission was exceptionally good. This is probably why they developed sensory organs that are not commonly found in cichlids living in the upper reaches of the lacustrine habitats.

SWAMPS

In strong contrast with the open-lake waters and the deep layers, the areas surrounding river mouths are often overgrown with dense aquatic plant thickets. In the murky, silt-laden water visibility is practically nil and fishes have to grope their way around

Many of the rocky bottom fishes are plankton feeders and spend much of their time up in the water column. Photo by Dr. Herbert R. Axelrod.

Other fishes are more closely tied to the bottom where they feed on the biocover. Photo by Dr. Herbert R. Axelrod.

much as they do in river marshes. The fishes coming from a river will settle in such a habitat without any trouble and there is need for very little adaptation on their part. The fishes in the basin area are thus very little differentiated, if at all, from pure riverine species.

SANDY BOTTOMS

Because erosion has been at work for so long and the slopes around the lake are usually steep, silt has covered the lake bottom since the early days, as is documented by the mile-high layer on the floor of the deep basins. The small particles of dust and sand rain down toward the deep, while landslides bring them down along the steep slopes so that any rocks falling down are eventually covered over with silt and sand. Ledges and every flat surface on a steep incline are also sanded over. Thus sandy bottoms, from gently rolling plains to the foot of rockstrewn slopes, prevail everywhere.

The first feeling of a diver exploring the vast expanses of the sandy floors is one of void. Seldom does one encounter a fish because they are living in tight concentrations wandering here and there over the barren grounds.

In the sandy areas a pair of *Lamprologus tetracanthus*, *L. modestus*, or *L. pleuromaculatus* can be seen busy around their nest. Funnel-shaped depressions, perhaps 4 meters in diameter and often half-filled with broken *Neothauma* snail shells, identify an abandoned *Boulengerochromis* crater nest. We might see the barely visible shapes of these giant cichlids in the haze of sand "dust". An occasional school of *Lamprologus callipterus*, always on the move, will pass by or, as we turn about, we will discover them following us, voraciously swallowing crustaceans that our flipper movements have stirred up from the bottom sand in which they lay buried.

As we approach a pair of *Boulengerochromis* and look

A rocky area with so few fishes is unusual. The diver/photographer has probably spooked them and they took refuge in the rock crevices. Photo by Dr. Herbert R. Axelrod.

A pair of *Boulengerochromis microlepis* cruising over the sandstone slabs. The individual with the less colorful pattern is the female. At about 70 cm, she is larger than the male. Photo by Pierre Brichard.

into the nest to see the development of the spawn, the two parents circle nervously about. If we stay quiet they get bolder and ram and bite our legs, our hands, and even tear at our hair. They abandon their defense against us as a boarding party of *L. callipterus* comes to the edge of the huge pit and heads toward the mass of perhaps 15,000 fry. Were they not repelled by the two doting parents before they reached the spawn, the *L. callipterus* would in a matter of seconds clean up the plate, although they are probably not more than 10 or 12 in number.

Something moves around an *Iridina* mussel shell half-buried in the sand. We discover it is a pair of *Lamprologus modestus*, another ubiquitous coastal cichlid, too small (it grows to a bit more than 12cm) to defend its nest in the open. Thus, using the mussel shell as a prop, they have dug a deep tunnel underneath it.

Lamprologus ocellatus and *L. brevis* are much too small to live in the open, much less to

Kavalla Island.

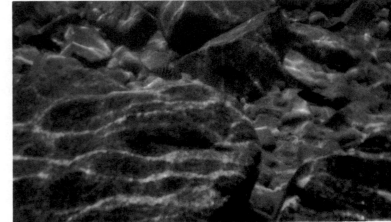

Loose rubble and individual uncemented rocks.

Massive calcite "reefs."

Calcite formations can reach large sizes.

Detail of cementing of small pebbles by calcite.

Broken edge of a calcite formation.

Bubbles of gas rising from the Lake bottom at Bemba. Photos by Pierre Brichard.

Exhaust vents of a thermal spring under the Lake.

spawn and defend their brood over open grounds. When they reach 2.5cm they have already reached sexual maturity. To survive on these hostile grounds they make use of their small size to settle in empty *Neothauma* shells and it is there that they breed and raise their offspring when no other shelter is available.

feeding, moving about, and even breeding. Some, like *Callochromis pleurospilus*, dive headlong into the sand and disappear in case of danger.

The typical sand-dwelling cichlids have developed their adaptations to their biotope to a very fine degree indeed, as we can note from *Xenotilapia*, whose shape and camouflage

The vast sandy bottoms extend all the way down to the deep regions and one might think that all sand-dwellers, as they wander about, could be found anywhere in the oxygen-bearing water layer. Far from it, as we discovered after more than 12 years spent in diving around the lake. Only in January '83 did we find *X. tenuidentata* (a species collected by M. Poll in 1947) for the first time at a depth of 40m. We might thus believe that some of the sand-dwellers are restricted to the upper reaches of the sandy plains, while other species live much deeper and never come up to the shallows.

There is nothing of course to prevent a pelagic fish from coming from the deep open waters over the sunken coastal bottoms and it is an exceptional sight to see the pelagic schooling *Perissodus paradoxus*, barely visible in the haze, speeding by over a 3m deep shallow plain. These pelagic fishes having been caught as well at 250 meters demonstrates how fluid the boundaries and rigid the categorizations set by man can be.

Callochromis pleurospilus dive headlong into the sand and disappear from view when danger threatens. Photo by Dr. Herbert R. Axelrod.

But the typical sand-dwellers are not solitary species. The best way to live, feed, and breed on barren, featureless floors when one is small is to bunch together. Schools of *Callochromis* and *Xenotilapia*, which live on the bottom, or those of *Aulonocranus, Cardiopharynx,* and *Lestradea*, which live in mid-water, number in the hundreds. Some species have developed strong gregarious instincts, responding to group stimuli for

are so good that it is difficult to discover their schools swimming just above ground in the golden haze of fine particles. They have extensive sensory organs on their sides (three lateral lines) that offer them protection against predators and they have special teeth set at such an angle that they can scoop sand to get at the shrimp buried within. Their very strong gregarious behavior provides them with additional protection.

ROCKY SHORES

Rocky shores are made up of a wide variety of rock formations sloping down at various angles toward the deep. Depending upon the amount of light penetrating from the surface, the influx of vegetable and animal plankton carried by coastal currents, the oxygen levels at various depths (which in turn depends on the layout of the coastline in relationship to the bearing of the main winds), as well as on the mineralogical nature, shape, and size of the rocks, each stretch of coast can be said to be different from the next.

If the relationship of the rock-dwelling fishes to their habitat is a very complex one and a very challenging problem to solve, it is also true when we look into several rock biotopes of a flabbergasting diversity.

What fishes are living on such a coast? There might be gregarious or solitary fishes, wanderers, or territorial fishes. Some build nests and raise their brood in the nestsite, others incubate their spawn in their mouth. Some feed on the vegetal carpet covering the rocks in the top layer, while others peck at the rocks to grab the tiny "bugs" creeping on their surface. Some other fishes station themselves off the slope, in midwater, so that they can get first try at the incoming phytoplankton or are interested only in the copepods and other crustaceans hopping about around them. Some predators attack other fishes and swallow them whole, but quite a few rip other fishes that are often diseased or weakened and wounded by the assaults of scale-ripping predators, to pieces.

Some species are restricted to the shallowest layer, while others seem to have colonized a much thicker layer. As for those fishes that usually live in the deeper parts of a rocky coast, our first dives around 40 meters and below have revealed that there appears to be a whole species flock of deep-living rock-dwellers that never come close to the surface. No wonder then if during his first explorations on

a typical rocky biotope a diver will be lost in a drama cast with intense activity by many players. In this respect rocky biotopes are very different from the offshore waters and sand floors, where the number of species appearing on a local "stage" is always very limited.

The variety of fishes on a slope can be amazing.

Rocky areas separated by sandy areas support a variety of fishes. Species of *Tropheus, Chalinochromis,* and *Xenotilapia* can be seen here. Photo by Pierre Brichard.

Inventories of the local populations that I conducted over the years in several biotopes often come close to or exceed 100 species, all living together or coming in contact occasionally with each other, on a stretch barely 50 meters wide. After getting used to the activity displayed by so many species, one starts to sort them out. Basically there are two flocks, as on sandy bottoms: (1) those fishes that live at ground level (or in the

recesses between the rocks) and (2) those that live off the bottom, in midwater.

Most of the time, except when breeding, fishes are positioned on the slope in the area where they feed, because feeding is a time-consuming activity. Predators have to wait for a prey to come by in order to attack, and they often fail.

Plankton-eaters take only minute quantities of food at each picking. This has to be repeated at short intervals so that the fishes are properly fed.

If the variety of fishes cohabiting a slope reflects the array of ecological niches that are at their disposal, their respective density is in direct relationship with the quantity of food supplies locally available. In this respect, the ready availability of drifting supplies supplementing those that

75

develop locally has a tremendous impact. Territorial nestbreeders, plankton pickers in midwater, wouldn't be so plentiful if they could not depend on the quasi-inexhaustible supply of drifting coastal plankton clouds. If along the Burundi coast there are many small areas that can support a school of 100,000 *Lamprologus brichardi* in addition to all the other species around it, it is because the school is fed by incoming zooplankton passing through the area in a never ending drift.

The distribution of the fishes along the coast will thus depend a great deal on their respective feeding habits. These feeding habits in turn are correlated with the species adaptations, not only in their dentition but also in their morphology and shape. A fish having to chase its prey around needs a streamlined body, while another feeding on drifting plankton remains much more stationary and can have a plump body endowed with banner-like filaments (*Ophthalmotilapia, L. brichardi*), and a rock-grazer like *Tropheus* needs maneuverability and a short body to attain the proper angle with the uneven surface of the rocks.

Food being so important with respect to the places occupied by the fishes and its indirect effect on their shape, we can try to sort the rock-dwelling fishes out according to their feeding habits into *feeding guilds*. We find an amazing variety!

1. *Insectivorous fishes* live close to the water's edge feeding on insects falling into the water and their aquatic larvae (mayflies). Ex. *Varicorhinus* (cyprinids).

2. *Herbivorous rock-grazers* feed mainly on the vegetal carpet of the biocover growing on the rocks. Their diet is composed also of animal proteins supplied by the "bugs" creeping among the algae. Ex. *Tropheus, Petrochromis, Simochromis, Tilapia tanganicae,* and *Asprotilapia.*

3. *Carnivorous biocover-peckers* live mainly on the "bugs" they pick from the algal carpet. The ratio of animal proteins to vegetable matter in their diet probably varies from one species to the next. Ex. *Eretmodus, Spathodus, Tanganicodus.*

4. *Carnivorous zoobiocover-peckers* specialize in picking crustaceans and probably insect larvae from the tiny crannies on the rough rock surfaces. Ex. Many *Lamprologus* (like *L. compressiceps* or *L. fasciatus*) and *Julidochromis.*

5. *Carnivorous zooplankton-pickers* live at ground level (Ex. *Lamprologus brichardi*) or in midwater (Ex. *Cyprichromis*) picking crustaceans as they "hop" by.

6. *Phytoplankton-pickers* feed mainly on the drifting vegetal organisms of the plankton in midwater. (For ex. *Ophthalmotilapia, Cyathopharynx,* and *Cunningtonia*).

7. *Snail crushers* or *rippers* are represented by several catfishes such as species of *Chrysichthys* and *Synodontis*. The largest cats also feed on crabs up to 10cm across.

8. *Bivalve shell crushers* feed on small bivalve mollusks. Ex. *Lobochilotes* and two *Lamprologus* species, *L. sexfasciatus* and *L. tretocephalus.*

9. *Aquatic plant browsers.* Ex. *Limnotilapia dardennei.*

10. *Sand-sifters* scoop mouthfuls of sand with their forward slanted teeth, sift it through their gills, and eat the crustaceans hidden in it. Ex. *Xenotilapia.*

11. *Diatom feeders* feed on diatoms and shrimp developing on decaying organic matter on the deep floors. Ex. *Trematocara.*

We thus have 11 feeding guilds of fishes living on food supplies provided by the rocky biotope or brought in to it.

There are, in addition, three guilds of fishes that prey on other fishes.

12. *Scale-rippers* have teeth that are set in such a way that they can seize the opposite edges of a scale on the side of a fish's body, apply pressure, and make it pop out of its seating in the flesh. Skin, mucus, and flesh are digested, bony scale structure is not. Scale–rippers will also attack open sores and wounds, thus helping dispatch a disabled fish. Ex. *Perissodus.*

13. *Macrocarnivores* will attack any fishes that they can swallow whole. There are many such predators on a rocky coast. For

example, in the dark recesses of the bottom all giant catfishes (*Chrysichthys, Bagrus, Heterobranchus, Malapterurus*, clariids, *Lophiobagrus*, etc.) and *Mastacembelus* eels, and in the open large *Lamprologus*, such as *L. lemairei, L. elongatus, L attenuatus, Cyphotilapia frontosa*, etc.

14. *Scavengers* feed mainly or preferably on dead or disabled fishes, which they dispatch. Ex. *Lamprologus callipterus*, although in several respects it is *omnivorous* like *Lamprologus modestus* and *L. mondabu*, or the larger *Telmatochromis*.

Of course feeding behaviors are not stereotyped. Even a herbivore will try to snatch eggs from a spawn or attack fry. It took me many years before I saw a *Tropheus* grabbing a disabled *Ophthalmotilapia* fry 2cm long and try to swallow it whole. On the other hand a *Julidochromis* that mainly picks a copepod or a shrimp from a rock anfractuosity will not let another one pass by in a drift. *Lamprologus brichardi* pecking at a rock to grab a copepod is not an unusual sight either.

Two feeding guilds have been identified in other African lakes but as yet have not been observed in Lake Tanganyika. The first includes *eye-ripping* cichlids specialized for tearing the eyes from other fishes. No such behavior has been identified among our lake fishes. Although several cases of missing eyes have been reported in shipments of *Julidochromis*, the fishes cannot be said to have specialized in this direction.

Lamprologus brichardi is a carnivorous zooplankton picker living at ground level. This is a pair preparing to spawn on a rock in an aquarium. Photo by Hans-Joachim Richter.

77

The second includes *paedophagous cichlids*, or species that have jaws and teeth allowing them to seize the mouth of female cichlids that are incubating fry, pry the mouth open, and then suck eggs or fry out. These have been reported from Lakes Victoria and Kivu, but not from Lake Tanganyika. But some omnivorous cichlids, like *Telmatochromis caninus* and *T. temporalis*, might pay special attention to the other fishes as they spawn and attack the eggs as soon as they are laid (ex. *Lamprichthys*). One might think that specialized paedophagous cichlids should have developed in lakes where there were no nestbreeding cichlid populations from which spawns could be preyed upon in the open and without much trouble. In Lake Tanganyika, with so many nests around, it is possible that such an opening toward specialization in preying upon incubating mouthbrooders provided less incentive.

If we compare the feeding guilds in the lake that are found on a rocky slope with those that have been identified around a tropical coral reef, we find very few differences indeed. There are a few guilds missing in the lake, such as the fishes that crush coral heads to get at the coral animals hidden within, or fishes that live on worms, because these organisms are missing from the lake's underwater landscape.

One cannot rule out that the lake jellyfish, *Limnocnida tanganicae*, is not preyed upon by some specialized fish.

If we sort out the fishes on a rocky slope according to the origin of their food, we discover that they either eat food that is produced locally and could thus become exhausted by overgrazing, like the phyto- and zoobiocover, or that brought in by currents that follow a pattern and are therefore very regular, but depend on the layout of the coastline and submerged slopes. The algal carpet on the rocks grows only so fast and it could happen that the density of rock-grazers and rock-peckers becomes too high; which might explain why herbivorous rock-grazers move about so much more than the fishes that feed on the crustaceans living on the rocks or that drift by.

The ratio of the respective variety and density of the fish flocks living on resources produced locally and those that live on the drifting plants and animals can be very variable. We will thus find a wide variety in the distribution of the two groups along a coast.

Basically there are places where:

1. Biocover and drifting plankton supplies are plentiful. In such areas we will find the highest concentrations of both fishes living on the biocover at ground level and fishes living off the bottom in midwater.
2. Drifting plankton input is low but the biocover is lush. In such cases rock-grazers and rock-peckers will be plentiful but there will be few fishes swimming in midwater.
3. Biocover is poor (silted over with sand for example) but there is a steady and abundant supply of plankton. This results in the depletion of rock-grazers and peckers and the boost of plankton feeders.
4. Biocover and plankton drifts are poor. The area can then support only a limited number of fishes from either group, except for ubiquitous omnivorous species.

Adverse local ecological conditions, like fermenting gases emitted by organic matter decaying in quiet coves or those emitted by thermal springs, can chase the fishes from the area. On the other hand, decaying organic matter in well-aerated areas, such as those at the foot of a slope, can boost production of shrimp and diatoms which in turn are fed upon by the fishes.

We can thus say that the food available on a rocky coast, consisting of supplies provided by the biocover and those brought in by the coastal plankton clouds, play an essential role in the pattern of fish distribution and the variety of fishes present on a coastline. This is why rock biotopes are never exactly identical to each other, nor homogeneous.

Because breeding behavior plays a major role in the ecology of the fishes, we can, once we have identified the behaviors, try to see if the occupation of the bottom is similar for nestbreeding cichlids and mouthbrooders. We discover that nestbreeders as a rule stay much closer to the bottom than mouthbrooders, and that they are more territorial and less prone to wandering than the

Telmatochromis dhonti (formerly *T. caninus*) is omnivorous and may pay special attention to spawning fishes so they can attack the eggs as soon as they are laid. Photo by Hans-Joachim Richter.

cichlids that use buccal incubation. Because they need a nestsite on which to breed and raise their brood and because nestsites are hard to obtain on the crowded bottom, many nestbreeding cichlids spend their entire lifespan in a very confined area (*Julidochromis* pairs may remain several years around the same crevice in a rock). The fry, once they have grown up, look around for a shelter of their own and, because of the dangers lurking behind every rock wall, don't go very far, settling for the first available suitable spot. The expansion of a colony of nestbreeders is often very slow, while mouthbrooding cichlids can take their brood along and travel quite a distance from the place they spawned. The spread of a colony of mouthbrooders toward new

grounds is thus potentially faster. This might explain why there are areas, such as the one along the corniche at Nyanza Lac, where nestbreeding cichlid species are very few.

In occupying a "new" rock habitat (such as the one created when man builds a pier or a road along the coast and scatters the rubble over a sand bottom), wandering mouthbrooders will be there first, because they can protect their fry in their mouth and do not need to stop very long for spawning. Nestbreeders, especially those living at ground level and raising their brood in a permanent nest amid the rubble, will be last.

The role played by ecological barriers in the distribution of fishes is very important to say the least. But a barrier can be fordable by

some species and not by others because each species has its own requirements toward a given type of substratum and ecology and will show more or less tolerance if the local conditions of the barrier are not to its liking. The distribution of species along the shores could thus not possibly be uniform for all of them.

In most habitats around the lake one finds species that are found everywhere as well as local species that do not appear elsewhere, even though the continuity of the layout of biotopes around the lake in the past should have helped them disperse. One might therefore guess that these species did not exist yet when the ways were open to coastal migrations. For example, if *Lamprologus moorii* lives exclusively to the

Julidochromis pairs generally do not roam far from the same crevice (their nestsite), many spending their whole lifespan in a confined area. This pair of *J. dickfeldi* is jawlocking in preparation for spawning. Photo by Hans-Joachim Richter.

Lamprologus moorii lives exclusively to the south of the Lukuga flats. It is one of several cichlids that treat that area as some sort of barrier to dispersal. This is the yellow form. Photo by Shuichi Iwai, courtesy of Midori Shobo.

south of the Lukuga flats and *Lamprologus toae* to the north, it is very probable that both developed in the lake after the flats became a barrier to their migration to the other part of the coast. There are several other examples of cichlids that did not cross the Lukuga flats.

Before starting to wonder why a species is not found in a particular biotope, one should investigate whether it belongs to the fish flocks in that part of the lake. Only systematic inventories along the various coastal biotopes and the identification of ecological barriers can provide us with clues as to how and when the fish flocks came about. As it is, the concentration of species on a rocky habitat can be very high at some times and very poor at others. On some of the best habitats, if one draws a circle 3 meters in radius and 3 meters high, one might count as many as 30 to 35 different species living in the area from the bottom to the surface.

The bottom crevices are occupied in strength by non-cichlid species, foremost among them catfishes and mastacembelid eels. Cichlids living in these recesses are *exclusively nestbreeders*. Mouthbrooding cichlids use the deep anfractuosities only temporarily for shelter, for spawning, or as youngsters. There are a few nest breeders that have become what one might call "cave-dwellers" (ex. *Lamprologus prochilus, L. furcifer, L. obscurus, L. schreyeni, L. savoryi,* and *L. niger,* more or less, and *Julidochromis).* On the other hand, mouthbrooding cichlids prevail in midwater off the slope, where only a few nestbreeders have gone to live.

If the nestbreeding cichlids depend much more on the availability of a given type of rubble on which to spawn and raise their brood, mouthbrooders might also eventually depend for their

distribution along a slope on bottom features for their spawning. Such are the *Cyprichromis,* fishes living off the slope and feeding on copepods drifting by. They spawn in midwater and don't need a nestsite, but when they look around for a place to release their fry at the end of the incubation period, they need overhanging rocks and shallow caves to do so. Because of their spawning rhythm, the schools are made up of females at various stages of incubation. *Cyprichromis* thus are found mainly near high rock pinnacles and uneven rock floors, steep slopes, and, much less frequently, over flat rubble.

Because a rock habitat on a steep slope goes all the way from the surface to the bottom of the oxygen-bearing water layer, we meet very different ecological conditions over a short distance. These range from the sunlit,

Spathodus erythrodon is one of the shallow-water species living no deeper than about two or three meters. It can be seen here why it is called one of the goby cichlids as it sits on the bottom. Photo by W. Staeck.

hyperoxygenated surface layer close to the water's edge, to the dark, quiet, poorly oxygenated deep slope waters where algal biocover can no longer grow. Although there are some species that are found from near the surface down to 30-40 meters or even deeper, most species have a rather restricted vertical range.

In the northern basin, in which most observations were conducted by our team, we discovered that some species live only in the top 2 or 3 meters (ex. *Spathodus erythrodon*), and some in 4-5m (*Eretmodus, Simochromis babaulti*, etc.). Many are not found deeper than 15m, but a few are (ex. *Tropheus, Petrochromis*). Some, on the contrary, are not found near the surface. *Cyphotilapia* doesn't appear to come

shallower than 7-8 meters (during the daytime at least), and *Lamprologus falcicula*, which is related to *Lamprologus brichardi*, does not live shallower than 8-10 meters, while the latter species is found from 2-3 meters down to 25-30m. Some *Trematocara* live quite deep, never having been collected in less than 60 meters at night, but their daytime station can be as deep as 200 m. As we progressively extend into the deeper habitats we will undoubtedly find many new species of deep-living cichlids and non-cichlids.

We have said that many sand-dwellers, such as *Xenotilapia*, come in contact with or venture among the rocks over stretches of sand intercalated within the rubble. They are not the only fishes

foreign to the rock biotopes to intrude. Many pelagic species at one time or another, but especially at mating season, come into the area and of course add to the local concentrations of coastal species, although their impact on the habitat and its dwellers remains to be assessed.

What should we remember about the fish distribution around the lake shores and in the open waters?

1. *Open waters* have a few species living gregariously in dense concentrations scattered here and there around the lake.
2. *Benthic* species, still poorly investigated, live in contact with the deep floors.

3. *Swamp-dwelling* species did not speciate much when they came in contact with the lake, which they avoid. They include an important group of non-cichlid fishes.

4. *Sand-dwellers* have specialized a great deal, whether they live on the bottom or in midwater. They live mostly in schools when small, although quite a few small species have managed to find shelters on the barren plains.

5. *Rocky habitats* overall are the biotopes where the fishes diversified the most and have become specialized in various ways. Those living in the rubble are mostly solitary species while those living off the slopes are gregarious. Non-cichlids predominate within the rubble, nestbreeding cichlids at its surface, while mouthbrooding cichlids prevail off the slopes.

6. *Ecological barriers* put a brake on or prevented colonization of new territories, according to the fish bonds toward a preferred type of rocky habitat.

7. A rocky coast is never entirely homogeneous as it is made up of stretches with different rock formations and with a variable layout with respect to prevailing currents and plankton drifts.

8. The distribution of the fishes on a rocky coast depends among other features on their diet (planktonic or locally grown foods), spawning behavior, specific bonds toward oxygen, light levels, shape of the rocks, etc.

9. Rock-dwelling fishes are very much confined to a specific level on the slopes, but soft bottom dwellers have also been found to live at a given level, either deep or rather shallow.

10. To be found on a given rock slope and not elsewhere, a species had to develop locally and be prevented from migrating past ecological barriers.

The fact that so many species can live together on a short stretch of slope can be explained by the number of ecological niches and the quantity of shelters available to fishes stemming from different backgrounds, as well as by the amount of food present that is capable of sustaining very large populations. In this respect a rock biotope around the lake is not very different from the coral reefs.

That so many localized species have already been found while the exploration of the biotopes can be said to have barely begun shows that my feeling that several hundred species are there waiting to be discovered is not overly optimistic.

Aulonocranus are not deep-water fishes, but they are always found over sand, even if only sandy patches between rocks. Photo by Glen S. Axelrod.

VERTICAL DISTRIBUTION OF COASTAL FISHES WITH RESPECT TO THE NATURE OF THE SUBSTRATE

Six types of slope bottoms showing the most frequent associations between the surface and about 30 meters are illustrated schematically with the feeding guilds most frequently found in each. Each of the vertical strips should be considered independently from its neighbors.

A rocky coast is composed of different types of rock and sand, ranging all the way from gigantic, monolithic rocks belonging to the core of the slope, to erratic boulders of more than 100 m³, to small ones about 0.5 m³, then loose but interlocked rubble with stones about 50cm x 50cm x 50cm or less, then more or less rounded pebbles smoothed out by wave action between 20 and 5cm across, gravel less than 5cm across, coarse sand less than 5mm across, and fine sand less than 1mm in diameter. After that comes mud and fine silt. They are, for the uninformed onlooker, distributed on a slope more or less at random.

B. *Feeding guilds of fishes living on or along a rocky coast.*
The letter code used is encircled.
Drifting plankton pickers
A. Mostly phytoplankton pickers
B. Zooplankton eaters (probably specialized more or less toward one category of plankton)
Biocover grazers:
G1. Mainly herbivorous biocover grazers
G2. Mixed phyto- and zoobiocover grazers.
Biocover peckers
P1. Highly specialized peckers on insect larvae and crustaceans (ex. *L. compressiceps*)
P2. Specialized pickers on insect larvae and crustaceans in drifting plankton (ex. *L. brichardi*) and also incidentally on the biocover.
P3. Unspecialized peckers (many nestbreeders).
Non-cichlid roamers, mainly insectivorous (*Varicorhinus*)
I. Top-layer insectivores
Carnivorous cave-dwellers
W1. Non-cichlids (catfishes, eels)

W2. Cichlids (*L. prochilus, L. furcifer, L. savoryi*, etc.)
Carnivores
C1. Close to the shoreline (*Haplochromis horei*)
C2. Deep (*L. profundicola, L. lemairei, Lates*, etc.)
Coastal scale-rippers
R. Excluding pelagic species (*Perissodus microlepis* and *P. straeleni*)
Mollusk-crushers
H. Bivalve shell crushers (*L. tretocephalus, Lobochilotes*)
N. Snail-eaters (catfishes)
Sand-sifters
S. Specialized sand-sifters for buried crustaceans (*Xenotilapia*, etc.)
Pelagic & semipelagic intruders
E. Scale-rippers or predatory (*Perissodus, Boulengerochromis, Lates*, etc.)
Benthic diatom feeders
D. Diatom, crustacean, benthic feeders (*Trematocara*, etc.)
Footnotes: The boundaries of each type of substrate are schematic as are the indicted depths around which the main flocks, not stragglers, are found. Only the daytime distribution of the fishes is shown. Nighttime distribution is as yet too poorly documented. Omnivorous and ubiquitous species are omitted.

Explanation of coastal strips:

COAST 1
A typical barren sand biotope, the shoreline eventually lined with reed stumps and a small beach, very seldom aquatic plants, and then only in quiet water. Normally a transition biotope toward an extensive sand coast or between two rock biotopes.

Explanation of Signs and Letter Codes
A. *Type of substrate:*

Sand bottom

Sand bottom with sparse rubble covered with silt

Sand floors with unsilted scattered rocks

Mixed grounds of sand and rubble

Pebbles

Big rubble in several layers, eventually with boulders

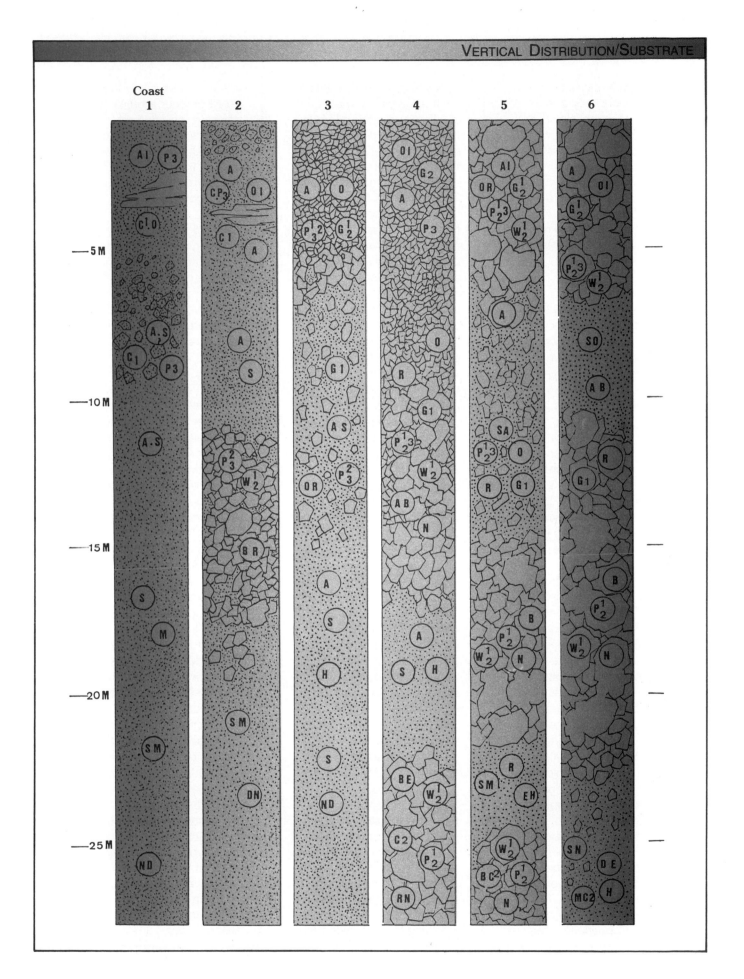

The top layer close to the shore harbors only a few species, roaming predators (cyprinids, *H. horei*), juveniles of sand-dwelling fishes, and zooplankton pickers like *Aulonocranus*. Deeper down there are sand-sifters, mollusk-crushers, and herbivorous fishes such as *Limnotilapia dardennei*.

COAST 2

Very much like the first superficially, with areas of loose scattered rubble well silted over and with poor biocover. Deeper down there are isolated outcrops of rock, most often sandstone slabs. This is too deep for rock-grazers, but there is a good population of cave-dwelling catfishes and eels and a few peckers on crustaceans. Further down the deep floors are composed of sand and are the grounds of the typical sand-dwelling species.

COAST 3

Clean pebbles, most often in several layers, line the shore, followed by scattered rubble with a good biocover but relatively few shelters. Deeper down, around 10 to 15 meters, the deep sandy floors rise to the coast. The first grazers appear between the surface and 15 meters deep, where the grounds are very favorable for drifting plankton pickers in midwater and the layout of the bottom offers excellent spawning sites. Small schools of sand-dwellers wander about the floor of scattered rubble. There is a considerable increase in the number of zoobiocover peckers, including very specialized species. *L. brichardi* often

appears in large numbers in such a biotope, but cave-dwellers are relatively few for lack of piled-up rubble and big cavities. Scale-rippers are common. When an extensive rock biotope is formed only of this type of layout and separated from the others by an ecological barrier, nestbreeding rock-peckers and plankton pickers might be entirely missing (Nyanza Lac corniche for example).

COAST 4

A typical very steep rocky slope layout, offering a wide variety of shelters and ecological niches, the slopes offering a vertical succession of rock falls and sand-covered ledges. The rubble piled up in a layer several meters thick offers overabundant shelter to cave-dwellers which abound (perhaps as much as 25% of the total population on the slope). Towering pinnacles of calcite-welded rock and big boulders rise from the rubble and offer additional shelters for juveniles of zooplankton pickers (*Cyprichromis*). Around 20-25 meters (sometimes more), new species of biocover peckers tend to replace those of the top layer. It is also around this level that intruders from pelagic roamers tend to appear along the coast. The area of the slope close to shore is composed of rubble, short stretches of pebbles in which juveniles of mouthbrooders and their parents, especially the rock-grazing cichlids, are often found. This type of habitat prevails, along with the two following layouts, along the

steep slopes lining the big escarpments around the lake (in the southern part of the lake: Chaitika-Chipimbi the Marungu; in the northern half of the lake: the Ubwari southern face, the Ngoma, the Ngombe).

COAST 5

Another typical rocky coast, similar to the preceding layout, but due to the nature of the rocks the shoreline is not lined with rubble but with very large rocks belonging to the core of the slope, which are slow to break down into smaller rubble. These rocks are full of deep cavities or fractured by crevices. Rubble and small patches of pebbles interlock on the bottom. Then comes a band of rubble scattered on the sand away from the wave action, and again a succession of steep cliffs, rubble layers several meters thick, and short stretches of sand. As on the previous slope, all kinds of rock-dwelling species are to be found there, along with pelagic species and scattered populations of sand-dwellers, among them sand-sifters. This type of slope shares with the previous one the big escarpments around the lake, as well as short stretches of rock biotopes in the northern basin, such as the one called "the spring" in Burundi, 30km south of Bujumbura, and Luhanga, south of Uvira in Zaire. It is also typical of the southern face of the Ubwari, and especially spectacular is Kavalla Island at the southern tip of the Ngoma escarpment.

COAST 6

Big boulders line the shoreline, but they fall directly on a sand bottom. The latter drops as a steep rubble slope disappearing in the deep. This is often the layout of bays and small coves along the big rock coastlines. Incoming breakers have smashed the rock core which fall in cyclopean piles along the shore. Sediments fill the bottom with a layer of sand. At the mouth of the cove the incoming waves repel the sand in the cove, and rubble from the main coastline core litters the floor in a very thick layer. This is a habitat where rock-grazing cichlids are fewer because there are few shelters for their young, most of those available being occupied by nestbreeding cichlids, and also because fine silt tends to cover the biocover. Typical of this layout are several coves around the Lufubu River mouth in Zambia, several coves on Kavalla Island, and the small cove at Kabimba on the Ngoma escarpment coast.

Many other variations of the layouts given here are of course to be found. Such is the entire north-facing coast of the Ubwari, lined with big rubble landing on a substrate entirely sanded over. There the rock habitat is only 10 to 20 m wide, but 40km long. Another is at Bemba (northern basin) were lava spills and thermal activity have produced coral-shaped outcrops entirely made of pyrite, as well as other types of lava-made ribbed rocks. Still another are the lava balls 4-5cm across lying in a thick layer at Mossi (Cameron Bay, Zambia) and agglomerated with sandstone along the beach, or the cyclopean boulders of Edith Bay lying on sand. Also there are the sandstone slabs lining the whole beachline, more than 35km, between the mouth of the Lukuga and Cape Bwana Denge and further north.

Seining along a sandy beach, these fishermen caught and killed a small crocodile. It will be sold at the local market. Larger crocodiles may tear the nets and are dangerous to the fishermen themselves. Photo by Pierre Brichard.

CHAPTER IV

SURVEY OF LAKE TANGANYIKA FISHES AND THEIR ENDEMISM

Until taxonomists reach a consensus about the advisability of using behavior as an important character in fish taxonomy species of *Sarotherodon* and *Oreochromis* have been retained in this list under the name *Tilapia*.

Despite the fact that Greenwood has placed the three species of *Limnotilapia* in synonymy with *Simochromis*, his point of view, based upon the morphology of the jaws, does not appear pertinent enough to overrule major differences in the ecology and ethology of the two groups of fishes. *Limnotilapia* are ubiquitous, live in deep water in schools, feed on aquatic plants as well as zooplankton and biocover, whereas *Simochromis* remain close to rocky bottoms, are very individualistic, and do not school. Moreover, they are restricted to the top 15 meter layer where they feed from the rock biocover.

The new classification by R. Allgayer of the *Lamprologus* group has not been followed. *Lepidiolamprologus* Pellegrin, 1903 has been rehabilitated. On the contrary, *Lamprologus nkambae* Staeck, 1977, has been put in synonymy with *Lamprologus kendalli* Poll, 1977. The sole difference between the two fishes being in the scaleless or scale-covered cheek, which is a feature varying from one specimen to the next as in several other species in the genus.

Haplochromis benthicola and *H. horei* have been kept under their previous identity despite Greenwood's revision of *H. horei*.

Tropheus annectens
Boulenger, 1900, was described by him because the two specimens he had collected (supposedly from Kalemie (previously Albertville)) had 4 spines in the anal fin instead of 5 or more as in *Tropheus moorii*. A new collection made at Kalemie by the author in 1981 failed to produce a single *Tropheus* with 4 spines. But in 1984 an intensive exploration of the southwestern coast uncovered two *Tropheus* close to Moba cohabiting on the same rocks. The two races could be distinguished by distinct color patterns and by the fact that one had consistently only 4 anal spines. (123 specimens were checked.)

Chalinochromis "dhoboi" and *C. "bifrenatus"* as well as *Lamprologus "cylindricus"* are names given to scientifically undescribed species and are in use only within the aquarium hobby.

At the end of 1985, 175 species of cichlids have been identified within the lake basin. They belong to 45 genera (47 if one takes into account the recent additions by Greenwood and Trewavas). Out of these 45 genera, 38 are found only in the lake and 7 include species that are also endemic to the lake and its basin, although other species in these genera are found elsewhere in Africa (*Astatoreochromis, Astatotilapia, Haplochromis, Lamprologus, Orthochromis, Tilapia,* and *Tylochromis*).

Of the 175 species, 171 are endemic to the lake. There of the four species that are not found only in the lake *Tilapia nilotica, Astatotilapia bloyeti,* and *A. burtoni* are found mostly in the northern sector of the lake, which means the neighborhood of the Ruzzizi delta. This restriction in habitat might point toward a recent acquisition when the Lake Kivu waters started to pour into Lake Tanganyika a bit more than 12,000 years ago.

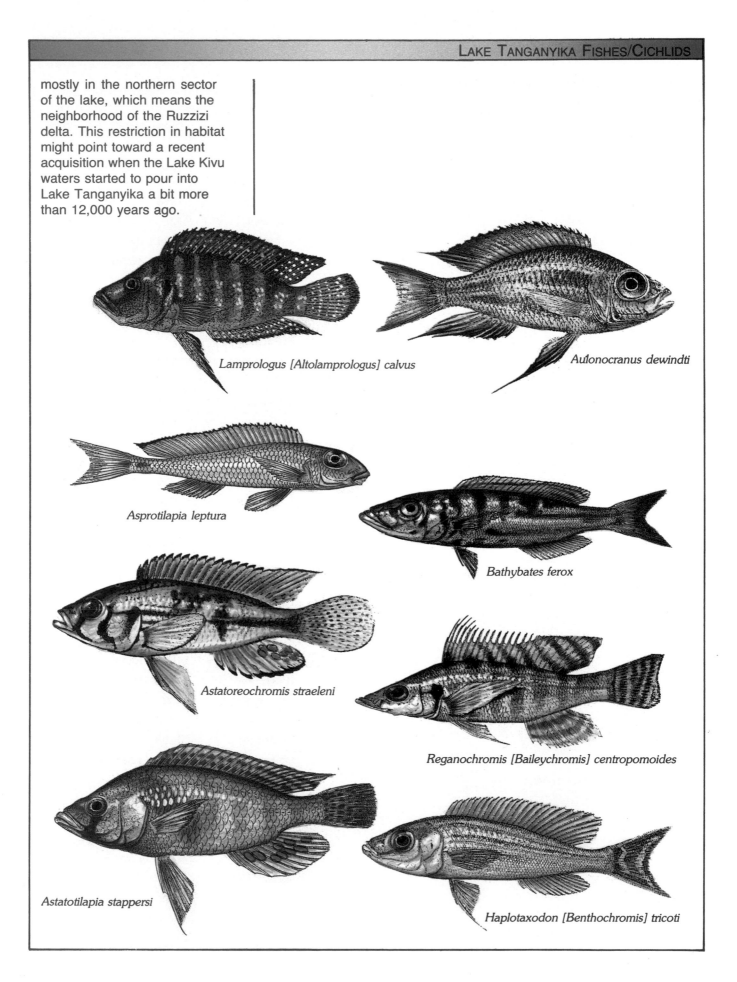

Lamprologus [Altolamprologus] calvus

Aulonocranus dewindti

Asprotilapia leptura

Bathybates ferox

Astatoreochromis straeleni

Reganochromis [Baileychromis] centropomoides

Astatotilapia stappersi

Haplotaxodon [Benthochromis] tricoti

LIST OF SPECIES as per 30 June, 1985*

1. CICHLIDS:

N	Genus and Species	Orig. Descr.	Date	Habitat
1	Asprotilapia * leptura *	Boulenger	1901	rock
2	Astatoreochromis straeleni *	(Poll)	1944	swamps
	vanderhorsti *	(Greenwood)	1954	rivers
4	Astatotilapia bloyeti	(Sauvage)	1883	swamps
	burtoni	(Günther)	1893	swamps
	paludinosa *	Greenwood	1980	river
	stappersi *	(Poll)	1943	swamps
1	Aulonocranus * dewindti *	(Boulenger)	1899	sand
7	Bathybates * fasciatus *	Boulenger	1901	Pelagic
	ferox *	Boulenger	1898	pelagic
	graueri *	Steindachner	1911	pelagic
	horni *	Steindachner	1911	pelagic
	leo *	Poll	1956	pelagic
	minor *	Boulenger	1906	pelagic
	vittatus *	Boulenger	1914	pelagic
1	Boulengerochromis * microlepis *	(Boulenger)	1899	sand
2	Callochromis * macrops macrops *	(Boulenger)	1898	sand
	macrops melanostigma *	(Boulenger)	1906	sand
	pleurospilus *	(Boulenger)	1906	sand
1	Cardiopharynx schoutedeni *	Poll	1942	sand
2	Chalinochromis * brichardi *	Poll	1974	rock
	popelini *	Brichard	MS	rock
1	Cunningtonia * longiventralis *	Boulenger	1906	sand/rock
1	Cyathopharynx * furcifer *	(Boulenger)	1898	sand/rock
1	Cyphotilapia * frontosa *	(Boulenger)	1906	rock
4	Cyprichromis * brieni *	Poll	1982	rock
	leptosoma *	(Boulenger)	1898	rock
	microlepidotus *	(Poll)	1956	rock
	nigripinnis *	(Boulenger)	1901	rock
1	Ectodus * descampsi *	Boulenger	1898	sand
1	Eretmodus * cyanostictus *	Boulenger	1898	rock
1	Grammatotria * lemairei *	Boulenger	1899	sand
3	Haplochromis benthicola *	Matthes	1962	rock
	horei *	(Günther)	1893	sand
	pfefferi *	(Boulenger)	1898	ubiquit.
2	Haplotaxodon * microlepis *	Boulenger	1906	pelagic
	tricoti *	Poll	1948	pelagic
1	Hemibates * stenosoma *	(Boulenger)	1901	pelagic

* denotes an endemic genus or species.

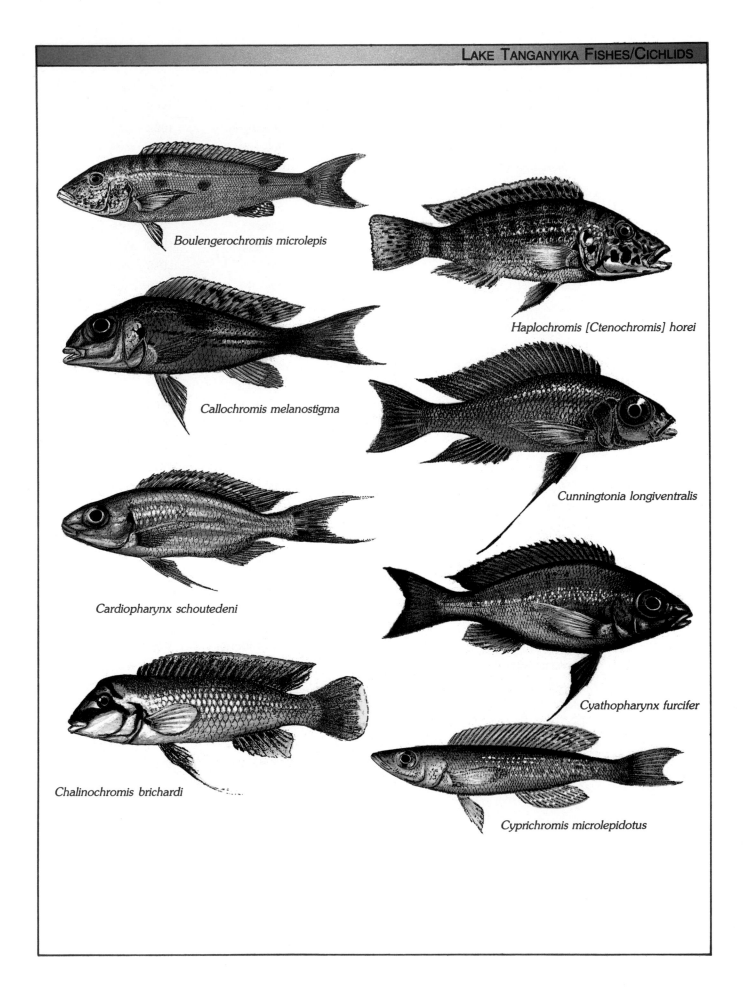

Boulengerochromis microlepis

Haplochromis [Ctenochromis] horei

Callochromis melanostigma

Cunningtonia longiventralis

Cardiopharynx schoutedeni

Cyathopharynx furcifer

Chalinochromis brichardi

Cyprichromis microlepidotus

5	Julidochromis * dickfeldi *	Staeck	1975	rock
	marlieri *	Poll	1956	rock
	ornatus *	Boulenger	1898	rock
	regani *	Poll	1942	rock
	transcriptus *	Matthes	1959	rock
42	Lamprologus brevis *	Boulenger	1899	rock
	brichardi *	Poll	1974	rock
	buescheri *	Staeck	1982	rock
	callipterus *	Boulenger	1906	ubiquitous
	calvus *	Poll	1978	rock
	caudopunctatus *	Poll	1978	rock
	christyi *	Trewavas & Poll	1952	rock
	compressiceps *	Boulenger	1898	rock
	crassus *	Brichard	MS	rock
	falcicula *	Brichard	MS	rock
	fasciatus *	Boulenger	1898	rock
	furcifer *	Boulenger	1898	rock
	gracilis *	Brichard	MS	rock
	hecqui *	Boulenger	1899	sand
	kungweensis *	Poll	1956	?/shells
	leleupi leleupi *	Poll	1956	rock
	leleupi longior *	Staeck	1980	rock
	leleupi melas *	Matthes	1959	rock
	leloupi *	Poll	1948	?
	lemairei *	Boulenger	1899	rock
	meeli *	Poll	1948	?
	modestus *	Boulenger	1898	Ubiquitous
	mondabu *	Boulenger	1906	Ubiquitous
	moorii *	Boulenger	1898	rock
	multifasciatus *	Boulenger	1906	sand/shell
	mustax *	Poll	1978	rock
	niger *	Poll	1956	rock
	obscurus *	Poll	1978	rock
	ocellatus *	(Steindachner)	1909	sand/shell
	olivaceous *	Brichard	MS	rock
	ornatipinnis *	Poll	1949	rock/shell
	petricola *	Poll	1949	rock
	prochilus *	Bailey & Stewart	1977	rock
	pulcher *	Poll	1949	rock
	savoryi *	Poll	1949	rock
	schreyeni *	Poll	1974	rock
	sexfasciatus *	Trewavas & Poll	1952	rock
	signatus *	Poll	1952	?

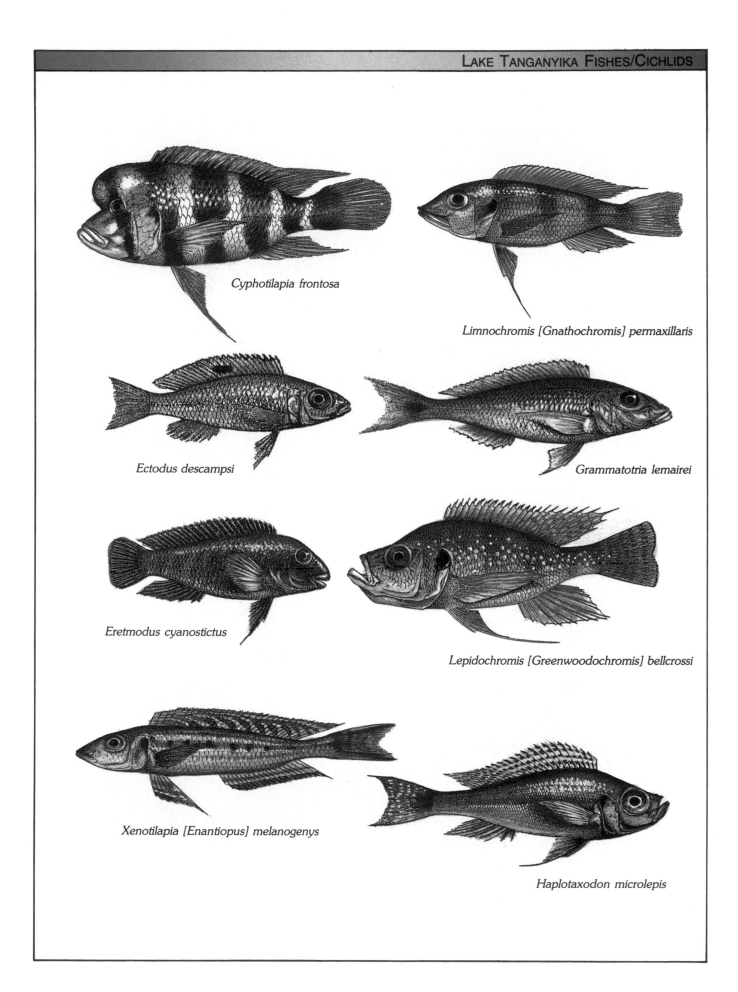

Cyphotilapia frontosa

Limnochromis [Gnathochromis] permaxillaris

Ectodus descampsi

Grammatotria lemairei

Eretmodus cyanostictus

Lepidochromis [Greenwoodochromis] bellcrossi

Xenotilapia [Enantiopus] melanogenys

Haplotaxodon microlepis

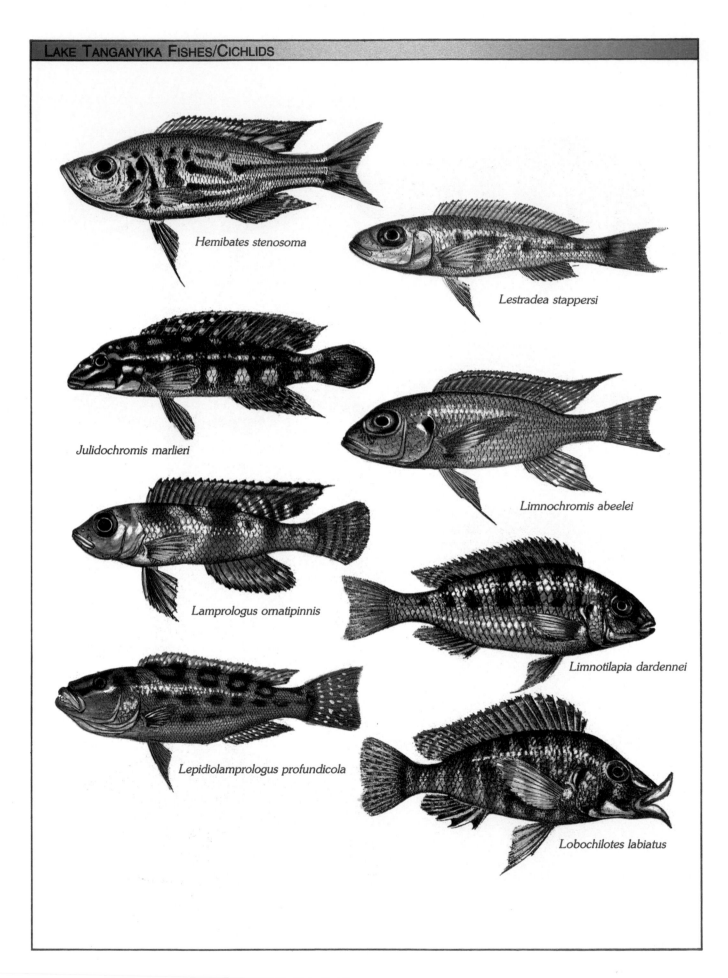

Hemibates stenosoma

Lestradea stappersi

Julidochromis marlieri

Limnochromis abeelei

Lamprologus ornatipinnis

Limnotilapia dardennei

Lepidiolamprologus profundicola

Lobochilotes labiatus

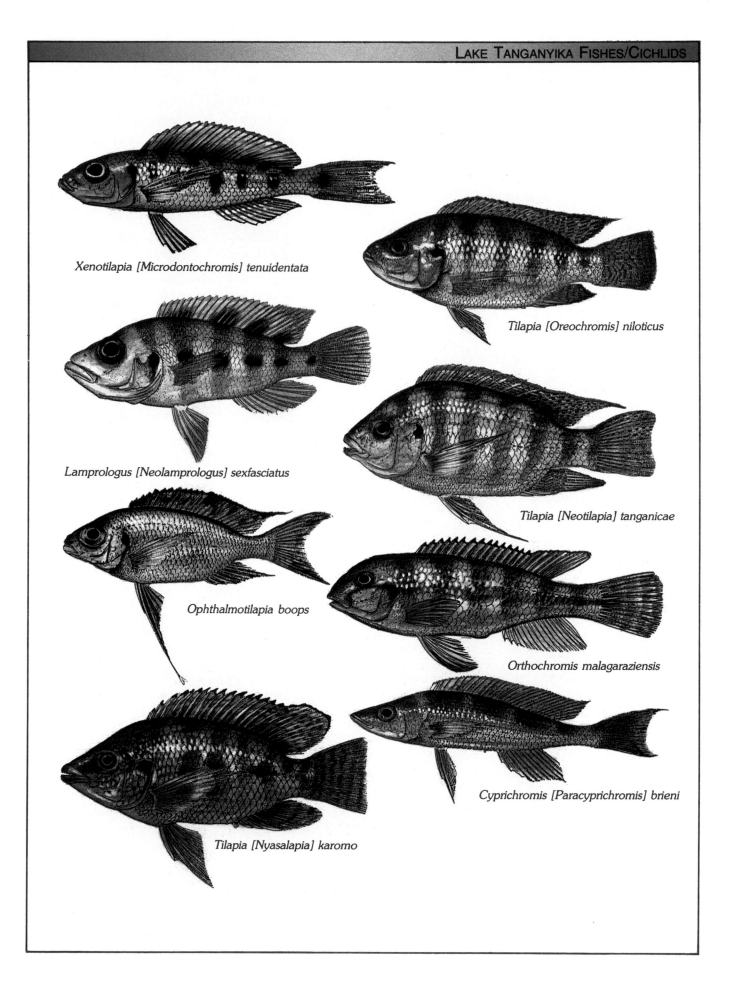

Xenotilapia [Microdontochromis] tenuidentata

Tilapia [Oreochromis] niloticus

Lamprologus [Neolamprologus] sexfasciatus

Tilapia [Neotilapia] tanganicae

Ophthalmotilapia boops

Orthochromis malagaraziensis

Cyprichromis [Paracyprichromis] brieni

Tilapia [Nyasalapia] karomo

		splendens *	Brichard	MS	rock
		stappersi *	Pellegrin	1927	river
		tetracanthus *	Boulenger	1899	sand
		toae *	Poll	1949	rock
		tretocephalus *	Boulenger	1899	rock
		wauthioni *	Poll	1949	?
6	*Lepidiolamprologus* *	*attenuatus* *	(Steindachner)	1909	ubiquitous
		cunningtoni *	(Boulenger)	1906	ubiquitous
		elongatus *	(Boulenger)	1898	ubiquitous/rock
		kendalli *	(Poll)	1977	ubiquitous/rock
		pleuromaculatus *	(Trewavas & Poll)	1952	ubiquitous
		profundicola *	(Poll)	1949	rock
2	*Lepidochromis* *	*christyi* *	(Trewavas)	1953	bottoms
		bellcrossi *	(Poll)	1976	bottoms
1	*Lestradea* *	*perspicax perspicax* *	Poll	1943	sand
		perspicax stappersi *	Poll	1943	sand
4	*Limnochromis* *	*abeelei* *	Poll	1949	bottom
		auritus *	(Boulenger)	1901	bottom
		permaxillaris *	(David)	1936	bottoms
		staneri *	Poll	1949	bottom
3	*Limnotilapia* *	*dardennei* *	(Boulenger)	1899	ubiquitous
		loocki *	Poll	1949	?
		trematocephala *	(Boulenger)	1901	?
1	*Lobochilotes* *	*labiatus* *	Boulenger	1898	rock
3	*Ophthalmotilapia* *	*boops* *	(Boulenger)	1901	sand/rock
		nasutus *	(Poll & Matthes)	1962	rock
		ventralis ventralis *	(Boulenger)	1898	rock
		v. heterodontus *	(Poll & Matthes)	1962	rock
1	*Orthochromis malagaraziensis* *		(David)	1937	river
7	*Perissodus* *	*eccentricus* *	Liem & Stewart	1976	pelagic
		elaviae *	(Poll)	1949	pelagic
		hecqui *	(Boulenger)	1899	pelagic
		microlepis *	Boulenger	1898	rock
		multidentatus *	(Poll)	1952	pelagic
		paradoxus *	(Boulenger)	1898	pelagic
		straeleni *	(Poll)	1948	rock
7	*Petrochromis* *	*ephippium* *	Brichard	MS	rock
		famula *	Matthes & Trewavas	1960	rock
		fasciolatus *	Boulenger	1914	rock
		macrognathus *	Yamaoka	1983	rock
		orthognathus *	Matthes	1959	rock
		polyodon *	Boulenger	1898	rock
		trewavasae *	Poll	1948	rock

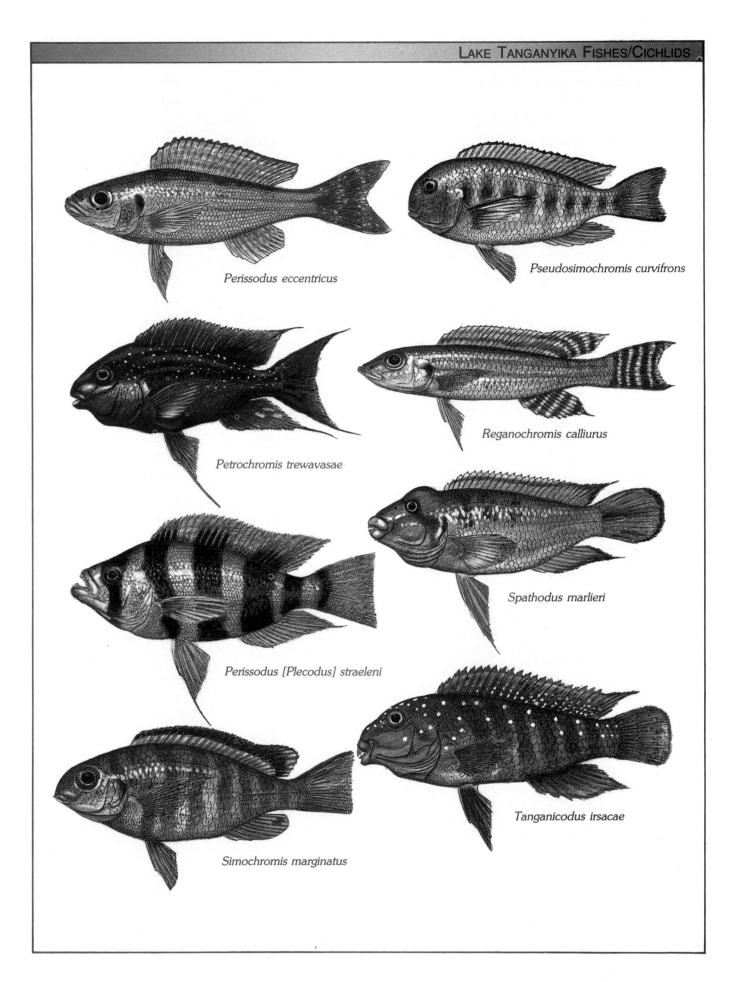

Perissodus eccentricus

Pseudosimochromis curvifrons

Petrochromis trewavasae

Reganochromis calliurus

Perissodus [Plecodus] straeleni

Spathodus marlieri

Simochromis marginatus

Tanganicodus irsacae

1	Pseudosimochromis * curvifrons *	(Poll)	1942	rock
2	Reganochromis * calliurum *	(Boulenger)	1901	sand
	centropomoides *	Bailey & Stewart	1976	?
5	Simochromis * babaulti *	Pellegrin	1927	rock
	diagramma *	(Günther)	1893	rock
	margaretae *	Axelrod & Harrison	1978	rock
	marginatus *	Poll	1956	rock
	pleurospilus *	Nelissen	1978	rock
2	Spathodus * erythrodon *	Boulenger	1900	rock
	marlieri *	Poll	1950	rock
1	Tangachromis * dhanisi *	(Poll)	1949	bottom
1	Tanganicodus * irsacae *	Poll	1950	rock
5	Telmatochromis * bifrenatus *	Myers	1946	rock
	burgeoni *	Poll	1942	rock
	dhonti *	(Boulenger)	1919	rock
	temporalis *	Boulenger	1898	rock
	vittatus *	Boulenger	1898	rock
4	Tilapia karomo *	Poll	1948	river
	nilotica	(Linné)	1758	swamp
	rendalli	Boulenger	1896	swamp
	tanganicae *	Günther	1893	ubiquitous
8	Trematocara * caparti *	Poll	1948	bottoms
	kufferathi *	Poll	1948	bottoms
	macrostoma *	Poll	1952	bottoms
	marginatum *	Boulenger	1899	bottoms
	nigrifrons *	Boulenger	1906	bottoms
	stigmaticum *	Poll	1943	bottoms
	unimaculatum *	Boulenger	1901	bottoms
	variabile *	Poll	1952	bottoms
1	Triglachromis * otostigma *	(Regan)	1920	mud bottom
5	Tropheus * annectens *	Boulenger	1902	rock
	brichardi *	Nelissen & Thys	1975	rock
	duboisi *	Marlier	1959	rock
	moorii kasabae *	Nelissen & Thys	1977	rock
	moorii moorii *	(Boulenger)	1898	rock
	polli *	G. Axelrod	1977	rock
1	Tylochromis polylepis *	(Boulenger)	1900	swamp
10	Xenotilapia * boulengeri *	(Poll)	1942	sand
	caudafasciata *	Poll	1951	sand
	flavipinnis *	Poll	1985	sand
	longispinis burtoni *	Poll	1951	sand
	longispinis longispinis *	Poll	1951	sand
	melanogenys *	(Boulenger)	1898	sand

98

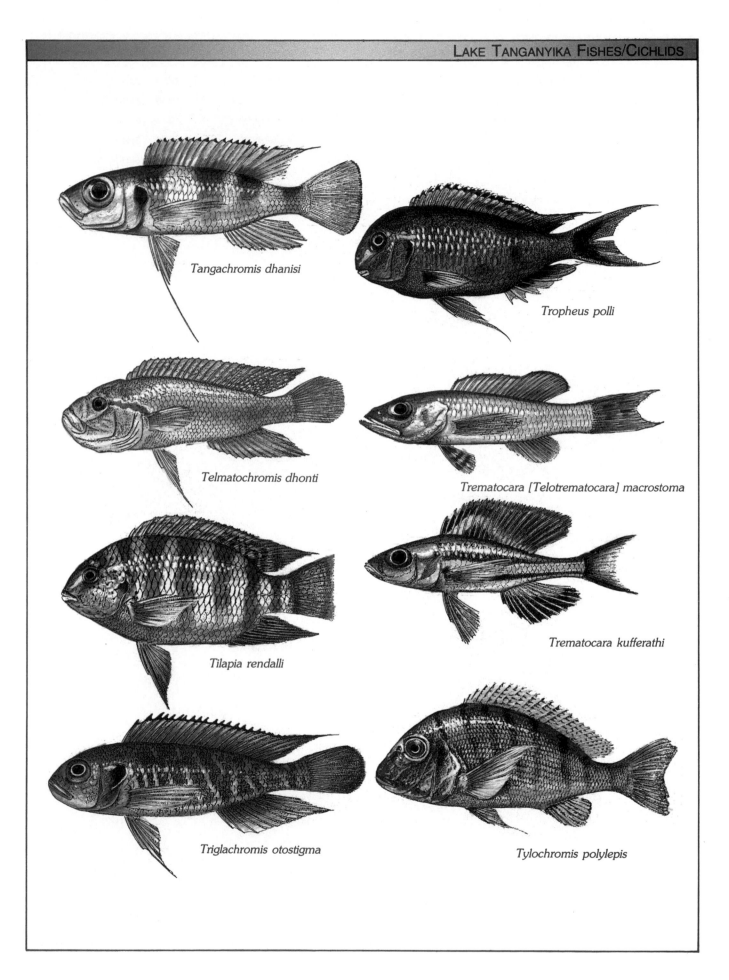

Tangachromis dhanisi

Tropheus polli

Telmatochromis dhonti

Trematocara [Telotrematocara] macrostoma

Tilapia rendalli

Trematocara kufferathi

Triglachromis otostigma

Tylochromis polylepis

nigrolabiata *	Poll	1951	sand
ochrogenys bathyphilus *	Poll	1956	sand
ochrogenys ochrogenys *	(Boulenger)	1914	sand
ornatipinnis *	Boulenger	1901	sand
sima *	Boulenger	1899	sand
spilopterus *	Poll & Stewart	1975	sand
tenuidentata *	Poll	1951	floor/sand

* denotes an endemic genus or species.

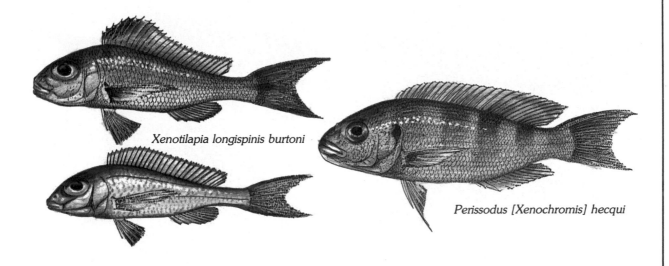

Xenotilapia longispinis burtoni

Perissodus [Xenochromis] hecqui

Since the completion of the basic text for this volume, Dr. M. Poll [*Mem. Acad. Roy. Belg.*, 45(2), 1987] has proposed a new scheme of classification for the cichlids of Lake Tanganyika. This involves the recognition of many new genera through the partitioning of more familiar genera. The color drawings accompanying the checklist have been identified according to the names used in the checklist, but the new generic names used in the Poll paper have been included in brackets where different. Whether these new generic positions used by Poll will be accepted by the scientific community and by hobbyists remains to be seen.

Tetraodon mbu is not a species endemic to the lake. It is an estuarine species found in the Malagarazi and no other river of the lake. Photo by Edward C. Taylor.

The 114 species belonging to families other than Cichlidae are distributed among 18 families involving 43 genera, of which only 8 are endemic to the lake basin. Thus, one might say that at the generic level the non-cichlids show much less endemism than the cichlids.

The eight endemic genera (*Limnothrissa, Stolothrissa, Bathybagrus, Lophiobagrus, Phyllonemus, Dinotopterus, Tanganikallabes,* and *Lamprichthys)* include only 12 species, all living in the lake proper and not in the basin affluents. Of the 12 species 9 live on rocky slopes, three in pelagic waters, and none over sandy bottoms. They all belong to families whose representatives have spread over most of tropical Africa.

As for the 35 non-endemic genera with their 102 species, they are distributed about evenly between endemic and non-endemic species, 45 and 57 respectively. Altogether the endemic non-cichlid fishes amount to 57 or exactly 50% of the total of these fishes.

If we look more closely at these families, we discover that less than half, especially the Clupeidae, Centropomidae, Mochokidae, Mastacembelidae, and Cyprinodontidae, in the non-endemic genera have colonized the lake proper. The other genera live in rivers or, when they come to live in the lake, have to go back to rivers for spawning purposes.

Putting together the endemic genera (totalling 12 species) and the endemic species from non-endemic genera (25) we find that 37 endemic non-cichlids live in and have adapted to the ecological conditions of the lake. In this biotope are found mainly rocky slopes and also somewhat pelagic waters, but practically no soft (sandy or muddy) bottoms.

If we turn toward the 57 non-endemic species we discover that only 15 have been found

The bichir *Polypterus ornatipinnis* is found in the swamps bordering Lake Tanganyika. Photo by H.-J. Richter.

in the lake and, as we just said, several, if not all, breed in rivers, or at least their estuaries. Some of them are found in many parts of Africa and it is very possible that they already lived in the rivers that were captured by the sinking grounds that were to become the lake. Of course some might have reached the Lukuga outlet on the upper course of the Congo River, climbed its falls and rapids, and thus finally reached the lake long after its birth. But it is very difficult to imagine how a slow-moving fish like *Tetraodon mbu* could have achieved such a feat, and then to explain how it happens to live only in one affluent across the lake from the Lukuga (Malagarazi) and in no other river around the lake. The same goes for *Polypterus ornatipinnis*.

As it is, although much less spectacular than with cichlids, the endemism displayed by the non-cichlids is by no means negligible, as made evident by the following chart.

Fish Flocks	Lake	Basin	Total
Cichlids endemic	167	4	171
non-endemic		4	4
			175
Non-Cichlids endemic	37	20	57
non-endemic	15	42	57
	219	70	289

LIST OF SPECIES as per 30 June, 1985★

2. *NON-CICHLIDS:*

N	Genus and Species	Orig. Descr.	Date	Habitat
LEPIDOSIRENIDAE				
1	*Protopterus aethiopicus*	Heckel	1851	swamps
POLYPTERIDAE				
2	*Polypterus endlicheri congicus*	Boulenger	1898	swamps
	ornatipinnis	Boulenger	1902	swamps
CLUPEIDAE				
1	*Limnothrissa ★ miodon ★*	(Boulenger)	1906	pelagic
1	*Stolothrissa ★ tanganicae ★*	Regan	1917	pelagic
MORMYRIDAE				
1	*Mormyrops deliciosus*	(Leach)	1818	rivers
1	*Pollimyrus nigricans*	(Boulenger)	1906	swamp/river
1	*Hippopotamyrus discorhynchus*	(Peters)	1852	swamp/rock
1	*Gnathonemus longibarbis*	(Hilgendorf)	1888	swamp/river
1	*Marcusenius stanleyanus*	(Boulenger)	1897	swamp/river
1	*Mormyrus longirostris*	Peters	1852	swamp/river
KNERIIDAE				
1	*Kneria wittei*	Poll	1944	torrents
ALESTIDAE				
2	*Hydrocynus vittatus*	(Castelnau)	1861	pelagic
	goliath	(Boulenger)	1898	pelagic
3	*Alestes imberi*	Peters	1852	coastal
	macrophthalmus	Günther	1867	coastal
	rhodopleura	Boulenger	1906	coastal
1	*Bryconaethiops boulengeri*	Pellegrin	1900	rivers
1	*Micralestes stormsi ★*	Boulenger	1902	Lukuga
CITHARINIDAE				
3	*Distichodus fasciolatus*	Boulenger	1898	Lukuga
	maculatus	Boulenger	1898	Malagarazi
	sexfasciatus	Boulenger	1897	coastal
1	*Citharinus gibbosus*	Boulenger	1899	swamps
CYPRINIDAE				
18	*Barbus altianalis*	Boulenger	1900	rivers
	apleurogramma	Boulenger	1911	rivers
	caudovittatus	Boulenger	1902	typical Congo River spec.
	eutaenia	Boulenger	1904	rivers
	kerstenii	Peters	1868	rivers
	lineomaculatus	Boulenger	1903	rivers
	lufukiensis ★	Boulenger	1917	Lufuku River
	minchini	Boulenger	1906	rivers
	miolepis	Boulenger	1902	rivers
	nicholsi	Vinciguerra	1928	Lukuga
	oligogrammus ★	David	1937	rivers

	paludinosus	Peters	1852	rivers
	pellegrini	Poll	1939	rivers
	serrifer	Boulenger	1900	rivers
	taeniopleura *	Boulenger	1917	rivers
	tropidolepis *	Boulenger	1900	rivers
	urostigma *	Boulenger	1917	rivers
	urundensis *	David	1937	rivers
4	*Varicorhinus leleupanus* *	Matthès	1962	coastal/river
	ruandae	Pappenheim & Boul.	1914	rivers
	stappersi *	Boulenger	1917	coastal/river
	tanganicae *	Boulenger	1905	coastal/river
6	*Labeo cylindricus*	Peters	1852	coastal/river
	dhonti *	Boulenger	1919	Lukuga
	fuelleborni	Hilg. & Papp.	1903	rivers
	kibimbi *	Poll	1949	rivers
	lineatus	Boulenger	1898	rivers
	velifer	Boulenger	1898	Lukuga
5	*Barilus moorii*	Boulenger	1900	river estuary
	neavii *	Boulenger	1907	coastal
	salmolucius	Nich. & Grisc.	1917	river estuary
	tanganicae *	Boulenger	1900	coastal
	ubangensis	Pellegrin	1901	river estuary
2	*Engraulicypris congicus*	Nich. & Gris	1917	Lukuga
	minutus *	(Boulenger)	1906	pelagic

BAGRIDAE:

1	*Bagrus docmak*	(Forsskal)	1775	estuaries
1	*Bathybagrus tetranema* *	Bailey & Stewart	1984	?
6	*Chrysichthys grandis* *	Boulenger	1917	ubiquitous
	graueri *	Steindachner	1911	ubiquitous
	brachynema *	Boulenger	1900	ubiquitous
	platycephalus *	Worthington & Ricardo	1936	ubiquitous
	sianenna *	Boulenger	1906	deep bottoms
	stappersii *	Boulenger	1917	sand/mud
1	*Lophiobagrus* * *aquilus* *	Bailey & Stewart	1984	rock
	asperispinis *	Bailey & Stewart	1984	rock
	brevispinis *	Bailey & Stewart	1984	rock
	cyclurus *	(Worth. & Rich.)	1937	rock
2	*Phyllonemus* * *filinemus* *	Worth. & Rich.	1937	rock
	typus *	Boulenger	1906	rock
1	*Auchenoglanis occidentalis*	(Valenciennes)	1840	ubiquitous
1	*Leptoglanis brevis*	Boulenger	1915	torrents

MOCHOKIDAE

7	*Synodontis dhonti* *	Boulenger	1917	rock
	polli *	Gosse	1982	rock

	granulosus *	Boulenger	1900	rock
	lacustricolus *	Poll	1953	rock
	multipunctatus *	Boulenger	1898	rock
	nigromaculatus	Boulenger	1905	swamps
	petricola *	Matthès	1959	rock
1	*Chiloglanis lukugae*	Poll	1944	torrents
	pojeri	Poll	1944	torrents

AMPHILIIDAE

2	*Amphilius platychir*	(Günther)	1864	torrents
	kivuensis	Pellegrin	1933	torrents

CLARIIDAE

1	*Heterobranchus longifilis*	(Valenciennes)	1840	ubiquitous
1	*Dinotopterus cunningtoni*	Boulenger	1906	rock
4	*Clarias liocephalus*	Boulenger	1898	swamps
	mossambicus	Peters	1852	swamps
	ornatus *	Poll	1943	swamps
	theodorae	Weber	1897	mud/coast
1	*Tanganikallabes* * *mortiauxi* *	Poll	1943	rock

MALAPTERURIDAE

1	*Malapterurus electricus*	(Gmelin)	1789	ubiquitous

CYPRINODONTIDAE

1	*Aplocheilichthys pumilus*	(Boulenger)	1906	swamps
1	*Lamprichthys* * *tanganicanus* *	(Boulengér)	1898	open/coast

CENTROPOMIDAE

4	*Lates (Luciolates) angustifrons* *	Boulenger	1906	open
	mariae *	Steindachner	1909	open
	microlepis *	Boulenger	1898	open
	stappersi *	Boulenger	1914	open

ANABANTIDAE

1	*Ctenopoma muriei*	(Boulenger)	1906	swamps

MASTACEMBELIDAE

12	*Mastacembelus albomaculatus* *	Poll	1953	rock
	cunningtoni *	Boulenger	1906	sand
	ellipsifer *	Boulenger	1899	rock
	flavidus *	Matthès	1962	rock
	frenatus *	Boulenger	1901	rock/rivers
	micropectus *	Matthès	1962	rock
	moorii *	Boulenger	1898	rock
	ophidium *	Günther	1893	sand/rock
	plagiostomus *	Matthès	1962	rock
	platysoma *	Poll & Matthès	1962	rock
	tanganicae *	Günther	1893	rock
	zebratus *	Matthès	1962	rock

TETRADONTIDAE

1	*Tetraodon mbu*	Boulenger	1899	estuary (Malagarazi)

In summary, we can say that in the lake endemic cichlids outnumber the other endemic fishes by nearly 5 to 1, but we find about 8 times more of the other fishes than cichlids in the affluents.

If we look into the distribution of each fish flock in the lake itself we can say that roughly their repartition is as follows:

Biotope	Cichlids	Non Cichlids Endemic	Non-endemic	Total
Pelagic	15	8	2	25
Deep floors	15	–	–	15
Rocky slopes	85	13	–	98
Sandy bottoms	25	5	–	30
Ubiquitous coastal	10	–	–	10
Ubiquitous	–	–	13	13
Unknown	10	–	–	10

Thus, an accounting of course not being taken of the many species that are still to be discovered mainly in deep waters and various rocky coastlines, one might say that at least 100 species of fishes, of which at least 50% are cichlids, can and do live together on rocky slopes. Only about 15% of the cichlids and about 10% of the other fishes have opted to live on soft bottoms. Non-cichlid fishes haven't been very successful when the opportunity was given them to colonize the lake waters, although 18 families with many genera existed in what became the lake basin.

At first the process was easy, as we can surmise, because when the depression started to fill in with river water there must have been mostly swamps at the bottom of the rift. The properties of the water were gradually changed through evaporation occurring during millions of years in the same way sea water became more and more saline.

Probably many lineages were progressively eliminated during this process, especially in the Cyprinidae, which are plentiful and ubiquitous in eastern Africa. For practical purposes let us say that only four species are regularly found in the lake. These still have to go back to river estuaries for breeding purposes.

What were the adaptations the fishes had to go through so that they could live in the expanding lake with increasingly mineralized water? Until the present it is very difficult for a fish coming from a river to survive in the lake, just as it is hard for a lake fish to wander into a river or the lake outlet and survive.

Rivers, swamps, or shallow lakes involve an array of ecological conditions that are basically different from the ones we now find in the lake. First and foremost they most often have murky waters that are heavy with fine mud particles, they are often very poorly oxygenated, more often than not they have a low pH, and their salt content is very low. They also reflect very quickly the seasonal climatic changes, alternating very dry seasons with long periods of heavy rains. Rivers and swamps often dry out more or less, leaving only a string of pools in which fishes try to survive, then a few months

later there is so much water they overflow.

By their physiology and behavior river and swamp fishes are for the most part well equipped to deal with such hardships and to survive amid their enemies or competitors. When they come into contact with the lake, their shortcomings, either physiological or behavioral, become apparent.

Professor Max Poll back in 1980 studied this problem and outlined the features of riverine fishes that help explain why many of them had so little success in Lake Tanganyika and several of the other Rift Valley lakes. To his outstanding work I can add only a few comments. By 1970 about 2,510 species of fishes had been found in African river systems and lakes, of which 1,238 belonged to the group called Ostariophysi, which thus represents 50% of the total African fishes, but much more if we exclude the many African cichlids from the Great Rift Valley lakes. Ostariophysi are typical of the river fauna, among which cichlids are in fact rather few. Non-cichlids represent 85 to 95% of all riverine fishes, with cichlids being outnumbered from 7 to 25 times according to the river system being explored. Ostariophysi include several families of characoids, cyprinids, gymnotids, and siluroids. Why have they been successful in rivers and not in Lake Tanganyika? Professor M. Poll finds an answer in their physiology as well as in their behaviors.

1. Many Ostariophysi have an excellent tolerance for low oxygen levels in slow moving water or murky waters carrying high amounts of decaying

organic matter being oxidized by aerobic bacteria. On the other hand many riverine fishes, and among them the Ostariophysi, have developed accessory breathing organs to assist their gills. At frequent intervals they rise to the surface to take in a gulp of atmospheric air. These trips are made possible by the rapid adjustment of their swim bladder to changes of pressure. Even when Ostariophysi do not have the accessory breathing system to regulate their intake of oxygen, all of them at least once, soon after they hatch from the egg, must go to the surface to fill their swim bladder for the first time. This trip to the surface can be made more or less safely by the young fish because their biotope is very murky and usually shallow. They can rise through the water column undetected by predators. The swim bladder of the Ostariophysi is connected to the gullet by a duct, which explains how the fish can increase or decrease the pressure inside the bladder quickly. That the accessory breathing organs are very useful is borne by the fact that many riverine fishes besides the Ostariophysi (for example, *Protopterus, Polypterus, Phractolaemus,* and *Channa*) use them as well.

2. Ostariophysi don't thrive in water with a high or variable level of salinity. Few live in brackish water or in inland waters with a high salinity, such as the Rift Lakes. They thrive in soft, acid water with a low pH and DH (especially the tetras).

3. Their eyes are often rather inefficient or even atrophied, but as they live in murky waters this is not very important. But their hearing is often outstanding. This is especially true of Ostariophysi which are endowed with the Weberian apparatus (basically a sound amplifier consisting of a string of small bones extending between the swim bladder wall and the inner ear). Through this connection sound vibrations reaching the swim bladder through the body walls are amplified and transmitted toward the ear. They give the fish information about what is happening around it beyond its visual range. The lateral line system of neuromast cells embedded in the scales on the fish's sides also provide information about disturbances in the flow of water along the body brought about by the intrusion of friends and foes alike. The need of very sophisticated detection means in riverine fishes is documented by the fact that, again, fishes other than Ostariophysi have developed them as well. For example, there are the mormyrids with electricity transmitting and receiving organs along the caudal peduncle with which they can emit and receive impulses. Mormyrids use this to compensate for very poor eyesight.

4. Olfactory senses are usually very well developed among riverine fishes. Catfishes and barbs have whiskers (barbels) with which they can detect food and many other riverine fishes emit chemical substances with which they can transmit signals.

5. Tropical fresh waters, such as rivers, respond quickly to seasonal climatic changes such as the dry and rainy seasons. During droughts rivers dry out entirely or are reduced to a trickle or overheated and muddy pools. Pollution of the remaining water increases to dangerous or toxic levels. Oxygen levels drop and fishes might suffocate. Food supplies are exhausted because the river has retreated from the grasslands along its banks. The density of fishes in the much contracted habitat is too high and deadly epidemics or parasitic diseases develop. The dry season is most often a dreadful ordeal for the river flocks.

In strong contrast the rainy season is a time of bounty. Decaying matter is flushed downstream by the rising water. Rivers and dried-out swamps creep back into the nearby flatlands. A new aquatic plant cycle starts and breeding grounds become available for the now fattened fishes. Breeding activity in rivers is thus very seasonal. Ostariophysi for their part lay large clusters of eggs, often several tens of thousands, which they often leave unattended on the bottom, stuck to aquatic plants or floating free. Losses through predation are very high, but they remain at an acceptable level because the murkiness of the biotope makes detection of the spawns difficult.

6. The camouflage of the riverine fishes, including the Ostariophysi, blends very well with their habitat. The patterns are usually plain and in shades of drab gray, brown, or purple. Few species display a strongly contrasting pattern. In virgin forests where water is laden with humic acid and brown in color (but often very clear), the robes of fishes (found also in savannah rivers) are much more colorful in shades of

107

bright brown, red, and purple. As a rule, aside from cichlids, riverine fishes have only one pattern, which doesn't change much through their various moods.

How do the cichlids fare in river biotopes? Having originated in the sea they thrive in rivers and freshwater biotopes with a high salt concentration and adapt to a fluctuating salinity. They don't thrive in rivers with a low pH and DH, as documented by the relatively small number of species found in the soft, acid waters of virgin forests. Only six such species have been found in the central basin of the Congo River.

On the other hand, they do not tolerate well fluctuations in the amount of oxygen provided by their habitat. Because their swim bladder is closed, it regulates the changes in pressure undergone when they rise or sink in the water column more slowly than other river fishes. Compensation for the pressure changes is controlled by two glands in the swim bladder, one for producing gas to fill the bladder, the other to absorb the gas when the fish swims down or up. This process, of course, is slower than the one used by fishes with an open duct bladder.

Cichlids cannot go to the surface readily for a quick gulp of air whenever they start to suffocate. They have to stay at the surface which is exhausting and might be dangerous when there are predators about. This is why, when the barometric pressure drops before an incoming storm, cichlids are often the first fishes to be seen by the hundreds gulping for air at the surface, as the water releases

the oxygen it had stored when the barometric pressure was higher.

The fact that cichlids are sensitive to oxygen levels is also documented by the species living in rapids. These suffocate very quickly when placed in water that is not strongly aerated. Proof of their oxygen requirements is also borne out from their respective populations in the Regina Rapids (17 species) and the nearby Stanley Pool swamps of the Congo River (9 species).

In contrast to most Ostariophysi, cichlids as a rule have excellent eyesight, but it is of little use in their murky river surroundings. Proof of this fact can be found in *Lamprologus lethops,* the blind cichlid living under the rock slabs of the Kinshasa rapids. It is the only blind cichlid in the world.

Cichlids also have well developed lateral lines of sensory organs and they can detect disturbances in the flow of water around them as well as many Ostariophysi.

Because of the seasonal changes in ecological conditions, riverine cichlids have to follow the breeding cycles of all riverine fishes. But their broods are much smaller than the average, which they compensate for by giving excellent care to their spawns, protecting them until the fry are capable of fending for themselves. Losses are thus kept to a minimum, but one might think that should a spawn be lost, and because of the long time it takes a pair of cichlids to raise its fry, there might be too little time to repeat the whole spawning process during the adequate season. Another advantage of cichlids lies in the fact that they

don't need to fill their swim bladder with atmospheric air, which saves the fry from the dangerous trip to the surface. Their bladder is filled by a gas-producing gland within.

On the other hand, most riverine cichlids are not very brightly colored, their colors involving mainly shades of green, purple, or brown (*Steatocranus, Leptotilapia, Orthochromis, Teleogramma, Lamprologus, Tilapia,* etc.). Only a few display bright colors, especially in clear virgin forest brooks, in arrays of crimson, red, purple, or even violet (*Pelvicachromis, Hemichromis, Nanochromis dimidiatus*). In such a habitat the capacity of these fishes to change their body color pattern according to their mood and thus transmit signals to friend and foe alike is put to better use than in the murky biotope of a large river.

With all these facts put together one can thus understand why it is that cichlids are outnumbered 7 to 1 by ostariophysans in the Congo River fluviatile system, and 11 to 1 if one takes into account all the inhabitants of this system.

M. Poll then turns to the Rift Lakes and analyzes how and why the main families of freshwater fishes managed their adaptations to an entirely different ecology. The high salinity of the main Rift Lakes was a handicap for the Ostariophysi, as can be deduced by the fact that they occupy in strength the surrounding rivers, but not the lakes themselves. Cichlids, for their part, occupied the lakes in strength and diversified as nowhere else, but not in the affluents. What are the ecological conditions that prevail in these lakes? Water is very transparent, its temperature is stable, and its volume acts as a buffer against seasonal variations. This is especially so in Lake Tanganyika (less in Lake Malawi) because it is close to the Equator, very deep, and laden with minerals that help keep the water crystal clear. Temperature in Lake Tanganyika is remarkably uniform and the lake, therefore, is practically not stratified into layers. Seasonal changes in the lake level are inconsequential (less than 1 meter) and do not destroy the coast's essential supply of vegetal matter, some of which is found in the phytoplankton but mainly in the rock's algal biocover.

Neither Lake Tanganyika nor, for that matter, Lake Malawi or Victoria receive the impact of seasonal changes in the supply of water from a major river which might bring about a drastic change of the water quality. All receive the water supply from their basin through a host of small rivers. Because they are oriented more or less on a north-south axis, both Lake Tanganyika and Lake Malawi are reached by the rainy season river output only very gradually.

Decay of organic matter in both lakes occurs in the deep layers and does not affect the fishes living in the upper reaches of the lakes. The top layer, 40 to 100 meters thick, is very well oxygenated.

Because of the seasonal plankton bloom there is a peak in sexual activity after the rains have started. But in coastal waters of Lake Tanganyika even in the midst of the long, dry season, there is no shortage of food. Fishes are

never starving and are capable of spawning more often and over longer spawning seasons than in unstable rivers and swamps. As they are much less affected by the dry season than in most fresh-water biotopes, the fishes are also less prone to seasonal epidemics and parasitic infestations, of which none has ever been reported from Lake Tanganyika (parasites do exist but do not reach epidemic proportions as does. *Ichthyophthirius* in Kinshasa). Because of the slow drop in oxygen levels with depth, cichlids can live deeper and adapt to low but stable oxygen supplies. They started living deeper perhaps as a way to avoid the crowded conditions prevailing on the upper reaches of a slope or perhaps also to go after a favored food supply.

Cichlids can live deeper because their fry do not have to go to the surface for a first intake of air. This is perhaps the single most important feature that prevented the Ostariophysi from becoming permanent settlers in the lake. Were they to spawn in the lake, how could their fry rise through crystal clear water and survive in the midst of so many predators! No wonder then that the Ostariophysi from Lake Tanganyika either keep going back to rivers to spawn or have undergone a drastic change in their physiology. By abandonning accessory breathing organs they don't need to go to the surface anymore. Some examples are the Clariidae of the lake (ex. *Tanganicallabes, Dinotopterus*). These fishes live and breed in the lake and do not go into rivers.

Survival of the brood is

perhaps the most vital problem the lacustrine species have to overcome. This is made evident by the variety of techniques used. In this respect cichlids fare much better than many other fishes because they are used to looking after their spawns until the fry can fend for themselves. Cichlids in several of the Rift Lakes developed the technique some of them used already in rivers—mouthbrooding. But if in Lake Malawi most cichlids, at least on rocky coasts, have not elaborated much and have spawns that are rather uniform in the number and size of the eggs as well as in the care afforded to the brood, in Lake Tanganyika we discover a remarkable diversity of the techniques used, not only for the breeding process but also in the number and size of the eggs. There is a tremendous diversity in the entire reproductive process among the mouthbrooding or nestbreeding cichlids. In this respect Ostariophysi can be said to have altogether failed to adapt.

On the other hand, in a biotope where visibility is outstanding, the poor sight of the Ostariophysi became a lethal handicap for many, even for those fishes that had excellent sensory organs. They were not at their best among cichlids that benefited not only from an excellent eyesight but also had well developed lateral lines. They developed the latter organs as is borne out by the number of lateral lines in *Xenotilapia* (reaching the unusual number of three lateral lines), by the variety of sensory cells and captors, and by the fact also that when the lateral lines

109

appear to be atrophied (as in *Trematocara*), they are compensated for by the cephalic drum-like cavities. More than any other African fishes, the cichlids are capable of displaying their moods by subtle changes in their patterns and colors. Again, in an environment in which fishes can see each other well, the cichlids started to display strong contrasting colors and a variety of patterns. Hence the bright colors of the Malawi cichlids and of many species in Lake Tanganyika. That the transparency of the water is directly linked to the brightness of the colors is borne out in Lake Tanganyika by the fact that whenever specimens of a race of *Tropheus* are found in parts of their habitat where water is cloudy, these specimens inevitably have a much drabber robe than their counterparts living in clear water a few kilometers away. Let us remember also that it is in the crystal-clear waters of warm seas that the marine percomorphs display their brightest hues.

The fact that in the lakes cichlids, because of the oxygen levels, could live deeper than in shallow rivers brought about other adaptations of their morphology. The dwindling light that prevails forced them to adapt. The eyes of the deep-living cichlids of Lake Tanganyika, such as those of *Trematocara,* are large instead of atrophied as in *Lamprologus lethops* from the Congo River. Simply because the transparency of the water is so good, vision remains sharp, as the eye has only to become better adjusted to dim light to keep its value for the fish.

Thus *Trematocara* and other deep-living fishes maintain strongly contrasted patterns so that they can make each other out and retain their social contacts.

As a result of the adaptability of patterns and because of genetic drift, new patterns tend to appear among isolated cichlid populations living along a coastline. This led to inbreeding and development of local geographical races (color morphs) which became distinct from the other populations. The number of cichlid species thus progressively expanded. Morphological, physiological, and behavioral adaptations became more and more complex as competition for shelters, nest-sites, food, etc., and predation among mainly carnivorous fishes increased. This is how, freed from the presence of other families of fishes and living under increasingly ideal conditions, cichlids multiplied in the Great African lakes and outnumbered all other fishes. In Lake Victoria this is by a margin of 4 or 5 to 1, in Lake Malawi by 6 or 7 to 1. In Lake Tanganyika, among the endemic species living only in the lake, cichlids outnumber other fishes by 5 to 1 (and probably many more cichlids remain to be discovered than non-cichlids), and among all species, endemic and non-endemic, by 3 to 1, although in affluent rivers of the lake they are outnumbered by non-cichlids 8 to 1.

This is, according to Prof. Poll (with a few additional features of my own) how one can explain the proliferation of cichlids and the penury of other fishes in the three major African lakes.

Young *Lamprologus tretocephalus* make a colorful display. Photo by G. Meola.

CHAPTER V

ECOLOGICAL BARRIERS ● IMPACT OF ISOLATION ● FORDS ● SPECIATION ● LOCAL INTRALACUSTRINE ENDEMISM

For a period of more than 20 years (1949-1971) I waded in many streams of the Congo River basin looking for tropical aquarium fishes to export. I collected them in many habitats, from virgin forest springs to savannah streams, including the river rapids in Kinshasa, swamps, and fetid mangroves. Some of the species I found in many habitats, more were to be found only in one. Some were spread about everywhere, others in a very special and limited area.

For example, a fish like *Xenomystus nigri* could be caught in virgin forest brooks among the dead leaves, from swamps in the Stanley Pool, and even in the foaming waters of the rapids. But the tiny *Barbus hulstaerti* could only be scooped out from around a leaf lying in the cool headwaters of a spring deep in the virgin forest where it seeps from the ground. As soon as there was more than a few cm of water the dwarf barb disappeared.

If we take apart the components of the biotope the two species occupy, we discover that *B. hulstaerti* lives in very cool (20-21°C), very soft (dH O°) water with a lot of humic acid and a pH of 4.5. As for *Xenomystus nigri*, it has a wide tolerance for pH and temperature, and a fair tolerance to hardness.

As soon as the water depth increases in the brook meandering through the stuffy forest, other fishes appear alongside *B. hulstaerti* that can prey on the fish. The temperature also increases to 23-24°C and the pH climbs to 5.5°. When the brook reaches a larger stream flowing in the sun, the temperature of the water gets even higher and *B. hulstaerti*, even without the many predators around, cannot acclimatize and disappears from the landscape.

In the Congo River temperatures average 26-27°C, pH is close to 6.5°, dH is still low but can reach 3°, and there are altogether about 130 species of fishes living in the river. *B. hulstaerti* is thus bonded, or infeudated, to its typical forest habitat and the ecological conditions that prevail in a very unusual

The dwarf barb *Barbus* (or *Capoeta*) *hulstaerti* has certain habitat requirements that must be met before it can survive. Photo by Aaron Norman.

The same Congo affluents that have been a barrier to the dwarf barb have at the same time been a vehicle for the spread of *Xenomystus nigri.* Photo by J. Elias.

aquatic microcosm.

If we turn back to *Xenomystus* we discover that the fish has spread not only over most of the Congo River biotopes, but also throughout most of western Africa. If the main affluents and the Congo River have been a *barrier* to the spread of *B. hulstaerti,* they have at the same time been a *vehicle* for *Xenomystus* in its attempts to colonize new territories.

The ambivalence of a biotope as a vehicle or a barrier for various species of fishes is a feature that we will find again and again in the coastal biotope of Lake Tanganyika. In fact, it is the adaptability of a species to new ecological conditions that lies in the background of its final range.

If two similar habitats are separated from each other by a biotope of a different type, such as rapids separated from each other by swamps, much will depend on the size of the barrier as well as on the different ecologies and their

degree of difference. Two populations of rapids-dwelling cichlids, one in the Stanley Falls in Kisangani, the other in the Regina Falls in Kinshasa, separated by 1,500km of swamps, are substantially different. For a fish used to living in the foaming waters of rapids, crossing a swamp is impossible because it cannot survive in such poorly oxygenated waters. If oxygen conditions are relatively good it might use a ford to cross the swamps, if they are not too broad.

A drastic change in the ecology provides a very efficient ecological barrier. It so happens that in the lower Congo, which for about 300km is a succession of impressive rapids, fishes from the right bank affluents are different from those from the left bank affluents. The first are related to the fishes found in the forest basin of the Kwilou-Nyari River; those from the left affluents are related to the Angolan fish fauna. These fishes are mostly small and

live in waters that harbor few predators. When they come into contact with the foaming waters of the Congo, they are not equipped to deal with such an immense volume of water travelling at more than 30km an hour that teems with a large variety of predators.

Sometimes natural phenomena bring about an accidental and episodic mix of two populations normally living well apart and in two different biotopes. Back in 1962-1963 I was very fortunate to witness the effects of a major seasonal flood of the Congo River and its affluents. All affluents from the smallest brook to the large streams became swollen by exceptionally heavy rains and carried away fishes that had been living in them, sweeping them into the Congo.

We thus started to discover *Bathyaethiops caudomaculatus* and a species of *Pelvicachromis* very similar to *taeniatus* in the Stanley Pool. They both came from small forest rivers upstream with a pH of 5.5, a very low 1° dH, and a high humic acid content. Their waters are practically devoid of *Ichthyophthirius* parasites (perhaps because of the low pH) and have few predators. Basically each of the fishes are the only ones of their family to live in the small rivers. When they were swept into the Stanley Pool they managed to settle down immediately and even to breed, as I could determine when I caught adults and juveniles for a couple of years. Then they disappeared. The question was why?

There were probably several reasons, a few of which we can outline here:

1. Competition from species belonging to the same type of fishes, such as characins for *Bathyaethiops*, or small cichlids for *Pelvicachromis*, that were indigenous to the Stanley Pool and well adapted to this habitat. The two intruding species had to compete in a new biotope with the species already existing there.

2. Pressure from predation, especially by fish-eating fishes that had become specialized for their task by developing faculties that are not found in small-stream predators. Against these predators the two intruders had few defenses (inadequate camouflage for example).

3. Parasites and diseases against which the two species had not developed inbred immunity. Forest fishes, when put into a tank, are especially prone to get and succumb to infestations of *Ichthyophthirius*, whereas savannah fishes usually display a higher degree of resistance.

The failure to settle down permanently in the Stanley Pool area was not so much due to physical or chemical properties of the water. The quality of the water, although it might have had a debilitating effect on the longterm resistance of the fishes, was not so adverse as to prevent them from living and breeding there for several years. Essentially, although several components of the intruder habitat had been present in the Stanley Pool, they failed because some of the other components in the swamps, competing species, predators, and diseases, could not be tolerated by the two species.

Another feature of the river fluviatile biotopes that I discovered while fishing in the Congo was the notion of the *fords*.

Some species, passing through "causeways" whose ecological conditions are more or less similar to the ones they live in on their original grounds, manage to cross "hostile" territories to reach another biotope of their liking. It probably happens by pure chance that they enter a causeway, because they can live on it, then progressively go further along the passage until it stops or reaches a biotope that they then colonize.

The swamps and low-lying islands of the Stanley Pool are only a few miles away from the Regina Falls rapids. Both habitats have a population of *Synodontis* that one finds in the other, but each also has a few species that cannot live in the other because the

Synodontis brichardi is a specialized rock-dwelling species requiring well-oxygenated water and feeding exclusively on a type of rock-encrusting algae. Photo by Dr. Herbert R. Axelrod.

ecological conditions prevailing in each are in many ways drastically different. One such species bonded to the rapids is *S. brichardi*, which clings to rocks, needs well-oxygenated water, and feeds exclusively on a type of rock-growing algae. It is a *specialized* rock-dwelling species that has never been found in swamps and muddy areas. It was thus a surprise to discover a small population of this fish at the downstream tip of the last island in the Pool where accelerating current has laid bare a rock promontory. One had to admit that the *Synodontis* had found a rock path on the mud-covered floor leading from the rapids upstream to the island. Further upstream along the island's long coastline (30km) the fish couldn't be found anymore. No other rapids-dwelling species were found on the promontory along with our species. The rock ford had thus not been used by other fishes which, in the rapids, share the habitat of the catfish.

As we progressed with our understanding of the forces at work on the riverine fishes we could not fail to discover that it was the variety of the habitats that multiplied the variety of the fish populations. Most of the many features of the fluviatile ecologies were, however, still hidden within the murky waters and frustrated our efforts at direct observation of the fishes in their environment.

At long last, when I switched my allegiance to the crystal clear waters of Lake Tanganyika and explored its slopes with scuba gear, could many unformulated riddles pop out into the open and I could

eventually begin to find answers.

In the lake, when fishes migrate it is not under the impact of a natural overwhelming force, as happened in the Congo River early in the sixties. It is because they have an opportunity to do so according to the layout of their biotope and their bonds to the biotope.

Physical and chemical properties of the water are basically everywhere the same in the lake within the top life-bearing layer. We can, however, say that on the surf-washed rocky shores oxygen levels are very high and that they dwindle with depth. Some fishes living on the fringes of the shore need higher oxygen concentrations than fishes always living deep on the slope. Some fishes have thus grown used to life in the very top layer, while others can move and migrate by using deeply sunken passageways that are unavailable to the first.

A ford can thus exist for one species and not for another. A *Julidochromis* can use a ford that is closed for a *Tropheus* simply because oxygen levels are too low for *Tropheus*. An ecological barrier can be open or closed for several other reasons. Its "porosity" will be specifically variable.

One of the main problems encountered by coastal fishes in the lake is the prevailing alternation of the three main types of littoral biotopes: sand, mud, and rock. We have seen that rock-dwelling fishes in the lake do not go to live, often not even for a short while, over a sand bottom. It all depends on each species' bonds toward its preferred habitat. If they are loose, rock-dwelling fishes will

wander over and eventually cross a sand or even a mud barrier. If they have become specialized to living on rock, they will not be capable of surviving on another type of habitat.

Infeudation to rock or to any of the other two habitats is variable from one species to the next. It might even be variable within a species, between juveniles and adults, because young fishes usually have requirements of their own that they no longer have when they become adult, such as a shelter into which only they can fit.

An excellent example of specialization is provided by *Tropheus*. It will never wander from its rocky bottom over sand for more than a few fin strokes. It will thus not enter areas of sand bottoms nor venture into open waters far from the slope. It needs a continuous belt of rock along the shore in the upper layer to be able to migrate.

The alternation of rock, sand, and mud floors along the coast has thus been a major obstacle to the spread of specialized species that are infeudated to a coastline. Their populations are fragmented in isolated biotopes.

But, as the bonds to their specific habitats are not identical from one species to another, some species will be more tolerant to changes in the ecological conditions they live in than others. These species will be found on a variety of habitats while the latter will be more localized.

The bonds that tie a species to a given type of biotope are many and difficult to assess. First we can say that a fish is well adapted to its habitat when the species survives and breeds there successfully, which means that, as far as we can see, the future of the species in this biotope is secure. The species has then found its *ecological niche*.

Sorting out the different components, or parameters, of any specific ecological niche is a difficult task, because if many are tangible features (such as physical or chemical properties of the water, i.e. pH, dH, temperature, organic matter content, luminosity, oxygen levels, etc., or of the surroundings, presence or lack of appropriate shelters and breeding sites, or nature and quantity of food supplies), many more cannot be measured so easily, or often cannot even be identified. Such items are the presence or lack of competitors and predators, the first in relation to room, shelter, breeding sites, and food, the second in relation to survival of the adults and their offspring. Also included are the abundance or relative lack of specific parasites or diseases that might cut down or obliterate a

The rock biotopes provide protection for many species, such as this *Tropheus duboisi*. But the patchiness of this habitat also serves to limit the rock-dweller's distribution somewhat. Photo by Dr. W. Staeck.

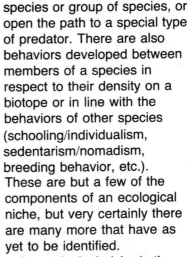

Other species of *Tropheus*, for example this *T. brichardi*, have similar rocky habitats. Photo by Glen S. Axelrod.

species or group of species, or open the path to a special type of predator. There are also behaviors developed between members of a species in respect to their density on a biotope or in line with the behaviors of other species (schooling/individualism, sedentarism/nomadism, breeding behavior, etc.). These are but a few of the components of an ecological niche, but very certainly there are many more that have as yet to be identified.

An ecological niche is thus not a territory (which is only a spacial entity) because the niche is made up of a set of physical components and a much more complex set of interspecific relationships, including behavior of the species one is studying and of the others.

Part of the ecological niche of a *Tropheus* is composed of its contacts with the other fishes that share the biotope. As a juvenile, *Tropheus* will suffer from the preying habits of some of the fishes living

Perissodus paradoxus is one of those highly specialized fishes that feed on the scales of other fishes. Photo by Dr. Herbert R. Axelrod.

within the recesses in which it hides. As prey for these predators the young fish will have an impact on their life. As an adult, *Tropheus* will have little impact on predators, except for the scale-ripping *Perissodus*. It will have little impact on nestbreeders as well, as it doesn't take breeding grounds from them, nor on the juveniles, which the fish will leave alone, and very little impact on their food supplies, because they are rather different. But a dense population of *Tropheus* might put pressure on *Simochromis* with which they are in direct competition in most parts of their respective habitats. Similarly, a *Tropheus* will, under normal conditions, not display aggressiveness toward a *Lamprologus,* but might engage in a very serious and prolonged fight with a *Petrochromis*. Also, when *Petrochromis orthognathus* occurs in a dense population, it might display schooling

behavior. If there are *Tropheus* nearby, the group of *Petrochromis* and the *Tropheus* move in a group. *Tropheus,* although they are basically asocial, will follow the *Petrochromis* in their moves and, for example, flee into the open as *P. orthognathus* most often does, instead of going into hiding amid the rubble. So the local schooling behavior of the *Petrochromis* has an impact on the change of social behavior of *Tropheus.* The presence of many species living together thus sets off a succession of interacting behaviors that reverberate on all the species.

Against the pressures of predation and competition, and the opportunities that open up because of the respective variety of each biotope, species tend to develop means to survive. They adapt and specialize. A species will make good the opportunities (for preying for example) or try to free itself from a pressure or

make it bearable. This involves finding a place on the slope where it can live, eating food whose supply is neglected or not exhausted by other fishes, finding a place to spawn and raise fry safely, and also to develop defenses against predators (morphological, physiological, or behavioral). It also involves a predator developing a better means to catch its prey. All this must occur within the context of potential change carried over from its ancestors. This means that a mouthbrooding cichlid will not return to nestbreeding if its breeding method is not successful under local circumstances, but the species will elaborate on mouthbrooding to make it safer. For example, *Tropheus,* faced with the competition for nesting grounds on the crowded bottom by a number of nestbreeding cichlids, has not abandoned mouthbrooding to become a nestbreeder, it has developed a better way to incubate its fry, i.e., mid-water fertilization, which is the most dangerous part of the process at ground level.

Each appearance of a new adaptation in any one of the species present on a biotope modifies the parameters of the other species. Ecological niches are thus fluctuating entities. New ones are arising on a coastline all the time and as the parameters are so many and so complex, no wonder then if we never will find two biotopes in the lake that are identical and two fish flocks that are similar.

As species specialize to adapt to life in a biotope that they share with other species, the more precise and delicate their tuning to local ecological

Mouthbrooding is one adaptation to the predator pressures of a rocky habitat. Here a pair of *Perissodus microlepis* release fry over a feeding area as they stand guard above them. Photo by Pierre Brichard.

conditions becomes. As a result, a very specialized species is less capable of adjusting to a strong change in its ecology. For a specialized species the ecological niche becomes very narrow and constricting. For a less specialized species its bonds toward a habitat are broader, less compelling; it might tolerate important changes in its surroundings. A "narrow" niche occupied by a fish that is specialized has a set of very precise parameters without which the fish cannot survive. A "broad" niche occupied by an unspecialized species has a set of parameters that can be found in a broad array of habitats.

Let us illustrate these notions with a few examples. *Lampologus compressiceps* is a typical rock-dwelling nestbreeding cichlid. Its body is very flattened laterally and heavily armored with very ctenoid scales; there are many long, sharp spines in the dorsal and anal fins. It is a sluggish fish with poor speed and stamina that lives in the open over rocks and feeds on crustaceans that it pries out from tiny crannies on the rock walls. It breeds in deep crevices with a very narrow slit as an opening. Its fertility is very low for a nestbreeder. It is a common species in most rock habitats but is missing from many; it is never very

abundant. From these features we can deduce that *L. compressiceps* cannot find a shelter or a breeding place over sand or mud bottoms.

The morphology of its mouth is perfectly adapted to pick a crustacean loose from a tiny hole on the surface of a rock but it is too narrow to successfully grab anything else in the open. Its requirements for a breeding place within a rock slab cannot be met in areas where rocks are silted over. Its low fertility requires that its fry escape from predation as much as possible, a condition that is not met on open floors.

117

The sluggishness of its movements prevents the fish from engaging in long migrations to other habitats. The parameters of the niche of the species are thus restrictive. But they have to be filled if the fish has a chance to survive. They can be found only on the most typical of rock habitats, which explains why the species is missing from many.

Lamprologus modestus is an ubiquitous nestbreeding cichlid whose morphology belongs to the mainstream *Lamprologus*, probably one closest to the fluviatile species of the genus. It is endowed with a fair amount of speed and stamina and is not as quickly exhausted after a fast run as *L. compressiceps.* It is an omnivorous species living on drifting zooplankton, fry of other fishes, crustaceans buried in the sand, etc. It breeds on sand and on rock bottoms as well. Its spawns involve several hundred fry. Its camouflage blends well with the landscape, especially over sand, of which many patches litter the bottom on rocky slopes. It withstands water loaded with a lot of organic matter and is thus to be found in very murky areas. It is one of the very few species, along with *Telmatochromis caninus,* to be found in rocky areas too poor in biocover and plankton drifts to support many species. This species is very tolerant of ecological conditions that are adverse to many.

Tropheus moorii is a mouthbrooding cichlid inhabiting rocks whose features have already been discussed. It requires clean, well oxygenated water and stable rubble on which biocover can grow. Rolling pebbles in which only their fry can hide are thus not the habitat of the adults. Although very mobile, the fish never wanders from its rock biotope. It spawns in midwater and keeps the fry in its mouth for more than a month. Their dentition is specialized to cut the biocover from the rocks, and their mouth is so shaped that it is very difficult for a *Tropheus* to grab food in midwater or pick it up from the bottom.

Tropheus are bonded to their habitat by the need to find rocks on which a carpet of biocover is growing, which means that they live in the top 15-20 meters of the oxygen-bearing layer. Their oxygen requirements have thus increased.

The need for the fish to release their fry in rock recesses that are appropriate to the size of the fry is another peculiar feature of their ecology. In this respect they are different from *Lamprologus compressiceps,* which can live deeper but are not on the lookout for a place in the midst of the biotope to abandon their fry at the end of their parental protection.

L. compressiceps and *Tropheus* are thus very specialized rock-dwellers, but the parameters of their respective ecological niche are in many respects very different. *L. modestus* is much less specialized and has a broader ecological niche. This is reflected in the range of the species in the lake, which includes most of the littoral biotopes.

Lamprologus modestus is an ubiquitous nestbreeding cichlid. It is omnivorous, living on zooplankton, fish fry, and invertebrates. Photo by Dr. Herbert R. Axelrod.

To sum up we can say that the variety of species in any one place reflects the number of ecological niches available there. The density of a species population reflects the number of territories that could be partitioned from the area in accordance with the parameters of its ecological niche. Unusually high densities of a species can result from the lack of competition by species that have an ecological niche very similar to the one occupied by the first, or from the lack of predators.

We were very fortunate to discover during our first dives with scuba gear a biotope in which the density of a species was much higher than anywhere else. To the south of Bujumbura, 8km toward the main rock coast in Burundi, huge sand shallows line the coastline for about 25 km. The sandy bottoms at the sand point of Ruziba are here and there broken by sandstone slabs scattered over the bottom, from which they rise by about 1 meter. The slabs are broken by long cracks and fissures, and in some places are full of holes like a swiss cheese, providing a number of shelters and narrow crevices in which large and small fishes can find hiding places. Altogether the slabs occupy perhaps 5,000 square meters of the bottom.

The slabs lie about one km offshore and are isolated from the main coast by a muddy bowl that will become in due time the central pool of a lagoon. They were laid bare under the action of waves, and although they lie between 2 and 3 meters deep, they are now being silted over at a fast pace by the building of the outside rim of the lagoon.

The sandstone building process is brought about by agglomeration of quartz particles cemented together by silica; this process is underway on most of the large beaches around the lake.

At Ruziba the slabs are separated from the main rock biotope by 15 km of sandy bottom, with several similar slabs of sandstone between, sunk deeper in the water. There is thus a broken-up ford of rock between the main coast and Ruziba. North of Ruziba, toward Bujumbura, if the sandstone belt exists it is buried in the sand and cannot be used by the fishes. In this area water becomes especially murky and opaque because of heavy sedimentation.

The rock-dwelling fish fauna at Ruziba is very poor and contains nothing like the variety found on the main coast. We find there all of the ubiquitous *Lepidiolamprologus* (except *profundicola*), *Lamprologus mondabu, L. tetracanthus, L. callipterus,* and *L. lemairei;* and among the rock-dwelling fishes *Julidochromis marlieri, Telmatochromis caninus* and *T. temporalis, Petrochromis fasciolatus* and *P. polyodon,* and juveniles of the ubiquitous *Limnotilapia dardennei.* The most typical rock-dwelling cichlids are missing, such as the rock-grazing *Tropheus, Simochromis, Eretmodus,* and *Spathodus,* as well as the specialized nestbreeders, such as *Lamprologus compressiceps, L. furcifer, L. brichardi, L. savoryi,* and *L.*

Lamprologus compressiceps is a specialized rock-dweller. Its strongly compressed body allows it access to narrower crevices than most other fishes of the same length. Photo by Dr. Herbert R. Axelrod.

schreyeni. We do not find either the plankton pickers of the *Cyathopharynx* guild, or the *Cyprichromis* or *Cyphotilapia*. We would not, in fact, have been drawn so often at first to these slabs had it not been for the extraordinary abundance of *Julidochromis marlieri*, which we needed to collect.

The slabs were also occupied by an exceptional abundance of non-cichlids, especially very large *Mastacembelus moorii* and *Malapterurus electricus*. These fishes occupied the largest holes and crevices in the slabs. Noteworthy was the lack of any non-cichlid juveniles. It is probable, but not documented, that these fishes breed in river estuaries and

that it is only later on, when they can fend for themselves, that the juveniles roam into the lake to find a cave to hide in.

Thus the *Julidochromis*, provided they keep to narrow slits and holes, are rather free from predation by their large enemies, which cannot slither into the narrow recesses.

In Ruziba, as opposed to the main coast, you never see a *Julidochromis* swimming in the open over the brightly lit stones; they cling to the rock walls. At first we pondered why *Julidochromis* had been the only typical rock-dwelling cichlids to occupy the slabs so far from the main rock coast. Successive expeditions increasing the range of our explorations provided us with information on the layout of the

bottom between Ruziba and the coast to the south as well as on the distribution of fishes along the coast.

Basically, the part of the lake lying north of the Ubwari peninsula is a shallow basin (by the lake standards) less than 200 meters deep on average. It is progressively being silted over by the sediments from the Ruzizi and many torrents, and by the silt brought by prevailing southerly winds. Since Lake Kivu found its outlet via the Ruzizi to Lake Tanganyika, perhaps not more than 10,000 years ago, sediments have already filled in about 70km of the lake's northernmost end, and the process is still going on at a very fast pace. In due time the entire northernmost basin will

An aquarium setup for any species of *Julidochromis* means plenty of rocks forming rock caves. This is one of the most recently described species of *Julidochromis, J. dickfeldi*. Photo by Hans-Joachim Richter.

be completely filled in and become sandy and swampy.

When the bottom along a rocky coast is deep, silt brought in by a torrent sinks to the bottom; if the bottom is shallow, silt accumulates in front of the river mouth and builds a delta.

The main rock coast in Burundi extends from Magara (40km south of Bujumbura) to Rutunga (about 20km south of Bujumbura). North of Rutunga we find a small mountain river delta about 2 km wide and with a small lagoon on its northern side. North of the delta we have a coast littered with scattered rocks half-buried in the sand and pebbles, then another small river swamp estuary.

We can reconstruct the history of the coastline from the available evidence. The main rocky coast at first extended much more to the north, then became fragmented by the building of the delta at Rutunga, which progressively became an ecological barrier between the two parts of the rocky coast. The one to the north of the delta was progressively silted over by sediments drifting to the north from the river mouth, hence the lagoon. It also silted over the rock and pebble shores to the north. Sediments from the Ruzizi outflow, from several small rivers on the coast, and from the river at Rutunga accumulated toward the north until they built the expansive sand floors around Ruziba.

Rock-dwelling fishes that had been isolated from the main population on the northern flank of the Rutunga delta saw their ecological surroundings progressively

Electric catfishes (*Malapterurus electricus*) are predators that grow to a fairly large size in the lake. This is a juvenile specimen. Photo by Dr. Herbert R. Axelrod.

deteriorate. Hiding places at the base of the rocks were invaded by sand, as were crevices and caves. Food growing on the rocks, algae, and crustaceans disappeared because of silt. Only scattered populations survived here and there, as long as their habitat provided them with their basic requirements. In most stretches of coastline they were already gone, by natural extinction or because some of them could migrate and in their desperate wanderings turned back to the coast to the south.

Today only *Tropheus*, *Simochromis*, *Petrochromis*, and *Eretmodus* (rock-grazers) are found along this coastline; all specialized nestbreeders have disappeared, as well as drifting plankton pickers.

Progressively they too disappear when one travels north toward Ruziba. Along the coast they have to cope with a predator they are not used to, *Haplochromis horei*, a sand-dweller against which the habitat provides no shelter. Juveniles of rock-dwelling cichlids find no shelter, adults do not find any breeding sites or a good supply of food.

Tropheus are the first to disappear along the coastline, and soon the other species also disappear from scenery that has become a sand habitat. As rock-dwellers disappear, one discovers the emergence of the typical sand-dwelling cichlids, like *Xenotilapia* and *Callochromis*.

What about the *Julidochromis*? *J. marlieri* lives from near the surface to about 40 meters in the north, perhaps deeper elsewhere. They feed on crustaceans buried in the biocover or hidden in the rock's tiny cracks. They need crevices to live in and breed, but they often wander over a sand patch in broad daylight (especially *J. regani*). Adults often forage rather far from their territory. Between Rutunga and Ruziba, because of their tolerance for reduced oxygen levels, they could use the ford provided by the sandstone slabs lying between and acting as stepping stones.

When the *Julidochromis*, alone among all rock-dwelling cichlids, reached the last slabs at Ruziba they found a wealth of recesses. In the labyrinths they could stake out a territory

With rock shelter so close at hand many species can afford to be conspicuously colored. This *Tropheus duboisi* exhibits striking white spots on its head and body. Photo by Glen S. Axelrod.

without being harassed by other species competing for room on the slabs. *Tropheus,* whose habitat goes from the surface to about 15 m deep, couldn't use the fords provided by the belt of sandstone, because they were too deep. *Simochromis* could use the slabs even less because these fish are seldom found beneath the 5-meter-deep layer. *Eretmodus,* which lives in the surf-beaten layer, has very high requirements for oxygen levels.

If *J. marlieri* had been capable of reaching Ruziba, they could not cross the murky waters beyond, toward Bujumbura, where a population of *J. regani* thrives in the rocks of the harbor and on sandstone flats lining the Ruzizi delta beaches. In these murky and opaque waters we

again find *Simochromis babaulti* and the two *Petrochromis* species as well as the two *Telmatochromis* species, but not *Eretmodus* or *Tropheus.* The rocks there are covered with a gelatinous slime and mud.

What can we extract from the facts? First, although they had been capable of fording the 15km distance between Rutunga and Ruziba, *J. marlieri* had not been able to colonize the Bujumbura harbor, perhaps because of the competition from a well established population of *J. regani,* but more probably because of the lack of stepping stones on the 8 km sand bottom. Two types of rock-grazers had reached and thrived in the murky waters of Bujumbura—but not *Tropheus* and *Eretmodus.* This showed

that the requirements for clear water of the latter were probably higher than those of the two first fishes.

Petrochromis had occupied the slabs at Ruziba but *Simochromis* did not. It was thus very probable that the latter had reached Bujumbura not via the slabs, but along the coast at a time when it was not so heavily silted over as it is now.

J. marlieri has multiplied on the Ruziba sandstone because it has little competition from closely related species and it is protected from predation by the lack of juvenile non-cichlids. As a result, on the 5,000 square meters of slabs there are perhaps as many as 50,000 *J. marlieri,* which is perhaps as much as 15% of all cichlids in the area, including those that live not in recesses but in the open. In the recesses *J. marlieri* is alone. In a bit more than a couple of years we managed to collect about 15,000 *J. marlieri* and did not witness a drop in their population. The study of the biotope of Ruziba was then abandoned because of the increased danger from a hippo herd nearby.

Telmatochromis, which do not live in recesses but in the open, were much more open to predation and their numbers are not significantly higher than elsewhere along the coast.

The wealth of information amassed at Ruziba was compared with the data that we were collecting on the main coast and provided us with our first clues on the various parameters in the ecological niches of several rock-dwelling cichlids. The parameters of many rock-dwelling species were many (we could guess), and minute differences between them would dictate the distribution of species on a coast.

We would, in the future, find more isolated habitats around the lake, and with the help of our Ruziba findings have some luck explaining the local ratios of the respective fish populations. Local varieties and species are today isolated in habitats that they could reach in the past because a ford was left open or opened when the layout of the coastline changed. The fishes could migrate and colonize and occupy new territories. What then happened depended on many features. If the grounds were practically devoid of predators and competitors the species could take hold and multiply. If competing and predating species were already settled, the intruder's destiny depended much on the adaptations they went through in order to survive.

Another parameter of an ecological niche is parasites and diseases. If we go back to *Bathyaethiops* and *Pelvicachromis* in the Stanley Pool, we remember that they had to face parasites and diseases in their new habitat that they did not have to fight in their original biotope. In the lake this is not often the case, from what we could see, but there is at least one endemic parasite in the lake that is not found everywhere. In the southern part of the lake *Tropheus* and other rock-grazers, like *Petrochromis*

Julidochromis marlieri has multiplied on the Ruziba sandstone. On 5,000 square meters of slabs there were as many as 50,000 individuals. Photo by Glen S. Axelrod.

trewavasae, are often infested with an endoparasitic worm that proliferates in their tissues. The fish become emaciated and eventually succumb to their parasites. This infestation spreads to perhaps 50% of the *Tropheus.* The problem is totally unknown in the northern part of the lake and I don't remember having seen one fish infested with the parasites. Surprisingly, when bred together in a pool neither the northern *Tropheus* nor their offspring were attacked by the parasites while adults of the southern races placed with them died along with their juveniles. It was not until the third generation that the southern *Tropheus* were rid of the deadly worms.

SPECIATION

Cichlids as a whole are very pliable fishes, and many of the species can readily adapt to a wide set of habitats. In Africa more than in any other continent they have formed many genera and species to become the most prolific family in the area. This would not have taken place if they had not colonized the various biotopes found in the African Rift valley lakes, especially in three of them, Lake Victoria, Lake Malawi and Lake Tanganyika. Most of the lakes are old and stable ecosystems with a wealth of various habitats in waters for which they had an affinity, born probably from their ancestral marine background.

The distribution of cichlids in Africa strongly suggests that most prefer hard, alkaline, and well-oxygenated waters, and do not like much the soft, acid, poorly oxygenated waters of rivers and swamps. In the latter type of water, their affinity for marine-type waters was put to rest.

Nevertheless, they managed to spread from the Dead Sea in the Jordan River valley to the southern tip of the African continent, through desert springs, savannah, virgin forest brooks, swamps, and even mangroves. They managed to withstand extremes of heat and cold. I have found *Hemichromis fasciatus* breeding in 45°C water and *Pseudocrenilabrus*

Pseudocrenilabrus multicolor is a colorful small mouthbrooder that has gained some popularity with aquarists. The female (lower fish) has her buccal area distended with eggs.

Phractolaemus ansorgii is one of the more unusual non-cichlid fishes inhabiting Lake Tanganyika. It is imported from time to time but is by no means common in the aquarium trade. Photo by Edward C. Taylor.

philander surviving under a thin coat of ice in a Shaba shallow pond.

Fishes other than cichlids show a remarkable degree of adaptability, such as the characins, cyprinids, and silurids all of which have an extensive range in Africa. Many more families are now localized; often they are much poorer in number of species.

Cyprinodontids are plentiful in the western virgin forest ranges but are scarce and much less varied in the eastern part of the continent. *Pantodon* is represented by a single species, as is *Phractolaemus*. Although the Notopteridae is represented by only two species, they cover most of central Africa.

In 1973 M. Poll reported that the 2,510 species of African fishes (to which about 50 newly discovered species have been added since then) had to be divided among 44 families. From the 44 families one could eliminate 5 families with about 60 species that live in the brackish waters of the mangroves. Of the 40 families of pure freshwater fishes 9 are very diversified:

Cichlidae	700 species
Cyprinidae	561
Mormyridae	202
Cyprinodontidae	178
Mochokidae (catfishes)	155
Clariidae (catfishes)	102
Bagridae (catfishes)	102
Alestidae	122
Citharinidae	82

These 9 families total more than 2,200 species, or 90% of all African fishes. Five of them total 65% (cichlids, cyprinids, mochokids, clariids, and bagrids) and these are the families that have mainly colonized the lake. If we examine the distribution of the other species within the remaining 35 families of freshwater fishes, we discover that most (23) are represented by 5 species or less. Some of the families have thus been rather sterile, while others were very prolific.

In some of the ancestral forms the smallest change in ecological conditions has brought about *speciation,* the phenomenon by which new species are created. In others a change in the habitat or isolation doesn't seem to have an impact on the future evolution of the species. The variety of the evolutionary line doesn't increase.

Some evolutionary lines respond very quickly to new ecological conditions, like the *Haplochromis* of Lake Victoria; in others the response appears to be much slower or there appears to be none at all. *Hemichromis bimaculatus* from the Niger River system for instance is identical to the same fish in the Congo basin thousands of km away.

We can thus divide the fishes according to their tendency to evolve into new forms, or speciate.

Ancestral forms can be:

-*Relict or fossil* forms. These forms, such as *Polypterus* and *Protopterus,* go back millions of years and are still spread over an immense area but have very few species, apparently remnants of a much more varied complex of species. They are very well adapted to the biotopes in which they still live (or they would have disappeared long

ago) but they appear to have come to the end of their evolutionary line.

-*Static or dormant* forms. These are represented by species that have appeared more recently and represent a step in evolution but did not evolve a large variety of forms. Their range can be widespread but their variety is low. Ex. *Pantodon,* notopterids, phractolaemids, etc., as well as several cichlid lines.

-*Prolific lines.* These are forms in which genetic drift has been at work, where evolution and speciation have occurred and often; still very active. In these lines it appears that the final products of evolution are not yet in sight.

In several groups, for example *Haplochromis,* speciation is so active that one might call it explosive. It is producing new species in any ecological niche that becomes available.

Speciation has been active but stopped a long time ago in the fossil forms, it has not been active in dormant lines, and it has been very active in the most prolific lines. In the

same family we may find lines that are dormant and others that are very active.

Hemichromis, considered as an archaic form of cichlid, is not represented by a large number of species and probably never was, as we can presume from the range of the present few species. The line could have diversified into many species as has *Tilapia* and *Haplochromis,* but for reasons which escape us it hasn't. The line apparently never came into contact with Lake Tanganyika, but we can speculate and wonder what would have happened to *Hemichromis* if it instead of *Lamprologus* or *Haplochromis* had colonized the lake.

If one looks at the distribution of cichlids in Africa one cannot fail to see that their variety in the forest, swamps, and savannah rivers of western Africa is not outstanding. Their occupation of forest streams is especially poor and there are seldom more than two or three species in any given area. *Tilapia* and *Haplochromis* fared poorly in forest rivers, but did much better in savannah swamps. *Serranochromis,* a southern form, is thriving in swamps, but is localized there.

But in the three main rift lakes cichlids now number more than 600 species, and I wouldn't be surprised if their final number after the lakes have been fully explored could reach 1,000 species. Rift Lake cichlids outnumber their fluviatile counterparts perhaps 10 to 1.

It is quite true that the colonization of the Rift Lakes by cichlids brought about their explosive speciation. But, once again, when speciation started in the lakes it stopped short in several of the cichlid lines,

Pantodon buchholzi is one of the evolutionally static or dormant forms. The range of this species may be great but the variety is low. Photo by Dr. Herbert R. Axelrod.

It would be interesting to speculate on what would have happened if it were *Hemichromis* and not *Lamprologus* or *Haplochromis* species that colonized the lake. Shown here is *Hemichromis lifalili*. Photo by Hans-Joachim Richter.

which then became dormant, while in others it accumulated new species in a cascade of successive adaptations.

One might raise the objection that with the dirth of fossil cichlids in Africa it is rather presumptuous to venture that some lines are dormant and others very active. One might answer that those lines in which morphological, anatomical, and physiological changes from the mainstem have not taken place in a wide diversity during a long period are probably dormant.

There are several such lines among the Lake Tanganyika cichlids. For example, let us take *Lamprologus*, which has proved to be very prolific and speciated much in the lake. 51 species have been identified so far, distributed among a dozen or more lines. Some of

the lines involve several species; others, like that of *L. prochilus,* include only one species, and a very specialized cave-dweller at that. The line to which *L. lemairei* belongs is another example of monospecific lineage. The *L. modestus* group and the *Lepidiolamprologus elongatus* lines each involve at least three species.

We can find other examples of uneven speciation among pelagic cichlids. *Bathybates* and *Perissodus* are represented by several species, *Hemibates* and *Boulengerochromis* by one each.

How fast speciation is at work is difficult to assess. One feels that at times it is a very slow process, but under other conditions and with another

line it might become a very fast process, as is documented by the Lake Victoria haplochromines. This is interesting because probably Lake Victoria is not as favorable an ecosystem as Lake Tanganyika, being in fact a recent oversized swamp whose shores, consisting mainly of reed and papyrus beds, bathe in a water not drastically different from the one found in rivers. Other swamp lakes like Bangweolo and Moero didn't witness speciation on such an explosive scale. Lake Victoria provided enough incentive to the genus *Haplochromis* to diversify in a very short time into more than 200 species, but they still look very much alike.

In Lake Tanganyika, whose lifespan covers millions of

Lepidiolamprologus elongatus pair zealously guarding fry. The specific name "*elongatus*" has unfortunately been applied to several cichlids. Hopefully, this mess now has been straightened out. Photo by Doris Scheuermann.

This fish has also undergone switches in genera. It is called *Perissodus straeleni* by some authors but *Plecodus straelini* by others. Each faction has their own arguments and may never reach an agreement. Photo by Dr. W. Staeck.

years, we find fish forms that are not only very diversified but have become so after having gone through major adaptations.

When we find a single species in a lineage, or only very few species, very different from any other, we might think that the pace of evolution, of genetic drift, after having been very active, has now slowed down.

If the lakes provided the cichlids with a strong incentive to speciate, so did some of the fluviatile biotopes as well. One example is the Regina Falls in Kinshasa, of which several parameters are similar to the ones we find on a rocky coast in Lake Tanganyika (high oxygen levels, turbulence of the water, many shelters and hiding places, and rich phyto- and zoobiocover). We find there an unusual array of cichlids: 3 species of

Nanochromis, 2 *Teleogramma*, 3 *Steatocranus*, 3 *Lamprologus* (of which one is a blind species), 1 *Orthochromis*, and one *Haplochromis*. Nine out of these 13 species are endemic to these rapids. This is indeed a spectacular variety for a river habitat which as a whole has not been very favorable to cichlids. We might wonder again how many more would have been found if one more of the lake ecological parameters had been found in the rapids, such as the hardness and high alkalinity of the water.

As it is the similarity of some of the parameters has already brought about an extraordinary level of convergence between some of the fishes in the rapids and the rocky coast in the lake. The best example is provided by *Steatocranus* and *Eretmodus*. Both share a

short, massive body, a small swim bladder (that doesn't provide the fish with much buoyancy), short but powerful fins, and gap-toothed jaws for grazing on the biocover.

In Lake Tanganyika the availability of rocky, sandy, and muddy coastal biotopes has provided the opportunity for the fishes to move toward a first step in speciation. Some of them started to adapt to habitats that they had not found in rivers, the wide open reaches offshore and the deep layers. These biotopes do not exist in shallow and narrow rivers.

Although the lake as a whole was a very stable ecosystem, the layout of the coastline could be modified by changes of the water level. A coastal biotope could be fragmented, or fused to another, or isolated from other parts of the coast. Each species flock that had speciated to adapt to one of the coastal habitats could thus evolve on its own, without contact with another population.

The variability of the layout of the coastline has been a major force in the speciation process in the lake. Again we might wonder what would the variety of the lake species be had the array of different types of coastal biotopes been more stable than it was.

Cichlids in Lake Tanganyika are more diversified than they are in any of the other Rift lakes. This has been attributed to the age of the lake, but also very probably to the alternation of the three types of habitats, which is more frequent than in the other lakes.

In comparison with a fluviatile ecosystem, the stability of ecological conditions prevailing in a large and deep tropical African lake compounds the effects of a large span of evolution. This theory calls for an explanation. A river ecosystem is bound to reflect the impact of changes in climatic conditions in its area. Thus, when evolution has given birth to more specialized fish-forms whose "tuning" to a given ecological niche is delicate, the future of these forms is much more fraught with the risk of extinction if the conditions having prevailed over their appearance happen to change. In the lifespan of a river such changes could happen more than once and each time the more specialized forms were more in jeopardy than the less specialized fishes. Each change could mean an evolutionary process under way could be cut short. The long-term fragility of a

riverine ecosystem might thus give an edge to the survival of fishes that are not too specialized and thus have a "broad" ecological niche. This is why the survival of a blind cave-dwelling *Lamprologus* in the Inga rapids of the Congo River is quite exceptional.

In a lake whose lifespan covers, without drastic changes in the ecology, several million years, evolution of a line can be a continuous process. It might bring forth highly specialized forms that will not become extinct by a change in the nature of their habitat. They, in turn, will keep evolving new and even more specialized forms in an unbroken chain of successive adaptations. This also is why there are more specialized cichlids in the African Rift lakes than in rivers.

The lack of expansive swamps on the coastline of the lake has not had a positive effect on the building of a huge flock of *Haplochromis* that find their preferred biotope in swamps. This fact could also account for the few *Haplochromis* in Lake Tanganyika.

Although the lake offered cichlids an exceptional opportunity to speciate, some of their behavior patterns could boost or hinder their potential for the appearance of new forms. *Sedentarism* and *nomadism* are two such behaviors.

Sedentarism stabilizes a population in its biotope. It thus promotes the setting up of pairs of breeders whose partners belong to a same filiation and favors inbreeding and the purity of blood in the local stock.

Nomadism, in its sense of mobility, tends to mix fishes

that belong to different ancestral backgrounds. It thus tends to promote the possibility of crossbreeding and the stability of the average genetic stock of the species across several local populations.

In the lake we can find several species that are very mobile, such as the rock-grazing mouthbrooders, but have diversified into local varieties and subspecies. *Tropheus* is an outstanding example of speciation with its dozens of local varieties. At first we might think that this contradicts our point that mobility promotes crossbreeding, but the fact remains that although there are so many varieties of *Tropheus* around the lake they belong, in fact, from what we have already been able to assess, to very few species. *Simochromis* and the goby cichlids *(Eretmodus, et al.)* are even less diversified.

Local races or varieties are the first step toward speciation. In spite of their mobility, each time one of the populations of *Tropheus* has been isolated for some time on a stretch of coastline it has diversified from its original stock. There are even differences in *Tropheus* distributed along a rocky coast that has not been fragmented by the appearance of an ecological barrier!

If we take a homogeneous rocky shore like the one stretching from Moliro to Nkamba Bay around Cameron Bay in the southern part of the lake and which encompasses about 60km of coastline, we discover that all *Tropheus* found on this coast belong to the same line, which I call "with a double V on the snout". Several color varieties, which I christened according to the

Although there are but few species of *Tropheus* in the lake, there are many, many different varieties. Above is an adult blue-faced form of *T. duboisi*, while below is a spotted juvenile. The spotted juvenile seems to hold true for *duboisi* throughout the lake. Photo by Hans-Joachim Richter.

locality where they were collected, are found along this coast. They are called: Moliro, Chipimbi, Mossi, Kachese, Sumbu, and Nkamba. All have a light colored snout and a double chevron extending from the upper lip to the eyes.

The first three varieties have a set of red dots extending along the caudal peduncle. This feature is the most developed in the Moliro *Tropheus* and reduced to a single dot in the Mossi specimens. The population at Chipimbi has some individuals that are the typical Moliro bright cherry red, as well as a few fish with a much darker blue-brown background color such as the one displayed by the Mossi specimens. In Mossi again we find some fish with the Kachese pattern and some colored like the ones in Chipimbi. But one will not find specimens sporting the olive green and orange robe from Kachese in Moliro.

Each part of the habitat was colonized by fishes that belong to the same line, but they soon started to feel the effects of speciation. The settlers that first colonized each part of the huge coast, because they were at first too few to carry with them the whole set of genetic characters, had a more limited genetic stock to pick from than that of their original population. They thus developed, under what is called the *founder effect,* into a population that was slightly different from the original population.

The founder effect is very strong when only a few fishes participate in the founding of a new colony. As *Tropheus* are solitary fishes, it is only piecemeal that they probably settled in the new habitat.

The impact of genetic drift and founder effect on speciation can be more or less neutralized when a trickle of successive immigrants manages to reach the colony after it has started. This genetic flow tends to maintain the similarity between the original population and the new colony. One can thus say that genetic drift and genetic flow are counteracting forces.

Around Cameron Bay, from Moliro to Nkamba, the main features of the lineage with the double V were kept by all the local varieties by the continuous immigration of individual fish from adjoining populations. But, although mobile, these individual immigrants could not in their aimless wanderings and during their lifespan cover more than a few kilometers. As we figure it, each *Tropheus* came by accident into contact with a population close at hand. Had these wanderings been so extensive as to cover the whole habitat, we would not today find the six varieties stretched out over the rock habitat. *All* the features of the original population would then have been preserved along the entire coast. As they didn't, this is why in each of the successive populations we see the similarities fade away with the distances involved.

Isolation by an ecological barrier is thus not the sole force leading to speciation on a coastline. It can be induced by the sheer size of a biotope.

We have seen in some of the pelagic cichlids like *Bathybates,* which are crisscrossing the open waters of the lake at will, that speciation has also been at work, although the fishes are not isolated.

SUCCESSIVE COLONIZATIONS

When two parts of a biotope are reunited into a single entity after a period of isolation, the respective fish populations can again be in contact and invade the other's territory. What will happen then between two populations of a line will reveal much about the degree of speciation they underwent for the time that they were isolated.

On the northwestern and northeastern coasts we discover isolated populations of *Tropheus duboisi* in the midst of a population of another *Tropheus, T. moorii* in the west and *T. brichardi* in the east, respectively. The *T. duboisi* population of the northwestern shore is very small, a few thousand fish scattered along a coast less than 1 km long. Their main population lies between the Burundi border and the Malagarazi River delta in the east. They are more than 150km apart and in between lies the 30 km wide swampy delta of the Ruzizi River. One might think that in past ages *T. duboisi* covered the entire range between the Malagarazi and Bemba, but that later on *Tropheus moorii* progressively became dominant and that the isolated population of *duboisi* in Bemba is heading toward extinction.

It is difficult to assess why *T. moorii* is dominant, but the respective fertility of the two species might give us a clue. In pond-raised specimens, during experiments covering hundreds of breeders, we discovered that the average fertility of *T. moorii* of the orange variety in Bemba was four times higher than displayed by *T. duboisi.*

We find another example of successive colonization on the northwestern coast. There, in different spots, we find scattered specimens of a green variety of *Tropheus* mixed with the black-tailed line of *T. moorii.* The green variety, always very much outnumbered by *T. moorii* in Uvira, occurs on the eastern coast of the Ubwari peninsula, along the coast south of Kasimia, and in Kalemie. All the green specimens found along the coast have an average of 6 anal spines, while *T. moorii* has only 5. In fact, the green specimens belong to a scattered population of *Tropheus annectens,* whose main range is the huge rocky coast south of the Ubwari peninsula.

We have thus on the northwestern coastline three different species of *Tropheus* superimposed on the same habitat.

It is not so difficult to sort out the facts behind the presence at Uvira of a few *T. annectens* when one knows that the Ubwari peninsula has been for some time a very large island separated from the mainland by a stretch of shallow bottom. The shallow bottom could have been used by *T. annectens* when they moved to the north until the progressive silting up of the shallows linked Ubwari Island (or Muzimu Island as it was called 100 years ago) with the main coast and established the path for future migrations of *T. annectens.* The populations of both *T. duboisi* and *T. annectens* north of the Ubwari could thus well be relics from a past occupation, faced with strong populations of *T. moorii* that had the upper hand.

When two varieties come

This vertically barred form of *Tropheus moorii* is one of the many color varieties of this species. A thorough examination of the species and varieties of *Tropheus* is desparately needed. Photo by Dr. Herbert R. Axelrod.

A very colorful variety of *Tropheus moorii* with red color in fins and on the body. Photo by Dr. W. Staeck.

The wide-band variety of *Tropheus duboisi*. Even though the patterns vary, individuals of the same species from different areas are still able to recognize one another and will interbreed. Photo by Hans-Joachim Richter.

This is probably a variety of *Tropheus brichardi*. Note the interesting distribution of orange and yellow spots on the anal fin. Photo by Dr. W. Staeck.

into contact after having undergone a certain amount of speciation, if they can no longer "recognize" each other as sexual partners they will not crossbreed. Specimens from each population will exclusively mate with partners from their own population. Hybridization will not normally occur in the wild (although it might be possible to force mating in captivity).

In cichlids, especially, visual signals displayed by potential partners are essential. In *Tropheus* the visual signals are provided not only by behavior but also by the color patterns displayed. The color patterns of *T. duboisi, T. annectens,* and *T. moorii* are very different, which explains why each species has remained distinct in the areas where they live together. In the south, a variety from Chipimbi will mate with one from Moliro because they still recognize each other by their color patterns.

SPECIALIZATION

At the local level speciation gradually increases the number of species living on a habitat and as variety increases new ecological niches appear to which the fish can adapt. Pressures from competitors and predators build up and fishes tend to adapt and specialize more. Ecological niches, as they multiply, tend to become more narrow and finely tuned and as a result they require even more specialization from the fishes. Once speciation has started, as long as the biotope provides shelter, food, and spawning grounds in ample and varied supply, it will become a self-propelled force increasing the pace at which new living forms appear.

The adaptations fishes go through are reverberated from one generation to the next. At each level of adaptation the new forms that are not fit for long term survival can survive for awhile, then will disappear because the ecological conditions for their survival are not met.

Speciation promotes those forms that at the time they appear are the best fitted. It not only acts on morphological or physiological features of a lineage, but on its behavior as well. Buccal incubation is a major behavioral adaptation in cichlids, probably born out of the need to protect spawns from predators. The various adaptations of buccal incubators in Lake Tanganyika are in strong contrast with the rather stereotyped behavior of mouthbrooders in the two other main lakes in the Rift valley. If *Tropheus* and *Cyprichromis* have started spawning in midwater instead of on the bottom, it is perhaps because the bottom was occupied by a number of permanent bottom-dwellers, like the nestbreeding cichlids, and that good and safe spawning grounds were hard to come by for fishes that did not have a permanent territory on the bottom: Thus spawning in midwater has become a specialized behavior of *Tropheus* and *Cyprichromis*. Such behavior belongs to the genetic patrimony of the fishes and is inborn in each as much as schooling or individualistic behavior.

Tropheus brichardi and *Tropheus duboisi* are known for their aggressiveness and *Tropheus brichardi* is more

Tropheus moorii is able to adapt its behavior to transient conditions. It is also less aggressive than either *T. brichardi* or *T. duboisi*. Photo by Dr. W. Staeck.

territorial than *T. moorii*. *T. moorii,* when their density is very high, can develop a measure of schooling behavior and, at times, move as a single body, which *T. brichardi* never does. The capacity for *T. moorii* to change its behavior according to fleeting conditions is in itself an adaptation and a specialization.

The threshold of intolerance is thus much lower in *T. brichardi* than it is in *T. moorii* and a feature of their genetic patrimony. Most behaviors are thus transmitted by one generation to the next, but some appear in a species under special circumstances to which an individual fish adapts

very quickly.

The fact that *Julidochromis marlieri* in its habitat swims sideways once it is threatened from above is a behavior that was transmitted through the genes. The fact that an adult *Julidochromis* captured and put in a tank learns in two or three days where its food is coming from and juts its head out of the water whenever somebody passes by is something that the fish living in the lake far from the surface did not do nor did it get the habit from its parents. *Julidochromis* transmit to their offspring an adaptability and a capacity to improvise that a clupeid will not display.

MUTATIONS

If speciation is a gradual process of evolution by which a species evolves by natural selection from among its stock of genes, mutation results from the sudden change of one of the genes.

Albinos do exist in African fishes, such as the well established *Protopterus aethiopicus* in the Stanley Pool.

In Lake Tanganyika I was very fortunate to discover a mutant of *Lates angustifrons* that is called the "golden Nile perch" locally because of its extraordinary color. The fish's scales are metallic gold, as if they had been plated with the metal. Normal *Lates* are dark colored with patches of paler scales. The mutant *L. angustifrons* was first caught 28 years ago, and since then about 30 additional specimens have been caught, among which were young adults, their size ranging from 2kg to 25 kg. A pair was even discovered among rocks busily spawning, which shows that the morph is reproducing itself and breeding true.

The color of the golden perch is strikingly different from the color and patterns found in the other *Lates* and intermediate shades are not found in the tens of thousands of these fishes that are caught every year. One has to admit that the new variety is a mutant and not the result of speciation. Because the fish has multipled, its short-term survival appears to be secure. Time will tell if, against the competition from the other *Lates* and the preying of the pelagic hunters, the glittering golden sheen of the mutant will not cause its ultimate demise. If not, we will see their

numbers progressively rise so that from the southern part of the lake, which is their present homeland, they might invade the whole lake like the other *Lates.*

PREFERRED NICHE

In any biotope there are slopes on which a species finds the optimal combination of all the parameters of its ecological niche. In other parts of the slopes the optimal conditions for the survival and well-being of the species are

Albinos exist in a number of African fishes, such as this *Protopterus ae-thiopicus* from the Stanley Pool. Photo by Edward C. Taylor.

not all found. It's possible that for a species needing very clear water there are too many particles floating around and settling down or that breeding sites are available but surrounded by hostile grounds without good shelters for the offspring, or that a competing species has already occupied the best territories, etc., etc.

Saturation of the best slopes of a biotope can lead individual fishes to migrate and try to find a territory in less favorable surroundings. It is a slow process for nestbreeding rock-dwellers, especially when they are very territorial, and one that is especially the task of the weakest, which means the juveniles, because the adult fishes, barring an accident, have already secured their own site. As the young fish radiate from their parent territory, they wander from rock to rock, repelled by the local occupants, until they reach the outskirts of the local area.

In some ways nature has provided the rock-dwellers with a safety valve. Many different species have overlapping territories and cohabit the same recess(*Julidochromis* with *Lamprologus furcifer, L. savoryi, L. brichardi,* etc.), but seldom do we find several specimens of the same species sharing a territory *(L. brichardi* is the most noteworthy). During this search for a territory the young fish go through the most dangerous part of their life, as they are exposed on foreign grounds to the full impact of predation.

When they survive their quest, mostly by sheer luck and not necessarily because they are the fittest, they will have to settle for the "second-choice" site, where life is more difficult—breeding fraught with danger, foraging for food in the open over longer distances, etc.

The less favorable ecological conditions prevailing in the area will leave the fish with a lowered fertility or less vitality. The habitat can still support the species, but the number of fish and eventually their size will not be up to the usual standards.

This happens in Cameron Bay with *Julidochromis dickfeldi.* The largest fish and largest population of this species lives around Moliro and Chipimbi. Down south its range extends to Kachese, where it cohabits with another *Julidochromis, J. regani,* and with *Chalinochromis brichardi.* In Kachese and Mossi *J. dickfeldi* are very few, small, and very drab in color. The habitat in Mossi is mostly like the one in Moliro and apparently suited to *Julidochromis.* In Kachese visibility is often very poor due to the drift of sediments from the nearby Sumbu stream, and

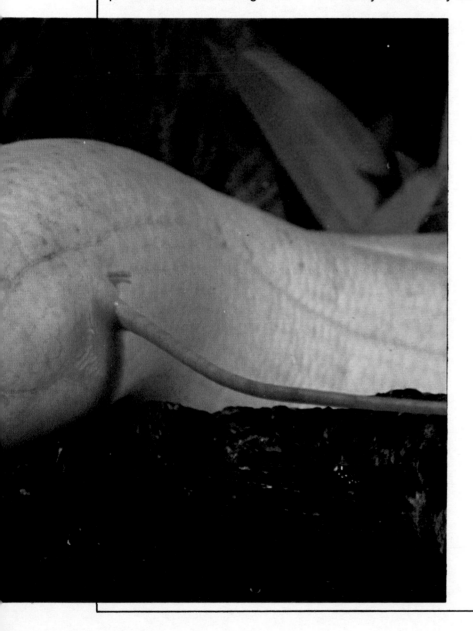

the species is outnumbered 10 to 1 by *J. regani* and *Chalinochromis*. *Chalinochromis* finds there its northernmost boundary in the area.

One might conclude that the trickle of *J. dickfeldi* travelling southward from the area of Moliro has reached in Kachese its ultimate boundary and ecological conditions. Among these conditions probably could be included the competition from the two other species, against which the undersized *dickfeldi* were at their limits of toleration. On the contrary, they have occupied the grounds in the area of Moliro and Chipimbi in strength, find optimal ecological conditions there, and attain a much larger size: *J. dickfeldi* in Kachese average about 6cm; in Moliro about 10cm. The ratio of populations between Moliro and Kachese is about 10 to 1.

Another example of the effects of a "second choice" niche is provided at the corniche of Nyanza Lac. Along the road that winds its way along the mountain slopes, a narrow fringe of rocks dumped into the lake by the excavation of the road lines the shore. It is barely 10 meters wide and rests on a huge sandy shallow extending for several hundreds of meters toward the open lake. It is there that we discovered *Tropheus brichardi*. The habitat is thus very limited and the nature of the rocks prevent the growth of a substantial biocover. Many typical rock-dwellers, namely all nestbreeders, are lacking.

On this rather new rock biotope (less than 50 years since the road was made) we find a few ubiquitous *Lamprologus* species, etc. *(L. mondabu, Telmatochromis caninus)*, but a wealth of mouthbrooding rock-grazers. It is the only place we know of in Burundi where *Eretmodus, Spathodus erythrodon* and *S. marlieri, Tanganicodus,* and three species of *Petrochromis* live together along with *Tropheus brichardi.* It is the best illustration as yet of the fact that nestbreeding cichlids are slower to occupy a new territory than mouthbrooders. Missing from the area are *Julidochromis, Lamprologus compressiceps, L. furcifer, L. savoryi, L. schreyeni,* etc. (all species that occupy similar habitats elsewhere).

Tropheus brichardi are there in numbers, although the biocover is poor and the surface area of the habitat is much smaller than average. Because of the narrowness of the habitat and its shallowness (not more than 3 meters deep), it is quite easy to go there periodically and make a catch of several hundred *T. brichardi* in a few hours—a very high yield indeed. All the specimens are less than 10 cm long.

The Nyanza Lac corniche belongs to the northern boundary of the species, which to the south extends to the Malagarazi delta. On all the other biotopes on this coastline *T. brichardi* reaches the usual size of all *Tropheus* species, which is between 15 and 16 cm overall length. Thus, in Nyanza Lac the *brichardi* managed to survive and multiply, at the cost of maximum size, on a biotope where food was scarce and had to be shared with other rock-grazers.

The other parameters of the biotope were apparently optimal, among them the lack of predation and competition from the specialized rock-dwelling nestbreeding cichlids and non-cichlids. The density of the *T. brichardi* population as a result was achieved at the cost of size and, perhaps in respect to their behavior, by an increase of their territorialism and aggressiveness.

ECOLOGICAL BARRIERS

Before deciding that a coastline acts as an ecological barrier for a species one has to be cautious. The components of the bonds of the species to its habitat have to be assured. The layout of the slope lining the shoreline has to be checked as in the case of the Nyanza Lac corniche to make sure that there are no fords that fishes could use to cross the barrier. From the appearance of the sand beaches and swamps around Ruziba nobody would think that sandstone slabs 1 km offshore would provide a rock biotope. One has to go underwater and check the layout of the slopes.

Most extensive sand beaches have a rocky belt underlining them in a few feet of water, sometimes for miles along the coast. This belt is made of sandstone, which is a very poor rock biotope for several reasons. Sandstone erodes quickly under the beating of waves and this erosion prevents biocover from taking a good hold on the rock. The base of these slabs is usually buried in the sand, crevices tend to broaden, and narrow recesses are at a premium. Many fishes do not tolerate very turbulent water, especially those that are used

Sandy areas are very effective barriers to rock-dwelling fishes. The more tied to the rocks is a species, the less chance it has to cross these barriers and become established in other parts of the lake. *Julidochromis dickfeldii* is one of the more rock-bound species. Photo by Hans-Joachim Richter.

to living in deep and quiet waters. *L. compressiceps,* with its high body and fins, would be smashed against the rocks in such an area. Thus, for many rock-dwellers a belt of sandstone slabs along a coastline doesn't help if they want to cross the open surrounding sand floors.

But the belt can be used by other rock-dwelling fishes if they can tolerate the local conditions. An adult *Tropheus* leaving a rocky coast can enter the belt at one end and gradually migrate along the sandstone slabs, even for miles, until it reaches another rock biotope, for example at the tip of a promontory or a cape. On its way the adult *Tropheus* will scratch a living from the biocover growing on protected walls, find a shelter against predators in the crevices, and manage very well in the pounding surf in which the fish likes to "spin around".

The porosity of an ecological barrier to transgression will thus depend on the requirements of each species as to its minimal ecological conditions. The minimal requirements of *L. compressiceps* are not met on a sandstone belt along a beach, but they are met for *adult Tropheus.* But barely, because the fish are not able to multiply by virtue of the poor habitat. When they mate the females have no place in which to abandon their fry with a modicum of safety. The few crevices are wide open, their walls land on sand, and there is a lack of deep and narrow labyrinths in which the defenseless fry can hide. In a hostile world, where the typical shoreline predator,

Haplochromis horei, is master and prowls, they are doomed.

Sandstone belts, lying in one or two feet of water between the mouth of the Lukuga and the coastline south of Toa Bay and the archipelago of Kavalla (40 km), have been used by *Tropheus* to colonize the pier in the harbor of Kalemie, which lies in a typical sand biotope.

The sandstone slabs can line a beach in an unbroken belt for miles, thus allowing the passage of a few adult *Tropheus* at a time, but are an ecological barrier for their young. The sandstone slabs are a *vehicle* for adult rock-grazing cichlids, they are a *barrier* for juveniles and nestbreeding cichlids.

The few examples with which we have tried to explain the problems linked to the distribution of coastal species show how complex the ecological factors can be.

INTRALACUSTRINE ENDEMISM

About all the cichlids in Lake Tanganyika are endemic to the lake, as well as about 50% of the non-cichlid fishes. We cannot expect, because of what has been said about the features of the coastlines and the ecological niches, that all these species will be found everywhere around the lake. In fact we discover that about 50% of the nestbreeding cichlid species are found only either in the northern or the southern half of the lake. This is due to the separation of the lake coastlines, in the west as well as in the east, by an expansive ecological barrier in their middle. The lake is virtually cut into two separate parts.

Fishes have thus been isolated into two major stocks with a high percentage of local species, especially apparent among nestbreeding cichlids, living north or south of the barriers. In each of the 2 major habitats we find "geminated" or "twin" species that have a closely related species in the other half of the lake. Some examples are:

	NORTH	SOUTH
Lamprologus	*leleupi*	*mustax*
L.	*tretocephalus*	*sexfasciatus*
L.	*mondabu*	*modestus*
L.	*toae*	*moorii*
Lepidiolamprologus	*pleuromaculatus*	*attenuatus*
Telmatochromis	*bifrenatus*	*vittatus*
Callochromis	*melanostigma*	*macrops*

Although the minimal requirements for life are not met for *Lamprologus compressiceps* on a sandstone belt along a beach, they are (barely) for *Tropheus* species. Factors such as this help explain the ofttimes patchy distribution of some of the cichlid species in the lake. Photo by Andre Roth.

Lamprologus furcifer is one of the specialized cave-dwellers but is found on all typical rocky coasts. A similar species, *L. prochilus*, has only a limited range. Photo by J. Shortreed.

And there are other examples among *Xenotilapia.*

The fact that rock-dwelling nestbreeding cichlids appear to be more confined in their range than mouthbrooders is probably due to their sedentary habits.

We discovered also that several species are found only very locally. Such is the case of *Lamprologus prochilus,* whose range extends probably only along the huge escarpment leading from Moba to Sumbu on the southwestern coastline. It is a cave-dwelling specialized fish that is never seen in the open and is only flushed out from under the rocks with the anaesthetic quinaldine. It shares the habitat of *L. furcifer,* another specialized cave-dweller. The latter fish is found on all typical rock coasts, but *L. prochilus* has only a limited range.

The specialization of the morphological features of this fish suggests that its lineage has gone through a long set of successive adaptations, of which the species is the sole remaining witness. It might thus be a kind of end-product of evolution. So, probably, is also the case with *L. furcifer.* Why then are the respective ranges of the two fishes so different?

Although *L. compressiceps* is a rather stable species in which local morphological differences are minor, the line to which the species belongs produced another "offspring", *L. calvus,* in the south, which lives on the same slopes.

There are many such examples of endemism along the coastlines. For example, *L. toae* is found on the coastline north of the Luguga outlet on the west coast, but not in

Burundi, although it is found again in Tanzania. *L. tretocephalus* has the same range and is not found on the main rocky coast of Burundi. *L. niger* is found from the archipelago of Kavalla to the Ubwari peninsula and nowhere else. *Telmatochromis burgeoni* is found around Kalemie and supposedly in Nyanza Lac, but nowhere else again. The distribution of *L. leleupi leleupi* is as yet very limited, less than 20km of coastline! In its habitat we do not find any *L. niger* and we don't find *L. leleupi* in the latter's range. It might thus be possible that the two species exclude each other because of competition for the same niche. On the main rocky coast in Burundi we find *L. schreyeni,* a dwarf cave-dweller. The species has not been found elsewhere.

If we put together all the data as yet collected on the ranges of species within the lake, one feels that it is among the cave-dwelling nestbreeders that local endemisms prevail. By their adaptations to a very special ecology they appear to be less fit to undertake migrations over hostile territories and settle in new areas.

Second after them, among the nestbreeders, are the fishes living close to the rocky floor, such as *L. leleupi, L. niger,* etc. They, too, appear to be often localized, but the wide differences in their respective ranges leave many questions unanswered.

Among mouthbrooding rock-dwellers, like *Tropheus, Simochromis, Petrochromis,* etc., there are relatively fewer localized species although some, like *T. polli,* are very localized as well. We find

Lamprologus leleupi has been split into several subspecies, each having a different color ranging from bright yellow to a darkish brown. Photo by Hans-Joachim Richter.

several species, like *Simochromis babaulti and S. diagramma,* that are seen everywhere, and species like *S. marginatum* that are more localized. And there are several other new species recently discovered on the west coast that have only a local distribution.

Several lines of mouthbrooding cichlids, especially among plankton pickers, have local races and species. Such is the case with *Cyprichromis leptosoma,* which is not found in the northern half of the lake. Such also are the various *Ophthalmotilapia* species, of which many local lines are only

color varieties, but among which there is little doubt that new species will be found.

Intralacustrine endemism appears to be strong in *Xenotilapia* and *Trematocara,* although one would think that crossbreeding among sand-dwelling roamers should be prevalent. As with *Tropheus* around Cameron Bay, the distances in the lake are so huge that although no ecological barriers exist to prevent passage from one half of the lake to the other (for sand-dwellers), local flocks do not wander so much that they could come in contact with flocks from the other half and crossbreed. The dimensions of

Chart of Intralacustrine Endemism in Coastal Cichlids

SPECIES	Northern	Central	Southern
Mouthbrooders			
Callochromis macrops macrops	–	–	*
m. melanostigma	*	–	–
pleurospilus	*	–	–
Tropheus moorii moorii	*	–	–
m. kasabae	–	–	*
annectens	–	* West	—
polli	–	* East	—
duboisi	*	–	–
Simochromis babaulti	*	*	*
diagramma	*	*	*
marginatum	*	–	–
margaretae	*	–	–
curvifrons	*	*	*
pleurospilus	–	–	*
Eretmodus cyanostictus	*	*	*
Spathodus erythrodon	*	*	*
marlieri	* East	?	* East
Tanganicodus irsacae	*	–	–
Cyathopharynx furcifer	*	*	*
Ophthalmochromis ventralis ventralis	*	?	–
heterodontis	–	?	*
nasutus	*	?	*
Ophthalmochromis boops	—	?	*
Cunningtonia longiventris	–	?	*
Cyprichromis brieni	*	–	–
leptosoma	–	–	*
nigripinnis	*	*	–
microlepidotus	*	*	*
Nestbreeders			
Lamprologus brevis	*	?	*
brichardi/pulcher	*	*	*
callipterus	*	*	*
calvus	–	–	*
caudopunctatus	–	–	*
christyi	–	*	–
compressiceps	*	*	*
falcicula	*	?	–
fasciatus	*	*	*
furcifer	*	*	*
hecqui	?	*	?
kungweensis	–	*	–
leleupi leleupi	* West	–	–

SPECIES	Northern	Central	Southern
l. melas	* West	–	–
l. longior	* East	?	–
leloupi	–	*	–
lemairei	*	*	*
meeli	?	* West	?
modestus	–	–	*
mondabu	*	*	–
moorii	–	–	*
multifasciatus	*	*	*
mustax	–	–	*
niger	* West	–	–
obscurus	–	–	*
ocellatus	*	*	*
ornatipinnis	*	*	*
petricola	* West	–	–
profundicola	*	*	*
pleuromaculatus	*	?	–
prochilus	–	–	*
savoryi	*	*	*
schreyeni	*	?	–
sexfasciatus	–	–	*
signatus	?	* West	?
stappersi	–	* East	–
tetracanthus	*	*	*
toae	*	*	–
tretocephalus	*	*	–
wauthioni	?	*	?
Julidochromis dickfeldi	–	–	*
marlieri	*	*	–
ornatus	*	–	*
regani	*	*	*
transcriptus	*	–	–
Chalinochromis brichardi	*	*	*
Telmatochromis bifrenatus	*	*	–
burgeoni	–	* West	–
vittatus	–	–	*
caninus	*	*	*
temporalis	*	*	*
Lepidiolamprologus attenuatus	–	?	*
cunningtoni	*	*	*
elongatus	*	*	*
kendalli	–	–	*

145

the habitat are such that there is a de facto isolation and inbreeding in the flocks.

Curiously enough, non-cichlids are much less localized. As yet I don't think that a single species of endemic non-cichlid fish has only been found on a small stretch of coastline. The endemic species of the lake, from the data so far collected, appear to be distributed everywhere. In contrast, it is the non-endemic species of non-cichlids that are more locally spread along the shoreline. In retrospect it is a normal situation. Once non-cichlids got used to the lake ecology they could spread because their bonds to a peculiar type of bottom were not too strong. Their average size is often quiet substantial, which is a help for long migrations.

Many endemic non-cichlids remain anadromous, which means that they go back to the rivers in which they were born to spawn. This entails that the fish are mobile and swift, like the cyprinid *Varicorhinus,* and that they can cover long distances.

As for the non-endemic non-cichlids, some of them live in the lake and are used to its habitats. Such is the case with *Malapterurus electricus,* which is found everywhere (but mainly on rock). They also have to go back to river estuaries to spawn.

Most non-endemics do not go into the lake proper. They live in its affluents, eventually venturing into the lake but remaining in small pockets close to the river mouth. Such are the fluviatile types of *Haplochromis* and *Tilapia,* except for *T. tanganicae,* which is found everywhere,

even over rocky bottoms on which they graze. The non-cichlids that are not endemic to the lake are thus yet too infeudated to their fluviatile biotope to be capable of a full colonization of the lake.

In summary: The proliferation of local species in the lake, some of them with a very restricted range, strongly suggests that speciation has been at work on isolated lineages belonging to forms that had colonized the whole lake earlier.

-It is likely that there were several lines in mouthbrooding and nestbreeding cichlids to start the speciation process in the lake, or that there were successive waves of colonization of the lake waters.

-Ecological barriers acted as filters to stem or slow down the fish in their aimless coastal migrations, each species feeling the impact of the barriers according to its bonds toward its preferred habitat.

-Isolation was not brought about only by the nature of the barriers and their size, but within a preferred habitat by its sheer size.

-Some species more than others are bound to have a restricted range because they are so infeudated to their ecological niche and behavior that they cannot cross any foreign grounds.

-Cave-dwelling cichlids, all of them nestbreeders, are probably the most localized fishes in the lake, with a few exceptions. To them we can add those fishes that have strong bonds toward a rocky floor and those nestbreeders that are living deep on a rocky slope, because in the deepest parts of a rocky slope most of the rocks are silted over.

With only one third of the

coastline superficially explored as yet, the odds are thus excellent that, given the prevalence of isolated populations, we will find among nestbreeding cichlids a large number of species that have not been accounted for as yet. I wouldn't personally be surprised if the number of *Lamprologus* and related species would in the long run exceed 100.

As soon as our team started to explore the second layer of the submerged slope along the Burundi coast, barely 20 km long and between 15 and 40 meters deep, we discovered three new species of nestbreeders in a little more than a year. This was without the use of quinaldine or any other tranquilizing drug.

Of the 84 species and subspecies included in this chart for which enough data are available about their distribution in the lake, 22 have been found only in the northern half, 19 only in the southern half, and 9 have been found on the central coastlines in the west or the east. It means that 50 species out of 84 have a restricted range and can be considered as intralacustrine endemics. More will probably be added in the future because, of course, it is mainly the fishes that are found everywhere along the lake coastlines that have been collected.

MAIN ECOLOGICAL BARRIERS IN LAKE TANGANYIKA

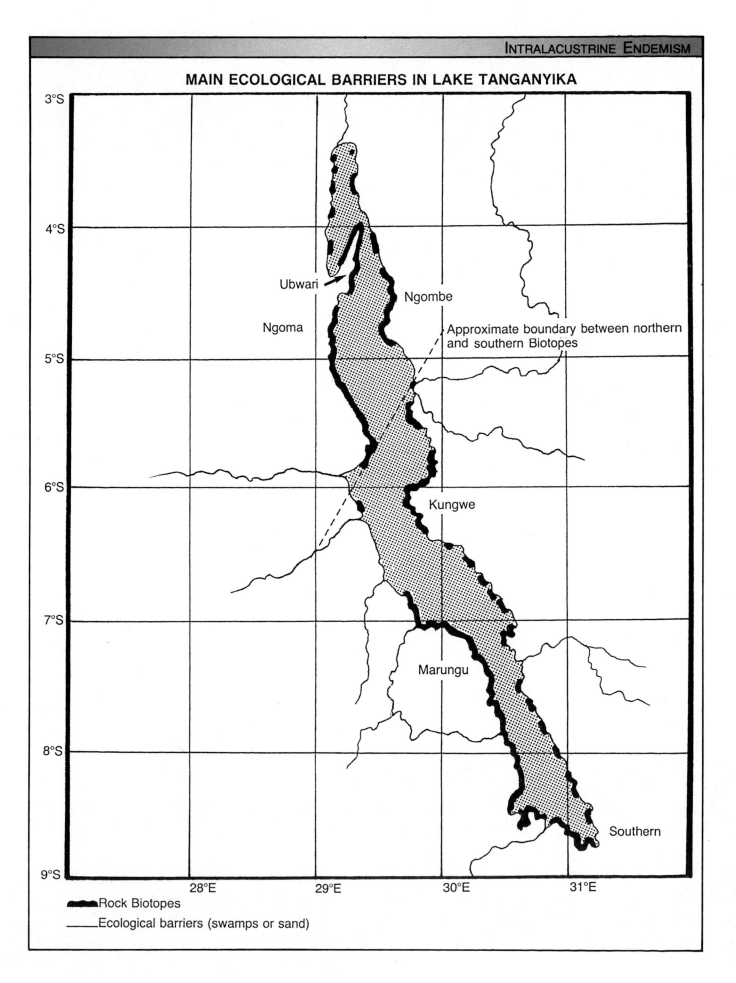

Ubwari

Ngombe

Ngoma

Approximate boundary between northern and southern Biotopes

Kungwe

Marungu

Southern

▰▰▰ Rock Biotopes

――― Ecological barriers (swamps or sand)

Tropheus duboisi, Kigoma, D.E. 4.50.

Tropheus duboisi, Malagarazi, D.E. 5.15.

Tropheus duboisi, Bemba juvenile, D.W. 3.35.

Tropheus moorii kasabae, Mpulungu, K.E.

Tropheus moorii kasabae, dominant male, Moliro, K.W. 8.10.

Tropheus moorii kasabae, dominant male, Moliro, K.W. 8.10.

Tropheus moorii kasabae, Kalambo, K. E. 8.38.

Tropheus moorii kasabae, Kalambo, K.E. 8.35.

CHAPTER VI

The TROPHEUS SPECIES COMPLEX AND THE GEOGRAPHICAL RACES

No other African cichlid genus has more captured the attention of scientists and amateurs alike as has *Tropheus*. This genus is often given as an example of the sophistication reached by buccal incubating cichlids. At the same time *Tropheus* is famous for the large number of local races that have developed around the lake. More than 40 have been identified so far.

I. *Tropheus* Species and Subspecies

Assessing a *Tropheus* with respect to its taxonomic status is not an easy task, because all of them look very much alike as their morphology is little differentiated. Only by putting two specimens side by side will one eventually be capable of noting a different curvature of the snout, or different proportions of the head and body, or the size and shape of the fins, as well as the basic color pattern. This will be possible only with specimens coming from different areas of the lake and belonging to two phyletic lines. When the two specimens belong to closely related races from the same coastline they cannot be told apart by their counts, measurements, or shape, only by their color pattern, which will, unfortunately, disappear in preserved specimens.

When the first *Tropheus* were described nobody suspected that there might be several lines and so many local races. As a result, nobody felt the need to devise an entirely new set of taxonomic parameters by which to identify these lines. Traditional measurements and counts were used. So it came

to pass that Boulenger described *T.annectens* as distinct from *T.moorii* because the first had 4 anal fin spines when the second had at least 5. The two species were put in synonymy by M. Poll and *T. annectens* "disappeared". *T. moorii* was then considered to have from 4 to 7 anal fin spines.

As the genus now stands there are five species and subspecies of *Tropheus*, which are:

-*Tropheus moorii moorii* (Boulenger) 1898
-*Tropheus duboisi* Marlier, 1959
-*Tropheus brichardi* Nelissen and Thys, 1975
-*Tropheus moorii kasabae* Nelissen, 1977
-*Tropheus polli* G. Axelrod, 1977

The Status of *Tropheus annectens*

This species was described by Boulenger in 1900 upon the reception of two specimens sent by Captain Hecq, then stationed on the lake at Albertville (now Kalemie), who allegedly caught them there. The two specimens each had 4 anal spines, which set them apart from the first specimens of *Tropheus* received by Boulenger. These had been caught by Moore at Kinyamkolo (Mpulungu) and had 5 or 6 anal spines.

To elucidate the case of this species, I went to Kalemie in 1981 and collected 65 specimens there, but failed to catch one with only 4 spines. 59 specimens had 6 spines, one had 5 spines, and five specimens had 7 spines. One could conclude that the two specimens caught by Capt. Hecq had not been caught at Kalemie. As it turned out, in

149

our two 1984 safaris we found a population of 4-spined *Tropheus* 90 km to the south at Mpala (which at the turn of the century was a very large station often visited by Capt. Hec). Because of this discovery *T. annectens* appears to be a valid species of *Tropheus* and should be rehabilitated.

Anal Spines in *Tropheus*

Within the percomorph fishes in general and especially the African cichlids, the number of anal fin spines displayed by a genus or a species is rather constant and displays a small range of variation. Most cichlid genera display three anal fin spines, especially *Tilapia* and *Haplochromis,* the two genera from which the buccal incubating species in the lake are thought to have developed. Only *Tropheus* among the 39 genera of incubating cichlids in the lake display more than 3 anal fin spines. This is even more remarkable in view of the fact that the cichlids most closely related to *Tropheus, Simochromis,* have only 3 anal fin spines.

Among the *Tropheus* only two species, *T.annectens* and *polli,* have consistently four spines; the other species, including many local races, have 5, 6, or even 7 spines. We have thought that among them some might display a mode of frequency in the number of anal spines, narrowing the range of variation of a taxonomically important character. Perhaps if several races shared an identical mode of frequency it could provide a clue to their phyletic relationships.

Obviously, we might expect that races distributed along one of the major biotopes in one part of the lake might display a similar number of anal spines and be more closely related to each other than to races living at the other end of the lake.

Color Morphs

Another clue to the kinship existing between several races could perhaps be found in their color pattern. *Tropheus* around the lake display a phenomenally wide variety of colors arrayed in different patterns. They were thought to be distributed around the lake more or less at random. Our experience with many races indicated that, contrary to the accepted theory, races belonging to the same genetic line displayed an identical layout of dark and pale areas on the body and that very often the layout was typical of the line. We discovered also that in each line the variety of colors found in the races was limited and that some colors were not found in any race belonging to the line being considered.

So it happens that in the northern basin the local races of *Tropheus* display a typical layout in their pattern, which is a black body (including all the fins), especially the hind part of the body and the tail. We might call the dominant color the *primary* color of the pattern. A *secondary* color, yellow, orange, brown, or red, covers the lighter areas, which are concentrated in the central part of the body in a broad or narrow vertical band that eventually disappears in the central part of the sides. There might be blotches of the secondary color on the head, or the secondary color could cover the entire head.

To this "black tail" line as we called it, we could ascribe a number of well-known races: the black and red from Burundi, the black from Uvira or Minago, the orange from Bemba, the Kiriza and the Caramba from the Ubwari, or the brown race of Rumonge. In this line not a single race or specimen displayed green or blue as a secondary color.

Identically, the "rainbow" line from the south, which involves several races, always has a green body (of various shades) and tail, a head dotted more or less strikingly with tiny whitish dots, and a broad pale area extending from behind the pectoral fins toward the rear. The secondary color is either yellow, orange, or pink. The dorsal and anal fins are either yellow, red, or blue, never black.

In the same vein, if one analyzes the color pattern of *Tropheus brichardi* races, one discovers that the background color is either green or brown, the secondary color being gray or yellow and exhibited as a large dorsal spot or as stripes on the body.

T.duboisi always have a dark blue head, a black body and unpaired fins, and a secondary color, either white or yellow, in a vertical band across the central part of the body.

There are other clues to be found among the various races, such as the multistriped mood pattern of adult fish found among the "rainbow" races. These are never seen on any of the northern races nor on those of *T.brichardi,* which might display a set of broad bars but not thin stripes.

In short, the layout of the dark and light colored patches on the body is typical of each line and the colors found in the

B.E.4.50 *Tropheus brichardi*. | *Tropheus brichardi*, blue-eyed Nyanza Lac variety. B.E.1.45

D.W.335 *Tropheus duboisi*, narrow-band Bemba variety. | *Tropheus duboisi*, narrow-band Bemba juvenile. D.W.3.35

M.UN.4.10 *Tropheus moorii*, broad-band Kiriza variety. | *Tropheus duboisi*, no-band Bemba variety. D.W.3.35

P.E.6.03 *Tropheus polli*. | *Tropheus polli*, barred form. P.E.6.03

line are limited by the genetic capital of the line. In fact, as they now stand most races found in the northern and southern parts of the lake and assigned to one of the existing species and subspecies belong to a corresponding line.

From the data gathered until then, we started back in 1976 to make a systematic survey of all races and to prepare a statistical study based on as many specimens of *Tropheus* as we could lay our hands on. The survey thus focused on two features:
–the anal spine count
–the layout of the pattern and the colors on the body of each race.

Using the two features we set out to discover the lines of *Tropheus* and check to see if they corresponded to the existing taxons.

Conclusion: Our theory about the clue provided by the dominant mode of frequency of anal spine counts among the *Tropheus* races with respect to the various lines they might belong to appears to be well founded. The variability in any one race does not exceed three spines (except in the so-called Kalemie stem) and one particular number of spines most commonly is represented by more than 90% of the local population.

Color Patterns of *Tropheus* Races

According to the layout of dark and pale areas on the body and the range of primary and secondary colors displayed by local races, which appeared limited in variety, we have identified several lines in combination with their respective mode of anal spine number.

Anal Spine Counts of the Races of *Tropheus*

1. Tropheus moorii moorii (line with the black tail)

Race	Specimens	Spines			
		4	5	6	7
Burundi (Magara)	27	–	27	–	–
Bemba (orange)	50	1	48	1	–
Ubwari (Kiriza)	14	–	14	–	–
Caramba (cherrycheek)	51	–	51	–	–
Rumonge	30	–	30	–	–
	172	1	170	1	0

Dominant mode: 5 spines (98.8%)

2. Tropheus species (Kalemie Line)

Race	Specimens	Spines			
Kalemie (own collection)	65	–	1	59	5
Kipampa	76	–	–	76	–
	143		1	135	5

Dominant mode: 6 spines (94%)

3. Tropheus duboisi

Race	Specimens	Spines			
Bemba	30	–	28	2	–
	30	0	28	2	0

Dominant mode: 5 spines (93.3%)

4. T. brichardi

Race	Specimens	Spines			
Nyanza Lac	97	–	1	96	–
	97	0	1	96	0

Dominant mode: 6 spines (99%)

5. Northern green Line

Race	Specimens	Spines			
Uvira (hybrids?)	52	–	25	26	1
Ubwari South	63	0	10	53	–
	115	0	35	79	1

Dominant modes: 6 anal spines (69%)
5 anal spines (30.5%)

6. Tropheus moorii kasabae

Race	Specimens	Spines			
Double chevron Line	49	–	2	44	3
Rainbow Line	52	–	1	47	4
	101	0	3	91	7

Dominant modes: 6 anal spines: (90%)
5 anal spines (7%)

7. Tropheus polli

Race	Specimens	Spines			
Mbulu	6	6	–	–	–
	6	6	0	0	0

Dominant mode: 4 spines (100%)

8. Mwerazi Line

Race	Specimens	Spines			
Masanza-Mwerazi-Zongwe	120	–	120	–	–
	120	0	120	0	0

Dominant mode: 5 anal spines (100%)

Race	Specimens	Spines			
9. Moba-Mtoto Line	126	126	–	–	–
	126	126	0	0	0

Dominant mode: 4 anal spines (100%)

M.W.3.33 *Tropheus moorii*, Kigongo-Makobola bronze variety. *Tropheus duboisi.*

X.W.3.30 *Tropheus* sp., Ulvira-Kalunda variety. *Tropheus* sp.

K.W.8.18 *Tropheus moorii kasabae*, Chipimbi Cape variety   *Tropheus moorii kasabae*, Chipimba Cape variety. R.W.8.18

B.E.4.15 *Tropheus brichardi*, Nyanza Lac variety. *Tropheus* sp.

So far the following lines have been identified:

1. The line "with the black tail" (corresponding to *Tropheus moorii moorii*)

This form has colonized the whole northern basin, including the Ubwari peninsula, but has not reached beyond Kasimia Bay. On the eastern coast it occurs down to Rumonge.

Dominant color: black - In all races at least the hind third of the body is black. Dominant color of all fins: black.

Secondary colors: lemon yellow, gold, orange, red, or brown, most commonly in the shape of a vertical stripe in the middle of the sides, but also hourglass-shaped, or as a spot on the back, and even with the whole forward part of the body of the secondary color (Bemba) or blotched (kash ekezi) or with spots on the head (Caramba).

Adults never possess a number of narrow vertical stripes. The central part of the pattern is most colorful.

Five anal fin spines.

2. "Green northern" or "Kalemie" line

Not identified as yet with any of the existing taxa, this form occupies a stretch of coastline between Mtoto and the Ubwari. This line is concurrent at each end of its range with other lines of *Tropheus*.

Primary color: green, gray, or chocolate brown.

Secondary colors: black, white, yellow and/or green. A specimen from the Kipampa race put into a white tank became entirely white save for a dark strip under the dorsal fin and a dark brown tail, all the other fins being yellow. Most races display a yellowish shade on the forepart of the head. The body is not striped or then only on the posterior half.

Six anal fin spines.

3. "*Blue head black body*" line (corresponding to *T. duboisi*):

The head is navy blue and the body black, most often vertically striped in white or yellow. Juveniles are the only ones among all the *Tropheus* species to be covered with white spots, which progressively disappear as the fish grow.

This form exists only on one part of the eastern coast, from around the Malagarazi estuary to the border between Tanzania and Burundi, and in the northern basin on one kilometer of the western coast around Bemba. It apparently does not exist anywhere else. The yellow-striped variety appears to be the basic color pattern, as white striped specimens from Bemba turn yellow in ponds.

4. "*Green northeastern*" line (corresponding to *T.brichardi*):

This form has colonized the northeastern shores of the lake from the Malagarazi to the Nyanza Lac area, and probably also the Ubwari, on which a race has been found belonging either to the Kalemie or to the *T.brichardi* lines.

Primary color: green or chocolate brown.

Secondary colors: yellow, gray, emerald green in stripes or an oblong spot under the dorsal. Tail dark green or chocolate, the other fins of an indistinct drab color more or less tinted in green or yellow. *T.brichardi* has a broader mouth than the other northern lines.

Six anal fin spines.

5. *The crescent tail* line (corresponding to *T.polli*):

This form has only four anal fin spines and a crescentic tail. It is found along the eastern shores of the Kungwe mountain range.

Primary color: green or shades of gray, or light purple.

Secondary colors: yellow, greenish, or bluish gray.

The pattern is plain or striped. A crescentic tail is also found in other *Tropheus* (Mwerazi line from the west coast), but they have five anal fin spines and apparently do not belong to the same line as indicated by other features.

6. "*Rainbow*" line (corresponding to *T. moorii kasabae*):

This form occurs between the area of Pala in southern Tanzania and Nkamba Bay in Zambia.

Primary color: green in several shades, most often jade or blue-green, sometimes bottle-green.

Secondary colors: yellow-orange-salmon or even pink (east coast of the Lufubu estuary), dorsal and anal fins yellow, blue, red, salmon, or even mauve. Tail always a shade of green. Pectorals shades of the secondary color.

The head is always covered with tiny white specks. The sides are always with a large paler area from beyond the pectoral fins and narrowing or disappearing toward the rear. A typical "mood" pattern is displayed most commonly by these *Tropheus* consisting of about a dozen thin (one scale width) vertical stripes always in shades of pale yellow, the first ones on the head.

Anal spine count: 6

7. "*double chevrons on the snout*" line (corresponding also to the present *T. moorii kasabae* taxon):

This form extends from

Telmatochromis sp. (probably new), Kalemie.   *Tropheus moorii*, Uvira, M.W. 3.25.

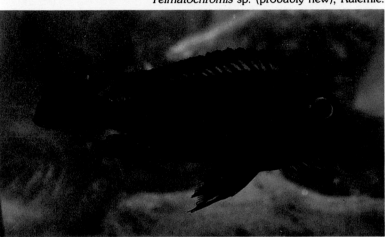

Tropheus moorii, Kachese, K.W. 8.30.   *Tropheus moorii*, Ikola, X.E. 6.42.

Tropheus moorii, Bemba, M.W. 3.35.   *Tropheus moorii*, Magara, M.W. 3.35.

Tropheus moorii, Caramba, M.US. 4.15.   *Tropheus moorii*, Cape Banza, M.UN. 4.05.

Nkamba Bay (where it is superimposed on the Kabeyeye race of the previous line) to Kapampa Bay (7°35′S).

Primary color: brown (chestnut, very dark olivaceous, or beige).

Secondary colors: yellow, orange, vermilion, cherry red, or brick red (Nkamba Bay and Sumbu). The tail is brown, usually very dark. The dorsal and anal fins are, according to the race, orange-blood red, vermilion, cherry red, or brick red.

The northernmost races (Lupota, Moliro, Chipimbi, and Mossi) display a series of red dots on the caudal peduncle. The typical feature of all the races belonging to this line consists of the two superimposed "V"s on the snout contrasting by their lighter shade with the color of the snout. This double chevron can be orange or red. The contrast of the secondary color with the background shade is especially strong just behind or around the pectoral fins and for the above-mentioned races on the caudal peduncle. Juveniles first start to display the adult pattern by the appearance of the double "V" on their snout.

As with the previous line, the pattern of the adults is more often than not about a dozen thin pale stripes, the first one on the head. They can disappear when the fish is dominated or upset.

Taxonomically the two forms belong to the subspecies *T.moorii kasabae*. It is obvious they are closely related but still appear, by the highlights of their respective patterns, to have split into separate lines.

Six anal fin spines
8. *"Mwerazi"* line (not

corresponding as yet to any taxon):

This form occupies an area north of the previous line on the southwestern coastline, from Kapampa Bay to a bit further than Cape Zongwe. It is morphologically different from the other lines in the areas to the north and the south by having only five anal fin spines and a large crescentic tail.

Primary background color: green in various shades from nearly black to olive green.

Secondary colors: yellow or whitish, disposed in an area extending behind the pectoral fins as a wide circular spot, eventually cut by a few thin vertical blackish stripes. All fins are either very dark or olivaceous yellow. The body in one of the mood patterns can be striped like the two previous lines. The head doesn't display any markings (dots or chevrons).

9. *"Moba-Mtoto"* line:

At first glance this form looks very much like the previous one, but checking the two patterns and the shapes reveals some differences. Among these are the fact that all the specimens of the Moba-Mtoto line display four anal fin spines, a feature that they share only with *T.polli*.

Primary color: green, striped with a dozen thin vertical stripes.

Secondary colors: all specimens from different localities extending from north of Cape Zongwe to Mpala have yellow as the secondary color. The dorsal and anal fins are yellow, and the tail is dark green and crescentic.

The line represents about 80% of the *Tropheus* found in its range, the other 20% belonging to the Kalemie line,

which are identifiable at once when seen underwater by their yellow snout and normally notched tail. These features are supported by the number of spines in their anal fin—6.

Dorsal Fin Spines

One of the many features separating *Tropheus* from their cousins is the number of dorsal fin spines. There are at least 20 spines in the dorsal of a *Tropheus* and less than 20 in a *Simochromis*. As far as we could check, in the 910 specimens of *Tropheus* that we surveyed only two had 19 dorsal spines. The mode of frequency among the northern lines *(T.moorii moorii, T.duboisi,* and *T.brichardi)* was distributed between 20 and 21, while in the south the mode of frequency was close to 21 with a trend toward 22 spines. Not a very significant difference in fact.

Conclusions: The survey proved that the two features, anal spine counts and the layout of the pattern, could be used to determine the lines between which the local races are divided. To determine the other features by which the various lines can be identified, one will have to develop new taxonomic guidelines.

II. Classification of the Races of *Tropheus*

An adequate classification of the more than 40 geographic races of *Tropheus* is sorely needed, with respect to the work on their phylogeny which has already been undertaken in various scientific institutions.

Tropheus moorii, Kiriza, M.UN. 4.10. *Tropheus moorii*, Kiriza, M.UN. 4.10.

Tropheus moorii, Kiriza, M.UN. 4.10. *Tropheus moorii*, juvenile, Kiriza, M.UN. 4.10.

Tropheus moorii, Mboko, M.W. 3.45. *Tropheus moorii*, Bemba, M.W. 3.35.

Tropheus juvenile, Kalemie, L.W. 5.55. *Tropheus* Kalemie, L.W. 5.55.

Various publications of lists of *Tropheus* color morphs compiled by amateurs have relied on crude bicolor pattern features in an attempt to sort them out. These lists have several defects, the basic weaknesses being (1) deciding about the shade of a color is a matter of taste, (2) the bicolor combinations, such as yellow/green, are too often repeated in different races, and (3) the lists neglect or are vague about the precise distribution of the races and their taxonomic status. More often than not they are simply identified by a senseless nickname given them by a dealer in tropical fishes.

The most damaging weakness of these lists is the fact that they do not prevent unscrupulous breeders from marketing hybrids between two races belonging to different lines produced in tanks as a fish found in the lake. This situation might in the short run make genetic studies of the lake *Tropheus* as well as research on the history of their colonization of the various lake biotopes well-nigh impossible. Thus we need a way of properly identifying the many races and the locality where they are found.

The following classification has been developed after careful consideration. Each race is identified by its locality and the geographical parameters. The code used to identify each race includes a first initial identifying the line the race belongs to and a second (and eventually third) initial locating the main coastline (east-west, south, or the Ubwari peninsula); then follows the latitude of the locality in degrees and minutes.

The lines are identified as follows:

Tropheus moorii moorii line:	Initial M
Lukuga (Kalemie) line:	Initial L
Tropheus duboisi line:	Initial D
Tropheus brichardi line:	Initial B
Tropheus moorii kasabae line:	Initial K
Tropheus polli line:	Initial P
Mwerazi line:	Initial W
Moba-Mtoto line:	Initial O
Unidentified line:	Initial X

The coastline is identified in the second and sometimes also third initial as follows:

Western coastline:	Initial W
Eastern coastline:	Initial E
Ubwari north shore:	Initials UN
Ubwari south shore:	Initials US
South coast (Zambia):	Initial S

To provide an example: The race found at Cape Caramba on the southern-facing shoreline of the Ubwari is identified by its code which reads: Caramba M (for *T.moorii moorii*) US (for south shore of the Ubwari Peninsula) 4.15 (for 4°15'S latitude). In short, *Caramba* M.US.4.15

Any two *Tropheus* races occupying the same biotope will be identified at once by their first initial, the remaining part of their respective codes reflecting the fact that they are found together in their locality. Should any line become identified and a new taxon created for it, its code will be at once updated by the simple expedient of the proper change of its first initial from X to the initial of the line.

Any hybrid that would eventually find its way into a scientist's hands will be immediately identified by the lack of its code (and locality it is supposed to come from).

List and Description of the *Tropheus* Lines and their Major Races—with their Codes

A. *Tropheus moorii moorii* Line

Eleven color morphs have been identified so far in the northern basin, all belonging to the same line, which might be identified as follows:

-5 spines in the anal fin.

-tail and caudal peduncle always opaque black.

-tail slightly concave and not very deeply notched.

-basic body color black (sometimes entirely), but more often with a secondary color pattern manifest as spots (rare), a vertical stripe or band, or one or two triangles opposite each other on the back and abdominal area.

-secondary color shades of yellow, orange, or red, eventually as a network of orange or red on a yellow background. The secondary colors are never green or blue.

Tropheus sp., Kalemie, L.W. 5.55.　*Tropheus* sp., Kalemie, L.W. 5.55.

Tropheus sp., Kavimvira, X.W. 3.30　*Tropheus* sp., Kalemie, L.W. 5.55.

L.W. 5.30 *Tropheus* sp., canary-cheeked variety.　*Tropheus* sp., juvenile, Kipampa, L.W. 5.40.

Tropheus brichardi, juvenile, Nyanza Lac, B.E. 4.15.　*Tropheus* sp., dominant male, Kipampa, L.W. 5.40.

-adult *Tropheus moorii moorii* never with any of their various mood patterns multistriped.

The various geographical morphs occur as follows from south to north and west to east in their biotope:

1. *Caramba* (M.US.4.15) code = M (for *moorii moorii*), US (for Ubwari southern face), 4.15 (for 4°15′S latitude)

Called the "cherry cheek" in the trade. Body entirely black, including the unpaired fins and pectorals. A large red spot present at the base of the dorsal fin and extending into the fin. Another irregular patch occurs behind the eye and extends onto the nape and cheek. Some specimens (rare) have another red patch on the abdomen. The spots can be cherry red, crimson, or brown-red, according to the place along the peninsula at which the fish are found. This form extends along the entire southern coast of the Ubwari from Cape Muzimu to Cape Karamba, or 50 km. Exported.

2. *Cape Banza* (M.UN.4.05)

Called the "Banza" by the collectors. A very variable subrace of the next variety. The body is entirely black with a thin vertical lemon yellow stripe extending from the vent to the dorsal fin (much like the better known "lemon-striped" morph of Mboko) or reduced to a spot on the back. Eventually with the upper half of the eye bright yellow. This race extends from Cape Muzimu on the southern face to the village of Manga on the northern face or on about 5 km of coastline. It has not been exported.

3. *Kiriza* (M.UN.4.10)

Body black except for a broad lemon-colored band on the sides extending from the belly to and into the dorsal fin,

where it is more orange.

The race extends from Manga near the tip of the peninsula on the northern face along the entire coast (over 35 km) where the rock coast blends with the swamps. There is no significant variation of the pattern although in this area the specimens are a bit more colorful, with a faint network of orange superimposed on the abdominal area. There is a slight sexual dimorphism, the band being broader in males than in females.

Called by Matthes the "Rubana" race, which has since been replaced by Kiriza because the original place has not been found anymore and Kiriza is easier to identify because of a falls high up in the mountains. The race is exported under the names Kiriza or "Kaiser" II because of its superficial resemblance to the *Kaiser* morph from Tanzania. The so-called "Baraka" race has not been seen at Baraka.

4. *Lueba* (M.W.4.00)

Originally found by Marlier and Matthes. In fact, this form is split into two populations approximately 3 km apart on two isolated and much silted-over rocky outcrops. It is separated from the previous races by Burton Bay and its swamps and sandy bottoms, and from the next race to the north by 20km-long sand flats and the estuary of the Kaboge torrent. Marlier and Matthes described the race as having a dark blue head and a black body, which leads me to think that it might be an offshoot of *T.duboisi*. Step by step I explored the coast for it between Mboko and south of Lueba. In fact, the race is black all over and the head is not blue. The two populations

appear to be restricted, each to an area not more than 50 by 50 meters of rock half buried in the sand. The population 3km north of Lueba is exceedingly variable, although probably not more than 100 specimens strong. Some fish are entirely black, others have a drop-shaped lemon yellow patch on the sides not more than about 1cm across; a few specimens sport two successive bright lemon yellow patches on the sides. The entire population at Lueba, also about 100 specimens strong, is, as far as I could see, entirely black.

It is obvious that the two populations are the farthest outposts to the south of the *Tropheus* from the Uvira-Mboko coastline, now isolated and on the road to extinction by the expanding sand flats of Burton Bay.

5. *Mboko-Kifumbwe* race (M.W.3.45)

This form is isolated on sand flats at islands one mile from the mainland, the main island being Mbuluma ("solitude" in the Babembe language). The group of three inlets is linked by a submerged causeway to the mainland, where scattered *Tropheus* are also found. The fish, all black except for a thin lemon yellow vertical stripe extending from the dorsal fin to the abdominal area, are probably not more than a few thousands strong. Exported from breeding stock (due to the small size of the population) under the names "lemon stripe" or "Mboko".

6. *Bemba* race (M.W.3.35)

This form is isolated from the previous race by the Suima estuary. It is found around cape Munene (meaning "white" in Babembe) on a 1 km long coastline. This purest orange morph of *Tropheus*

Tropheus moorii moorii, Bemba orange variety. *Tropheus moorii moorii*, Magara red/black variety.

Tropheus moorii kasabae, Kamba Bay variety. *Tropheus moorii kasabae*, Kamba Bay variety.

Tropheus moorii kasabae, Kapemba Bay variety. *Tropheus moorii kasabae*, Chaitika Cape rainbow variety.

Tropheus moorii kasabae, Kachese variety. *Tropheus moorii kasabae*, Kapeyeye.

provides an example of cohabitation of two *Tropheus* species on the same grounds. It is found on this coast along with *T.duboisi* (code D.W.3.35).

The posterior part of the body is always black, most often also the anterior part and the head, although specimens can be found with the front half of the body entirely bright orange. A broad band of orange-vermilion extends across the body from the back (and in part also the dorsal fin) to the belly.

There is a fair-sized population with a high density at the cape. The name "Bemba" was applied to a village found nearby which has now disappeared, but is kept here as it has been in extensive use in scientific papers. The *orange Tropheus* is exported.

7. *Makobola* race (M.W.3.33)

This is in fact a melting pot of varieties with very changing patterns, from narrow and faint banded orange morphs mixed with specimens nearly entirely black to blotched yellow or copper–orange and black, distributed from Makobola to Uvira. For simplification it is coded M.W.3.33-3.25.

The future evolution of these hybrid populations might eventually result in a few better defined patterns through isolation as the coast is increasingly fragmented by the Makobola-Kivovo torrent estuaries. As yet the Makobola race is not isolated from the Bemba biotope and there is still evidence of a gradual evolution, although contact has become more and more difficult along the coastline between Uvira and Bemba. This race is not exported.

8. *Uvira* race (M.W.3.25)

This is an all black variety with here and there a few specimens displaying a dark orange band on the sides. More often they are a "smoky" black. The specimens have the upper half of the eye brown, which is typical of *Tropheus* on the west coast of the basin. The "Uvira" or "black" morph has not been exported due to its drab colors.

9. *Magara* race (M.E.3.35)

This is the first *Tropheus* to have been exported by myself from the lake in 1958 and again, afterwards, in 1971. It is called the "Brabant" or the "Burundi" *Tropheus* as well as the "Red" *Tropheus* in the trade.

It is all black with two triangular patches on the sides placed opposite each other, one on the back and the other in the abdominal area. The triangles are golden yellow with a network of crimson red superimposed. The two triangles form an hourglass shape. In some of the mood patterns either yellow or red color prevails and only the triangle on the back might eventually persist. The size and contrast of the patches are thus variable depending on the moods.

The race extends (with few variations) on about 20 km of coastline between Kabezi (south of Bujumbura) and Cape Magara, which are separated from each other by the growing barrier of the Mutumba estuary. The habitat of Magara is the southern boundary of the major rock biotopes on the east coast of the basin. Further south there remain at first only pebble beaches and isolated rock outcrops on a coast invaded by sand shallows.

10. *Minago* race (M.E.3.50)

For 10 km south of Cape Magara small populations of *Tropheus* persist on a coastline that for them becomes more and more difficult to live on. The local race is entirely black and resembles much the one found in Uvira. There is a faint dirty yellow hue in the abdominal area. Called the "black" *Tropheus*, it has seldom been exported, although in areas closest to Cape Magara the populations are fairly large, but scattered.

11. *Rumonge* (M.E.3.55)

This form has not been collected since 1960 when it was first seen by Matthes. Due to the proximity of *T.brichardi* at Nyanza Lac on the other side of the ecological barrier between the two localities, I was in doubt about its status as Matthes had not been aware of *T.brichardi* at the time (described in 1973). We organized a new collecting expedition in 1984 to ascertain the status of the local race at Rumonge. The pattern and number of anal fin spines are similar to those of *T.moorii moorii*. The background color is black. There are faint brown dorsal and beige brown ventral areas, the former extending into the dorsal fin. In a white storage tank, the beige brown area extends to the entire anterior part of the body, including the head, as in the other *moorii*.

What can we make out from the distribution of *T.moorii moorii* races around the northern basin? Foremost, perhaps, we see the repetition of very similar local varieties at various points of the biotope, where it is evident that they developed separately from each other, reverting to or maintaining a similar pattern. Such are the two "black" morphs at Uvira and Minago, or the "thin lemon stripe" races at Mboko and Cape Banza, or the "Brabant" of the Burundi coast and the "cherry-cheek" race of the southern face of the Ubwari. It might become evident in the future, but is still too early as yet to think that under ecological pressure a *Tropheus* line will repeat patterns within its genetic drift potential in separate populations. The few "tear-drop" pattern specimens found north of Lueba might in this context be related to the "Kirsch-fleck Tropheus", found along the Kungwe mountain range south of the Malagarazi, whose taxonomic status is as yet unknown.

Another feature that we should remember is the spread of *T.moorii moorii* on both sides of the muddy Ruzizi delta. Since we know that *Tropheus* avoid mud and have never been found in areas such as a river estuary, and since we also know that the Ruzizi alluvional plain and delta are rather recent, probably not much older than 10,000 years, one can imagine as a likely explanation of the spread of the *moorii moorii* line that they previously occupied solid rock biotopes that are now buried underneath the alluvional deposits. All this occurs much farther to the north than the present shoreline, along the Kamaniola escarpment.

B. *Kalemie* Line (or *Lukuga* line)

This line extends from the Mtoto-Mpala area, from about 6°45'S, to a little above 5°20'S. It can be identified by several features:

-anal fin spine mode 6 or 7; dorsal fin spine mode 21-22 (?), caudal normally forked and not exceptionally long, never black.

-body pattern: usually chocolate or mouse gray, plain anteriorly, often striped posteriorly. Secondary colors: yellow-green. The stripes on the body are yellow, sometimes green, although in one race they are black dorsally and white underneath. In all races there is a yellow or more or less orange area on the head involving the snout, cheek, opercle, and postopercular area, but never the sides. The paired fins as well as the dorsal and anal fins are often various shades of yellow or brownish. Kalemie race specimens have been found with ocelli on the pelvic fins. There is often a bluish metallic sheen on the soft parts of the dorsal and anal fins.

So far six local races have been identified. From south to north they are:

1. *Mtoto-Mpala* race (L.W.7.00-6.45)

The body is plain (without any stripes), brownish on the anterior part and more yellow posteriorly. A strong yellow hue is present on the snout. It is yellowish on the underparts and more chocolate dorsally. The unpaired fins are dark chocolate, yellowish posteriorly. The pectoral fins are a dirty yellow.

This form is sympatric in its habitat with another race of *Tropheus* sporting only 4 anal fin spines.
See Moba 0.W.7.00 .

2. *Kalemie* race (L.W.5.55)

This form was found at the pier at Kalemie. The body is dark brown, dirty yellow underneath. The head is dark brown with an indistinct orange hue. A few very short stripes with a faint green or orange hue run down from the back, but are most often reduced to one or two scale lengths. The dorsal and anal fins are dark brown with a blue sheen on the soft part, and the caudal fin is olivaceous, the upper lobe often with an orange tip (vermilion in juveniles). Juveniles are very similar to those of *T.brichardi* from Nyanza Lac (striped green or brown on a yellow background, dorsal and anal fins yellow to vermilion). The pelvic fins are often with ocellated spots. This race is sold under the name of "Kaniosha" or "Kalemie".

3. *Kipampa* race (L.W.5.40)

This form is very variable depending upon the place where it is found. The race was named after the coast along which the best-looking specimens are found. It is separated from the previous race by the extensive sandy bottoms north of the Lukuga, but which could be crossed along the sandstone slabs that line the beaches for more than 40km. The race exists from Cape Bwana Denge to near the Kabimba Bay. It most commonly has a light chocolate body with a yellow ventral area. The posterior part of the body is striped with alternate bands of brown and yellow. All fins are more or less yellow or orange, but not very bright in color except for juveniles. The tail is darker brown, but never black. The most beautiful population at Kipampa extends only along a 300-meter-long shoreline.

Specimens there are pale mouse gray in front, the posterior part of the body alternately banded in yellow and a dark band that is black dorsally but becoming pure white ventrally. All fins except the tail are a bright canary yellow. This is probably the best looking *Tropheus* in the whole lake. It is dangerous to collect because of the presence of a huge crocodile population. It is exported under the name of "Kavalla" or "Kipampa".

4. *Kabimba* race (L.W.5.30)
This form follows the previous race along the shoreline, from Kabimba Bay to 10km from the bay to the north. Thus it has a narrow range. The two races are not separated by a barrier. The "Kabimba" *Tropheus* is basically a mouse-gray fish, the color of all fins being canary yellow, but a bit less bright than the fins of the small Kipampa population. The tail is dark gray, not black. There is a conspicuous yellow patch on the cheek and temporal area, hence the nickname of the race is "canary cheek," under which it is being exported. This is an outstanding and very colorful race.

This color variety of *Tropheus moorii* was one of the first imported for aquarists, and it still retains much of its initial popularity. Photo by Burkhard Kahl.

165

5. *Kabogo* race (L.W.5.25)

This form extends north of the previous race, from the Muzimu River (10km north of Kabimba) to the Kabogo River, 20km north of Kabimba. Thus it also has a 10km range. It does not have the typical yellow patch on the head of the "canary cheek" and resembles much the race called Kipampa, but for the fact that the fin colors are much paler and of a ghostly white-yellow tinged with pale green. By far this race is not as beautiful as the two previous races. It is not exported.

6. *Zinzia* race (L.W.5.20)

This form extends north of the Kabogo River and the Zinzia Creek, which means it occurs between 5°20'S and 5°10'S. The body is dark green with a faint but broad yellow spot under the dorsal fin. All fins are olivaceous green, the pattern being much drabber than in the other races of the line located north of the Lukuga.

Conclusions about this line

This line is distributed over a long coastline, covering about 200km. Most populations are separated from each other, especially between Mpala and Cape Bwana Denge, by long stretches of sand bottom through which passage was provided by unbroken fords of sandstone slabs. Aerial photographs to which the author had access show that the sand bottoms were built by convection currents and not by alluvional deposits from nearby rivers, which might have brought in muddy waters that were unfordable by *Tropheus*. Small rock promontories jut out from the coastline every few km, which apparently provided a marginal type of rocky habitat for the *Tropheus* in their migrations along the coast by the sandstone slabs.

The Mtoto-Mpala population is mixed with the Moba line, which have a different morphology and basic pattern. In the north of the "Lukuga" line range, from the islands of Toa Bay toward the north, the local populations are increasingly mixed with another *Tropheus* with again another morphology, the foremost difference being a much higher, longer, and crescentic tail, the tail of the Lukuga stem being short and simply notched (such as the one found on *Tropheus moorii moorii*). The pattern of this new line is also different.

The second peculiarity of the Lukuga races' distribution along the Ngoma coast is the fact that it was enough to have a small barrier intercalated by the alluvions of a mountain torrent (like the Muzimu or the Kabogo Rivers, covering only a few hundred meters of coast) to bring about a change in the pattern of the local populations. In fact, the Kipampa race pattern extends toward the Kabogo barrier, with differences in shades only, and the Kabimba canary-cheek *Tropheus* is a local, if spectacular, reminder of the trend in the stem to have a yellow patch on the head.

C. *Tropheus duboisi* Line

Only three populations of the famous *T.duboisi* have been found around the lake. We had hopes of finding more populations during our 1984 explorations along the Ngoma range south of the Ubwari and along the Marungu range south of Kalemie, which brought us to the border between Zaire and Zambia. We found none, so that one might guess that the three populations already identified in the past are the only ones of this lineage.

All three share the following features:

-anal fin most often with 5 spines, rather seldom with 6.

-dorsal fin with 21 (80%), 20 (10%), or 22 spines (10%).

-mouth narrower than in other *Tropheus* and with fewer teeth.

-head covered with a dark blue skin, the body otherwise entirely jet-black except for a median vertical white or yellow stripe extending from the dorsal fin to the abdomen. Juveniles dotted all over with pure white spots that disappear before they are fully adult.

Two main races have been recognized in the aquarium trade, namely:

1. "White-banded", occuring north of the Ubwari Peninsula and Malagarazi.
2. "Broad yellow-banded", found nearby in the Malagarazi estuary.

It appears from pond-raised products of the first race that the color of the band might depend on the diet or the chemistry of the water in which the fish are raised.

Narrow White Band

1. *Bemba* (D (*duboisi*). W.3.35)

This form cohabits with the orange morph of *Tropheus moorii moorii* on a very narrow stretch of coastline, less than 1km long, around Cape Munene (the village of Bemba no longer exists). This relict population, more than 160km from the next one in Tanzania

The white-banded varieties of *Tropheus duboisi* occur north of the Ubwari. Photo by Glen S. Axelrod.

on the east coast, involves a few thousand fish at most.

2. *Kigoma* (D.E.4.50)

This is the largest population as yet found of *T.duboisi,* and occurs from the border between Tanzania and Burundi, south of Nyanza Lac (4°30'S), to the Malagarazi (5°10'). Distributed along nearly 100km of coastline, this population involves tens of thousands of fish. It shares its habitat with *Tropheus brichardi.*

Broad yellow band

3. *Malagarazi duboisi* (D.E.5.15)

The broad-banded *T.duboisi* has been found around the Malagarazi River delta. Its full range is as yet unknown. By mid-1983 a new very broad-banded yellow variety, whose whereabouts have not been identified, has been exported from Tanzania. It doesn't appear to be significantly different from the specimens previously exported. *

The broad yellow-banded variety is found mostly in the estuary of the Malagarazi. Photo by Pierre Brichard.

N.B. Claims of the discovery of a red-banded *duboisi* have not been substantiated.

Several varieties of *Tropheus brichardi* have been exported from the lake from time to time. The species range lies between 4°15′ and 5°10′S latitude. Photo by Dr. W. Staeck.

D. *Tropheus brichardi* Line

Several varieties and races of this species, sometimes difficult to identify and locate, have been exported from time to time. The main habitat appears to be north of Kigoma to the south of Rumonge. The northernmost habitat appears to be the "corniche" of Nyanza Lac from which the first "shoko" were identified and exported.

So far the range of *T.brichardi* lies between 4°15′ and 5°10′. The main features of this species are:

-anal fin with 6 spines (99%), exceptionally 5 (1%); dorsal fin with 20 (40%) or 21 (60%) spines.

-mouth broader and lower than in other species.

-body background color varying from green to chocolate brown with lighter yellow, slate gray, or light green bands; with a pale yellow or light green oval dot under the dorsal fin.

Other newly found geographical races, for example striped yellow on a jade-green background in the posterior half of the body, have been localized on the Ubwari peninsula and might link *Tropheus brichardi* to *T.annectens.*

The anal fin spine count of *T.brichardi* puts this species well apart from *T.moorii moorii.* Dominant males of *T.brichardi* display an off-white, yellow, or sky-blue iris that is typical of the species.

Although more local races might eventually be discovered among the *brichardi* as yet exported, none appear to be living south of the Malagarazi delta, which appears to have been a major barrier to their spread toward the southern part of the lake.

Three well-identified races have been exported so far:

1. *Nyanza Lac* (B.E.4.15)

This form is called in the trade the "shoko" *moorii* (sic). It is identified by an oval yellow, off-white, or jade green spot on the back. The variability of the patterns along the Nyanza Lac corniche is often exceptional. On two sides of a cove the regular type of "shoko" *moorii* and another sporting alternate bands of chocolate brown and bright emerald green stripes can be found.

The Nyanza Lac *T.brichardi* never exceed 10cm in length. They cover a coastline about 10km long, but not exceeding 10 to 15 meters in breadth. It has been exported since 1972.

2. *Kigoma* (B.E.4.50)

This form extends along the Ngombe mountain range, south of the border with Burundi, to near the Malagarazi delta. It has green stripes on a pale yellow or slate gray background, with local insignificant variations. It has been exported since 1973, but fell out of favor.

3. *Karago* race (B.E.5.15)

This is also called the "Malagarazi" or "blue-eyed" Tropheus. It has a green body and head and a lemon yellow area on the abdomen. The pectorals and tail are green and the dorsal and anal fins are yellow-green, sometimes highlighted with orange-red. The eye is sky blue. It has been sporadically exported since 1973.

D2. Unidentified Line

Before leaving the *Tropheus brichardi* line and the races related to the northernmost basin of the lake, a few words should be said in respect to several local populations of an

The chocolate-striped variety of *Tropheus brichardi*. This is a young individual of the original Nyanza Lac type. Photo by Pierre Brichard.

This *Tropheus brichardi* variety has a yellowish dorsal fin and is of the Nyanza Lac type found around Kigoma. Photo by Pierre Brichard.

as-yet-unidentified line of unclear taxonomic status.

1. *Ubwari* race (X.US.4.15)

This form has been nicknamed the "Ubwari green" by our team who found the population in 1981. It is distributed unevenly between Cape Banza and Kasimia Bay, thus over 60km of the southern Ubwari shores, where it is superimposed on the Caramba race of *T.moorii moorii*. The maximum density of this peculiar race is found midway down the peninsula and is weaker at Caramba. The background color of the body is green, usually jade-green, the hind part of the body being striped alternately with green and lemon yellow; the ventral parts are lemon yellow. The eyes are blue-green. The dorsal and anal fins are green; the tail is long and crescentic. Six anal fin spines are found in 84% of the specimens.

2. *Yungu* race (X.W.4.50)

This is the race whose population at first is sympatric on the Lukuga races from Toa Bay on toward the north, then is alone from 5°00′S on the coast toward Kasimia Bay. The body is bottle green with many pale yellow stripes. The tail is crescentic and very long and high. This important population extends over a 130km coastline and is especially dense in the northern part of its range. It is unidentified as yet, having been discovered only early in 1984.

3. *Kavimvira* race (X.W.3.30)

This form has the body very dark green, the ventral area paler and suffused with yellow. The tail is very dark green, nearly black. There are one or two thin yellow stripes on the sides, sometimes fusing together in the shape of a "Y". The dorsal and anal fins are of a drab, indistinct dark color with a greenish hue. From its anal spine counts (about 50% of the specimens have 6 spines, the other half have 5) it is not impossible that this population is a natural hybrid between the "Yungu" line and a local *moorii moorii*. It is heavily outnumbered in its habitat by the latter.

E. *Tropheus moorii kasabae* Lines

The southernmost shorelines of the lake, between the borders of Zambia with Zaire and Tanzania, have yielded a set of local *Tropheus* races that have been ascribed by Nelissen to the subspecies *Tropheus moorii kasabae*. From the features found on the various races that have since been identified, it looks as if they should be divided into two distinct lines, if not by taxonomic parameters—which are rather similar—at least by the pattern featured in their respective robes. Nelissen had identified, I am made to understand, faint taxonomic discrepancies that lead him to believe there might be two lines in the subspecies. The two lines thus cannot be easily identified from preserved specimens, but mainly by their respective colors and patterns. They share taxonomic features that set them well apart from the *Tropheus* in the northern part of the lake. Their basic pattern is also strikingly different. Most specimens in the lake have a striped pattern; but this is a mood pattern that can disappear entirely when the fish is excited or dominant or, to the contrary, dominated, frightened, or sick.

The stripes, numbering from 10 to 12 and only a half-scale or at most a single scale wide, start on the head and extend to the end of the caudal peduncle. In all races from both lines they are usually pale yellow or cream-colored. They start on the back and extend down the sides of the body to eventually blend into the lighter underparts.

The two lines that I identified among the southern races have a different distribution along the southern shorelines, one to the west the other toward the east. They come into contact in Nkamba Bay but do not hybridize, indicating that they belong to different strains.

a) **The line "with a double chevron on the snout"**

All races from this line are identified by the two "V"-shaped chevrons, contrasting with the background color, that cross the snout. The color of the chevrons is always that of the secondary color. All juveniles of the races belonging to this line always start to develop their adult pattern with these two chevrons.

The line occurs from Nkamba Bay on the southern coast to Kapampa Bay in the northwestern part of Cameron Bay, approximately 7°35′S or 130km. Kapampa Bay is the boundary between this line and the *Tropheus* from the Mwerazi, which belong to another line.

The lack of strong ecological barriers along this long coast helps explain the spread of the line, even past the sand bottoms brought in by the small Sumbu River.

Refusing the strong temptation to multiply the number of races according to minor pattern differences, we can subdivide the line into its major geographic races. There

Tropheus moorii kasabae comes from the southern shores of the lake. This subspecies can be divided further into several races or color varieties. Photo by Dr. W. Staeck.

are six. From north to south they are as follows:

1. *Luhangwa* race (K.W.7.35)

From south of the barrier of Kapampa Bay, 20km long, and composed of alluvions falling down from the coastal hills to Luhangwa Bay, the *Tropheus* are a beige green, very pale in shade, striated with thin pale yellow stripes. The chevrons are pale red, with the northernmost population having them pale brown.

Between the gillcover and the pectoral fin extends a thin vertical stripe widening on the thorax and spreading toward the chin and lower part of the jaw. This patch is brown in the northern part of the race's range, but carmine red in Luhangwa Bay. The underparts of the body have a reddish hue. The dorsal and anal fins are medium brown in the north, reddish in Luhangwa. The body of the Luhangwa specimens is darker than those close to Kapampa

Bay and might have a dark orange cast rather similar to the one found in the Kachese race down south.

2. *Lupota* race (K.W.7.55)

The olivaceous brown color of the Luhangwa race has turned reddish brown in Lupota Bay specimens, and the color becomes dark brown further down toward Moliro. The double "V" on the snout and the sides turn a deep purple red, as does the postopercular area, chin, and chest. The sides behind the pectoral fins are suffused with the same bright color over a broad area. This circular patch is in many specimens vertically crossed by two or three thin but very dark stripes. The dorsal and anal fins are brick-red.

It looks as if the race has been collected lately and exported under the nickname of "rotkehl" (meaning redthroat in German). By the color pattern, this race is a preview of the most famous of the

double "V" races—the "Moliro" *Tropheus*.

3. *Moliro* race (K.W.8.10)

The features already noted in the previous race are increased in the Moliro race. The red has become a bright cherry red that covers the whole body of dominant or excited males and females. The background color is a dark brown, the sides and the head being of a lighter shade and covered with a crimson hue as are the underparts. The caudal peduncle, brown in color, is horizontally striped with a fluorescent cherry red bar about 3cm long on a large adult fish. The dorsal and anal fins are cherry red all over, with a black area on the soft part of the fins. The tail is a very dark brown. The race is known in the trade after its locality. It has been exported since 1976.

4. *Chipimbi* race (K.W.8.18)

This is a race to which the Mossi population can be

added as the two display only minor differences between the majority of individual fishes. A minority in each population displays an affinity with another population, which is Moliro for the Chipimbi *Tropheus* and Kachese for the Mossi specimens.

It is very similar to the Moliro fishes, from which they differ by a much darker color, the brown being suffused with a dark blue gleam, and the red instead of being vermilion or cherry red is blood red. The dorsal and anal fins are also a fluorescent blood red.

The red stripe on the caudal peduncle of the Moliro *Tropheus* has retracted into individual spots, most often three, but sometimes only two. At Cape Chipimbi and further down the coast (at Mossi) only a single spot is left. The Mossi specimens are even darker than those at Chipimbi, the background color sometimes appearing dark blue, but basically it is still a dark brown; the red has become less conspicuous and darker.

We can thus say that between Moliro and Mossi, on approximately 30 km of coastline, the highlights of the *Tropheus* pattern are focused on the forward half of the body, including the head with its red sheen and chevrons, when the fish meet each other head-on. But when they are seen from the side the focus is on the red head, the unpaired fins, and the red area of the caudal peduncle.

This is the dotted head type of *Tropheus* sp. from the Lufubu River, Kabeyeye Cape region. Photo by Pierre Brichard.

5. *Kachese* (K.W.8.30)

This variety extends from the southern side of Ndole Bay toward Cape Kachese and perhaps beyond toward Sumbu, although the permanent murkiness of the water has not allowed us to find the boundary of this race. The first Kachese *Tropheus* to be exported came from the cape area, the remaining part of Ndole Bay being mostly sandy. Altogether the race doesn't inhabit much more than 2km of coastline.

The upper half of the body, the head, and the caudal peduncle range from olive green to brown, with an orange cast on the abdomen and caudal peduncle, the latter in fact being the remainder of the red spots or stripes of the Moliro-Chipimbi races. The dorsal and anal fins are fluorescent orange, the soft parts of which are mixed with black. The tail is olive brown or green.

6. *Sumbu-Nkamba* (K.W.8.30)

This form extends for about 35km between Sumbu Island and part of Nkamba Bay, along the southern shores of Cameron Bay. It displays a further evolution of the pattern of the Kachese type toward a reddish brown instead of olive color. The dorsal and anal fins are brick red, the abdomen and caudal peduncle lighter colored (orange-brick red).

This race is in contact with but does not hybridize with the Kabeyeye race of the second southern line in Nkamba Bay.

b) **The *"Rainbow"* line**

This form has 5 main races as follow: (from west to east): Kabeyeye-Chaitika-Mpulungu-Kalambo-Kala. They cover about 150km of coastline with several ecological barriers, none of them very large, in between (sand beaches and Lufubu estuary, the mud and sand flats of Hoare Bay, etc.). Several additional races might eventually be found in the future north of Kala in Tanzania.

All races display a head spotted with more or less contrasting, basically cream-colored tiny flecks. The sides

172

display a large lighter oval area in the abdominal region that is eventually elongated toward the caudal peduncle and shaped like a long triangle. The tail is never very dark, but usually olive green or brown. Pectorals yellow, orange, or green. The basic color is shades of green or blue.

1. *Kabeyeye* (K.S (south). 8.30)

The dominant color of this form is jade or olive green superimposed on a yellow background. A bright yellow triangle extends from the pectoral fin area toward the tail. The dorsal and anal fins are yellow as are the pectoral fins. The head, back, and tail are olive-green. The vertical stripes are pale yellowish white.

The Kabeyeye morph extends west of the Lufubu around Cape Kabeyeye toward the Inango Peninsula and Nkamba Bay where it is in contact with the Sumbu-Nkamba race of the line with the double chevron on the snout.

The bright orange spots on the anal fin stand out more sharply against the brownish black background of this *Tropheus moorii.*

This beautiful red variety is a *Tropheus* sp. of the double "V" type from Chipimbi Cape. Photo by Pierre Brichard.

2. *Chaitika* (K.S.8.35)

This form extends from the Lufubu River eastward past Chaitika Cape (from where most specimens have been fished). It is called the blue rainbow in the trade, so nicknamed because of the many color shades found on the first specimens. The best specimens are found close to the river mouth but are extremely difficult to catch as they live on megalithic rocks. These specimens display a magnificent light blue-jade green color, covered by a network of pink; the head is pastel strawberry, the spots and flecks creamy-white. The dorsal and anal fins are mauve or violet, the pectoral fins salmon colored. The tail is pale olive green or brown, the tips of the lobes orange-red.

The loss of colors is already noticeable at Cape Chaitika, one km to the east, especially in the dorsal and anal fins, which are sky blue striped with purple, the network of secondary color also being

purple. The degradation of the color pattern of the rainbows toward the east is very fast. Two or 3km east of the cape the secondary colors have become drab and the dorsal and anal fins are a grayish blue. The degradation of the colors increases toward Mpulungu, where the next race is the next normal evolutionary step of the Chaitika race.

3. *Mpulungu* (K.S.8.45)

This form was called the "E 1" variety when first exported in 1977. It is now called the "lemon fleck" or "zitron-fleck" in the trade. The fish is a dark olive green or bottle green depending on its mood. The sides behind the pectoral fins (which are dark green) display a broad lemon colored area. This patch is what remains of the broad triangle displayed by the Kabeyeye and Chaitika races. The dorsal and anal fins are poorly colored (a faint blue-gray). The vertical stripes, often running only halfway down the sides, are lemon colored. The specks on the head are barely visible.

The trade name is a poor choice as two other races around the lake have already

Although somewhat somber in appearance, this variety may turn out to be quite attractive if the red coloration in its fins becomes enhanced. Photo by Dr. W. Staeck.

This attractive color variety is a member of what has been referred to as the rainbow line. The reddish fins and yellow lateral blotch make a nice contrast. Photo by Dr. W. Staeck.

been found to sport lemon-colored patches on the sides, and we will see that the Mwerazi line has two, if not three races sporting yellow patches on the sides.

4. *Kalambo* race (K.E.8.35)

Several subraces have been identified that have developed around the estuary of the Kalambo River (boundary between Zambia and Tanzania). It is a development of the "rainbow" line and of the race found around Mpulungu, but the colors are much brighter. The background shade is blue-green, the secondary color being salmon on the sides, occurring as a broad patch behind the pectoral fin. The dorsal and anal fins are blue or purple, becoming paler posteriorly. The spots on the head, which were very faintly visible on the specimens from Mpulungu, stand out better and are cream-colored. This fish is known in the trade as the "papagai" or "parrot" morph. It has been exported since 1979 and is one of the best "rainbows" yet.

5. *Kala* race (K.E.8.08)

This race was introduced for the first time early in 1983. It is the last population of the "rainbow" line to have been found and the one extending most to the north on the eastern shores of the lake.

The highlights of the "rainbow" races are the large triangular light colored areas extending from behind the pectoral fins toward the rear on a background that is always green in color. The secondary color, reflected in the patch, is yellow in various shades, mixed with pink or red (Chaitika-Kalambo-Mpala). The spots on the head probably play a role in signaling the mood of the fish, but appear to be less significant than the very visible chevrons of the "double chevron" line. Altogether the focus of the pattern displayed by the "rainbow" races appears to be the area just behind the pectoral fins.

As with the northern stems of *Tropheus* we can note that in areas where the habitats are marginally capable of sustaining a *Tropheus* population (Kavimvira-Minago, Kasimia Bay, Mpulungu, etc.) the local specimens display a loss of intensity and contrast in their colors. Apparently *Tropheus* living in very clear waters display the brightest and most contrasting colors, while those living in muddy or

This is one of the southern races (Kamba Bay) of *Tropheus* sp. with the double "V" marking. Photo by Pierre Brichard.

Another rainbow variety with less yellow and more red. The rainbow appelation generally is attributed to *Tropheus moorii kasabae*. Photo by Dr. W. Staeck.

heavily sedimented habitats have a darker and drabber color pattern.

F. *Tropheus polli* Line

A single race of this species has been found along the Kungwe mountain range. Discovered and described by G. Axelrod, *T.polli* has only 4 anal spines and is distinguishable at once by its huge crescent-shaped and filamentous caudal.

1. *Mbulu* race (P.E.6.03)

The color of this race is most often pale purple or brown, sometimes with a greenish cast. The pattern can be striped or not; this feature is said to be involved with sexual dimorphism. Two stripes descend from the top of the head toward the chin. The posterior part of the dorsal and anal fins is decorated with small orange-red spots; the upper tip of the dorsal can also be orange-red. The dorsal fin has between 20 and 21 spines.

This race shares its habitat with another *Tropheus* whose taxonomic status is as yet unclear, although it has been exported in substantial

Tropheus polli, often referred to as the swallowtail *Tropheus*, was discovered near Bulu Point and Bulu Island and named after Dr. Max Poll. Photo by Glen S. Axelrod.

quantities since 1977 ("Kirschfleck" *Tropheus*, see Kungwe X.E.6.10).

T.polli has the same low number of anal fin spines (4) as the *Tropheus* described by Boulenger in 1900 as *annectens*. As we have already pointed out the two Boulenger specimens apparently did not come from Kalemie, but they don't look like *T.polli*.

The *Tropheus* that we found in Moba also have 4 spines and a crescentic caudal, but the lobes of the tail do not end in filaments like that of *polli*. Still, they could belong to the same line. But then how can we explain how or why the two races occupy grounds opposite each other across the central part of the lake, which is at that point 1400 meters deep?

G. Mwerazi Line

This line, which we discovered in June 1984, has not been studied as yet, but at first glance appears to be basically different from any we have seen in the lake so far, and especially from those that live near its biotopes. It is at once distinguishable by two features: it has 5 anal spines and a striking color pattern. The caudal fin is also longer, higher, and much more crescent-shaped than in the northern and southern *Tropheus* lines.

The background color is very dark green (it looks nearly black) and all unpaired fins are rather drab, dark yellow or green. Behind the pectoral fin extends a very large circular patch in strong contrast with the surrounding area. This patch is either lemon yellow or

a ghostly white and is the only contrasting feature of the color pattern. The average size of the Mwerazi *Tropheus* is well above normal and probably near 14cm, with many specimens reaching about 16cm. Juveniles were very dark and were still incubated at a size of about 25mm, which is also unusual.

The line extends north from its boundary with the "double chevron" line in Kapampa Bay to an area above Cape Zongwe. Its color pattern is drastically different from the "double chevron" line, as are its anal spine counts and shape of the caudal fin. It is also different from the *Tropheus* in the northern part of the coastline (Moba) that have only 4 anal spines and a different pattern (striped). There are three races distributed from 7°30′S to 7°10′S.

1. *Masanza* race (W.W.7.30)

The body of this form is dark green. The specimens observed in the lake at first appear to be black all over except for a very conspicuous broad circular patch 3 to 4cm across just behind the pectoral fin. This patch is a pale yellow, pale green, or nearly white. All fins are blackish, the dorsal and anal with a faint bluish sheen; the pectorals are dark gray or dark green. The head doesn't show any unusual features. Not exported.

2. *Mwerazi* race (W.W.7.25)

The color of this morph is jade or emerald-green, with a large lemon-colored patch behind the pale green pectorals. The dorsal and anal fins are a rather drab yellow. The head is of the same color as the body, but a bit paler. Not exported.

3. *Zongwe* race (W.W.7.15)

This form is not strikingly different from the Mwerazi with respect to the background color and the lemon patch on the side, but the head is rosy, especially on the cheeks. Not exported.

H. Moba Line

Mistaken at first for a Mwerazi race, the population at Moba is distinguishable by the fact that the *Tropheus* there all have only 4 spines in the anal fin, and there is a striped pattern. The caudal fin is large, long and high, and crescentic. The basic color is green striped with pale lemon yellow.

This race has been coded O.W.7.00 and is called the "Moba". It ranges from Moba Bay toward the north. From Mtoto on toward Mpala it lives on the same slope as another

Tropheus that has 6 anal spines, a normally forked caudal, and a color pattern featuring a plainly colored body with a yellow snout. (see the race called "Mtoto-Mpala." L.W.7.00-6.45)

I. *Tropheus* Morphs whose Line and Taxa have not been Identified

Several color morphs have been introduced in the trade that have been included in color morph lists as *Tropheus moorii* though they have never been properly identified as such and perhaps do not belong to this taxon. Few if any

The red (bright red in life) blotch on the side of this *Tropheus moorii* adds some color to an otherwise drab variety. Photo by Dr. Herbert R. Axelrod.

Without specific collecting data the exact designation of the various races of *Tropheus moorii* cannot be ascertained. Once the color fades almost all identification clues are gone. Photo by Dr. Herbert R. Axelrod.

This is the Kapeyeye variety of the rainbow line of *Tropheus moorii kasabae*. Photo by Hans-Joachim Richter.

specimens are available for study and one will have to wait until a sizeable collection of specimens has been assembled to know exactly what they are.

The main races of the various color morphs have been temporarily coded "X" when the exact locality where they are found is known. There has been no attempt made to classify races whose habitat is still shrouded in mystery, or color morphs that obviously are only a local variation of one of the main races. It is useless to increase the number of code names for races that are not substantially different from one of the main

geographical races so thus identified. In case of need, one could add the name of the secondary population after the existing code.

1. *Kungwe* race (X.E.6.10)

This form is known as the "kirschfleck" *Tropheus* in the trade. It appears to cover most, if not all, of the coastline along the Kungwe mountain range, which means at least 70km. Several local varieties have been reported from the area between 5°55'S and 6°30'S.

The basic background color is dark chocolate brown. Usually one, but very often two, irregular tear-drop patches are present on the sides, one behind the other.

The patches basically are yellow, but in the best specimens they are suffused with pink, which becomes crimson around the edges of the patches. Unpaired fins are dark brown.

The best specimens are spectacular, one of the best looking *Tropheus*. From the layout of the pattern, reminiscent of the "Lueba" race of *Tropheus moorii moorii,* one would be led to think that the "Kungwe" race belongs to this species. However, specimens have not been made available for study, although the fish have been exported sporadically since 1977.

2. *Ikola-Karema* race (X.E.6.42)

This is one of the best known races of *Tropheus*. It is known in the trade as the "Kaiser" *Tropheus*. Occurring south of the previous race one would be led to think that they don't belong to the same line, as they are very different in color pattern. The status of the "Kaiser" has still to be defined, as only very few specimens have been made available for study, even though the fish has been exported in large numbers since 1974. The four specimens that were available for study displayed 5 anal spines like *Tropheus moorii moorii* from the northern basin.

The "Kaiser" morph pattern consists of a black background that is especially strong in the rear part of the body, the abdominal area and the sides above being orange. The best specimens have the front part of the body, including the head, orange as well, but in most the head is black. Many specimens have a faint network of black suffused in the orange area, which is then a bit off-color and drab. Nice specimens of the "Kaiser" are outstanding and rather like the "Bemba" race (orange *Tropheus*) of *T.moorii moorii.* There is a distinct possiblity that they belong to this species, but one will have to wait until more preserved specimens are available in order to form an opinion. It would be quite remarkable that a population of *T.moorii moorii,* separated from the main habitats by biotopes occupied by *T.brichardi,* could exist so far in the south.

3. *Utinta-Mpimbwe* race (X.E.7.05)

The *Tropheus* reported from this area do not appear to belong to the same line as the "Kaiser", but no specimen is available for study even though the morph has been known for a long time in the trade. The color is said to be dark brown or blackish, the chin and underparts being yellow or faintly orange.

This race seems to have a wide range along the southeastern shoreline south of the previous race and should include several local varieties. Not exported anymore.

4. *Kipili* race (X.E.7.25)

No specimens available.

5. *Murago* race (No code.)

The locality of this race, which I discovered in 1981, cannot be revealed at the request of the collectors who financed the exploration. Its taxonomic status is still not defined for lack of preserved specimens.

The body is entirely green, darker on the head and dorsally. The whole body, but mainly the head is covered with tiny mother-of-pearl specks; these increase with age. Juveniles are not spotted.

Thus we have a pattern exactly opposite to the one displayed by *T.duboisi* during its growth. The juveniles and the adults are vertically striped, the creamy yellow stripes narrowing with age. The vertical fins are green with the dorsal and anal suffused with yellow. The underparts are lemon yellow.

A full adult "Murago" is perhaps one of the two or three most beautiful *Tropheus* in the lake. It is very difficult to catch in its habitat, which does not hold a large population, and occurs only along a short coastline. This race will probably become available through breeding. Half of the first few fish collected died from internal parasites.

The Chipimbi Cape variety of *Tropheus moorii* is similar to the Moliro fishes but is darker and the red is blood red rather than vermilion. Photo by Dr. W. Staeck.

Tropheus moorii moorii (Line M)
M.US	4.15	Caramba	Cherry cheek. Caramba
M.UN	4.05	Cap Banza	Yellow eye. Banza
M.UN	4.10	Kiriza	Kiriza. Kaiser II
M.W	4.00	Lueba	Lueba
M.W	3.45	Mboko	Yellow stripe. Gelbstreif
M.W	3.35	Bemba	Orange. Bemba
M.W	3.33	Makobola	. . . .
M.W	3.25	Uvira	Uvira black
M.E	3.35	Magara	Brabant. Red. Burundi
M.E	3.50	Minago	Minago. Schwarz.
M.E	3.55	Rumonge	Rumonge. Braun. Brown.

Tropheus duboisi (Line D)
D.W	3.35	Bemba	White-stripe. Weiss band
D.E	4.50	Kigoma	White-stripe. Weiss band
D.E	5.15	Malagarazi	Broad band. Yellow *duboisi*

Tropheus brichardi (Line B)
B.E	4.15	Nyanza Lac	Shoko *Tropheus*
B.E	4.50	Kigoma	Kigoma
B.E	5.15	Karago	Malagarazi. Blue-eye *Tropheus*

Tropheus moorii kasabae (Line K)
K.W	7.35	Lunangwa	. . . .
K.W	7.55	Lupota	Rotkehl. Lunangwa
K.W	8.10	Moliro	Moliro
K.W	8.18	Chipimbi	Chipimbi. Mossi
K.W	8.30	Kachese	Kachese
K.W	8.32	Sumbu-Nkamba	Sumbu. Nkamba
K.S	8.30	Kabeyeye	Yellow rainbow. Kabeyeye
K.S	8.35	Chaitika	Rainbow. Chaitika
K.S	8.45	Mpulungu	E 1. Zitron Fleck
K.E	8.35	Kalambo	Papagai. Kalambo
K.E	8.08	Kala	. . . .

Tropheus polli (Line P)
P.E	6.03	Mbulu Island	*polli.*

Lukuga-Kalemie line (Line L)
L.W	7.00	Mtoto-Mpala	Yellow nose. Gelb Nase
L.W	5.55	Kalemie	Kalemie. Kaniosha
L.W	5.40	Kipampa	Kavalla
L.W	5.30	Kabimba	Canary cheek
L.W	5.25	Kabogo	. . .
L.W	5.20	Zinzia	. . . .

Mwerazi/Tropheus (Line W)
W.W	7.30	Masanza	. . . .
W.W	7.25	Mwerazi	. . . .
W.W	7.15	Zongwe	. . . .

Moba/Tropheus
O.W	7.00	Moba	. . . .

Undetermined Tropheus
X.US	4.15	Ubwari	Green Ubwari
X.W	4.50	Yungu	Yungu
X.W	3.30	Kavimvira	. .
X.E	6.10	Kungwe	Kirschfleck
X.E	6.42	Ikola-Karema	Kaiser
X.E	7.05	Utinta	. . .
X.E	7.25	Kipili	. . .
. . . .	. . .		Murago

CONCLUSIONS

Most of the major rock biotopes of the lake have now been at least superficially explored and thus most of the *Tropheus* lines and the races they are broken into have now probably been identified. A few races with limited ranges will probably still be found along the eastern and western shorelines between the areas that have been explored less superficially.

Forty-seven races have been accounted for so far that appear to be distributed among at least 9 lines. Each line should normally be given at least subspecific status in due time. We would thus wind up with 9 or 10 species and subspecies of *Tropheus*.

Of the 47 races included in the present classification 8 have not been identified with respect to a line, but 39 can already be properly coded and localized. These 39 races are distributed unevenly among the various lines with some lines appearing to be richer than others.

T.moorii moorii: 11 races and subraces.

T. "Lukuga": 6 races, perhaps 3 more?

T.brichardi: 3 identified races, probably at least 2 or 3 more.

T.duboisi: 3 geographical races.

T.moorii kasabae: 6 races for the "double chevron" line; 5 races for the "rainbow" line.

T.polli: as yet only one race has been identified.

T. "Mwerazi": 3 geographical races have been found.

T. "Moba": as yet only one race has been found.

Until 1983, of the coastlines extending along the western

Cases of Superimposition of Two *Tropheus* on the same Biotope		
Locality	*Latitude*	*Lines*
Bemba	W3.35	M + D
Kigoma	E4.50	B + D
Ubwari	US4.00-4.30	M + X
Ngoma	W5.20-5.40	L + X
Kungwe	E6.00-6.40	P + X
Mtoto	W6.45-6.30	L + O

shores of the lake for about 800km, only the shores lying in the waters of Cameron Bay in the southern part and those lying north of 4°30′S had been explored (amounting to about 220km). Today, only 90km of coastline essentially composed of sand bottoms and lagoons remain unexplored by our team. The exploration of nearly 500km of rocky shores has now been successfully completed by our team in just two years time. The remaining gap, between Kalemie and Mpala, 90 km wide, has probably been colonized by the Lukuga line of *Tropheus* and might yield a few more local races.

I don't think that a single important race has been overlooked as the coastal shores were systematically sounded every 5 or 10km along the whole megabiotopes of the Ngoma and Marungu ranges. By 1983 we had completed the exploration of the *T.moorii moorii* line in the northern basin and its approaches along the Ubwari. The present distribution of the various races belonging to this line tells much about the way the habitat was colonized, how it looked at first, and how the intercalate ecological barriers came into being. As barriers appeared in the midst of what

until then had been homogenous rock biotopes, they isolated local *Tropheus* populations on remaining fragments of the biotopes. Until then the *Tropheus* could wander along the coast and freely mix. From then on the isolated populations evolved separately into local races by inbreeding. And as we look to clues on how they became isolated and find answers to our quest, the whole geological history as well as the future of the area suddenly come into focus.

CHRONOLOGY OF COASTAL EVOLUTION IN THE NORTHERN BASIN

Three lines of *Tropheus* have been identified so far within the basin lying north of the Ubwari peninsula. The first is *Tropheus moorii moorii* of the black-tailed line distributed between Rumonge on the eastern coastline and the entire western coast of the basin, including both sides of the Ubwari peninsula. The second line is the one of *T. duboisi* that exists only at Bemba on the western coast. The third involves the green *Tropheus* race found near Uvira alongside *T. moorii moorii*. This race doesn't correspond as yet to any taxon.

Geographical features

The tip of the peninsula at Cape Banza is separated from the east coast by a scant 20km, the floor of which lies, with approximately a 350-400 meter depth, on top of a high submerged cliff diving down on a steep slope toward a deep basin 1300 meters below the present surface of the lake. Before the lake entered the northern basin this very high cliff had been the northernmost boundary of the lake. As this cliff is about 900 meters high it must have taken the lake untold millennia to rise along its slopes.

During this period the rock-dwelling fishes could travel the whole length of this biotope and along the foot of the peninsula to reach habitats on the coastlines across the lake from east to west or vice versa.

The cliff reached the east coast somewhere around Nyanza Lac, at the border between Burundi and Tanzania. North of this area we now find the ecological barrier between Nyanza Lac and Rumonge, a coastal plain lined with sand beaches, swamps, and even a very old virgin forest. Apparently the rock-dwelling fishes could not cross the barrier once the lake started to submerge the cliff and flood the northern basin. None of the typical fauna from Nyanza Lac appears to have crossed the barrier. To give but a few examples: *Ophthalmotilapia ventralis*, *Petrochromis macrognathus*, *Tropheus brichardi*, *Lamprologus tretocephalus*, and *Tropheus duboisi*. All these species are found on the Ubwari and on the west coast of the northern basin. Thus one has to admit that they

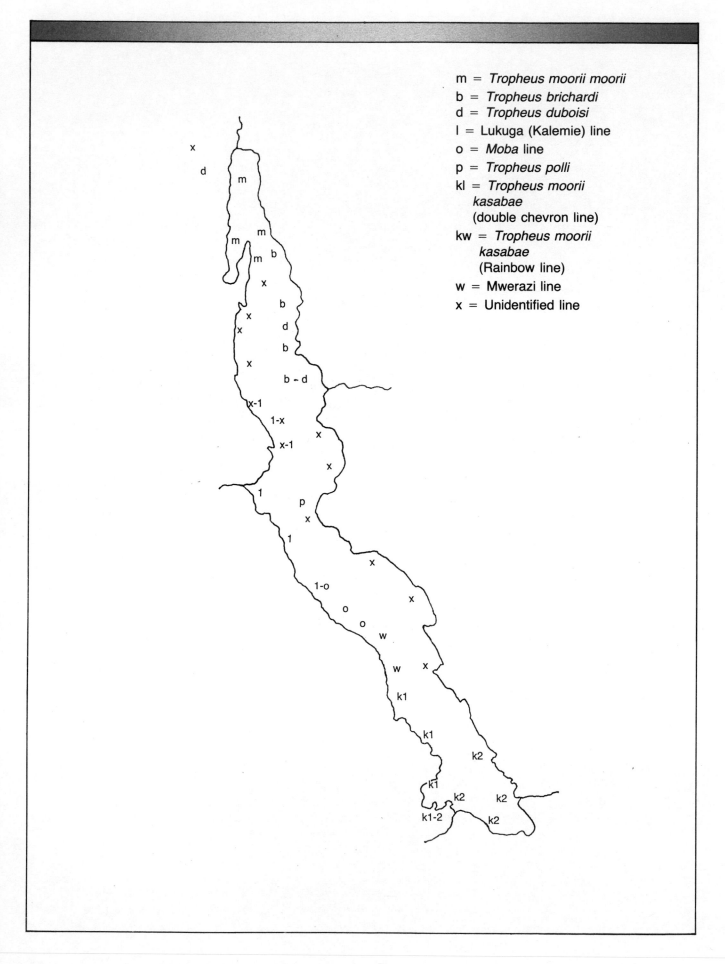

m = *Tropheus moorii moorii*
b = *Tropheus brichardi*
d = *Tropheus duboisi*
l = Lukuga (Kalemie) line
o = *Moba* line
p = *Tropheus polli*
kl = *Tropheus moorii*
     kasabae
     (double chevron line)
kw = *Tropheus moorii*
     kasabae
     (Rainbow line)
w = Mwerazi line
x = Unidentified line

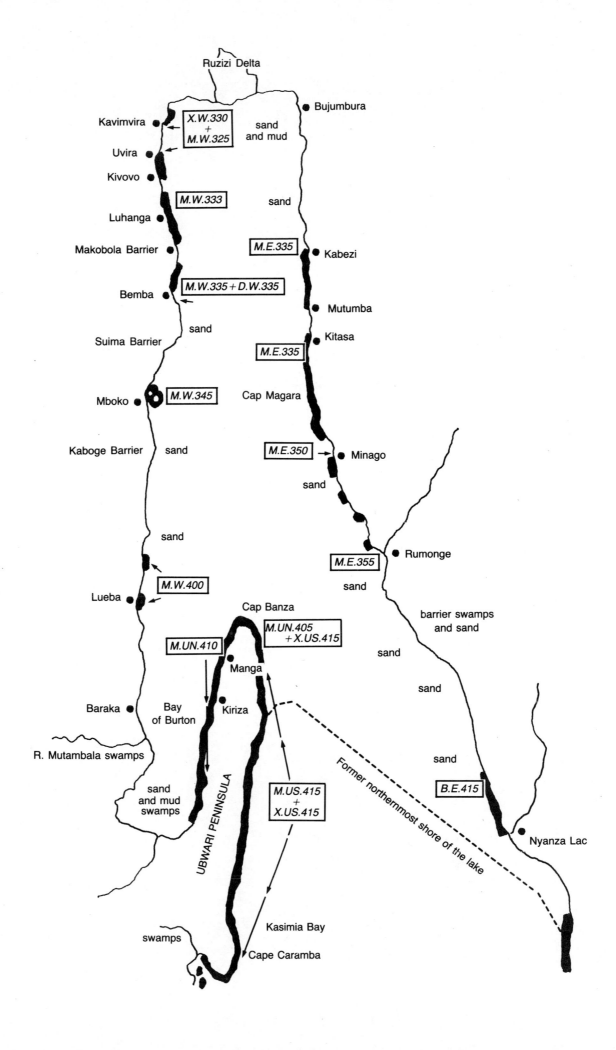

Ruzizi Delta

Bujumbura

Kavimvira $X.W.330 + M.W.325$

sand and mud

Uvira

Kivovo

$M.W.333$

sand

Luhanga

Makobola Barrier

$M.E.335$

Kabezi

Bemba

$M.W.335 + D.W.335$

Mutumba

Suima Barrier

sand

Kitasa

$M.E.335$

Mboko

$M.W.345$

Cap Magara

Kaboge Barrier

sand

$M.E.350$

Minago

sand

sand

Lueba

$M.W.400$

$M.E.355$

Rumonge

sand

Cap Banza

barrier swamps and sand

$M.UN.405 + X.US.415$

$M.UN.410$

sand

Manga

Baraka

Bay of Burton

Kiriza

sand

R. Mutambala swamps

Former northernmost shore of the lake

sand and mud swamps

$M.US.415 + X.US.415$

$B.E.415$

UBWARI PENINSULA

Nyanza Lac

Kasimia Bay

swamps

Cape Caramba

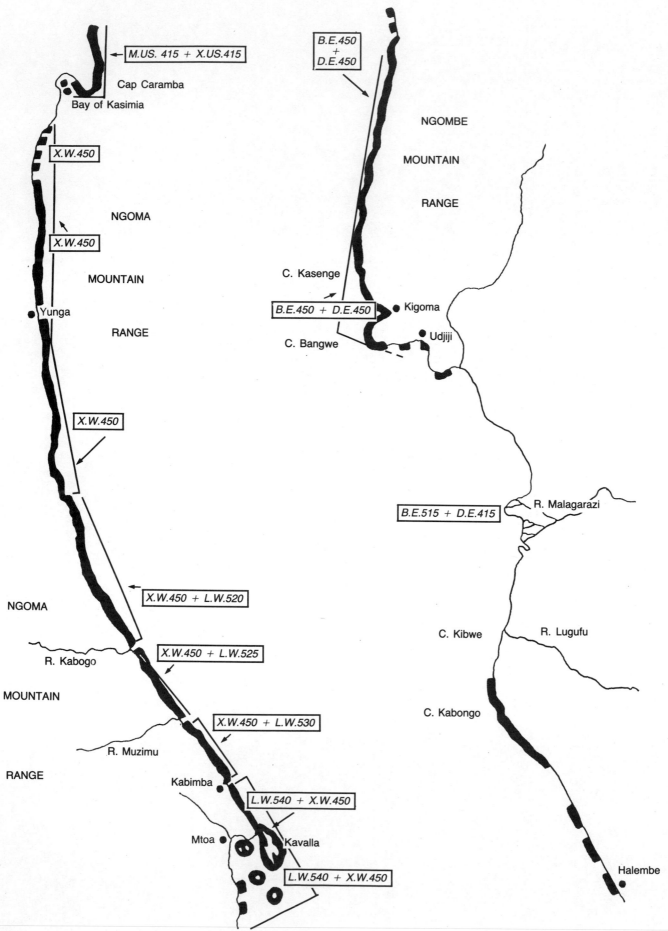

M.US. 415 + X.US.415

B.E.450
+
D.E.450

Cap Caramba

Bay of Kasimia

X.W.450

NGOMBE

MOUNTAIN

RANGE

NGOMA

C. Kasenge

X.W.450

B.E.450 + D.E.450   Kigoma

MOUNTAIN

Udjiji

Yunga

C. Bangwe

RANGE

X.W.450

B.E.515 + D.E.415   R. Malagarazi

NGOMA

X.W.450 + L.W.520

C. Kibwe   R. Lugufu

R. Kabogo

X.W.450 + L.W.525

MOUNTAIN

X.W.450 + L.W.530

C. Kabongo

R. Muzimu

RANGE

Kabimba

L.W.540 + X.W.450

Mtoa   Kavalla

Halembe

L.W.540 + X.W.450

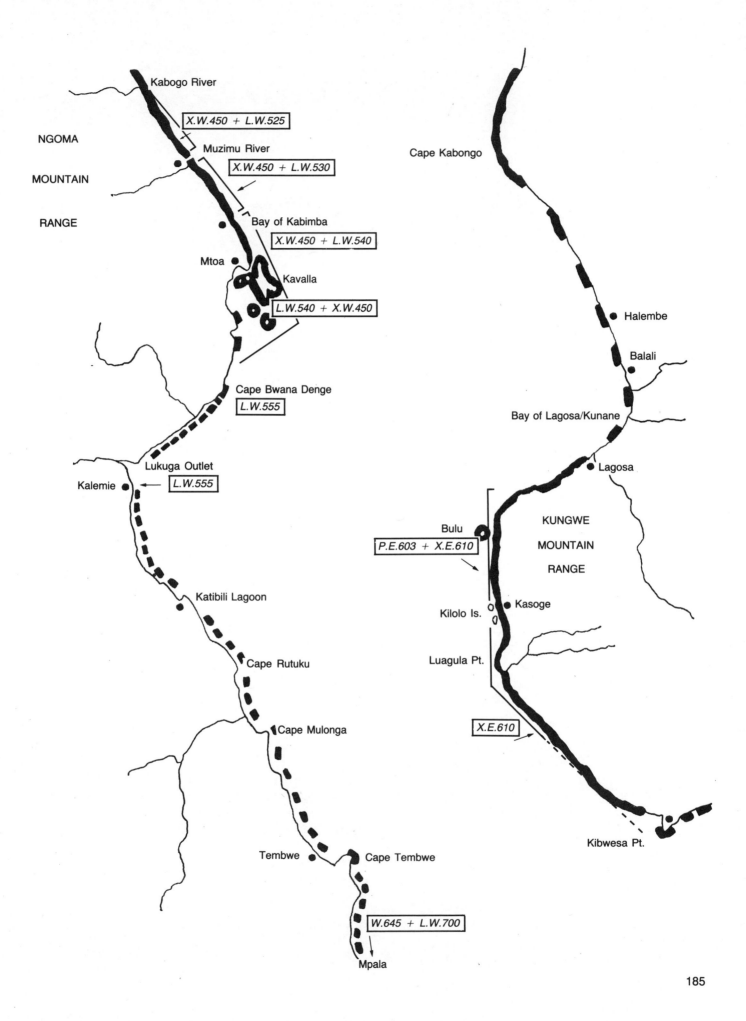

Kabogo River

NGOMA

MOUNTAIN

RANGE

X.W.450 + L.W.525

Muzimu River

X.W.450 + L.W.530

Bay of Kabimba

X.W.450 + L.W.540

Mtoa

Kavalla

L.W.540 + X.W.450

Cape Bwana Denge

L.W.555

Lukuga Outlet

L.W.555

Kalemie

Katibili Lagoon

Cape Rutuku

Cape Mulonga

Tembwe

Cape Tembwe

W.645 + L.W.700

Mpala

Cape Kabongo

Halembe

Balali

Bay of Lagosa/Kunane

Lagosa

KUNGWE

MOUNTAIN

RANGE

Bulu

P.E.603 + X.E.610

Kilolo Is.

Kasoge

Luagula Pt.

X.E.610

Kibwesa Pt.

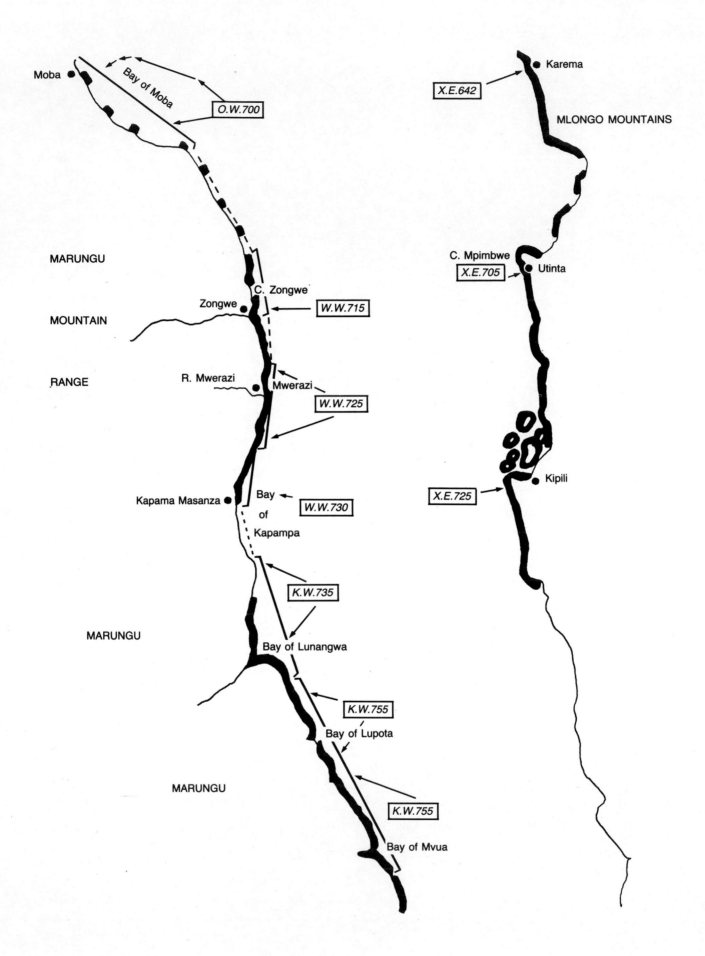

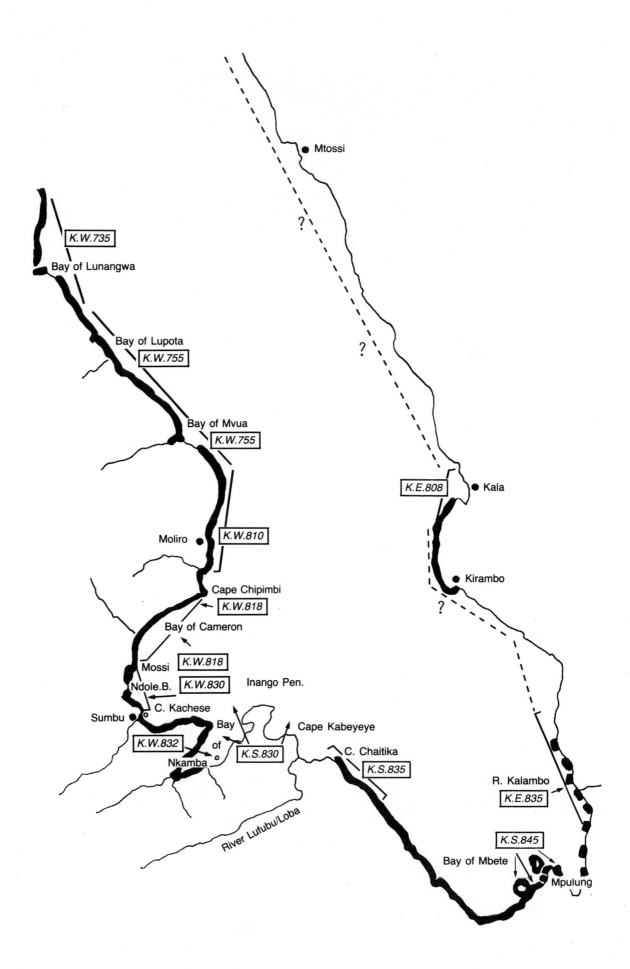

Mtossi

?

K.W.735

Bay of Lunangwa

?

Bay of Lupota

K.W.755

Bay of Mvua

K.W.755

K.E.808   Kala

Moliro

K.W.810

Kirambo

Cape Chipimbi

K.W.818

?

Bay of Cameron

Mossi   K.W.818

Ndole.B.   K.W.830   Inango Pen.

C. Kachese

Sumbu   Bay   Cape Kabeyeye

K.W.832   of   K.S.830   C. Chaitika

Nkamba   K.S.835

R. Kalambo

K.E.835

K.S.845

River Lufubu/Loba

Bay of Mbete   Mpulung

This juvenile *Tropheus moorii* is of the western type Chipimbi Cape variety. Photo by Pierre Brichard.

probably came into the basin via the west coast.

As for *Tropheus moorii moorii,* none appear to exist on the southern approaches to the basin, thus one cannot decide where they came from.

As the lake water crept northward and eroded the topsoil of the slopes they opened up new rock biotopes, with the rock-dwelling species colonizing the new territories thus made available. At that time the Ruzizi River was apparently not yet descending from Lake Kivu as this lake didn't exist yet, and the huge alluvial plain of the river had probably barely started its buildup. The depth of the alluvial deposits in this plain is several hundred meters , and several strata show that at times it had been flooded by the lake. So, in all likelihood the northern basin at its widest expansion reached the foot of what is now the Kamaniola escarpment, damming Lake Kivu 80 km to the north of the present Lake Tanganyika shores. Around these hills

rock-dwelling cichlids could extend around the coastline and, starting from the west coast, reach the eastern shores. There they followed the rock biotopes. Some of them reached Rumonge, where they were stopped by the Rumonge-Nyanza Lac barrier we already mentioned.

With the increased buildup of the Ruzizi delta the rock slopes lying in the north were buried by silt and disappeared, and the western and eastern rock biotopes were split apart. Rock-dwelling fishes from the Uvira coast could no longer be in contact with their cousins on the eastern shoreline.

This is how we can understand how we find *Tropheus moorii moorii* on both sides of the delta, but not in Nyanza Lac, nor *Tropheus brichardi* in Rumonge.

As for *Tropheus duboisi* they must have come from the Kigoma-Nyanza Lac area, as none are found anywhere on the central west coast; they probably followed the path to the north of the peninsula

isthmus until they reached Bemba, and perhaps further up the coast. They were in competition with *T.moorii moorii* and perhaps lost the battle for survival, as they are now restricted to less than 1km of shore at Bemba in what we might call a fossil or relict population.

Many species from the west coast are not found on the east coast of the basin. For example there are *Lamprologus tretocephalus, L. leleupi, Julidochromis ornatus,* and *J. transcriptus.* Probably they came in too late, after the Ruzizi delta was built, or they might have migrated but were wiped out which, everything considered, doesn't appear very likely.

As for the green *Tropheus* line there are two possibilities: (1) they belong to the *brichardi* line and probably came from the eastern coast via the cliff and the Ubwari; or (2) they belong to the line from the coast south of the Ubwari (the Ngoma) and crept around the Ubwari or across its isthmus.

L. leleupi probably came from the Ngoma as this coast harbors a sizeable population of this species. They do not exist there any more or never reached the Ubwari itself. If they were wiped out on the Ubwari it is perhaps due to the presence of a huge population of *Lamprologus niger* on this peninsula. As a rule where you find *L. niger* you don't see any *L. leleupi.*

Let us go back to *Tropheus moorii moorii.* Once they had colonized the whole basin, they split up into local races. This was easy since their expansive rock biotope soon started to disintegrate. The basin is narrow, rather shallow, and more than one hundred rivers and mountain torrents empty their alluvions ripped off the steep mountain slopes. The Mutambala, among others, started to build the peninsula isthmus, and from a long rocky island made a peninsula. Since then it has kept on filling the very large Burton Bay. The Ruzizi delta steadily crept southward. All stretches of rock biotope lying in rather shallow floors started to be silted over, leaving at best a narrow ribbon of rubble along the shore. Local small torrent estuaries were increasing their girth so much faster because the floor in front of them was rising with incoming sediments. As they did, they fragmented and further isolated the rocky habitats on which rock-dwelling fishes were trying to survive.

If we can envision that at first the rock biotopes circling the primitive northern basin had a girth of about 280 to 300km from the isthmus of the Ubwari to Rumonge, none of the present day rock habitat biotopes there are more than 15km long, except for the one on the northern coast of the Ubwari, a mere ribbon 10 to 15 meters wide lying on a sand floor and following the coast of the peninsula for about 35 km. Altogether in the northern basin not more than 70km of rocky shores now subsist.

As we follow the coastline from Cape Banza northward we can note the *Tropheus* populations, their isolation, and the intercalate ecological barriers, and see if they fit with the picture we have just drawn of the basin history.
Cape Banza: rock, 5km long. Yellow striped *moorii*.

An adult dotted head rainbow variety of *Tropheus moorii kasabae*. The rainbow varieties cover about 150 km of coastline with several ecological barriers included. Photo by Pierre Brichard.

Manga: small brook swamp with reeds, a few hundred meters.
Peninsula main coast: 30km rock. Uniform Kiriza *moorii*.
Bay of Burton: swamps and sand (the isthmus). No *Tropheus*. 15km
Baraka/Mutambala estuary: no *Tropheus* on 25km.
Lueba: a few rock outcrops in the middle of sandy bottom. A few dozen *Tropheus* (black or yellow-patched morph).
Kaboje estuary: 18km long. No *Tropheus*.
Mboko: sand floors with three islets and a submerged reef; a few hundred yards. The lemon striped *moorii*.
Suima estuary: sand. No *Tropheus* for 20km.
Bemba: first typical rock bottoms since the Ubwari. *Tropheus moorii* orange morph.
Makobola-Lunangwa-Kivovo: a succession of small barriers and rocky habitats, the only typical one being at Luhangwa. A melting pot of *Tropheus* with features from

the orange morph at Bemba and the black race at Uvira.
Ruzizi: a 30km long barrier creeping slowly southward, with a fringe of marshes and coastal lagoons. The output of alluvions from the Ruzizi and small rivers crossing the delta toward the lake is filling the northern basin at an increased pace, its action being felt along the coastlines of the eastern shores 25km to the south.
Kabezi biotope: a very marginal habitat, very silted over, with a small population of the *Brabant* or *Burundi Tropheus* along with a few other rock-dwelling cichlids. This biotope has been cut off from the next rock biotope to the south by the birth of an intercalate barrier.
Mutumba estuary: a 2 km wide delta with a coastal lagoon has fragmented the Magara rock coast. North of this delta rocks have disappeared under the silt with sand beaches littered with pebbles and gravel remaining. South of the

This is the juvenile of the dotted head rainbow variety of *Tropheus moorii kasabae*. This one is from Kamba Bay. Photo by Pierre Brichard.

barrier the bottom of the lake is deeper and not yet much raised by silt.

Magara biotope: extending for approximately 15 km, the Magara coastline is the only major rock habitat remaining more or less intact between the Ruzizi and Nyanza Lac on the northeastern coast. All the Burundi rock-dwelling cichlids are found there, but there are a few noteworthy missing species with respect

to the western coast of the basin.

Minago barrier: after Cape Magara the coast becomes sandy, with only sandstone slabs linking its various stretches, as the coastal shelf is shallow. Rock promontories offer marginal habitats, such as at Minago. This situation prevails until Rumonge, over 40km.

Rumonge habitat: rock outcrops submerged in

shallow waters provide a biotope on which *Tropheus* can survive—mostly scattered rubble. The last *T.moorii moorii* is found on this marginal biotope. In Minago they were all black, with a faint yellowish cast on the sides. In Rumonge the color is very dark, but one can still see a faint beige-brown triangular patch below the dorsal base, and the upper rim of the eye is brown, as in many morphs of the *moorii moorii* line. In a tank specimens turn brown-beige on the anterior half of the body and remain black on the posterior half, which is a typical feature of the line.

Rumonge-Nyanza Lac barrier: no rock-dwelling cichlids are found south of Rumonge till the Nyanza Lac corniche, 25km to the south. The barrier consists of a huge sand floor traversed by several small river deltas, one of which is occupied by a virgin forest.

This darker individual is also a dotted head rainbow *Tropheus moorii kasabae* belonging to the southern race. Photo by Pierre Brichard.

Conclusions

We can reach the following conclusions about the history of colonization of the northern basin and its future.

1. At first the basin was settled by fishes migrating along the western coastline north of the Ubwari and not by direct access along the eastern shores because of the Rumonge-Nyanza Lac flat grounds, which did not provide rock habitats.

2. The rock-dwelling species coming into the basin belonged to populations that at that time were common to the two shorelines of the central

basin south of the Ubwari.

3. Once they had occupied both sides of the northern basin via a rocky coast now lying buried under the Ruzizi delta, the delta started to build up and the two opposite coastlines became separated.

4. As the alluvions poured into the basin through so many rivers, the northern limit of the lake is progressively being pushed back southward and at the same time the deepest floors of the basin are being filled in. Thus additional sediments from all the rivers on the east coast pour onto shallow floors and the rock biotopes disappear at an increased pace.

5. River deltas are abuilding because the lake bottom in the area has risen, and the effects of the barriers intercalated within the rock biotopes increasingly fragment and isolate the fishes in their habitats.

6. The last step of evolution in the basin is made clear at places like Kavimvira near Uvira, Lueba and Mboko on the coast of Burton Bay, Minago, and Rumonge, where a few rock outcrops remain in three or four feet of water surrounded by miles of sand floors.

7. My guess is that in perhaps as little as one thousand years the Bujumbura harbor will be one or two miles inland and that it will not take 100,000 years to have all of Burton Bay, except for its tip, entirely lined with swamps and filled with sand. It will take much less time to have the remaining rock biotopes of Luhanga, Bemba, and Magara silted over. Then nothing will

The double chevron or double "V" can clearly be seen on this *Tropheus moorii kasabae*. The color of the chevrons is always that of the secondary color. Photo by Dr. W. Staeck.

remain of the 200-300 km of rock megabiotope in the northern basin and the rock-dwelling fishes there will be a thing of the past.

As it is we can conclude from this survey of the local races of *Tropheus* in the northern basin that by carefully studying the layout of local lines of rock-dwelling cichlids with respect to the existing topography of the coastline in their area, we can have our attention awakened to the presence of ecological barriers that explain why the change of races occurred.

If we improve on the quality of the survey that we developed, it might be that by applying the same principles we might one day understand

how, in other parts of the lake, according to the local layout of the coasts and the other geographical parameters, we might one day discover how the coastal fishes colonized their biotopes, why they evolved as they did, and how the biotopes themselves evolved during long past ages.

This is one of the major reasons why we should be very careful to properly identify all the *Tropheus* races and lines, as well as several other species, and have to investigate how they relate to each other. As for the taxonomic principles and methods used to identify the *Tropheus* taxa, it is obvious that they have to be improved upon.

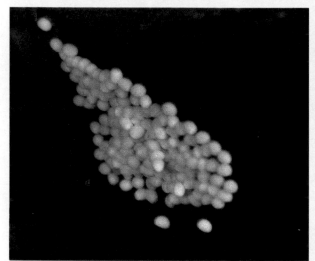

Lamprologus leleupi is one of the more popular of the Lake Tanganyika *Lamprologus* species. It has been successfully spawned, as can be seen in these photos by Hans-Joachim Richter. Shown are the eggs (center left), the moment of hatching of the fry (center right), and hatched fry with a good deal of yolk yet to be utilized.

CHAPTER VII

REPRODUCTION AND BREEDING BEHAVIOR

Only recently has the exploration of Lake Tanganyika biotopes provided information on the reproduction of the lake fishes, and much remains to be discovered about the breeding behavior of many species.

The lake is unique among the Rift Valley lakes in the composition of its fish fauna, which includes three main populations: nestbreeding cichlids (about 60 species identified so far), mouthbrooding cichlids (about 110 so far), and an important flock of non-cichlid fishes (110 so far).

Beyond the fascinating discovery of specific breeding behavior we thus find in the lake an exceptional opportunity to see how the three populations have adapted their breeding behavior to similar ecological conditions. We can also see how the presence of the other populations has had an impact on fishes belonging to a particular population. This is especially of value as nestbreeding cichlids and mouthbrooders most often share common breeding grounds. Finally, we can perhaps add valuable information on the distribution and spread of coastal species around the shorelines and discover whether their breeding habits could have an impact on their range.

Because the lake is a very different ecosystem from the fluviatile biotopes they were used to, riverine fishes had to adapt many of their ways when they started to colonize the lake waters. Among their behavior patterns breeding was essential, as it secured the future of the species in the lake.

For many fishes, especially among those belonging to non-cichlid families like cyprinids, several catfishes, etc., the adaptation of their breeding behavior simply proved too difficult and they remained in rivers, or when they could roam the lake they had to come back to rivers for breeding purposes. These fishes are *anadromous* in the sense that, like salmon, they have to leave the habitat where they roam as adult fishes to spawn in the river where they were born.

Some of the cyprinids in the lake leave the lake at breeding time, migrate to the mouth of a river, and climb a succession of cascades and falls to spawn in the cool headwaters, sometimes 1000 meters high in the mountains. Whether they die after they spawn or whether they return to breed in the same river where they were born, like the salmons, is not known.

Anyhow, one can say that even when they can roam the lake at will the anadromous species have not achieved a perfect adaptation to the lacustrine ecology, even when, like *Mastacembelus frenatus,* found on every rock coast around the lake, in all other ways they have adapted to local conditions.

Though as a whole the non-cichlids still breed in rivers, there are several lines that have succeeded in adapting their breeding pattern to lake conditions. Such are most *Synodontis,* most *Mastacembelus,* centropomids (like *Lates*), and clupeids. The one cyprinodontid living in the lake, *Lamprichthys tanganicanus,* also lays its eggs on rock

The breeding grounds of *Lates angustifrons* were discovered by Pierre Brichard in early 1977 in the southern part of the lake. The photo is of a relatively young individual. Photo by Dr. Herbert R. Axelrod.

shelves in the lake.

Behavior depends a great deal on the habitat the fishes have opted for in the lake, and we will thus deal successively with the pelagic open water dwellers, the fishes that live over sand bottoms, and end with the incredibly complex world of littoral rock slope dwellers.

PELAGIC SPECIES

Very few of the pelagic roamers (three *Lates* and one clupeid) are thought to breed and spawn in open waters, leaving their eggs and fry to drift in plankton clouds. Most pelagic species, and this includes the cichlids, have thus to migrate to the coastline for breeding grounds where they will raise their fry (when they do look after their brood) and where they will leave them afterwards.

Living in open waters searching for local concentrations of food (plankton or other prey) and avoiding roaming hordes of predators is fraught with risk. Pelagic species have therefore all taken to schooling. The schooling behavior is for them the most essential, the one on which their survival will depend, and the one behavior from which all other behavior will proceed. Spawning cannot be an individual affair breaking the cohesion of the school. It has to be an activity to which all the individual fish in a school have to devote themselves at the same time and in the same place.

Community breeding comes quite naturally to pelagic species by a *de facto* process linked to their mobility. As a school is roaming, weaker (younger) fish lack stamina

and speed in which to follow at the same pace and they disappear from the main body. Usually fishes of the same size can keep up with the school. This means that schools are made up of individual fish having reached a similar growth stage. Pelagic schools are always segregated by size, fishes belonging to different age groups living apart from each other. The urge to breed is felt by all members of the school at the same time of the year, especially so because the output and availability of food are seasonal in open waters, and as a cohesive body they can head toward the shores, find a breeding place, and start spawning.

Because conditioning leading to spawning readiness depends on seasonal climatic cycles, the spawns of all pelagic species are also seasonal. They breed once or twice a year (commonly in November/December but clupeids spawn a second time in May/June) following the start of the rainy season.

Being all born at the same time and place, the fry of a species grow up together, and as a result schooling for them comes quite naturally from their birth on. They will form a homogeneous school of their own at once. The threadlike fry of clupeids, barely 1 cm long, already move about in incredibly tight schools.

The breeding grounds of *Lates angustifrons* were discovered by the author in February/March of 1977 in the southern part of the lake, i.e. Nkamba Bay, along a beach cut by a small river mouth. These breeding grounds are the only ones of any pelagic species as yet discovered around the lake. They are used year after year, and one might thus think that the *Lates* return to breed to the place they were born. Perhaps there are other breeding grounds around the lake but as yet none has been identified. Given the importance of *Lates* for the commercial fisheries around the lake, their breeding grounds at Nkamba Bay should be protected.

The shoreline used for breeding grounds is shallow and the bottom sandy. Female *Lates* by the thousands begin to lay their eggs, several tens of thousand per female, over the whole area. Soon, under the push from the waves the entire shore is lined with a fringe of eggs and fry several inches wide. They are scooped up by aquatic birds and by local fishermen who relish this kind of caviar, which they call "silver pearls". Millions of eggs and fry with their yolk-sac thus disappear daily.

As soon as they are laid the eggs are abandoned by the adults and left unattended. As they start to hatch, the littoral waters around the breeding grounds display broad patches of agglutinated fry. These fry will remain in coastal waters feeding on the local zooplankton until they are several cm long, but thinning out under the impact of predation. As they grow they go deeper until one day they leave the inshore waters and start their endless roaming in the open. The other species of *Lates* breed offshore and their eggs and young fry are found drifting along with the plankton clouds.

Among the clupeids, *Limnothrissa* breeds inshore but *Stolothrissa* breeds far from shore, an adult female being capable of laying as many as 35,000 eggs, whose negative buoyancy causes them to sink slowly in the water column.

Stolothrissa fry, growing amid the plankton clouds, feed on the tiny crustaceans therein. The thread-like clupeid fry, initially totally unpigmented, move about in dense schools like a pale offcolor mist, from which individual fish are difficult to make out. A school can be 100 meters long, 10 to 15 meters wide, and 2 or 3 meters thick and involve millions of densely packed fry.

The more coastal *Limnothrissa* breeds during the entire rainy season but with two peaks, one at the beginning of the season and the other at its end. *Stolothrissa,* the more pelagic of the two species, breeds twice a year, the first time in December/January, the second from April to July, in coordination with the heaviest rains.

The pure red/black northern variety of *Tropheus moorii.* This is one of the first *Tropheus* varieties from the lake to be spawned. Photo by Dr. Herbert R. Axelrod.

Pelagic cichlids are found incubating and releasing their fry in coastal shallows. So far there are no reports of a mating having taken place either in pelagic waters or along the coastline. Whether spawning takes place in open waters and then the fish start their migration toward the coast or they first start their migration and come to breed inshore is not elucidated. From the evidence gained by capture of the fry, it looks like the breeding seasons are well delimited and triggered by the planktonic explosion following the first rains.

We can thus say about the breeding behavior of all pelagic species that they are seasonal and cyclical, that a whole school (and perhaps all the schools in an area) is engaged in sexual activities at the same time. Breeding activity among schooling and roaming species is *synchronized.*

COASTAL SPECIES—NON-CICHLIDS

Most of the breeding behaviors of fishes belonging to families other than the cichlids have not been much documented. Most often we can only guess that a fish breeds in the lake when we often find its young fry in one of the lake biotopes or, to the contrary, that it doesn't breed in the lake when we never, or very seldom, find one of its juveniles there.

The lungfish *Protopterus* breeds from November to April, at the height of the rainy season, when waters in the swamps are high. The male digs his tunnel-shaped nest in the mud between reeds. He attracts the female to the nest where she lays between 3,000 and 5,000 2mm long eggs and then is chased away by the male. The male, alone, takes care of the spawn. When the eggs hatch, the fry are about 3cm long and are provided with external gills whose length depends on the amount of oxygen in the water. When the water is oxygen poor the gills are long, when rich the gills are shorter. The larvae also have an adhesive disk with which they cling to aquatic plants.

Mormyrid reproduction is very poorly documented but it appears that they also seem to breed in March-April in the northern part of the lake.

Catfishes breed from October to May, with a peak from February to April. Breeding grounds include coastal and estuary swamps for species such as *Auchenoglanis, Chrysichthys,* and other large species. *Clarias* also breed outside of the lake. *Chrysichthys sianenna* breeds on sand in the lake. Other lake spawners include *Dinotopterus* (clariid) (on gravel and pebble floors), *Tanganikallabes* (clariid) (under the rocks), *Lophiobagrus* and *Phyllonemus* (both bagrids). *Synodontis* (except for one species which breeds in rivers) breed under a rock shelf, upside down, laying a few hundred eggs that are stuck on the ceiling of the recess. The fry remain upside down in the cave when they hatch browsing on the biocover. The riverine *Clarias* lay about 4,000 eggs among the plants, where they hatch in about 36 hours.

Ctenopoma muriei breeds in swamps. It doesn't build a floating nest like other anabantoids, and the eggs float about among the plants and hatch in about 24 hours.

Engraulicypris minutus is the only cyprinid living and breeding in the lake. It breeds in coastal lagoons and bays from December to March. Juveniles remain in the shallows and feed on drifting zooplankton. Then they head for the open waters where they mix with the clupeids.

The large alestids and citharinids (*Distichodus* and *Citharinus*) breed in large swampy rivers, but also in the lake in similar biotopes (coastal swamps, mud bottoms, and lagoons).

Two *Mastacembelus* breed in rivers (*M. frenatus* and *M. cunningtoni*) although the second also breeds on sand bottoms. So does *M. ophidium,* the curious viper-headed spiny eel whose 6-8cm long fry are seasonally found in quiet sandy coves. Most of the other mastacembelid eels appear to breed in rocky areas. One of my friends, Chris Blignault, during one of his night dives once observed in the beam of his light the strange sight of hundreds of *Mastacembelus* fry, knit together in a tight ball 20 cm across, slowly "rolling" their way through the darkness.

LITTORAL CICHLIDS

The three types of littoral biotopes are so different that it is little wonder that the breeding behaviors of sophisticated fishes like the cichlids have developed various techniques in line with the diversity of the habitats. We thus need to explore them individually.

Spawning of *Polypterus ornatipinnis*.

The pair break away at times for a rest.

During breaks the fish may cruise about the tank.

Activity is relatively high at this time.

Encounters between the pair may precipitate another spawning bout.

They may spawn at intervals over a period of days.

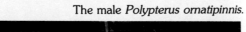
The male *Polypterus ornatipinnis*.

The female *P. ornatipinnis*. All photos by Dr. Hiroshi Azuma.

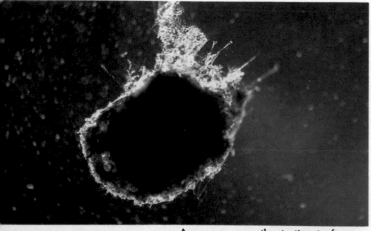

An egg apparently starting to fungus.

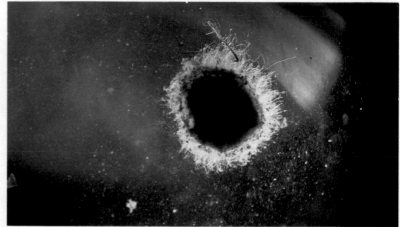

The egg well covered with fungus.

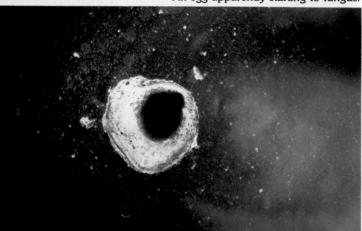

A backlighted egg shows that development is commencing.

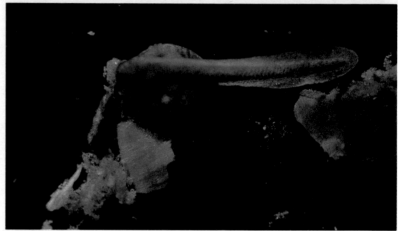

A newly hatched *P. ornatipinnis* fry.

Note the large yolk sac of this newly hatched fry.

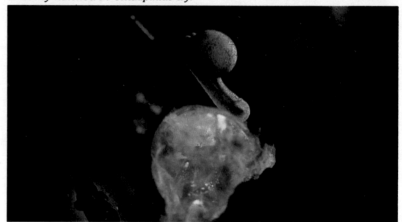

An empty egg case and newly hatched (still poorly developed) fry.

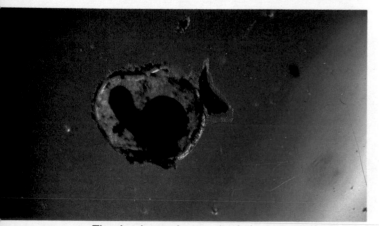

The developing fry can clearly be seen inside the egg.

Three eggs with developing embryos. All photos by Dr. Hiroshi Azuma.

A group of *P. ornatipinnis* eggs.

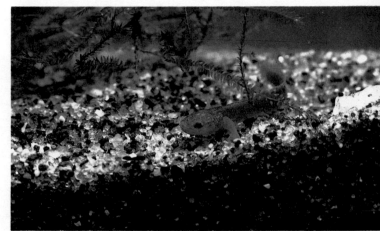

A young juvenile with the characteristic adult shape.

At the end of 20 days the external gills begin to recede.

A more ventral view of the 20-day-old juvenile.

Six-month-old juveniles are about 70mm long.

The pattern of the adult is already established at six months.

The juveniles can be fed small live fishes and bloodworms.

The juveniles look like miniature adults. All photos by Dr. Hiroshi Azuma.

Tilapia guineensis occurs in brackish lagoons in countries bordering the Gulf of Guinea. In this nestbreeder both parents share in the guarding of the fry. The fry are moved occasionally from one pit to another. The photos show various aspects of the color repertoire of this species. Top: Sexual display. Middle: Fright pattern. Bottom: Guarding pattern. Photos by Dr. J. Voss.

Swamp-dwelling cichlids

Due to the steep gradient of most of the lake shores there are few swamps around the lake, a feature by which it is very different from Lake Malawi (Nyassa) and Lake Victoria. The swamps of the lake are always built by sediments brought in by a river. Most typical are the Malagarazi and the Ruzizi deltas, to which might be added the Katibili lagoons on the west coast, the lower course of the Lufubu in the south, and the Mutambala delta in Burton Bay. Many more mountain torrents build small swamps at their mouth, but they are seldom large enough to host a large population.

Cichlids living in swamps belong to fluviatile lines such as *Tilapia* and *Haplochromis*. As they live in a habitat not much different from the river biotopes, there is little pressure for the fishes to change their ways. Most are mouthbrooders, and some of them, like *Astatotilapia burtoni* and *A. bloyeti,* or *Astatoreochromis straeleni,* are famous for the egg-dummy ocelli found on the anal fin of the males. They use these decoy egg ocelli to help the female fertilize the eggs she has already taken into her mouth by attracting her toward the male genital vent as it is exuding the sperm. Trying to pick the ocelli up in her mouth, she takes the sperm in and thereby fertilizes the eggs.

These swamp-dwelling fishes do not venture into the lake much or stray far from the river mouth. They have never been found breeding far from a river and certainly not in a rocky habitat.

Sand-dwelling cichlids

The extensive sand bottoms that rise from the deep lifeless floors to the shores of shallow coastlines and the foot of steep slopes are usually flat or with a small gradient and barren. In quiet, shallow waters aquatic plant thickets can grow (even as dense bushes) and provide good breeding grounds. But this is definitely not a swamp-like biotope for the lack of an essential feature: the murkiness or lack of transparency of the water. In the lake the water around aquatic plant thickets is often crystal clear. Except for these areas sand-dwelling fishes do not have many shelters in which to hide and breed. They have to rely on ingenuity to make use of the few that are available.

The most specialized sand-dwellers, like *Xenotilapia, Callochromis,* or *Aulonocranus,* are mouthbrooders that also school. Everything taken into account, the sandy bottoms are rather like offshore open waters because they provide little, if any, shelter but give ample room to wander about.

A *Tilapia guineensis* parent tending eggs in the nest, removing infertile ones and seeing that the others get sufficient aeration. Photo by Dr. J. Voss.

The fishes have to search for their food as well and the schools are thus mobile. Not all of the sand-dwelling fishes live on the bottom. Many roam in midwater, much like the pelagic species. Because of their mobility the schools are also composed of fishes having reached approximately the same age. We thus have all the ingredients leading to simultaneous, synchronized spawning sessions.

For the first time since we started the study of breeding behavior among lake fishes we discovered nestbreeding cichlid species on the bottom. *Boulengerochromis,* a semipelagic coastal genus and and the giant of the cichlid world, breeds there along with several large *Lamprologus* species. The most typical sand-dwelling species are *L. tetracanthus, L. attenuatus, L. pleuromaculatus, L. cunningtoni,* and occasionally *L. elongatus.* They are all solitary and monogamous.

One species at least, *L. callipterus,* schools in small packs. How *L. callipterus* breed has long been a mystery for us. We had never been able to witness their mating nor had we seen their fry. Then one day I discovered a small pack of the fish standing guard over a pile of broken *Neothauma* shells in which the fry took shelter. The snail shells are fortuitous for sand-dwellers, and many use them as breeding grounds or hiding places for their fry. Among these are spiny eels, *L. tetracanthus, L. callipterus,* and a group of dwarf *Lamprologus* species. I am sure in the future more species will probably be added to this list.

Xenotilapia sima is one of the small schooling cichlids that wander along the bottom. A school is generally composed of fish approximately the same age and size. Photo by Dr. Herbert R. Axelrod.

The small schooling cichlids like *Xenotilapia* species wander along the bottom. As the school is composed of fish of approximately the same age, they grow and reach sexual maturity together.

Seasonal conditioning through zooplankton supply increase is probably less marked than in pelagic species (because the sand bottom dwellers pick crustaceans buried in the sand) especially near the coast, but cannot be ruled out. As it is, the breeding impulse doesn't break the cohesion of the school, which is the foremost requirement for the survival of the fish, and they breed together on common spawning grounds called an arena.

Females rise a bit above the bottom, while the males start digging a crater in the sand. The males in full nuptial garb

(with *Xenotilapia* this is often an iridescent rainbow of pastel colors) entice the females to the nest where one by one they get their eggs, which they are already carrying in their mouth, fertilized.

Males are thus polygamous, but very often we discover that a female, after having had her eggs fertilized by one male and risen off the bottom, returns to another nest and repeats the operation with another male. We thus have an unusual case of polyandry and polygamy.

While females incubate their brood in the mouth, the whole school resumes its wanderings. Several weeks later they will stop, and the females will release and discharge their fry, now more than 1cm long, on the bottom at the same time and at the same place. Probably there are several such sessions in a school. This may be by accident, i.e. females were not all ready to release their fry or they were prevented from doing so, for example by the approach of predators, but this is only a guess.

Several predators in the lake are keenly attuned to the breeding behavior of other species. They are attracted to the various rituals by the behavior of fishes ready to spawn and follow in their tracks. When the spawning fish belong to a very peaceful species with few defensive means, the freshly laid eggs or the fry can be wiped out in a few seconds as they are released.

In a school of *Xenotilapia* from which we captured incubating females during a dive, the broods spewed out by the females had always reached the same level of development. When the females simultaneously release their fry, the young fish, being all together, can again, as with pelagic juveniles, form a school of their own and thus stand a chance to survive.

Several times we witnessed simultaneous breeding in the same arena in the lake that had been picked by a school of *Xenotilapia ochrogenys* and another of *X. melanogenys.* The males of both species were busy digging out their nests quite peacefully without interfering with each other. The nests of the two species were mixed over the whole area, so that the females of the two schools were circling together. They could recognize the nests built by their own males because they had a different shape. *X. ochrogenys* builds a circular nest made up of a row of small sand cones a few cm high, while *X. melanogenys* digs a regular saucer-shaped crater. No signs of aggressiveness were noted during the whole process between the two schools, intent on their own activities.

The schooling behavior of *Callochromis* is well documented, but their spawning behavior is not, at least at the schooling level. Only individual males or small groups of males have been seen spawning in shallow waters near the shore. There are always stragglers—fishes that have been separated from the main flock—that try to form a new school of their own and follow the breeding rhythm. *Callochromis melanostigma,* *C. macrops,* and stragglers from *Aulonocranus* schools have been seen digging their crater nest, about 30cm across in quiet areas, with females hovering above the area. At the end of buccal incubation the release of the fry was made in a quiet cove very close to the shore and near the surface for *Aulonocranus.* These species are also polygamous, but we could not ascertain if the females also eventually got their eggs fertilized by several males and were polyandrous.

The status of several species and genera of the *Limnochromis*-group of genera, i.e. *Triglachromis,* *Tangachromis,* *Lepidiochromis,* and *Cyprichromis* (which are rock-

Callochromis pleurospilus is a very attractive fish with delicate pastel colors. The schooling behavior of *Callochromis* is well documented — the spawning behavior is not. Photo by Dr. Herbert R. Axelrod.

dwellers), has not as yet been fully documented. Some of the species live over mud and sand bottoms and are known to be mouthbrooders (*Limnochromis auritus, L. pfefferi,* and *Cyprichromis* species) but it is as yet not sure, although we might surmise that the others are buccal incubators as well because they school and are mobile.

Although *Triglachromis otostigma* is now a rather common fish in aquarists' tanks it has never been bred in captivity. No incubating females were ever captured, no breeding sessions have been witnessed in the lake, and the breeding behavior of this well-known species is not yet elucidated.

Limnochromis auritus breeds in captivity. This enabled us to discover that both parents incubate their eggs, a rare behavior among the lake cichlids (also found in *Haplotaxodon*).

Some of the sand-dwelling nestbreeders have fascinating spawning behaviors that are also a forecast of what we will discover on rock biotopes.

Boulengerochromis, a benthic littoral genus often found in pairs in shallow water 3 to 10 meters deep, probably breeds in much deeper water. Young fry up to 15 cm long are often found by the thousands in dense schools roaming in very shallow waters (about 1 to 2 meters deep). The fish is a roaming predator, but apparently doesn't embark on the long migrations of the pelagic fishes.

The nest is often enormous, from 3 to 4 meters across and more than 50cm deep. Within the crater the pair digs a

Cyprichromis leptosoma is known to be a mouthbrooder. Perhaps the bright yellow tail is a sexual character. These are schooling fish that are quite mobile. Photo by Glen S. Axelrod.

number of smaller pits only 25 to 30cm in diameter. There might be as many as a dozen or more of these pits in a crater. In one of them the pair will lay from 10,000 to 15,000 eggs about 1mm, at most, in diameter.

If the nest is dug near rocky outcrops the pair will often dig the nest close to a rock wall, the rock wall blocking the approach of any predator from this side. When the pair is satisfied that this protection will be very efficient they might abandon digging the big crater altogether and will simply dig the small pits. As the pair never seems to move the eggs or fry from one pit to another, one might wonder if the fish dug the pits until they were satisfied with the layout, or as a defensive measure and with a purpose!

It is a fact that predators at ground level, when they approach the area, cannot deterimine in which pit the spawn has been laid unless they come to the rim of the pit, which is a very dangerous thing to do with the parents around and very watchful. It is especially dangerous to ground-level predators such as *L. callipterus, L. modestus, L. mondabu,* etc. who are much smaller than the huge *Boulengerochromis.* This multiplication of the pits might thus well be deliberate, so as to confuse and delay the enemies.

It is also a fact, as we have already said, that the ground level predators are often attracted by the behavior of spawning pairs, and the *Boulengerochromis* nests are often surrounded by a ring of predators ready to jump the nestsite. They are always very wary and jumpy when they approach the crater and start exploring the pits.

The breeding pair is stationed well off the bottom, about one meter high, and they tolerate nobody close to the crater—as a rule nobody ever tries! We have often been attacked during our observations. Although the fish were obviously wary of our looming shapes, much bigger than anything they had ever seen in the lake, they didn't hesitate to come and bite our goggles if we were lying down or our limbs and hands if we were standing. We got scratched and I can tell you—it hurts! The brood is thus very well protected.

The eggs are laid in one of the pits. What is unusual is that they are laid in layers, often nearly one finger thick, the more than 10,000 eggs covering a patch barely 20cm across. When the fry hatch the adults do not aerate them. The fry do not remain still but wriggle their tails in a very fast vibration so that the whole mass seems to be shuddering. Thus they provide their own aeration. The doting parents can therefore concentrate on the essential defense of the brood. They circle the fry, one standing guard two or three meters from the crater, the other one often disappearing in the gloom several meters away, then returning.

The yolk of the freshly hatched fry is very small so that they soon become very active. The tiny babies, barely 3mm long, start to hop up from the wriggling mass and, unable to swim, fall back. Not long after that they become free-swimming and leave the nestsite in a tight swarm *under parental guard* to start their wanderings.

I had written in my first book that, along with my daughter, I had once met a swarm of the fry already abandoned by their parents when they were perhaps 15mm long. I had been a bit hasty when I presumed that the *Boulengerochromis* fry were left to their fate much earlier than other nestbreeders raised in the open. I was mistaken. It is in fact much later that the parents abandon their fry, as we eventually discovered when we saw juveniles about 8cm long still under parental guard. What had happened? Probably by sheer luck our first sighting had been of a swarm of fry that for one reason or another had been abandoned.

I have talked at some length of the extraordinary behavior of these fry that had accumulated in a hemispheric dome close to one meter in diameter hovering just above the bottom. So close were the tiny copper-colored bodies pressed together that not a single fry swam even as little as 5mm from the mass and the surface of the dome glittered like brass in the sunlit shallows. Seen from a distance of a few meters the dome was opaque as if made of solid metal. Not a single fish was to be seen around and one might think that predators were frightened by this strange glistening shape slowly swirling about in the haze.

We tried to disrupt the array of fry, thrusting our arms through the quasi-solid mass. We could manage to punch a hole through it, but we never made it explode in panic. Never did one fry even try to get away from his brothers. Playing with our hands we thus spent several minutes bending

Limnochromis auritus is a mouthbrooder in which both parents incubate their eggs. This is rare in the lake cichlids. Photo by Dr. Herbert R. Axelrod.

the symmetrical dome into ectoplasmic shapes in a kind of eerie ballet until at last, sorry to disturb the fry anymore and with our hearts bursting with wonder, we let them go.

In several ways the behavior of *Boulengerochromis* as adults or fry illustrates the versatility and sophistication of the nestbreeders' spawning behavior.

The other sand-breeding cichlids also spawn and raise their brood in a nest. Depending on their size, they will, if large, be able to raise them in an open crater nest and, if small, will utilize on the barren bottom whatever shelter they might find to protect the breeding site from predators. This might be a sunken tree limb or anything under which or within which

they can spawn.

The largest roaming *Lamprologus* and *Lepidiolamprologus* species, such as *L. tetracanthus*, *L. cunningtoni*, *L. pleuromaculatus*, etc., will dig a crater nest that might be more than one meter across and from 20 to 40cm deep, in which they will lay several thousand eggs. Larger craters are probably old *Boulengerochromis* nests that are reused by other nestbreeders after they have been abandoned by their builders.

Smaller nestbreeders, like *L. modestus* and *L. mondabu*, do not reach a size commensurate with an open crater, which they would be unable to defend. So, whenever they breed in a sand plain (they also breed near

rocky areas) they dig a narrow tunnel under a stone or an *Iridina* mussel shell.

The smallest of all sand-dwellers are the dwarf *Lamprologus*, the so-called shell-dwellers like *L. brevis*, *L. ocellatus*, *L. kungweensis*, and *L. multifasciatus*. Although perhaps not all truly shell-dwellers, the fact remains that they breed in the empty *Neothauma* shells, even when they spend most of their time wandering over the sand plains.

Lepidiolamprologus pleuromaculatus is a large roaming cichlid that constructs a crater nest more than a meter across. Photo by Richard Piken.

In captivity *Lepidiolamprologus pleuromaculatus* spawns in caves or other secretive places like broken flower pots. This one is guarding young. Photo by Richard Piken.

The *Neothauma* snail shells, always empty, are found in piles up to one meter in diameter. They are always broken, as if they had been crushed, and they rest on a bed of broken shards. They are heavily encrusted with a calcite layer and appear to have been dead for a long time. The snails live in deep water and during more than 10 years of underwater observations only twice did I find a live snail. How come then that the empty shells are found in rather shallow water (but deep enough not to have been rolled about by the surf) and in areas where the live snails are not found? We spent a lot of time pondering the problems, until two clues provided us with a possible explanation. First: stomach contents of several big catfishes revealed that they eat crabs, fishes, and also shellfish. The largest *Synodontis* are known to feed on *Neothauma. Synodontis multipunctatus* is one of them. Second: we discovered small patches of freshly broken bivalve shells in piles on big rock slabs 15 or 20 meters deep. As the bivalve molluscs live in sand and not on rock, one had to conclude that a fish had taken the small shells to a rock and there had a kind of picnic. The shells were thus transported from the collecting grounds to a site where they were systematically crushed and eaten.

As for the *Neothauma* shells, although more or less broken they provide one of the few protected breeding grounds for sand-dwelling fishes in a very barren biotope. Even the much larger *L. callipterus,* which, unlike the dwarf species, cannot slither into a shell to spawn, probably lays their eggs inside of the shells and remains over the pile and thus assumes their protection.

L. callipterus fry, from our observations, remain over the shell piles until they are at least 4cm long. We could check this fact when we saw the oversized fry trying to hide head-on within a shell that they had outgrown, thinking they were hidden when half of their body was still sticking out from the shelter.

Many of these shell-breeders, such as *L. brevis* and *L. ocellatus,* swim around in schools several hundreds strong, as might be expected from small species living in the open. Because of this schooling behavior, one has to believe that there are areas where the empty shell piles are a very common feature, allowing adults in search of breeding grounds to find enough shells to breed in without disrupting the cohesion of their school.

The tiny *L. multifasciatus,* sexually mature when only 2cm long, was found breeding in a small pile of shells along with *L. callipterus* on a ledge of a rocky slope where obviously the *Neothauma* shells had to be carried. *L. multifasciatus* is probably one of the few species using the shells as permanent dwellings.

L. ornatipinnis, which grows to more than 10cm, is probably using the shells only as temporary breeding quarters and not as dwellings, the fish often being seen swimming in the open, off the bottom in midwater. With so many species using the shells we tend to believe that they are used in rotation. By our reckoning some 13 species utilize the *Neothauma* shells as breeding grounds.

The diversity of breeding behavior among nestbreeders living on the sand plains is therefore amazing and the ingenuity they display in finding breeding sites suited to their specific needs provides a preview of what we will discover when we enter the rocky habitats, where they are in their preferred biotope.

Coastal shallows gradually slope toward the deep floors of the lake, which become increasingly difficult to explore with scuba gear. The time spent in underwater exploration is reduced by the need for an increasingly long decompression process when returning to the surface. The need for electric lights to find one's way around the deep, dark areas scares the fishes away or disturbs their activities. Data on breeding behavior is thus very hard to come by. Little is known about the breeding behavior of benthic species like those of *Trematocara,* which we never could witness in more than 10 years spent underwater. They are only known to be mouthbrooders and schooling fish.

The shell-dwelling species of *Lamprologus* use the shell as their spawning site. Here a female *Lamprologus* "magara" enters the shell to spawn while the male waits. Photo by H.-J. Richter.

The male *L.* "magara" entering the shell to fertilize the eggs. Photo by Hans-Joachim Richter.

ROCK-DWELLING CICHLIDS

Nestbreeders:

As we leave the sand bottoms and come into contact with a rock-strewn slope we witness a quantum leap in the variety and density of the fishes living in the area. Everywhere mouthbrooding cichlids such as *Tropheus, Simochromis,* and *Petrochromis* skim in and out of the rubble in an endless ballet, while overhead the dark shapes of mouthbrooding plankton pickers of the *Cyathopharynx* guild move slowly about. Deeper down they are replaced by the streamlined bodies of *Cyprichromis* and the eerie bulk of *Cyphotilapia.* Mouthbrooding cichlids are everywhere and nestbreeding cichlids appear to be heavily outnumbered.

But beware! The nestbreeders are hidden in the rubble, and only a few specimens here and there of *L.modestus, L. compressiceps, Julidochromis, Telmatochromis,* and the ubiquitous *L. brichardi* schools can be discerned at first glance. If nestbreeders forage in the open, many more remain in the rubble, where they live on their own. The layout of the slope, with the tormented relief of broken rock, provides the fishes with countless shelters in which they can live.

Nestbreeders as a rule are carnivorous, even if this entails only for some of them feeding on the flesh of tiny crustaceans, like shrimps creeping on the surface of the rocks or buried in the sand, or the small "bugs" hopping about in the drifting plankton clouds. For species used to living at the surface of the rubble or within it, there is little need to forage far. They can pick their food from nearby rock walls or as it drifts by. They can afford to be sedentary and remain in a small part of the rubble. They are territorial, and it is in their territory that they will breed. The bottom-dwelling nestbreeders will thus secure a place on the bottom, but because the slopes are so crowded the territory is usually very small and more often than not will be reduced to the nestsite and little else.

209

A female *Julidochromis marlieri* tending eggs.

Julidochromis marlieri is a cave spawner that will deposit eggs on the roof of the cave. The eggs are then guarded by the parents. Photos by Jorgen Hansen.

Territories often overlap and we can find places, like a small cave barely one meter across, where several species live together. They will each have a slit, a hole, or a ledge in which they will retreat in case of danger or when they breed, but the cave or anfractuosity will be shared by all as common grounds. We may find a cave with one or two pairs of *Julidochromis marlieri,* one of *L. savoryi,* and a harem of *L. furcifer,* not counting the several pairs of *L. brichardi* that will come in and breed near the opening of the cave. If we add to these cichlids several catfishes, like the ubiquitous *Lophiobagrus,* and the spiny eels, we have an approximate idea of the breeding activity in the rubble and the density of occupation.

The caves, except for the individual breeding recesses, are thus the common ground on which the fry from all species will have a gradual contact with the outside world. Apparently they will cohabit in relatively good harmony, as we will find the juveniles of *J. marlieri* and *L. furcifer* distributed on the same vertical cliff. Afterwards they will emerge from the cave and try to find another piece of rubble under which to hide and stake out a territory. The whole slope, therefore, is partitioned in a mosaic of territories and breeding grounds, densely occupied by the bottom-dwellers. The cichlids that live off the slope, be they nestbreeders or mouthbrooders, must secure a nesting place on a substrate that they do not control permanently, and that they do not know as intimately as the bottom-dwellers. Although some of them, such as the cruising predators, are quite capable of overpowering the resistance of local dwellers and securing a nesting place, many appear to have problems, considering the number of mouthbrooding cichlids that have taken to mating in midwater.

Some of the cave-dwellers display a fierce aggressiveness when they repel an intruder. An example is *J. ornatus,* the *Julidochromis* with the worst temper. A female *J. ornatus* (or a male for that matter) barely 6cm long will immediately repel a *Tropheus* 15 cm long and much more bulky. The *ornatus* will tolerate the *Tropheus* coming to the entrance of the hole in which the brood is raised, but it will head for a *L. toae* perhaps 30 cm away, and chase it away never giving the fish a chance to come near the spawn. Mouthbrooding rock-grazers like *Tropheus, Simochromis,* and *Petrochromis* apparently are used to being repelled from the holes they try to sneak in, and they seldom persist.

Another species of *Julidochromis, J. ornatus* spawns in a similar manner. This one utilized the protected side of a rock. Photo by Hans-Joachim Richter.

As a rule monogamic pairs of bottom-dwelling nestbreeders form early in life. *Julidochromis* and *Chalinochromis* pairs are often already formed when the male is 4 cm long and the female about 5cm (especially in the largest species, such as *regani* and *marlieri).*

Once the pair has selected a nest site in the rubble they might, as far as we could see, remain bonded to it for the rest of their lifespan. I have encountered stable pairs of *Chalinochromis* under the same stone over a period of several years. They could be identified because they were exceedingly large or had a body feature that could help identification. The permanency of such bonds explains why, in captivity, pairs are often difficult to form. The fish are forced to accept a new mate and replace the original bond with its old mate (which was broken by their capture). Such new bonds are more fragile, even after the pair has started spawning. At the slightest disturbance in the tank the stronger mate will often turn against the other and kill it. Nestbreeders at ground level on a rocky slope are very sedentary and territorial and often form stable bonds with their partners.

The fry, raised in the safest part of a narrow territory, only very progressively will venture out of the nest and start to distribute themselves on the nearby rock walls. As soon as the fry from *L. furcifer* or *Julidochromis* emerge from the slit in which they were born, they are no longer attended to by their parents and are on their own. The strongest and boldest are the first to move out, then gradually the others

follow. The process from the time they hatch till they go out on their own is a long one. It takes several weeks for them to grow accustomed to the outside world and learn how to survive. At first, when in danger, they head back to the place where their parents stay and the pair provide them with parental protection. But with genera like *Julidochromis,* the pair is often busy raising the next batch(es) of fry and can do little to protect the oldest fry, which are already outside of the crevice. There is seldom a lull in parental guarding during the day, even when they share the responsibilities, one staying within the nest or close to the swarm, while the other stands watch outside of the slit.

For fishes living more at the surface of the rock and not so much within it, like *L. leleupi, L. mustax, L. niger,* etc., we will eventually discover the fry in the open, in a small "clearing" between the rocks; but they never ascend above the rubble.

Many nestbreeders live some distance away from the bottom. Examples are *L. brichardi, L. toae, L. moorii, L. sexfasciatus, L. tretocephalus, L. compressiceps,* and *L. calvus.* They breed more or less like the pure cave-dwelling nestbreeders. But there are specific differences in behavior patterns and nest sites. They seldom spawn deep within the rubble but usually select a place near the opening in a cave or anfractuosity.

The fry will rise from the rocks and progressively stay 10 to 20cm from the bottom. *L. compressiceps* and *L. calvus* breed for their part in narrow horizontal slits, barely 1cm

high, in which they alone among all the cichlids are capable of entering by slanting their body sideways. The slit is so narrow that only baby eels and dwarf *Telmatochromis* (like *T. bifrenatus)* can enter the hole, and these fishes are not very dangerous for the breeders as they are much smaller. In more than 10 years of diving I have still to see my first swarm of *L. compressiceps* swimming in the open. The fry, when they are discovered swimming low amid the rubble, are already about 3cm long and are single.

The fry of several rock-dwellers, such as *Julidochromis* species and *L. furcifer,* cling to the walls and are never found swimming off the rocks. The mottled garb of baby *Julidochromis* and the slate gray of young *L. furcifer* provide them with excellent camouflage and they are often very difficult to distinguish against the rocky background.

Most other fry of bottom-dwellers when they become free-swimming will develop in the open under their parents' watchful eye. As they grow, their swarm, at first very dense and concentrated, expands and becomes thinned out through predation.

In the early days, when menaced by an intruder the fry head back to the hole where they were born and disappear from view. The way the swarm disappears into the hole is often paradoxical. One would expect the outermost fry, i.e. those closest to the menace, to retreat first toward its center, so that, from fry to fry the signal would be transmitted to the most remote. The fact is, very often the fry at the periphery of the swarm are not the ones that move first, but

A *Julidochromis dickfeldi* male.   *Julidochromis dickfeldi* courting.

Further courtship in *J. dickfeldi*.   The courtship is nearing its final phase.

Spawning in this pair has actually commenced.   Spawning continues. All photos by Hans-Joachim Richter.

Lamprologus moorii is another cave spawner. The eggs can be seen on the roof of this cave. Photo by K. Oki, Midori Shobo.

L. moorii guarding fry. This species can also be found in a bright yellow color. Photo by S. Iwai, Midori Shobo.

those that are near the bottom and the nest opening. They start to enter the nest and the signal is transmitted then through the swarm to the fry located at the rim of the swarm. The whole bunch gives a viewer the funny impression that they are being sucked into the nest. From a ball initially barely 30 cm across, the swarm in a few weeks expands to more than one meter in diameter. For cruising *Lamprologus,* the swarm can reach between 3 and 4 meters across.

The guarding parents are always very watchful, with the sharing of responsibilities following a pattern that, except for minor details, is found among all nestbreeders. One adult stays close to the fry, the other has the long range patrol. The one closest to the

fry stays closer to the ground, while the second opts for a high station providing a panoramic view of the neighborhood.

The amount of energy spent by nestbreeders in the protection of their fry is enormous. The day is spent repelling intruders and boarding parties and they often combine their efforts. It often happens that while both parents are engaged in the pursuit of an intruder the fry fall prey to marauding predators. The rate of attrition by predation is thus very high. An example will illustrate the effects of predation. A swarm of baby *L. tretocephalus,* just free-swimming and 5mm long, was seen one day on a rocky slope. They were about 300 strong. Fifteen days later the fry, still under the guard of the

Fry of some mouthbrooders are released over feeding grounds as the parents keep a watchful eye. At the first sign of danger these *Perissodus microlepis* fry will dart for their parent's mouth. Photo by Pierre Brichard.

pair, were discovered again on the same spot. By that time the fry were a bit more than 15mm long. There were no more than 20 or 25 left.

One can guess that on average only between 10% and 15% of nestbreeder spawn survive by the time the fry are left on their own.

As the swarm expands the young fish gradually lose visual contact with their parents, either because they become lost in the many passageways in the rubble or because they remain too far away from each other in the open. The largest and boldest fry become self-supporting and the time eventually comes when they leave the flock. The swarm melts away at an ever-increasing pace. The respective receptivity to stimuli that kept parents and fry together becomes blunted by the expansion of the swarm. Finally they part.

It would be a mistake to think that among the nestbreeding cichlids in the lake the fry are "abandonned" by their parents. More exactly, barring an accident, one can say that it is the fry who take their leave.

Far from being the passive objects of the parental care the young fry are often seen participating in an active way in their own defense against predators. We have seen the fry of *Boulengerochromis* assemble in a dome and thus secure a measure of safety when the parents are gone. We have seen the fry of *L. tretocephalus* head for shelter when the parents are out of sight. The eldest fry of *L.*

brichardi, barely 2cm long, start to participate in the defense of a nursery along with the adults. Among cichlids this is the only case as yet recorded of an altruistic behavior. In mouthbrooders like Perissodus microlepis the fry are often released on the top of a boulder so that they can start picking their food from the biocover. When the parents have to leave the fry on their own for a short while, the young fish descend to the surface of the rock and lie there until the parents are back and danger has disappeared. These examples show that the fry of nestbreeders can to some extent take care of themselves and adopt defensive measures of their own.

As a rule nestbreeding cichlids are monogamous, but there are exceptions (ex. L. furcifer) in which the males assemble a harem of females under a big boulder. This behavior is not unheard of in the genus. Lamprologus mocquardi and L.werneri of the Congo basin are also polygamous, the males controlling a large territory in which the females have individual nest sites.

On the other hand, nestbreeders with a very mild temper, such as L. brichardi, have communal breeding grounds in the superficial rubble in which several parents raise their respective broods in a kind of nursery. All the fry are looked after by the parents and even the oldest fry, as we have said, participate in the defense according to their feeble means. They even get out of the hole they are in to face an intruder, although they don't have a spawn of their own to defend.

When they have to leave, the swarm spreads out over the walls of the rocks nearby. Juvenile nestbreeders living at ground level have to secure a territory. This is not an easy nor safe task on the crowded slope. They thus skim along the walls entering recesses slowly to explore them very cautiously. They are on the alert and more often than not are repelled by a tenant. Thus they creep from rock to rock until they have found a vacant place. There are not many available at any given time and this is perhaps what brought Julidochromis to adopt sequential spawns and other species, like L. savoryi, L. leleupi, and other nestbreeders, to spawn only a few dozen eggs at a time.

Julidochromis have a double spawning rhythm. The first consists of the deposition of several hundred eggs once every 4 to 6 weeks. Only J. regani most commonly appears to have this breeding cycle. The second rhythm, called sequential, involves laying small batches of a dozen eggs (or for the smallest species even less) nearly every day.

As a result of sequential spawns there are usually a number of eggs and fry in the nest that have reached different levels of development at all times. The fry often stay in the nest until they are pre-adult, at which time they are expelled by the brooding pair. Eggs, freshly hatched babies, and juvenile fish nearly 4cm long can cohabit in a J. marlieri nest with no harm being done by the eldest fry to the latest brood. The consequences of this behavior and the small broods of the other nestbreeders are far-

reaching for the fry and for the adults as well.

1. The pair is bonded to the nest site by the permanent breeding activity, as there are always fry to be watched over in the nest. This means that sequential spawns increase the sedentarism of the species and curtail its migratory potential. It also increases the bonds within a pair of breeders.

2. As the fry are at different age levels, there is a constant trickle of pre-adult fish coming out of the nest and spreading throughout the neighborhood in search of a shelter. This search by a few fish at a time for a vacant place is likely to be more successful than if many juveniles had to divide among themselves the few shelters available on the slope at any given time. Many would not find one.

The sequential spawns of Julidochromis fit perfectly with the dirth of available shelters on the crowded slope. They increase the odds for survival of the fry once they have left parental guard. Not that Julidochromis always spawn sequentially. J. marlieri can switch from normal spawns of several hundred eggs to sequential spawns of a dozen and vice versa, although we don't know as yet what the trigger is that brings about the change of rhythm. J. transcriptus, the smallest species of Julidochromis and the most cave-dwelling as well, probably breeds only in sequential batches. Chalinochromis for their part do not breed sequentially.

The number of eggs laid by the other nestbreeders is highly variable. The average appears to be well over one hundred eggs and can reach

Julidochromis transcriptus is the smallest species of the genus and most secretive as well. It spawns in sequential batches. Shown are an adult and the spawn. Photos by Hans-Joachim Richter.

two or three hundred in several species such as the largest *Julidochromis, Chalinochromis, Lamprologus sexfasciatus* and *L. tretocephalus,* the large *Telmatochromis,* etc. Several species lay much fewer eggs as we have seen with *L. savoryi, L. leleupi, L. mustax,* etc. The spawn of the dwarf *L. schreyeni,* a secretive cave-dweller, probably doesn't amount to more than two dozen eggs from what we could see.

The eggs of nestbreeders are very small and never more than one mm across, even for the largest species like those of *Boulengerochromis.*

Some of the nestbreeders' behaviors are very unusual and unexpected. This was illustrated one day by a pair of *L. sexfasciatus* busy defending its brood from several predators on a coast in the southern part of the lake. As I came by I could notice the pair standing guard over the still very small fry in the usual nestbreeder fashion, one high up off the bottom, the other one down by the swarm. They were very active, having to repel successive intrusions from nearby fish, such as *L. callipterus, L. modestus,* and other bottom-dwelling fishes. High above loomed the dark shapes of *Lepidiolamprologus elongatus* and of plankton pickers like *Ophthalmotilapia* and *Cyathopharynx.* As I watched, a pattern in the behavior of *L. sexfasciatus* progressively became apparent. Their defensive maneuvering against the various marauders was not stereotyped, but appropriate to the nature of the intruder. Whenever the fish standing watch above detected an *L.*

elongatus high above the bottom still several meters away from the swarm and although apparently still unaware of the presence of the fry, the *L. sexfasciatus* rushed in a headlong attack and chased after the powerful cruising predator until both disappeared in the haze. *L. elongatus* was not tolerated less than three or four meters from the brood.

On the ground foraging predators were approaching the nest site as well, and as they came near they were headed off by the other mate. *L. callipterus* could approach perhaps to one meter from the fry. The *L. sexfasciatus* facing its opponent would make a short but determined advance to meet it perhaps over 20 or 30cm, and the intruder would retreat in panic. An intruding *L. modestus,* the same size as the *L. callipterus,* would be tolerated much closer to the nest, perhaps as little as 50cm before being repelled in the same way, although both fishes had come over the same grounds and had been detected early.

Time went by as the intruders were repeatedly beaten back with the same pattern of behavior, which, perhaps, would have not been very significant but for the fact that at the same time several *L. brichardi* were swimming quite peacefully and without being bothered by the breeding pair right in the middle of the swarm. The *L. brichardi* didn't pay any attention to the fry, nor were the latter displaying any panic. The fry simply moved a bit away when the *L. brichardi* came too close (which means perhaps 5 or 10 cm).

One had to admit that *L.*

sexfasciatus had a knowledge of the potential threat to their brood represented by each of the intruding fishes. That they *knew* that *L. elongatus* was a powerful and swift predator that they could not prevent from preying on the fry if they allowed it to come too close to the swarm. They knew as well that *L. callipterus* is a voracious marauder always intent on mischief, but a coward, that *L.modestus* is a rather timid forager, and that *L.brichardi* would do no harm to their fry even though the latter were barely 4mm long.

One had to conclude that some of the rock-dwelling nestbreeders are quite capable of sorting their neighbors out with respect to the danger they represent to their broods. On a slope teeming with potential predators, such a knowledge would enable the breeding pairs to fit their defense to the nature of their enemies and to economize on the energy they would have spent had their defense been stereotyped and indiscriminate.

It happens all too often that the tremendous care the nestbreeding cichlids devote to their spawns is wasted. This happens when the two adults are overwhelmed by a simultaneous attack from different quarters by several predators. Even *Boulengerochromis* spawns can be devastated by a number of predators when the parents are busy chasing other predators away. It happens very quickly (in a few seconds) with the intruders gorging themselves on the hapless fry or eggs. Even when the parents return and chase the marauders away in time to save a small part of the spawn, they appear to lose interest in

Lamprologus leleupi is a popular aquarium species. In the top two photos a spawning cave is defended against an inquisitive *Haplochromis*. In the middle pair of photos a pair of *L. leleupi* court before spawning, then one parent guards the entrance to the cave. At right are the eggs themselves. Photos by Dr. Adam Kapralski.

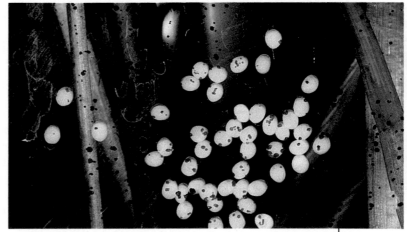

the remaining brood and soon abandon their efforts.

Sometimes the pair doesn't even try to defend its brood, as I discovered when I saw a pair of huge *L. elongatus* rise above the tree limb under which they had spawned and without a move witness the invasion of the breeding grounds by a school of *Xenotilapia flavipinnis*. It was as if the mates had realized that they couldn't halt the progress of the several hundred fish in the school, although, probably because of the peaceful and shy nature of the invaders, the powerful *elongatus* might well have been capable of chasing them away. The fact is that they didn't even try. The *Xenotilapia*, by the way, didn't bother the fry.

Care as they might to protect their brood during daytime, nestbreeders sleep at night, and in the lake this is when the many non-cichlids (catfishes and spiny eels) come to life and start creeping out from the recesses and prowl. All nestbreeders, even those that by daytime remain off the bottom, come to rest with their brood on the substrate. This is when the broods probably are most depleted by predators. Even if the parents are awakened by the approach of an intruder, they can do little in the dark to protect their offspring and it is by sheer luck that the latter, and often the breeding pair itself, escape annihilation.

To sum up our survey of nestbreeding cichlids on a rocky coast we can say that:

1. The breeding grounds are fragmented into a mosaic of territories that are occupied permanently by the bottom-dwellers that spawn in them.
2. Nestbreeders living off the bottom have to find a territory that they will occupy until the brood is raised and ready to start swimming in the open.
3. The fry of cave-dwelling nestbreeders at all times remain in the rubble and do not appear in the open. Only fry from species living in the open start swimming in the open after a while. This is commensurate with their size.
4. Nestbreeding cichlids are usually monogamous, but there is at least one species (*Lamprologus furcifer*) that is polygamous. The bonds between the mates in cave-dwelling nestbreeders (monogamous or polygamous) seem very strong and eventually might cover their entire lifespan.
5. The protection of the fry is difficult although the fishes often have developed very uncommon behavior patterns to secure the success of their spawns, selective defense *(L. sexfasciatus)*, pooling of resources *(L. brichardi)*, etc.
6. The problems of inserting the fry at the end of the parental guarding period into the overcrowded slopes are very difficult to solve. The uneven development of the fry in a spawn help to spread this insertion over several days. But several fishes are especially well adapted to this problem by spawning small sequential batches of eggs instead of periodic large spawns. Individual fry can thus secure a territory more easily.
7. There are many examples of the active role the fry themselves assume in their protection and in their abandonment of the parental protection.
8. The bonds of bottom-dwelling nestbreeders toward their territory and nesting place are very strong, especially among those fishes living in the rubble. Their sedentarism is very high, especially for species that have sequential spawns. As a result they often develop qualities that wandering species don't have, such as the knowledge of the topography of the slope on which they live (*Chalinochromis-Julidochromis*) and probably of their neighbors. Because of sedentarism a species will colonize new territories and newly opened slopes very slowly. The gradual expansion of a species on a slope will allow nestbreeders to live and breed apart from each other at different levels. This is exemplified by the color patterns of *Julidochromis marlieri*. The specimens living deeper than 15 meters are systematically much darker than those living near the surface. This is not a temporary adaptation to different light conditions but a permanent character of the various lineages.

The variability of the nestbreeding cichlids' spawning behavior and the sophistication of their approach to the difficult problems stemming from the dense occupation of the rock-

Prespawning courtship can be rough at times, leading to the battering of one of the spawning partners. A sure sign of the breakdown of compatibility is one of the partners departing the scene with clamped fins as seen here in *Lamprologus leleupi*.

strewn floors are illustrated by their successful colonization of these coastal biotopes. Nowhere else in the lake or in the rivers are so many nestbreeding cichlids living together. They could not have if their spawning behavior, a major feature of the specific ecological niche, had been more stereotyped.

Pseudocrenilabrus multicolor spawning in typical mouthbrooder fashion. The eggs are laid and the male follows to fertilize them. The female is approaching the eggs, ready to take them into her mouth. Photo by Hans-Joachim Richter.

MOUTHBROODING CICHLIDS

As we leave the nestbreeders and come to the mouthbrooding cichlids we cannot fail to discover major differences, not only of course in their behavior but in their occupation of the biotope and the problems they have to face as well.

There are no true cave-dwelling mouthbrooders, meaning by this that we don't find mouthbrooding cichlids living permanently in deep recesses. All mouthbrooders living at ground level dart in and out of the rubble anfractuosities and spend much time in the open, in midwater, or on the surface of the rocks. They go into the

rubble to take shelter, occupy a nest site for a mating season, get the fry used to the rubble, or to release them at the end of the incubating period. But they don't have a permanent dwelling in which they live, a territory that they own deep in the recesses. When they select a spawning site, which is not always the case, their occupancy of the site is short—at most a few days.

Some species, like those of *Tropheus* and *Cyprichromis,* don't even breed on the substrate, but in midwater. The fact that a rock-grazer like *Tropheus,* living at ground level and with such strong bonds toward the rubble, has abandoned use of the

A female *Lamprologus leleupi* tending the remaining eggs from her clutch. The empty egg cases of those that have hatched can be seen. In the bottom photos the female guards the eggs (left); the fry after hatching have fallen from the wall to the bottom of the cave (right). Plenty of yolk is provided for their first days (bottom). Photos by Hans-Joachim Richter.

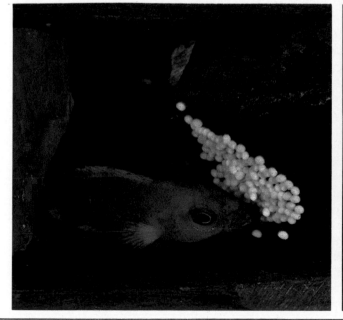

substratum to breed might be considered a very significant clue to the problems encountered by mouthbrooders when they look around for a safe breeding site on the ground. The bottom is occupied in strength by nestbreeding cichlids and a host of non-cichlid rock-dwellers. Most have a permanent dwelling, are very territorial, and often are predators. To some extent all are carnivorous. Some are keen spawn predators. The amount of predation is especially high on eggs and juvenile fishes, because of the prevailing promiscuity.

Several marauders, such as *L. callipterus, L. modestus, L. mondabu, Telmatochromis caninus,* and *T. temporalis,* can be considered as ravenous spawn eaters. They are keenly aware of any unusual behavior. Of course any spawning activity, with the antics the mates engage in, attracts their attention.

A good illustration of their behavior is provided by the spawning sessions of the semi-pelagic *Lamprichthys tanganicanus,* the endemic cyprinodont, itself a predator of juvenile fishes swimming in midwater. *Lamprichthys* mate and spawn along a rock ledge on which the female deposits a row of crystal-clear, transparent eggs. The eggs are invisible even at close quarters and can be made out only by the magnifying-glass effect they have on the biocover underneath.

As soon as a pair of *Lamprichthys* dives toward the rock and starts embracing side to side, the above-mentioned species that happen to be in the neighborhood lineup behind the spawning fish

which are totally oblivious of anybody nearby. As the eggs are laid they are gobbled up at once and without fail by the predators. I often wondered at the "stupidity" of the *Lamprichthys,* at the apparent waste, and at the predators who knew very well what was going to happen and what the *Lamprichthys* were up to.

Thus, on a rocky slope bottom it is very difficult indeed to spawn and raise a school successfully. Nestbreeders manage because they are sedentary and have picked a safe place: the safest place within their territory to live, to breed, and to raise their brood. They aggressively defend their brood and have fine-tuned their skills through their evolution for this essential task. They are at least two strong to defend the spawn for once they have selected a nestsite the mates pool their resources to defend the nest.

The evolution of mouthbrooding cichlids has changed this order of priority. The nest has become much less important because at best it is only used, if at all, for a short while (a few seconds or a few days). It has often dropped so low in the order of priority that it has become a mere part of the mating ritual.

What has become essential for mouthbrooders is how to get the eggs duly fertilized and safeguarded against predators into the mouth of the female as soon as possible. In some species the eggs are even already stored in the female's mouth before she looks around for a male to fertilize them. This calls for a physiological adaptation of the egg which remains perceptive to spermatozoa for an extended period.

If nestbreeders have to pool their resources and thus form strong bonds between the mates to safeguard the brood (some, like *L.brichardi,* even pool the resources of several adults), mouthbrooding females don't need the males after mating and the latter can thus devote their spare time to mating with other females. A few mated mouthbrooders remain together to raise the brood (ex. *Perissodus*) but this happens only with species attaining a small size as adults and having large spawns to look after. Mouthbrooders thus tend to be polygamous.

Because the fry are not raised in a nest but taken along in the mouth, mouthbrooders are not bonded to a site or territory in the same way as nestbreeders. Because they are less territorial, their wandering potential is much higher and they are as a rule much less sedentary. The fry are taken along during their incubation and are released at its end in a place where the female happens to be at that time, which can be far from the one where she originally mated.

Mouthbrooders can thus colonize new territories, reaching newly available biotopes much faster than the territorial and more sedentary nestbreeders, especially when the latter are cave-dwellers.

On the other hand, many mouthbrooders depend less on the substratum on which to build an appropriate nest. Mouthbrooders can thus eventually occupy and multiply in a biotope in which nestbreeders cannot reproduce for lack of proper nest sites. There are thus places around the lake where *L. compressiceps,*

Lamprichthys tanganicanus courting.

Selecting a nestsite.

Courting intensifies.

Spawning becomes imminent.

The eggs released are virtually invisible.

In the wild spawning attracts egg predators.

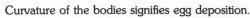
Curvature of the bodies signifies egg deposition.

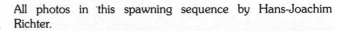
All photos in this spawning sequence by Hans-Joachim Richter.

Julidochromis species, and *Telmatochromis bifrenatus* cannot settle permanently for lack of breeding places, while *Tropheus* and other rock-grazing mouthbrooders can.

These several features of the mouthbrooders' ethology explain why they were the ones that went into pelagic waters, while the nestbreeding cichlids remained bonded to the shorelines.

On the genetic level, the wandering potential of the mouthbrooders along the coasts tended to favor crossbreeding and there are relatively few species in each line *(Simochromis-Tropheus, Petrochromis-Tilapia, Eretmodus-Spathodus-Tanganicodus, Cyathopharynx-Ophthalmotilapia,* etc.) although the number of local varieties is high. In nestbreeders, with a slow expansion potential, inbreeding brought about the birth of a large number of species in the most sedentary lines.

To be able to raise their fry in the mouth and have them develop without the help of outside food supplies, mouthbrooders have to pack their eggs with enough supplies to carry the embryo through several weeks of growth in full autonomy. Eggs of mouthbrooders are thus always larger than those of nestbreeding cichlids. The egg of *Eretmodus* is approximately 2mm long, the one of *Tropheus* from 5 to 6mm, or about 5% the length of a young adult fish. Those of *Boulengerochromis* are about 1 mm long or 1/300th to 1/400th the length of a young adult. It contains a very small

amount of yolk, just enough to carry the growth of the embryo through the few days before the egg hatches. Soon afterward the fry will have to find food in the outside world.

Mouthbrooders' eggs have a very large yolk that will sustain the growth of the embryo and later on of the fry for several weeks. The energy packed by the mother into the egg will see *Tropheus* fry through several weeks of development and growth. During the time it takes for a *Tropheus* fry to absorb its yolk, it will remain as helpless as the fry from a nestbreeder, i.e. until it becomes free-swimming, but it is safeguarded in the mother's mouth.

Because she packs so much energy, in the form of proteins, into her eggs, a female mouthbrooder will lay, as a rule, much fewer eggs than nestbreeders of similar size. A *Tropheus* female 10cm long can lay from 10 to 20 eggs; a female *Julidochromis* 10cm long can lay between 150 and 300 eggs.

Those few mouthbrooders that lay a high number of eggs, for example *Perissodus microlepis,* have eggs barely larger than those of nestbreeders. Another example of this rule is provided by *Lobochilotes,* a mouthbrooder reaching a large size. Comparable to *Boulengerochromis* among nestbreeders, *Lobochilotes* perhaps lays as many as 400-500 eggs, *Boulengerochromis* up to 15,000.

When they do spawn large batches of eggs, mouthbrooders have to reduce the period of incubation, because they cannot hold all their fry for as long a time as

the species that lay only a small number of eggs. The fry, having exhausted the scant supplies provided by a smaller yolk-sac, have to start feeding earlier in the biotope.

There are thus wide differences in the length of incubation time provided to their fry by the various species. Some, like *Tropheus,* in the various aspects of their technique have brought mouthbrooding to its most sophisticated development. Other species in various aspects of their technique have remained rather close to nestbreeders. *Perissodus straeleni* is a mouthbrooder but it uses a nest site in the rocks where it lays several hundred eggs over several days and both parents look after and raise the fry, mostly in the open.

In terms of the energy that the female nestbreeders and mouthbrooders store in their eggs, which they must take from their own energy source, one might thus think that for similar sized females the energy spent must be rather similar. A mouthbrooding female probably spends less energy when she is incubating her brood than a nestbreeder as she looks after the growing fry, but she has a problem feeding herself. Nestbreeders often take turns while watching the eggs and fry, to feed and forage. Thus mouthbrooding females often are more or less emaciated at the end of the incubation time.

In terms of the efficiency of the two modes the evidence is difficult to assess. But we have a few good clues. The survival rate of the fry of mouthbrooders is definitely better than that of

Sarotharodon karomo spawning. The spawners circle over the spawning site with the female depositing eggs, the male quickly fertilizing them, and the female snatching them up. In the lower photo the female is waiting to pick the eggs up while the male is busy fertilizing them. Photos by Hans-Joachim Richter.

nestbreeders. We have seen that at the end of the parental care, nestbreeder's spawns at best had a survival rate of 10 to 15%, not counting the eggs already preyed upon while they were still hidden in the nest.

If we take the most sophisticated of all mouthbrooders, *Tropheus,* and we accept that between 15 and 20 eggs are laid in an average spawn, we discover that on average it releases from 10 to 12 fry at the end of the incubation time. This gives a survival rate between 50 and 80% of the spawn, which is enormous.

But in absolute terms, the picture is not so bright for mouthbrooders. A *Tropheus* female will release 10 or 12 fry on the bottom and a *L. tretocephalus* female as many or a bit more. At first glance the result will be more or less identical. But a pair of nestbreeders can breed between 8 and 12 times a year, which means that between 100 and 250 juvenile fish will be injected into the biotope over a 12 month period, while a female *Tropheus,* spawning between 6 and 8 times a year, will release about 100 juvenile fish at most. In pond-raised *Tropheus* breeding in "natural" conditions, which means with a lot of predators around (frogs, snakes, monitor lizards, and other species of fish), the average is about 50 fry per female. In this respect mouthbrooding doesn't appear to give buccal incubators much of a lead over nestbreeders in the lake.

Aside from *Tropheus,* which is exceptional in the small size of its spawns, most mouthbrooders lay between 25 and 50 eggs, even when they are large fish such as *Cyphotilapia* (whose females reach 25 cm and lay at most 50 eggs), with an average spawn of about 30 eggs.

When the mouthbrooders come near the end of buccal incubation they face a much more serious problem than nestbreeders. They have to release their fry into the biotope. To nestbreeders this occurs by a gradual process in an area with which the fry have become acquainted. When the fry are fit they leave their parents.

For mouthbrooders it has to be done in a place that they don't *own* and do not usually control. In short, on foreign grounds. Their fry, contrary to those of nestbreeders, have only had, at best, episodic contact with the outside world. For some of them this occurred toward the end of their incubation, when their mother released them at an accelerating pace to graze on the biocover. They could in that manner get acquainted with the world they would soon be living in and learn about it. What mouthbrooders gained in safety for their brood by keeping them in the mouth they have perhaps to lose by delivering their fry poorly equipped to deal with a world they don't know well, if at all. In fact, several mouthbrooding cichlids living in midwater, like the *Cyathopharynx* guild, do not seem to release their fry periodically before they abandon them in the open. Their fry come all of a sudden into contact with a world they don't know at all.

That the problem is a very serious one for mouthbrooders is illustrated by the behavior of *Tropheus* females. These not only keep their fry in the mouth for several weeks after they have become free-swimming, but spend more and more time in the rubble to *train* them in the open areas between the rock walls.

More clues are provided by the drifting plankton pickers such as *Cyathopharynx, Ophthalmotilapia,* and the species of *Cyprichromis.* The first two fishes release their fry, which they spawned in a crater nest on the bottom, at the surface of the water close to the shoreline. Paradoxically, *Cyprichromis,* which didn't need the bottom substrate and a nest but spawned in midwater, opted to go to the bottom to release their fry in deep recesses or under overhanging rocks.

The need to find an appropriate place in which to release the fry brought mouthbrooders to a variety of options and was of such urgency that it might have had an impact on the distribution of the species around the lake. To illustrate this point let us remember what we know about the respective ecological niche of *Cyathopharynx* and *Cyprichromis.* Both live off the bottom in schools with unsynchronized spawns— which means that they are bonded to the shoreline. *Cyprichromis* breed in midwater, *Cyathopharynx* in sandy or rocky areas. *Cyprichromis* release their fry in rock recesses or under overhanging rocks (which means that the relief of the rubble must be very uneven). *Cyathopharynx* release their fry off the bottom near the surface and in the open. They are found mostly over rocky slopes but just as well may be seen over rather steep sandy

Ophthalmotilapia ventralis spawning sequence.

The female expels an egg in the center of the nest.

The egg has been picked up. Her throat is already distended from other eggs.

She backs up and prepares to pick it up.

A school of brooding females in the wild. All photos by Ad Konings.

The male patrols his territory.

The female mouths the bright yellow tips of the male's ventral fins.

slopes where *Cyprichromis* are not found. *Cyprichromis* remain restricted to the rubble areas with deep recesses to hide the fry in and do not wander far from these places.

Thus one might say that the lack of synchronism in their spawning and the need to find the recesses in which to release their fry are both essential features of their ecological niche and have restricted their distribution to a part of the coast. The distribution of *Cyprichromis* could have been like that of *Cyathopharynx,* i.e., could have been broader, had it not been for the need to remain near the towering areas of rubble.

There is one last aspect of the comparison between mouthbrooding cichlids and nestbreeders that needs to be investigated, namely, the minimum size at which it is possible for a cichlid to be a mouthbrooder.

The smallest nest sites that a biotope can provide a nestbreeding cichlid with are the empty *Neothauma* shells, barely 3cm across, or the very narrow slits and cracks of rocks. *Lamprologus multifasciatus* can breed when they are only 20mm long. The dwarf nestbreeders can of course produce only so many eggs, probably no more than two dozen, and their eggs are very minute.

Given the energy stored in an egg by a mouthbrooder there might be a limit to the miniaturization of the fish. This could be because it would exhaust the females too much, because they would be too minute to defend their fry at release time, or, even moreso, because they would be without a territory of their own in which

to deposit their fry. The number of fry reaching the end of incubation could also be so small (probably only one or two) as to not be worth the effort for such small fish.

The smallest mouthbrooder found in the lake so far might well be *Spathodus erythrodon,* ready to incubate when it is only 35mm long, or the elusive semibenthic *Xenotilapia tenuidentata,* which incubate at the same size, although both species reach 6cm or more.

Thus, with the available information at hand we might think that the minimum size at which nestbreeding cichlids become sexually mature is about 50% of the one needed by mouthbrooders.

After having given much attention to the comparative merits of the two cichlid breeding modes, let us now investigate the various groups of mouthbrooders on a rocky coast.

They can be split according to the area they occupy with respect to the bottom—off the slope or at ground level. Mouthbrooders at ground level are composed of rock-grazers, including the genera *Tropheus, Simochromis, Petrochromis, Eretmodus, Spathodus,* and *Tanganicodus,* and two scale-rippers, *Perissodus microlepis* and *P.straeleni.*

Altogether there are fewer than 20 species in any given place. They are very individualistic and often very aggressive at the intraspecific level. They usually breed among the rubble, except for *Tropheus,* which often breed in midwater. Males entice the females by a short display across the path of the females, during which they shiver. The female lays one or several

eggs, takes them into her mouth, and bumps the male in the side with her head causing him to emit sperm with which she gets her eggs fertilized. The operation is repeated with the next batches of eggs. It is entirely possible given the short time spent in one spawning session and the mobility of the partners, that a female can get her spawn fertilized by several mates. Polyandry cannot be ruled out.

Although egg-decoy anal ocelli adorn the anal fin of males in several species (*Simochromis* and *Petrochromis*) they are not a feature found in the pattern of male *Tropheus, Eretmodus, Spathodus, Tanganicodus,* or *Perissodus. Tropheus* males do have light colored spots on their anal fin (and in *T.annectens* even sometimes on the ventral fins) but they don't duplicate the enormous eggs of the species in shape, color, or size, and thus are not egg dummies.

The anal ocelli of *Simochromis* and *Petrochromis* could be considered as egg-decoys. All species are endowed with at least one anal ocellus that, through the lateral moves of the fin, can bulge out of the fin plane and mimic an egg in relief. This is brought about by the thinness of the ocellus membrane in comparison with the thicker tissues of the rest of the fin. Such bulging ocelli may exist even in the dorsal fin of *Petrochromis.* Although *Simochromis* and *Petrochromis* have brought to a very high degree of perfection the duplication of an egg by an ocellus, it is paradoxical that they don't appear to be paid any attention by the female. She will bump

Removal of unwanted gravel is one of the first prespawning chores for this *Lamprologus ocellatus*.

A possible site is excavated under the shell by the female.

This site was worked for a while, then abandoned in favor of the shell itself.

The shell is chosen, and the female enters to clean before depositing her eggs.

She deposits her eggs inside the shell.

He then prepares to fertilize the eggs.

The male waits outside the shell until she finishes.

The female guards the eggs. All photos by H.-J. Richter.

the male most often in the abdominal area and will not try to pick up or swallow the egg-decoy ocellus.

The fact that very often female mouthbrooders in the lake already have their eggs in their mouth when they approach the males might help explain the paradox of finding fish with the most sophisticated egg-decoy ocelli remaining useless. The fact that *Tropheus,* the fish that brought mouthbrooding to a high degree of perfection, doesn't have any ocelli might hint that the egg-decoys of *Simochromis* have become a mere part of the nuptial garb of males and have lost their primary function.

The fact that all *Simochromis* species are endowed with anal ocelli, one of which is always capable of bulging, while *Tropheus* lack the ocelli, is one of the reasons why the two genera should be kept apart and not put into synonymy, as had sometimes been suggested.

Buccal incubation takes a long time, extending over several weeks in *Tropheus, Simochromis,* and *Petrochromis.* No data are available about the breeding behavior of *Eretmodus* or other goby cichlids as none apparently have been raised in captivity.

Once they have been released (by the time they have reached about 11 to 12 mm), the fry of the first three genera grow rather slowly. In the wild they probably don't reach sexual maturity before they are about one year old, perhaps a little less. They remain hidden in the rubble until they are 3 or 4 cm long and then start to appear on top of the stones. They will not rise

from the bottom and start large scale wanderings in the open until they are adult. The fry are very active and restless. They display much animosity toward each other and engage in mock fights very early (a rare behavior among nestbreeder fry).

Perissodus are very different from the rock-grazers as we have already said. Their breeding behavior is very close to that of nestbreeders as they form monogamic pairs in which the two mates take care of the fry, produce very large spawns, etc.

The second group of rock-dwelling mouthbrooders involves all the species that live off the bottom in midwater schools. Their breeding behavior is very different from that displayed by pelagic and sand-dwelling schooling mouthbrooders in at least one essential aspect. Because the biotope provides a steady and bountiful food supply, they don't have to roam to feed. They find it on the spot. Whenever they need a nest to breed they find one on the nearby slope. They don't have to migrate like pelagic species to far-off breeding grounds. They are thus much more static and less mobile than roaming or wandering species.

Because they remain in the same area and can breed without a seasonal cycle, their fry, once they become free-swimming and large enough, join the main flock. The schools of these mouthbrooders are thus composed of some fishes that have reached sexual maturity and others that have not. As a result, the spawns are not concentrated within a short period and cyclical.

The spawns of coastal plankton pickers are more asynchronic and there is a constant mating activity going on between the mature males and females ready to spawn. This is probably the single most important feature that prevents coastal plankton pickers from roaming far from shore and becoming pelagic. They are bonded to the coastline because a school never stops its breeding activity, due to the lack of uniformity in the size of the fish and their lack of cohesion.

Not that the lack of synchronism in spawning sessions is identical in all species. The spawns of *Cyprichromis microlepidotus* are more synchronized than those of *C. nigripinnis,* and perhaps more seasonal as well. The spawns of *C. leptosoma* are not synchronized. *Cyathopharynx, Ophthalmotilapia, Cunningtonia, Cyphotilapia,* and *Limnotilapia dardennei* are not synchronized and not seasonal.

The information gathered during ten years of diving is very poor with respect to sightings of the spawning activity of these species. Few have been observed mating, but the lack of breeding cycles has been deduced from the countless captures of females and males displaying various stages of egg and fry development and sexual maturity.

In *Cyathopharynx* the technique is similar to the one used by the sand-dwelling *Xenotilapia.* The males in full mating regalia are busy building their nests, cleaning the ground, and rising to meet a female that is circling slowly

A spawning sequence of the shell-dweller *Lamprologus brevis*. The top two left photos are of the "Magara" form, as indicated by the golden spot behind the eye. Photos by H.-J. Richter.

in midwater along with immature males and juveniles . One of the females in the school dives toward the male and follows him to the crater nest. *Ophthalmotilapia ventralis* females already have their eggs in their mouth as they descend. The male enters the nest and ejects the sperm on the floor of the crater while the female remains outside and waits. The male leaves the nest across from where the female stands. She then dives into the crater and starts taking mouthfuls of the sperm, which by then has been diluted in the water and is nearly invisible. She then turns back and rises again to join the school above. *At no time are the two mates in physical contact with each other.*

Not even for a few seconds can the predators that might have been around have a chance to prey on the eggs. They were already secure in the female's mouth.

To avoid predation on the eggs when they are laid in a nest, the lake mouthbrooders have split the spawning session into two distinct episodes. In the first step the female spawns, then keeps the eggs in her mouth for quite a while before she is selected by a male and only then gets them fertilized. *Cyprichromis* for their part don't even go to the bottom. They mate in midwater.

The fact that two very different fishes like *Cyprichromis* and *Tropheus* avoid laying their eggs on the substrate, and that females of the *Cyathopharynx* guild safeguard their eggs first, are illustrations of the dangers surrounding spawning on the substrate for many mouthbrooders.

This is probably also the reason for the strange behavior of *Cyathopharynx* males when they build their crater nest on a rock-strewn slope. They pick the highest boulder in the area, sometimes two meters above the bottom, and start bringing sand to the top of the rock to build their crater nest. More or less it is 30cm across and about 5cm high, the central area being the bare, cleaned rock. Not every kind of sand is picked from the bottom. Fine sand won't do, probably because the walls of the crater might collapse if the water is churned up by rough weather. Only the coarse variety of sand will be selected by the male, even if it entails venturing a few meters more from the boulder to find the proper quality.

All males in the area select the top of boulders on which to erect their craters, even when sand patches do exist on the bottom nearby. We often wondered at the incredible work involved in carrying four or five pounds of material from the bottom to their lofts, until we were able to witness the endless stream of males rising to meet the egg-carrying females circling above. Those males who were occupying the highest nest sites had the shortest distance to travel toward the females and could thus fertilize more spawns during their breeding activity. As the highest loft was also the first to be occupied by a male, one could assume that the dominant males had first pick, and that afterward newcomers had to settle for lower boulders. The second fact that one should remember from the *Cyathopharynx* behavior is that on a densely populated rock-strewn slope, with many

predators at ground level, the craters on top of the boulders could be in a safer place than if they had been on the bottom. And anyhow, these were the only nest sites in the area not occupied by nestbreeders. In fact, I don't remember having seen one predator living normally around the rubble close to any of the hundreds of *Cyathopharynx* lofts I was able to explore during my dives.

Not that the males will breed exclusivly on top of boulders. In areas where rocks are missing from the scenery, or when their tops are too close to the surface, males dig their craters in the sand to provide a quiet mating place. This is especially true in quiet coves. One can see them looking after their nest along with the males of *Aulonocranus* and *Callochromis macrops melanostigma,* with which they mix.

Ophthalmotilapia build their nests on small sand patches between the rocks, and I have never seen them use the top of high boulders.

The eggs of the plankton pickers are smaller than those of *Simochromis, Petrochromis,* and, of course, *Tropheus.* They appear to lay about 30 to 40 eggs at most. So do the *Cyprichromis* that breed in midwater without any nest preparation.

We have mentioned already that the *Cyathopharynx*-type related species abandon their fry near the surface along the rubble lining the shoreline. The fry are seen swimming in troops 10 to 30 strong that probably come from a single female. At the time of their release they are not more than 15mm long. The species to which the fry belong are at this size difficult to identify as the pickers all look very much alike.

Cyprichromis live at various depths according to the species. *C. brieni* (in the past incorrectly known as *C. nigripinnis*) has the shallowest habitat and can be found between 10 and 20 meters, but the fry are found hidden in the dark recesses between 6 and 8 meters deep. This is a feature common to all mouthbrooder fry—they are found higher up on the slope than the adult fishes. This is perhaps due to a problem of the higher amounts of oxygen needed by the fry, which the incubating female cannot provide in her mouth but which she can find in the upper part of the specific habitat. Even *Tropheus* fry are found close to the shore, in a few cm of water, where they get water that is hyperoxygenated by the pounding surf.

The breeding behavior of *Cyphotilapia* is poorly documented in the wild. Females are much smaller than males, reaching only about 25cm against the 35cm of a fully grown male. They incubate between 20 and 30 fry in their mouth, often less. Spawns of up to 55 eggs, however, have been reported from captive females.

Incubating females when captured from a single school have eggs and fry at various stages of development in their mouth, indicating that they don't have a synchronized breeding pattern. Spawning occurs throughout the whole year. As might be expected from fishes living off the slope and wandering little, they follow the normal coastal schooling fish breeding cycles. Again fry are often found much shallower than the adults, and

These *Pseudocrenilabrus multicolor* fry are large enough so that they no longer seek refuge in the female's mouth. Here they are being fed newly hatched brine shrimp, which they avidly consume.

can already be seen at 8 meters depth, when the main flock of adults is found mainly around 25 meters or more.

Fry remain mostly hidden from view under rock ledges and in small recesses until they are about 4 to 5 cm long. At depths of 25 meters or more they can be seen, already schooling, with juveniles of their specific predator, *Perissodus straeleni* (which breeds in the same area), in their midst.

235

Limnotilapia dardennei is another fish whose spawning sessions have never been sighted in the lake. The adults live in deep water in schools near *Cyphotilapia* but are much more elusive. Schools of juveniles 5 to 15 cm long are a common feature of shallows, mostly over sand or even in aquatic plant beds but also over rocks. Adult fishes are seldom seen in shallow waters and then mostly single incubating females, which, one might think, are looking for a place to release their fry.

Lobochilotes is a fish with a wide ecological niche, as it is often found on mixed grounds of sand, pebbles, and rock. But again, as with the two previous species, adults live in deeper water and young fry much shallower. *Lobochilotes* differs, however, by the size of its spawns, which may reach several hundred eggs. Because of this feature, and also because the fry are kept over a period of several weeks in the female's mouth, she has to release them often to feed on sand patches and graze from the rock biocover. Females busy watching over the fry are often seen in shallow water, and I remember having once disturbed a very large female as she was standing guard over fry about 10mm long. She tried to collect them all, couldn't, departed in a hurry, and disappeared toward deep water. Perhaps a hundred fry remained on the bottom. I went past and then waited. After one or two minutes she reappeared, sped toward the area where the fry were waiting, gobbled them up, and disappeared for good toward the deep.

This episode illustrates the strong bonds of a mouthbrooding female toward her brood. The instinct to safeguard them in the mouth is so strong that it is not rare to see a female captured and brought up with decompression problems, and wait until she is really suffocating before she releases the fry.

This is especially true with *Tropheus* and *Cyphotilapia*, but much less so with *Cyprichromis*. It is enough for a female from one of the four species of this last genus to bump against a net to start dropping their eggs and fry. When several females are trying to escape in panic from a net they often all start to release their broods in a cascade of eggs and freshly hatched fry. There are thus degrees among mouthbrooding cichlids in the tenacity with which females cling to their fry.

SEXUAL DIMORPHISM

There is very little dimorphism between males and females of nestbreeding cichlids. Males are often a bit more elongated than the more plump females and the color pattern is perhaps a bit brighter in males than in females, but the differences are not much. It is therefore very difficult to tell them apart unless one takes a look at the genital papillae. That of the male is much narrower than the broader female orifice. The sex of a *Julidochromis* is easy to identify: males have a long and slender tube protruding behind the vent; in females it is much shorter and in the shape of a broad "V".

Contrary to nestbreeders, the sexual dimorphism of mouthbrooding cichlids is often spectacular. Females are usually much less flamboyant than males, but there is at least one exception to this rule in the lake. The *Haplochromis benthicola* female is much more colorful than the male. In this species females are vermilion red with patches of metallic golden and black scales; the male has a dark brown background color with a few horizontal rows of metallic blue scales on the sides.

One might wonder why nestbreeders do not show much difference in shape and garb between the mates while mouthbrooders display so much.

Nestbreeder pairs or harems are formed usually for a long period of time in the lake and remain together. Pairs often form early in life and remain bonded during the lifespan of many species. The mates thus know each other well and don't need visual signals to identify each other in the biotope. Also most are sedentary and territorial.

Among mouthbrooders the fishes are polygamous and wander about a great deal. They are less territorial and partners usually don't know each other individually as their bonds are episodic. As the mates approach sexual maturity females need to identify those males that are ready to fertilize their eggs. This is when the males develop the gaudy colors. Unripe males can be told from sexually mature specimens because their pattern, in itself much more colorful than that of females, lack the special sheen produced by the excitement of sexual maturity. Sexual dimorphism from one

The first eggs of this *Lamprologus brichardi* were placed on the dark stone, but something made them change to the lighter colored stone in mid-spawning. Photo by Hans-Joachim Richter.

species to the next manifests itself in various ways. Without tending to be exhaustive, we will examine the patterns of a few species.

It is often difficult to say at first glance if a fish is a male or a female because there are so many color varieties around the lake, the colors of both are often very bright, and the patterns depend so much on the moods of the fish. Still, there are morphological features by which one can tell the sexes. *Tropheus* males can be identified by:

1. The genital vent. This is much smaller in males than in females, in which it is a broad circular papilla.

2. The mouth. This is more curved and more narrow in the male; the mouth of a female is straight and broader. The profile of the male snout is also more convex.

The patterns of male and female *Tropheus* are rather similar, except for the patterns of dominant males. Males of most varieties have spots on the anal fin which are seldom found on females or, if so, they are much less obtrusive. Some males of the Kalemie *Tropheus* even have several circular spots of contrasting color on the ventral fins as well.

Simochromis males always have ocelli on the anal fin, one of which is a bulging egg-decoy. The colors of males are definitely brighter than those of females at breeding time, but less so when they are not sexually ripe.

Sexual dimorphism in *Petrochromis* is not much apparent, aside from the striking ocelli on the anal fin (in some species on the dorsal fin as well), one of which is also always of the bulging type and

Simochromis males always have ocelli on their anal fins. One of these is the three dimensional egg decoy described in the text. This is *S. babaulti*, one of the typical shallow-bottom grazers. Photo by Pierre Brichard.

an egg-decoy. Females are often striped, males not.

Eretmodus, Spathodus, and *Tanganicodus* display little sexual dimorphism, except for the slightly brighter colors of the males, which are endowed with more and brighter blue spots than females. Males can grow cephalic humps.

Coastal scale-rippers such as the two *Perissodus* have a pattern in which males and females are very similar, probably reflecting their monogamy and perhaps formation of long-lasting pairs.

Sexual dimorphism in *Cyphotilapia* is striking. Females remain much smaller than males while the latter grow an increasingly protuberant hump between the occiput and the snout. The head of males is suffused with a deep blue sheen and the body stripes are more sharply outlined in males than females.

Limnotilapia adult males have anal ocelli.

Sexual dimorphism reaches its peak among the various plankton pickers living in midwater. In all species the males sport very gaudy colors when they reach sexual maturity, while females as a rule have a much drabber pattern from which all flamboyant shades are excluded.

In the *Cyathopharynx* guild males of all species are endowed with very long, filamentous ventral fins that often reach the caudal fin. They are 7 to more than 10 cm long. In several species the tips of the ventrals are forked or spatulate and the tassels a rich yellow, orange, or white, according to the species and the locality. The female's fins are much shorter and a pale gray; they are not forked. The colorful spatulated tassels are not used as egg decoys.

The variety of colors in the male's repertoire is large from one species to the next, and even between local populations within a species as well.

Cyathopharynx males are the most colorful fishes in the lake, the most common color variety being a deep blue with prussian blue filamentous unpaired fins striped with jet black and bright yellow. The fish is incredibly beautiful and straight out of a fairy tale. The colors of the fish could hardly be expected to become even more striking, but by raising ever so slightly the plane of their scales the males manage to suffuse the deep blue of their sides with glitters of metallic green that pass through like a flash. *Cyathopharynx* males in Nyanza Lac are sometimes a deep brown with an orange head; those in Cameron Bay are often velvety black with oblique stripes of metallic red alternating with emerald green. Females of all the *Cyathopharynx* varieties and of related genera are always silvery gray, sometimes with darker stripes and patches.

We can now conclude our survey of reproduction with this summary:

1. When food supplies are abundant and steady fishes breed several times a year. When food supplies dwindle and increase periodically breeding is concentrated in one or two annual cycles.

2. Non-endemic species belonging to families other than cichlids engage in seasonal anadromous migrations toward spawning grounds in rivers. The endemic

Cyathopharynx furcifer males are endowed with very long ventral fin rays tipped with a bright yellow tassel. The body and head colors are quite variable from one population to the next. Photo by Glen S. Axelrod.

This appears to be a pair of spawning *Cyathopharynx furcifer*. The male is leading a female to his nest. Photo by Glen S. Axelrod.

Lamprologus furcifer spawns in rocky crevices. The eggs of this species are bright green in color in contrast to the orange or yellow ones of most mouthbrooders. Like most nestbreeders, this species is territorial. Photo by J. Shortreed.

species of these families mostly breed in the lake.

3. Schooling is essential for the survival of a roaming species. All other behavior, including breeding, stems from this necessity. Breeding of these species is thus synchronized in pelagic and sand-bottom schools, for the former after a seasonal migration toward coastal spawning grounds.

4. Only mouthbrooding cichlids and not nestbreeders have colonized the pelagic waters.

5. Wandering and schooling cichlids not only synchronize their spawning activity but also the release of their fry at the end of the incubating period.

6. On a rocky coast the constant availability of food and shelter has made possible the individual survival of the fishes. Thus they can become the owners of a territory in which they breed and raise their successive broods. Breeding activity is not cyclical or seasonal.

7. Nestbreeding cichlids tend to be individualistic, i.e. territorial, sedentary, and aggressive toward all intruders. Mates often build strong and permanent bonds and polygamy is rather rare. Mouthbrooding cichlids do not control a piece of ground as much or as permanently as nestbreeders. They tend to be less individualistic, territorial, and sedentary. Their aggressiveness is limited more to conspecific males. They are most often polygamous, and sometimes even polyandrous.

8. Mouthbrooders do not permanently occupy the deep recesses that are dominated by non-cichlid species and a few nestbreeders. At ground

level nestbreeders dominate mouthbrooders, because they control the bottom. Thus mouthbrooders have to spawn without a site, or when they have one it is only temporary. The nest of nestbreeders plays an essential part in the success of their reproduction. For mouthbrooders it is a prop of the mating ritual at best. Mouthbrooders dominate the midwater column off the slope.

9. All fishes on a rocky coast breed where they live and juveniles of all species have to insert themselves into the biotope once the parental protection stops. Thus the problem of securing the safety of the fry at this dangerous step in a crowded biotope is a very difficult one to solve. It has led to a wide variety of behaviors. In some cases this problem might even have had an impact on the distribution of a species between the biotopes.

As the fry of the wild *Lepidiolamprologus elongatus* grow they have to insert themselves into the biotope once the parental protection ceases. In an aquarium the aquarist provides protection even beyond this stage. Photo by Doris Scheuermann.

Taxon	Social Behavior		Ecology		Sedentarism		Breeding mode	
	Gregarious	Individualistic	Open	Shelters	Territorial	Vagrant	Buccal Incubation	Nesting
SAND								
Callochromis	x		x			x	x	
Xenotilapia	x		x			x	x	
Cardiopharynx	x		x			x	x	
Aulonocranus	x		x			x	x	
Limnotilapia	x		x			x	x	
Lamprologus								
and Lepidiolamprologus								
attenuatus		x	x			x		x
pleuromaculatus		x	x			x		x
tetracanthus		x	x			x		x
ocellatus	x		x			x		x
brevis	x		x			x		x
modestus		x	x			x		x
callipterus	x		x			x		x
cunningtoni		x	x			x		x
Boulengerochromis	x	x	x			x		x
Perissodus paradoxus	x		x			x	x	
Limnochromis auritus	x		x			x	x	
ROCK								
SCHOOLING FISHES					Static			
Ophthalmotilapia	x		x		x		x	
Cyathopharynx	x		x		x		x	
Cyphotilapia	x		x		x		x	
Cyprichromis	x		x		x		x	
Lamprologus brichardi	x		x		x			x
LONERS								
Eretmodus		x		x	x		x	
Spathodus		x		x	x		x	
Tanganicodus		x		x	x		x	
Tropheus		x	x		x		x	
Simochromis		x	x		x		x	
Petrochromis		x	x		x		x	
Lobochilotes								
furcifer		x		x	x			x
savoryi		x		x	x			x
schreyeni		x		x	x			x
prochilus		x		x	x			x
leleupi		x		x	x			x
niger		x		x	x			x
obscurus		x		x	x			x
Julidochromis		x		x	x			x
Chalinochromis		x		x	x			x
Telmatochromis	x		x		x		x	
Perissodus microlepis		x	x			x	x	
straeleni		x	x			x	x	

The female *Telmatochromis dhonti* (formerly *T. caninus*) generally guards the immediate vicinity of the nest. The male will patrol the border area and act as the first line of defense. Photo by Hans-Joachim Richter.

Tilapia mariae with two-week-old fry. Predation is heaviest in the wild on the younger stages of the species. Parental guarding is exercised at least until the young have a fighting chance at survival. Photo by G. Marcuse.

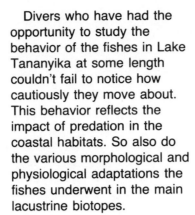

Divers who have had the opportunity to study the behavior of the fishes in Lake Tananyika at some length couldn't fail to notice how cautiously they move about. This behavior reflects the impact of predation in the coastal habitats. So also do the various morphological and physiological adaptations the fishes underwent in the main lacustrine biotopes.

Fear and wariness are more often displayed than any other behavior. Other moods, such as those linked to sexual activity, are more readily perceived by a diver because they are more spectacular—the fishes often being in their best attire—and perhaps also because the diver has been trained by the study of captive fish to look for the breeding behavior. But altogether, along with conspecific dominant/submissive social contacts, breeding behavior occupies only a small part of the overall activities going on in a habitat.

Fishes in the wild spend much of their time looking for food after which they rest (by daytime if they are nocturnal, at night if they are diurnal), sometimes in a deep trance-like slumber. They are at that time barely aware, if at all, of what is going on around them. Still they remain linked to a subconscious alarm system that will trigger uncontrolled evasive actions.

Fear in all its degrees, caution and awareness, fright and panic, will always remain the background of their life. It will put a stop to any of the other activities in which the fish is engaged, even the last the fish are likely to abandon—looking after their spawn.

The impact of fear on fishes has been overlooked by many ichthyologists because most of their observations could only be conducted on captive fishes. In a tank the sampling of fishes living together includes only at best a few individuals chosen for their relative compatibility and of course their natural behavior is distorted or muted. This is more true with fishes having an elaborate ethology (like cichlids) that are capable of adapting to the artificial conditions of a tank, even when they are a far cry from what they used to live in.

They respond to the crowded conditions and the promiscuity by establishing hierarchies and feeding and pecking orders, and display permanent aggressiveness in stereotyped rituals that the inferiors cannot escape and that in natural surroundings would have been very fleeting and less destructive. For example, they fight for food at preset times when in the wild they would have fed whenever they chose.

The oversimplified and distorted social relationships in captivity provide very little insight into the impact of fear in the lacustrine biotopes. The *modus vivendi* reached by fishes in a tank does not reflect the reality of the biotope they elected to colonize in the lake.

The intricate relationships between predators and prey are as yet poorly documented, but we can already reach some conclusions if we look for clues provided by the morphology and behavior of the fishes in their various habitats.

A predator roaming in open water in quest of its prey has

specialized in ways to be successful (and the prey likewise) other than a predator lying in the rubble waiting for a prey animal to pass by (and the prey to frustrate the attack). There is thus more similarity between predator and prey from a similar biotope than between two predators living in different biotopes.

Specializations call for morphological, physiological, and behavioral adaptations.

PELAGIC WATERS

The open waters of Lake Tanganyika constitute a homogeneous biotope whose main features are:

1. A total lack of shelters for fishes to hide in.
2. Lots of space in which to move at will.
3. Exceptional transparency of the water.
4. A thick layer of well oxygenated water in which temperature is remarkably stable.
5. The quiet of the water column as soon as one leaves the surface, which is undisturbed by strong currents.
6. One might add the various climatic conditions prevailing from one end of the lake to the other and the staged bloom of the plankton.

Let us start with this last feature.

Clupeid concentrations in large schools are linked to local concentrations of the plankton on which they feed. Because of the staged plankton bloom, plankton-eating fishes (and their predators) had to undertake long-distance forays in search of the planktonic clouds. To do so safely in open waters they

had to bunch together as well as to gain speed and stamina. Bodies are slender and laterally compressed, the tail fin is powerful, and the other fins are used not for propulsion but for equilibrium. Hydrodynamism has been increased by fragmentation of squamation in order to reduce drag and make the body as smooth as possible. The mouth is always large and the teeth set to grab.

One might believe that in all likelihood, because of the place their relatives occupy in rivers, the lacustrine clupeids were the first fishes to colonize the open waters of the lake. Probably at first they were not followed by other species, which still had to adapt to the new ecology, so that clupeids could multiply fast (they did so quite recently, in less than 20 years after they were introduced by man into Lake Kivu).

Their occasional contacts with the shorelines and their predators increased in frequency. Progressively some of the predators left the coastal waters to follow their prey into the open and progressively became specialized pelagic species. Probably the ancestors of today's open water predators are to be found among fishes that were already roaming in the water column above the sandy floors.

The adaptations of the pelagic hunters called for a very fine tuning of the qualities that had brought them there in the first place.

As schooling and speed had to be maintained, every other behavior ran second and evolved from the gregarious instincts. Rigorous timing and

coordination during the moves kept the school together. Signals had to be emitted, received, and re-transmitted by the members of the school, although only very few were near any commotion and could assess its source. Thus pelagic fishes rush blindly to attack or flee, not because all have detected a prey or a predator, but because they perceive their neighbors' moves and duplicate them.

The larger a roaming school is the less its cohesion becomes manageable, especially so in time of stress. On the other hand, the smaller the fish in a school the more vulnerable they are. Small fishes, like the clupeids, tend to offset this handicap by bunching together more than larger fish do. So it happens that a school of clupeids in the lake barely 25mm long can number as many as 50 million fish and perhaps more. A school that size would be unmanageable and lose its cohesion had the clupeids not lost most of their capacity for personal decision making. Strays from the flock are thus doomed; this is more true if they belong to species in which gregarious bonds are the strongest.

Because most pelagic predators are also hunted by larger species, all had to adopt schooling to be able to roam in the open waters. There are thus no individual pelagic predators with the sole possible exception of the giant *Hydrocynus* (tiger fish) which are thought to be coastal.

The ecological features that were the background against which the fishes evolved are:

The *transparency* of the water, which led to the

Cyphotilapia frontosa is one of the deeper dwelling cichlids. Mouthbrooding females stay at the same level as the school, where there is increased safety. In fact, other females may descend to these depths while brooding eggs. Photo by Pierre Brichard.

development of good eyesight, efficient as well in the dazzling light of the surface areas as in the darkness prevailing at the bottom of the oxygen-bearing layer 250 meters deep. As the fishes move up and down through the water layers they must be capable of adjusting to the changing light conditions rather quickly.

This applies to the oxygen levels, which drop sharply within the layer, as well as to pressure. Pelagic species must be capable of adjusting their hydrostatic equilibrium quickly to the depth they reach. The fact that *Perissodus paradoxus* has been found as shallow as 2 or 3 meters deep and also at a 250 m depth tells us much about the capacity of its swim bladder to absorb punishing variations in water pressure.

The *stability* of the water column, i.e. being usually devoid of any strong turbulence, helped the pelagic species develop finely tuned sensory organs able to detect changes in pressure that might give advance warning to the approach of other fishes.

All pelagic species except for clupeids (which probably have other means at their disposal) are endowed with long lines of neuromast sensors on their sides. Pelagic cichlids in particular developed two lateral lines on each side numbering more than 100 sensors. Very seldom do coastal species develop as many, and then only when they live off the slope (in midwater). Coastal cichlids living in the surf-beaten and very turbulent area of a slope have only short, or even

atrophied, lateral lines. It is thus worth investigating the lateral lines of pelagic cichlids.

The lateral line sensory neuromasts are distributed in two parallel lines of scales running along the sides of the fishes. The upper one (superior lateral line) extends from the upper end of the gill slit to the base of the tail; the lower one extends from the tail base toward the abdominal area. They overlap more or less in this latter area.

The number of neuromasts in each line depends on the number of scales in the longitudinal rows. The scales may be small and many or large and few. The length of the lines also varies from one species to another.

The extent of overlap between the two lateral lines provides us with a clue about the development of this sensory system. Pelagic fishes, among them some of the cichlids in the lake, have long lateral lines with a very high number of neuromasts.

Pelagic species are always on the alert and jumpy. The smallest incident, such as a leaf falling on the surface or the shadow of a cloud passing by, is enough to trigger a panic in a school of clupeids. And well they might be, because if encounters between predators and their victims are very sporadic and far between, they reach a paroxism of apocalyptic proportions. In a few seconds predators have to make the most of a fleeting opportunity before the prey manage their escape.

In this respect *Lates* are the most spectacular. A school of these huge fishes, several hundreds strong and averaging perhaps as much as 50kg each, provides a never

Schools of predatory *Lates* roam the lake waters in search of prey. A 40 cm *Lates* can swallow a dozen or more clupeids whole. Photo by Heiko Bleher.

forgotten sight in its savagery. The two schools (predators and prey) merge in a maelstrom of frenzied activity. Jaws agape, the perches zigzag through the school of clupeids and gorge themselves. A 40cm long *Lates* can swallow whole a dozen or more clupeids, each 10 to 12 cm long, at one time and by the time the surviving clupeids have disappeared in the blue haze, wind up with more than a pound of the fish in its stomach.

In spite of their losses clupeids thrive in the lake because their fecundity is very high and their losses are quickly made good.

Moreover, fishes like *Lates,* unless they survive to reach a size that puts them beyond the reach of any predator, fall prey themselves to the other predators around. *Lates* suffer enormous losses during the first weeks after they hatch from the egg because their spawns are seasonally concentrated on a few breeding grounds around the lake. Their eggs and young fry are decimated by aquatic birds, fishes, and man, if not

by their own breed. Adult clupeids themselves will not be above eating *Lates* fry when they happen to find them in planktonic clouds.

The fact that preying in open waters depends so much on the chance detection of a school moving nearby gives full meaning to the well developed sensory organs of pelagic fishes.

Camouflage of pelagic species that lack any shelter in which to hide can delay or deflect the moves of the predator, which is important for very fast-moving fishes as soon as visual contact has been established.

Clupeids sport the plain silvery olive-green pattern typical of the family. As juveniles they are mere pale gray filaments, nearly totally unpigmented. When, during a dive, we happened to meet one of their schools, we realized what was happening only by the fact that all of a sudden the strongly contrasted background of a rock slope was progressively being blotted out in a kind of gray mist and we found ourselves engulfed in the midst of

thousands of the young fry.

The robe of the adults is very different. If the bodies of the young fish don't reflect any light, the adults with their silvery sides are constantly glittering in the sunlight. Seen from below they blend with the surface light; when detected from above their olive-green backs blend with the dark blue of the deep. The shimmering of thousands of silvery bodies flickers on and off as the fish are caught in the sunbeams playing under the surface and then disappear in the blue. When a school breaks in explosive panic, light is reflected everywhere by the shiny bodies and the enemies can only rush forward blindly.

The larger size of the predators gives them an advantage in top speed, and one might think that camouflage is less important for them. But predators fall prey to other predators and in a struggle whose outcome will be decided in a few split seconds between fishes with lightning-fast reflexes, predators need to be discovered as late in the game as they can manage. The ability to confuse an enemy when a delay can put the success of its maneuvers in jeopardy is essential. No wonder then if pelagic predators, each according to its ancestral background, have adopted camouflage patterns.

The lake *Lates* sport muted blotched color patterns reminiscent of the one displayed by their river relative *Lates niloticus.* Pelagic cichlids have two main patterns.

Bathybates and *Hemibates,* when under stress, have adopted patterns that are very similar to the ones found in marine tuna fish. The

Without doubt the tigerfish, *Hydrocynus vittatus*, is one of the fiercest predators in African waters. The teeth are well suited to capturing prey. Photo by Shuichi Iwai.

The garb of *Perissodus microlepis* makes it very difficult to spot in the greenish haze of a plankton cloud. Photo by Dr. Herbert R. Axelrod.

background color is silvery gray or light blue, often overlaid with dark blue vertical and horizontal stripes and spots in specific arrays, all of which might be found together on a fish at one time. One can imagine how these unusual patterns breaking the body outlines and endlessly repeated in an onrushing school can dazzle the prey.

Other cichlids, like *Perissodus* and *Haplotaxodon,* have a more discrete garb, usually plain yellow-green with a few glittering pale blue scales on the sides and a spot on the caudal peduncle. They are very hard to make out in the greenish haze of a plankton cloud, and the black spot on the tail might well serve as an eye-dummy to help confuse the prey.

The teeth of pelagic fishes are always sharp and set in multiple rows. Some species have their teeth hinged so that they lie flat when the mouth is shut and are raised when the jaws open. *Perissodus* have their teeth embedded in thick gums which retract and bare the teeth when the fish bites.

Let us conclude this review by emphasizing the extraordinary road taken by cichlids when they colonized the open lake biotope. Their evolution took them well beyond the path of river species.

Myers used to say that the teeth of *Perissodus* were supralimital. He meant by this that their shape was beyond the scope of what could be considered as normal evolution in cichlid teeth. What would he have said had he examined the teeth of *Perissodus eccentricus,* where the teeth in one specimen can be oriented toward the left and the next one toward the right of the jaws. When a school of *P.eccentricus* (known as scale-

rippers) attacks, some of the predators will hit the sides of their victims from the left, others from the right, as they pass by. In any case, as they try to elude their tormentors, the victims will always be in the proper position to be attacked.

In spite of the many gaps in our knowledge of the pelagic species, this example provides us with a clue regarding the exceptional sophistication that the struggle for life has reached in the open waters of the lake.

Preying in open waters, we might now conclude, is the work of very specialized schooling predators during brief, sporadic, and frenzied encounters with schooling prey, themselves very well adapted to the ethology of their predators.

SANDY BOTTOMS

Because of the breakdown of rock along the shores and the input of alluvions by rivers, sand plains cover most of the lake bottom. Underlining or fragmenting the coastal rocky biotopes, the sand habitats are by far the largest biotope of the lake. Few features break the monotonous expanses of the gently rolling plains or mounds on the bottom.

If the water column above the bottom is in many respects very similar to the offshore biotope, the presence of the sand bottom gives a new dimension to the ecological conditions of open waters.

We will thus find, as in pelagic waters, fishes living together and roaming over vast distances as well as fishes that live on the bottom. As the sunken plains extend from the coastal shallow habitat to the lower boundary of the oxygen-carrying water layer, we will find fishes on the bottom that live in shallow water only, other species living deeper, and some fishes living as well in shallow water as deeper down.

Predation on or by the various fish flocks—pelagic, benthic, semipelagic, and sand-dwelling—will thus be more complex than what we could observe in the offshore waters.

The ecological conditions will be mainly (1) bright sunlight in shallows with a gradual decrease in light toward the deeper layers; (2) oxygen levels high near the surface with also a gradual drop with depth; and (3) transparency usually good, except in front of river estuaries and over shallows.

Preying and defense against predation will be very similar to what happens in the open waters and the protagonists will use the same tactics that were developed by pelagic species, i.e. schooling and sporadic attacks by roaming and fast-swimming predators on elusive prey. This is also true because many if not most of the pelagic species' juveniles grow in coastal waters where they prey on local species. Thus we found *Bathybates* fry mixed with schools of *Xenotilapia*, and *Perissodus paradoxus* with schools of juvenile phytoplankton pickers like *Ophthalmotilapia*.

As the water column above the bottom is a natural extension of pelagic waters, pelagic species intrude into the biotope and are found along with schools of semipelagic species like *Aulonocranus, Cardiopharynx, Lestradea,* *Grammatotria,* and *Boulengerochromis,* which roam in midwater above the bottom.

These semipelagic species display, on a smaller scale, the same morphological adaptations as their pelagic counterparts. At ground level wandering species living on the barren plains display strong gregarious impulses and try to avoid detection by predators. They blend with the fawn-colored sand turning to gray with depth and have developed a morphology and behavior giving them a measure of security. Included are typical sand-dwellers like *Xenotilapia* and *Callochromis.* Aside from schooling and moving about they have little in common with the fishes that roam the water column above them.

They wander about in a sedate manner and, although not sluggish, are incapable of sustaining a high speed for long. Their best defense is to remain as unobtrusive as they can, in which they are also helped by their elongate and low profile and by their color pattern.

Their slender body throws very small shadows on the sand, which in shallows brightly lit by the sun is a definite advantage. Many fishes are revealed on a barren landscape by their shadow. Seen from ground level they are barely outlined against the greenish background as their scales transmit few reflections. From above they cannot be discerned against the sand as their colors do not strongly contrast with it.

Rock-dwelling species like this *Lamprologus leleupi* must be provided with a rocky environment with caves. Sand-dwellers would not prosper in such a setup. Photo by Hans-Joachim Ricter.

Sand-dwelling cichlids need open sandy areas in their aquarium. *Callochromis pleurospilus* will actually dive into and hide under the sand when danger threatens. Photo by Pierre Brichard.

They are mostly pastel-colored in shades of fawn, beige, and pink, the only bright colors displayed being yellows and light blue. All of these colors become grayish in the deeper water layers and the fishes are then even more difficult to detect. Over shallows they blend with the particles of sand dust lifted from the bottom in a golden haze by incoming waves.

All sand-dwellers have large eyes and are endowed with a well-developed lateral line system. *Xenotilapia* even has developed a third lateral line extending along the caudal peduncle, a very unusual feature among cichlids.

Callochromis pleurospilus avoid their enemies by burying themselves in the sand in a lightning-fast motion and can remain hidden for several minutes before reappearing on the bottom. *Haplochromis horei* also does this as well as one other cichlid from Lake Malawi.

As they roam over the sandy plain toward the deep regions, *Xenotilapia* will come into contact with *Trematocara*.

Oddly the latter have not as developed a lateral line system as the sand-dwelling fish. Theirs is even atrophied. But the fishes are endowed with another detection system consisting of rows of drum-like cavities pitting the facial bones, again a unique adaptation in the lake cichlids.

The camouflage of *Trematocara* appears to be linked with the depth at which most live. Their sides are silvery and the unpaired fins transparent except for occasional black stripes or spots. A deep-living *Xenotilapia, X. tenuidentata,* also has a silvery pattern with a few thin vertical stripes on the sides. Another deep-living species, *Ectodus descampsi,* has a salmon pink body and a large ocellus on the dorsal fin.

Who are the enemies of these fishes? We can only guess because observations have been few and far between and most clues are provided by the inventories of captures.

On the shallow plains solitary *Lepidiolamprologus,* like *L. cunningtoni, L.*

attenuatus, L. pleuromaculatus, Lamprologus tetracanthus, etc. are found. Then, deeper down, small packs of *Boulengerochromis* and of *L. callipterus* occur. Still deeper is the group of species related to *Limnochromis (L. abeelei, Gnathochromis permaxillaris)* about which we know very little as they live deep indeed. They have been found between 100 and 200 meters, too deep for us to visit them.

At night the sand-dwellers are preyed upon by the mastacembelid eels, slithering over the vast expanses of the sunken plains, and by foraging catfishes, such as *Auchenoglanis,* and perhaps large specimens of *Synodontis.*

One might think that the flat sand bottoms cannot be inhabited by dwarf species. But we find several among the smallest cichlids in the lake there. *Lamprologus brevis, L. kungweensis,* etc., have managed to occupy the sand bottoms, over which some of them are found roaming while others occur in small colonies in the *Neothauma* shell piles. The small species apparently manage to survive in a world crisscrossed by schools of predators many times their size.

Like *Trematocara,* the rare *Gnathochromis permaxillaris* has developed a very unusual detecting organ in the shape of a large scale on each side of the snout. This scale is perforated by a circular row of pits the physiology of which has not been investigated as yet.

How can we summarize predation on the barren sandy bottoms?

- The fauna is diversified into several fish flocks displaying a wide variety of adaptations to peculiar ecological conditions.
- Predation is more diversified than that which is found in open waters. There are solitary predators, but schooling predation still prevails.
- Predation is still sporadic, but less stereotyped than in pelagic waters.
- The availability of shelters on the bottom has allowed several very small species to survive on the barren expanses of the sand bottom. One has succeeded in using a featureless biotope for its protection by burying itself in the sand in case of danger. Others might have developed the same behavior.
- The development of unusual detection devices by cichlids living on the sunken plains is one of the major aspects of the ecology of this type of biotope.

COASTAL WATERS

As we rise along the slope toward the coastal shoals and look for predators and their prey, we discover a wide variety of fish flocks according to the type of shoreline: beaches, swamps, and rock biotopes.

Beaches

If the gradient of the slope is a gentle one, we gradually pass from the deep bottoms to the surf-beaten shoreline and wind up on a beach. The line of breakers harbors only a few fishes, but they are typical of the area.

Foremost among the predators is the cyprinid

Many species build their nests in the sand. *Boulengerochromis microlepis* build crater nests that they defend. This pair built their nest at 4 meters depth near sandstone slabs. Photo by Pierre Brichard.

Varicorhinus. These are swift, shark-shaped cruising predators that move in small packs seldom more than a dozen strong, more often by twos and threes. They are devold of any teeth, but a hard cartilaginous ridge on their jaws is enough to cut and crush their prey. They average 25cm (but often reach 60 cm) and are voracious feeders against which the juvenile fishes living close to the water edge have little defense.

A voracious cichlid, *Haplochromis (Ctenochromis) horei* is also typical of the area. Solitary, swift, and powerful, this largest haplochromine in the lake attacks on sight, after a very short but fast rush, the juveniles of other cichlids in the area, especially those of *Xenotilapia, Callochromis, Aulonocranus, Limnotilapia,* and of the semipelagic species.

Camouflage colors for the cyprinids and *H. horei* are a greenish yellow which blends well with the aquatic plant beds, of which there are many in the shallows.

Aulonocranus, which lives more in midwater, are among the most colorful fishes around, with a pale blue body and horizontal rows of yellow scales. *Aulonocranus* adults roam in schools over the brightly lit shallows.

The development of lateral lines on fishes living close to the beaches reflects their ecology. The lines are short in *Haplochromis horei;* they are long and above average in species of genera like *Aulonocranus, Cardiopharynx, Grammatotria, Lestradea,* etc. All these fishes have keen eyesight and large eyes, especially *Aulonocranus.*

Coastal swamps

Over shallow bottoms in front of a river mouth the incoming waves carry away the mud so that the transparency of the water can be very good between the stems of aquatic plants and reeds that grow on a clean sand floor. This is an unusual condition for a swamp. No wonder then if the lacustrine species enter the rim of the estuaries and mix with typical swamp-dwelling species.

Predation activities are thus complex as shelters and hiding places become available and ambushes increase. As we penetrate within the swamp visibility gradually drops to a few inches and the lacustrine species that depend essentially on sight for their survival are now facing the riverine fish flocks that depend more on olfactory detection.

Ambush predators dominate the area. Foremost among them are catfishes (such as clariids, bagrids, especially *Chrysichthys, Malapterurus,* and large *Synodontis*) as well as *Polypterus, Protopterus,* etc. They have poor eyesight but a keen olfactory sense with which to track their prey. They lie in the mud or hide among the plant stems.

Cruising predators, sometimes schooling, sometimes single, swim in midwater. These include large *Alestes,* large barbs *(Barbus),* and *Barilius.* Among them and often venturing into the lake, even along the rocky coasts, swim the most powerful and fearsome characoids in the world, the two species of *Hydrocynus,* most commonly *H. vittatus,* less often the giant *H. goliath.*

The first species reaches a length of about 80cm and is often found in the midst of a school of *Tilapia tanganicae.* The second reaches two meters in length and 80kg in weight, with interlocking, needle-sharp teeth 3cm long, and is the most spectacular freshwater killer. It has happened to one of us, while diving at Bemba, that when he lifted his head from the rock he was investigating   found himself staring into the cold eyes of an *H. goliath* nearly his size. The terrifying beast, displaying the frightening array of its barracuda-like teeth, remained motionless for several minutes investigating the intruder and wondering whether he was worth a bite.

Occasional attacks on swimmers or fishermen wading waist-deep in the shallows have been reported from the Congo River and always left the victim dead or maimed for life. A very large *H. goliath* in a single stroke can tear out several pounds of flesh and can swallow a 25 pound fish whole in one bite. They are incredibly swift and probably the fastest freshwater fish in the world. It is more than a match for any South American piranha.

The deeper we enter the swamps the less we discover lacustrine species. They are progressively replaced by the typical river fish flocks that are better adapted to the murky, poorly oxygenated waters. Predation there has become an individual activity calling for stealth and ambush, in which the river species excel.

Rocky Coasts

From pelagic waters to the swamps, we have seen predation diversify and become more complex. Morphological, physiological, and behavioral adaptations of the various fish flocks that are directly linked to predatory activities among predators (as well as among their victims) follow the same pattern. As we

Catfishes, like this *Malapterurus electricus*, are the foremost ambush predators of the lake. They usually have poor eyesight but a keen sense of smell.

reach the shores and slopes lined with rocks the diversity of the rock–dwelling fish flocks and of predatory activities reaches its zenith.

In strong contrast to the open waters or barren bottoms, where contacts between fishes belonging to different species are very sporadic, on a rocky slope they never stop and are a direct result of the prevailing promiscuity. As many as 100 different species can live together, some of them in very dense populations. Again, as with the sandy bottoms, we have to split the local fishes into two groups. The first involves the fishes living at ground level, amid the rubble. The second involves those fishes that hover in midwater off the slope.

At ground level schooling is very difficult, at least for wandering or roaming species, because the uneven surface of the rubble, towering piles of rocks, or large boulders, obstruct the view and hinder the transmission of signals (acoustical or visual) maintaining a school's cohesion.

With availability of many types of shelters suited for a wide variety of fishes, the problem associated with schooling has thus given an edge to the trend toward individualism and sedentarism over those leading toward schooling and gregariousness. Small species do not have to find safety in numbers.

On the contrary, the ecological conditions prevailing off the slope in midwater are similar to those met in open waters (space, lack of shelters, plankton clouds) and the fishes there have to school, even when they don't move much.

Because of the promiscuity, the fishes living on the bottom or in midwater are always in visual contact with each other, and many of the neighbors are either predators, and as such specialized, or potential predators if given a chance, and are more or less unspecialized with respect to predation. On a rocky slope there are relatively few "professional" killers, and many "amateurs" will prey only under a specific set of circumstances and only on certain prey.

Another point is that on a rocky slope all of the fishes breed *in situ.* Even mouthbrooding cichlids must release their fry into this habitat at the end of the breeding cycle. The rubble is a nursery for the growing fry of the local species, except for a few fishes, living off the slope, that release them into the water column nearby (*Cyathopharynx* breeding guild).

Thus, beside the swamps, about which we know too little to know the answer, a rocky habitat is the only one in the lake where adults and their brood live together.

Many fishes that are not predatory on other adult fishes will voraciously swallow their eggs and fry. I don't know of a single fish in Africa that will miss the opportunity to gobble up the eggs of another species, or even its own. I have seen *Ophthalmotilapia,* a phytoplankton picker, *Tropheus,* a biocover grazer, and even elephant-nosed mormyrids grab fry.

The simultaneous presence on a slope of specialized predators and of the many occasional predators has had a double impact:

-It has a cumulative effect on the frequency of the attacks.

-The multiplicity of methods used by predators, according to their nature, has prevented the development of stereotyped patterns of

Chrysichthys species are also ambush predators. Some of the species grow quite large. This is only a juvenile. Photo by Dr. Herbert R. Axelrod.

predation against which the prey could reply by developing a simple defense. From this aspect has stemmed a wide variety of morphological and behavioral adaptations found in rock-dwelling fishes.

Finding an adequate defense is even more difficult for rock-dwelling fishes because the rubble shelters such a large population of predators that are not found in such a high density, if at all, in the other biotopes. Among these must be included non-cichlid species, foremost among them being the

reaching barely 20cm; bagrids by *Bagrus docmac,* reaching 100cm, and by *Lophiobagrus,* not exceeding 5cm; *Mastacembelus* by *M. moorii,* growing to more than 80cm, and *M. platysoma,* not exceeding 12cm. Thus all hiding places, from the largest cave to the narrowest crevice, are within reach of at least one predatory species.

-Most of the non-cichlids are nocturnal, but some of them prey by daytime; thus they take turns in their predatory forays.

-Non-cichlids are often so

1. There is an abundance of shelters and the broken relief of the rubble brought about conditions at ground level leading to individualism, sedentarism, and promiscuity between a variety of fish-flocks. Their density, the promiscuity on the same grounds of adult and young fishes, and the overlapping cycles of nyctemeral activity between cichlids and non-cichlids are another set of components leading to around-the-clock predation.

2. The water column above the bottom is a different kind of habitat whose ecological conditions are very similar to those prevailing in open pelagic waters. There fishes school together, but as the schools move about only a little, fishes of different species are very often in contact with each other and with their fry. As a result, fishes on a rocky slope never live in peace, are always menaced, and have to be watchful to deflect or blunt an attack by a predator.

Many catfishes are nocturnal predators, so the danger to the cichlids continues 24 hours a day. This is *Auchenoglanis occidentalis,* which feeds mostly on items (including fish eggs and fry) that it sucks up from the bottom.

catfishes, such as *Bagrus, Chrysichthys, Auchenoglanis, Heterobranchus, Dinotopterus, Tanganikallabes, Lophiobagrus, Phyllonemus, Synodontis,* etc., not to forget *Malapterurus* and the spiny eels. They bring a new dimension to predation on the slopes for several reasons:

-Most of the families are represented by giant or large species and dwarfs. For example clariids are represented by *Heterobranchus,* reaching 150cm, and *Dinotopterus,*

large that no fish, especially those among the cichlids, can be said to be free from their activities. But the preferred habitat of the many non-cichlids around is rubble; they prey at ground level and not in midwater. Cichlids living off the bottom in schools would thus be free from predation by non-cichlids were they not coming back to the bottom at night to rest.

If we summarize the ecological conditions presiding over predation on a rocky coast we might say that:

Predators use different methods in order to secure prey. Headlong chases by specialized predators prevail in midwater, while ambush or a stealthy approach prevail among the rubble, where unspecialized foragers complement their diet with incidental predation on weakened or young fishes.

Occasional predation is very common, especially so among cichlids because so many have a carnivorous diet, even if it consists mainly of water "bugs" (crustaceans), such as shrimps, copepods, and cyclops.

Many zooplankton or zoobiocover pickers will not disregard the opportunity to prey on another fish's spawn. Omnivorous species like *Lamprologus modestus* and the two largest *Telmatochromis* species will look around more systematically for prey as they forage among the rubble.

But occasional predators, being less systematic in their approach to predation, will display less boldness and give up an attack more quickly than specialized predators. Numbers play an important part in the brazenness displayed by occasional predators. Two or three fishes belonging to different species eventually will display more individual aggressiveness than if they were single, because they compete.

Any uncommon behavior by a fish, like the jerky moves of a wounded or sick fish or, to the contrary, its apathy, attracts the attention of neighbors. Typical in this respect are the foraging *Lamprologus* species like *L. modestus, L. mondabu, L. tetracanthus,* and foremost among them *L. callipterus,* the only predator on the rubble roaming in small, fast packs. Timid and easily overcome by a show of resistance when single, they become very aggressive and even rash when their pack circles around a prey.

The combination of efforts by occasional predators, even from different species, is what brings about the downfall of so many spawns when the stubborn defense of their nest by the parents is overwhelmed. Even fish as powerful as *Boulengerochromis* pay a very

heavy price to predators when young.

DEFENSIVE MORPHOLOGICAL ADAPTATIONS

The lake fish fauna, especially cichlids, and among them nestbreeding species, offers a unique opportunity to discover adaptations that answered pressures born from predation. We can compare, for example, the lacustrine *Lamprologus* with their river relatives from the Congo basin.

Basically, fishes can

Hydrocynus vittatus, attaining a length of up to 80 cm, may commonly be found in a school of *Tilapia tanganicae*. Photo by Heiko Bleher.

respond to predation in two different ways. They can try to flee or they can develop passive defenses. Which means that they can develop streamlined, often torpedo-shaped bodies endowed with lightning-fast reactions, they can develop spines, or they can cover themselves with stronger scales that will increase the difficulty for a predator to grab them.

The scales of the first group (those that flee) are smaller and smooth, and everything that might increase drag has

been discarded. On the other hand, fishes belonging to the second group have deep or plump bodies and high fins, often with flowing filaments that slow them down even more. To the first group belong the cruising *Lamprologus* and *Perissodus,* to the second belong several *Lamprologus,* like *L. moorii, L. compressiceps,* or *L. brichardi.*

A fish like *L. moorii* is rather slow, moves about very little, but swims off the bottom in full view of the many predators around. It can do so because it has a rather formidable temper and is aggressive, but also because it is one of the most "armored" fishes in its biotope. The scales are large and bristle with several rows of raised denticles at their posterior edge. The scale-plating extends along the rays of the dorsal, anal, and caudal fins, so that they are covered with a strong sheeting. The scales on the back ascend along the base of the dorsal fin so that it can fold safely into a hard and well protected groove. So does the anal, which has a higher than average number of spines that are very strong and long. The fish thus has powerful static defenses and can hold its own against most predators, even though it is slow moving and lives exposed in the open.

L. calvus and *L. compressiceps* have added to these rather formidable defenses with a few adaptations of their own. Foremost among them is their capacity to curl their body sideways so that their tail reaches their snout. In this position the scales of the outer side of the body project from the body curve and their bristling denticles are fully raised. The fishes then become a very unpalatable ball of spikes, something that even the hungriest predator might think twice about before grabbing. They are also difficult to swallow because of their shape.

The capacity to raise the scales away from the body is common among cichlids, but some species appear capable of raising them higher than others. The size of the denticles, their shape, and their number also exhibit a wide variety. Riverine *Lamprologus* in this respect are much less remarkable than the most typical rock-dwelling *Lamprologus* in the lake.

There are, of course, different levels in the two main trends (toward speed or toward armoring), and some of the *Lamprologus* species, especially *L. callipterus,* are very mobile fish, although their sides are covered with the bristling spikes of strongly ctenoid scales. Aside from this feature they don't display any armor on their fins, nor do they have many more spikes in their anal fin than is average in the genus.

In midwater, where most predators and prey animals are excellent swimmers, *L. lemairei* displays a trend toward armoring. Contrary to the form of the cruising *Lamprologus,* this species is very short and stocky with an enormous head and cavernous mouth, more or less the counterpart among nestbreeding cichlids of the mouthbrooding *H. horei.* Hovering in midwater and not moving much, it has developed strongly ctenoid and very erectile scales, but the fins are left unprotected by the scale plating. Were they so protected the fish would have lost its capacity to "jump" its prey.

Among the mouthbrooding species living on the plankton clouds drifting off the slope, we find the same ambivalent trends toward speed or static defense. *Cyprichromis* species are slender and can move fast while *Cyathopharynx* and the other related genera are plump fishes whose body sprouts banner-like filaments that can only increase the drag when they move and thus slow them down. But none of them display the armor plating due to strong squamation that appears to be exclusive with the *Lamprologus* and related genera.

PHYSIOLOGICAL ADAPTATIONS

If *Lamprologus, Julidochromis, Chalinochromis,* and *Telmatochromis* have increased the number of spikes and reinforced the protection given by their scales, all of them have also developed the sensory organs that the riverine relatives use

Lamprologus compressiceps has some unique defensive adaptations, including curling up head to tail to resemble a ball of spikes. Photo by Glen S. Axelrod.

to detect what is happening around them. In this respect the coastal species of nestbreeding cichlids appear to have a more sensitive lateral line system than the mouthbrooding species occupying the same part of a rock biotope.

The time has now come to investigate this system more in detail. Each scale included in a lateral line system is perforated in its center by a pit. A tuft of uneven hairs anchored at their base in neuromast cells grows in the pit and is in contact with the water around the fish's side. The hairs bend under the pressure of turbulence in the water and their movements activate nerve cells that then emit weak electric impulses. The impulses are transmitted via the nerve underlining the lateral lines toward the inner ear and the brain. There they are decoded and provide the fish with a sensation of anything that is disturbing the normal flow of water along its sides and to the rear, beyond its sight.

The scale pits in cichlids are normally covered with a tube or tubule bisecting the scale more or less horizontally. It is open in front and sometimes at both ends. It is in these that the tuft of sensory hairs lies protected.

Although pored (or open pit) scales have been reported on in some *Lamprologus* species (by M. Poll) only tubuled scales have as yet been included commonly in the lateral line counts that are given in taxonomic determinations of cichlid species; pored scales have usually been overlooked or disregarded.

During a systematic investigation of the morphology of the lake *Lamprologus* and their related genera I discovered several disturbing features in their lateral lines. In each species the fry first start with all sensory scales of the open pit type in both the upper (superior) and lower (inferior) lines. After a few weeks at most tubules start to appear and grow first on the upper line, at its anterior end near the head, and they spread progressively toward the posterior end of the line. Then tubules appear on the lower line, but at its posterior end, near the tail first, and they then spread toward the head and progressively cover the open pits. The two lines of tubules thus grow in opposite directions.

When the tubule building activity reaches the posterior end of the upper line, tubules on the lower line stop growing as well. But as the tubules appeared on the lower line later than on the upper line, all open pits in the second line are not yet covered with a tubule, leaving several pored scales in front of the tubules. Depending on the species, their number can go from 10 to 25, often outnumbering the tubules on the lower line.

The tubule activity can stop before the end of the first line is reached and in this case as well a few pored scales are to be found behind the row of tubules. In this case the total number of sensory scales (pored + tubuled) is rather constant in a given species, although the recorded count based on the tubules alone

Many species of the nestbreeding *Lamprologus* have better developed sensory organs than species of some other genera that are mouthbrooders. This is *L. tetracanthus* with young. Photo by Glen S. Axelrod.

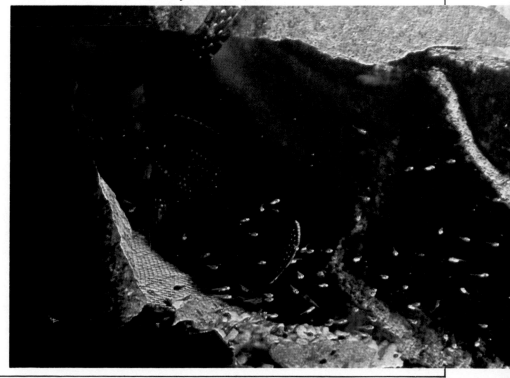

can show wide discrepancies.

In most mouthbrooding cichlids from Lake Tanganyika (with the sole exception of the *Cyathopharynx* guild) the lateral lines are composed exclusively of tubuled scales, and occasional pored scales can be considered as individual accidents, not a systematic feature.

If one compares two cichlids, one a nestbreeder the other a mouthbrooder, sharing the same part of a rocky slope (like a *Julidochromis* and a *Tropheus*) one discovers that the tubuled lines are proportionately of the same length.

However, if one adds the pored scales to the lateral line count in *Julidochromis* we can conclude that the lateral line system of the nestbreeder appears to be much more developed than the one in *Tropheus.* This holds true when we systematically compare all the cichlids living near the bottom from the surface to the depth as yet explored.

Off the bottom, if we compare the species belonging to the *Cyathopharynx* guild with the cruising *Lepidiolamprologus* that share their habitat, we discover that they have about the same number of sensory scales and that the first group includes species that, as mouthbrooders, are the only ones to have open pit scales in a long row in front of their tubuled line. The length of the lines, especially that of the lowest, appears to be more or less linked with the station occupied by the species in respect to the rubble and the depth at which they live.

The *Eretmodus*-group have very short lateral lines, which

might be due to the fact that they live in the very turbulent water of the surf-beaten slope. If we examine fishes living in the recesses, and there are only nestbreeders there, we discover again that the lower line has many pored scales. If we look into the shell-dwelling species, like *L. brevis* and *L. ocellatus,* we discover that these fishes, reported to have an atrophied lower line, have in fact few tubules but many pored scales and that their lateral lines are not as short as was reported in their determination.

When we turn back to riverine relatives of the two cichlid groups we again find that the riverine *Lamprologus* have pored scales in addition to their tubules; but riverine mouthbrooders, such as *Astatotilapia burtoni* and *A. bloyeti, Astatoreochromis straeleni,* and the few species living in the lake like *H. horei* and *H. benthicola,* lack the pored scales.

If we consider again the sensory scales of the lacustrine nestbreeders we find an amazing variety in their morphology. Depending upon the species the lateral lines can be uninterrupted or broken. Some sensory scales might be missing in a row or a few tubules might be replaced by pored scales, as if the growth of the tubule had miscarried. In the row of open pits a few tubules might appear as if there had been a last spurt of tubule-building activity. Individual specimens often display such "malformations" but some species appear to be systematically unable to build a complete and normal lateral line.

The shape of the tubules or

of the pored scales is not uniform and there are specific shapes. The tubules can be pear-shaped or cylindrical or be very protuberant or deeply embedded in the scale tissue. The pits can be wide open and crater-shaped or very narrow openings. Most usually there is only one pit on a pored scale, but several species have more than one and *L.leleupi longior* can have as many as four!

Considering that there are so many variations in the size and shape of the neuromast scales, that the two lines grow in opposite directions and not simultaneously, and that there appears to be a link between the development of the lines and the location of its ecological niche, it is strongly suggested that the systematic study of the lateral line system is long overdue and that it might reveal fascinating aspects of the nestbreeders' adaptations to their biotopes.

In another still poorly investigated field, the cephalic sensory network, more adaptations appear to have been developed in nestbreeders than in most mouthbrooders. All *Lamprologus* and related genera have either a pitline of sensory neuromasts or a sensory canal with gaping holes below the eyes. They have similar holes on the jaw that gape more or less wide open according to the species. Several appear also to have a line of neuromasts extending across the nape of the neck.

The variability of the infraorbital openings and of the other cavities again strongly suggests an array of adaptations in organs that might be vital to the fish. Unfortunately, as with the lateral lines we will have to

Lamprologus brevis may have few tubules but many pored scales in the lateral line. The lateral line is not as atrophied as some people have reported. Photo by Hans-Joachim Richter.

wait for a thorough investigation before we can assess its importance.

CAMOUFLAGE PATTERNS

Rock-dwelling fishes swim against a very different background depending upon whether they live in midwater with its blue-green haze, against the mottled scenery of the rubble, or in the deep and dark recesses of the lake. Camouflage patterns are thus displayed in a variety of patterns and colors.

Fishes living in midwater often sport an olive, light blue, or blue-green shade with few contrasting stripes. The cyprinodont *Lamprichthys* female has a color pattern reminiscent of the clupeids—blue-green on the back and very silvery on the sides. The male is decorated with several rows of sky blue scales making him one of the most gorgeous fish in the lake. Male *Cyathopharynx* have a dark blue, jet black, or burnt sienna body that can turn entirely iridescent at times of excitement. *Ophthalmotilapia ventralis* has a background color that can be white, very pale blue, cream, or very pale gray. *O. nasutus* males have a variety of background colors depending on the place in which they are found. They can be vermiculated dark olive on a yellow background, entirely gray, or bright orange with the tassels at their pelvic fins pure white, lemon yellow, or bright orange.

The females of all these fishes are always unobtrusive and pale silvery gray, with no other colors other than darker blotches or vertical stripes.

Once again we might think, as is the case for the riverine *Astatotilapia* and

Astatoreochromis straeleni, that the gaudy pattern of the males makes them the more obvious target of predators swimming in the area and that the females carrying the progeny of the species escape their attention.

In the deeper layers the garb of the species living off the slope follows two different trends. We find first that the *Cyprichromis* males are again much more colorful, especially their finnage, than the females. But both genders carry a slightly contrasted dark pattern in shades of bronze, copper, and browns that make them poorly outlined in the twilight. On the other hand, the banded blue and pale bluish white pattern of *Cyphotilapia frontosa* makes them the most visible fish in their habitat. The pale bands reflect whatever light remains in the area.

The banded pattern of *Cyphotilapia* is shared by other fishes, among which are *Lamprologus sexfasciatus, L. tretocephalus,* and juvenile *L. savoryi.* The fact that the blue and white banded pattern is shared by mouthbrooding and nestbreeding cichlids appears to give special value to this type of camouflage in midwater as well as against the mottled background of rock walls.

Which reminds me of the day when as a fish collector in the Congo Basin I found a *Lamprologus* —a single specimen among the several million fish I had collected— that had exactly the same colors and pattern of a *L. tretocephalus* in the Stanley Pool. Although the unique specimen of this unidentified species was subsequently lost, it was certainly not one of the lake fishes, but a riverine

species having adopted the same kind of camouflage.

The crimson red color of female *Haplochromis benthicola* and the bright yellow color of *L. leleupi* have the opposite effect of camouflage when these fishes venture to or come to live on the sunlit bottoms of the upper part of the slope. But the brightness of these colors disappears in deep water, where most live, the *Haplochromis* turning a dark brown, the *Lamprologus* a dirty gray-beige.

There are no definite color patterns among the rock-grazers. *Simochromis* are generally pale in color, varying from pale blue or steel gray to olive green with faint bands, spots, or stripes. Few of them are found deeper than 5 or 6 meters and one might think that this pale camouflage, much lighter in shade than that of *Tropheus,* suits them well in sunlit shallow water.

Tropheus, in their many local varieties, display strong contrast between the dark and light colored areas. Fright mutes the contrasting colors by darkening the lighter areas. These also contract in size.

In the southern races the thin vertical light-colored stripes typical of the carefree mood of the fish and the gaudy patches of color displayed by the fish when they are courting disappear when they are frightened. In northern races the bright bands of red, yellow, or orange and the oval yellow or green dorsal patches of *T. brichardi* contract and become suffused with black or dark gray. Darkening of the pattern helps *Tropheus* hide in dark recesses.

Petrochromis also have a wide variety of color patterns

The male *Lamprichthys tanganicanus* is decorated with many rows of sky blue spots making him a gorgeous fish. In the natural environment, however, it is not so very conspicuous. Photo by Aaron Norman.

and one might say, perhaps, that the species most often seen swimming in the open display shades of green, yellow, and blue, while those like *P. trewavasae* that live more in the darker recesses have predominantly dark patterns.

The small biocover grazers of the *Eretmodus* guild have pale beige and golden yellow colors suffused with light blue spots, as might be expected from fishes living in the dazzling light reflections of the shoreline pebbles.

Beige in various shades is probably the most common color of the nestbreeders living on a rocky slope, especially among foraging species and cruising *Lamprologus.* Among the former we can include *L.*

modestus, L. mondabu, L. tetracanthus, L. callipterus, L. brichardi, L. petricola, L. mustax, L. falcicula, and *Chalinochromis;* among the predators are *Lpd. attenuatus* and *Lpd. pleuromaculatus.* But we also find a few fishes that stand out in their habitat as if they didn't need camouflage, such as the jet black *L. moorii.* In the dark recesses the colors are also very dark, and fishes like *L. prochilus, L. furcifer,* and *Julidochromis* are often difficult to discover under the rocks.

There is also a variety of shy fishes that do not come out often from the recesses, like *L. obscurus, L. niger, L. schreyeni,* and *L. savoryi.* The trend toward a pattern well adapted to the specific ecology

is so well defined among the nestbreeders that one can say by observing the normal shade of the pattern whether a fish lives more in the open or, on the contrary, prefers to stay in the dark passageways of the rubble.

On the other hand, there are some of them that, instead of camouflaging themselves at all times, abandon their camouflage, for example when they are looking after their spawn. An example is the *Lpd. elongatus* pairs that are silhouetted by the darkening of their pattern against the background of open water. Dark shapes are most commonly given a wide berth by other fishes.

Dark colored fishes such as this *Lamprologus furcifer* are often difficult to spot in the dark rocky recesses or caves. Photo by J. Shortreed.

SCALE-RIPPING AND STEALTH

After a look at the adaptations of ambush predators in the maze of the rubble and at ground level, the sporadic predation by unspecialized foragers, and the long distance chases of midwater predators, we now reach a special type of predation that is very common on the rocky slopes—scale-ripping.

Scale-ripping perhaps evolved from the habit of many fishes of attacking wounded or diseased fishes, tearing from their body chunks of skin and meat and to dispatch them peacemeal. Scale-rippers digest the skin and the soft chunks of flesh ripped away along with the scales, but cannot dissolve the bony structure of the scale with their stomach juices. *Petrochromis*

often are skin-grazers when they are hungry and I have seen riverine fishes, such as several *Synodontis* species, scrape each other to death. I remember having once had a young *Distichodus sexfasciatus* barely 4cm long in my tank that took to nibbling at the skin of a 20cm long prize *Synodontis angelicus.* The pinpricks couldn't possibly hurt the catfish much, or so I thought, but after perhaps 15 days it died from its multiple wounds.

The teeth of all scale-rippers in the lake are very specialized—the leaf-shaped crown is set transversly at the end of a long stem. The teeth remain buried in the gums as long as the jaws are closed, but as soon as a *Perissodus* opens its mouth to bite the gums retract, baring the pickax-like fangs. The tooth

blades of the opposite jaws lock around the edges of a scale and as the jaws close and pressure is applied on its edges the scale bulges and then pops out of its seat.

Sometimes two or three scales are ripped off in one stroke, leaving a broad wound on the victim's side. This might attract successive attacks by scale–rippers or scavengers like *L. callipterus* and the weakened fish will eventually succumb under the onslaught.

P. microlepis is probably the most common long-distance predator in most rocky habitats. It doesn't move much but hovers above the ground watching for anything that might lead to an attack. It will often disregard several nearby fishes and rush at another fish several meters away. Seldom have I seen one miss its victim, but it is not always

The densely packed tricuspid teeth of this *Petrochromis polyodon* indicate a biocover scraper or picker. Photo by Glen S. Axelrod.

The teeth of *Petrochromis polyodon* are usually visible at all times, no doubt some modification for its feeding behavior. Photo by Dr. Herbert R. Axelrod.

certain that it can succeed in getting a scale loose. It is certainly one of the swiftest fishes to be seen on a rocky coast.

Juvenile *P. microlepis* swim in loose schools amid juvenile swarms of *Lpd. elongatus,* other cruising predators and the *Cyathopharynx* group of species. The juvenile *Perissodus* look very much like the young *Lpd. elongatus* on which they appear to selectively prey (at this stage of their life). But as adult fish, contrary to what has been reported, they appear to be less selective, and few *Lpd. elongatus* carry the scars of the scale-ripper's attack although the density of *P. microlepis* is rather high (one might say at least one per 100 square meters of bottom area). In my opinion they attack at random, as opportunity dictates their choice.

The fact that a scale-ripper's morphology and color pattern are very much like that of its victim's is not a unique adaptation.

Another scale-ripper, *Perissodus straeleni,* uses mimicry and stealth to such a point that it can be considered as the epitome of the sophisticated adaptations reached by this very special group of cichlids. It stands apart from the other fishes in the genus by its squat and high body, all the others being streamlined and torpedo-shaped. Thus, *P. straeleni* cannot hope to match their speed and has to rely on stealth to approach its victims at close range.

Probably no other freshwater fish in the world can match *P. straeleni* in its talent for mimicry. The basic colors of the fish are dark beige or pale chocolate suffused with glittering sky blue scales covering the whole body. As the scales are very small the overall impression is that of a brown body covered by a bluish haze. According to the mood of the fish one of the two colors prevails. There are a few darker broad vertical bands on the sides.

The capacity of this fish to adapt its pattern to its moods or to the pattern displayed by its expected victim results from a complex set of changes that occur in its basic pattern. First, and in this respect *P. straeleni* is like many other fishes, the contrast between the bands and the background color can be enhanced or muted, resulting in a strongly banded or a plain pattern. At the same time the dominant background color can become light brown, beige, chocolate, jet black, light blue, deep blue, or pure white, and the glitter of the metallic sheen can disappear entirely. The width of the bands can increase or become narrower, or the bands can split into narrow stripes revealing the background color between, or the bands can disappear entirely.

P. straeleni seems to be capable, in fact, of adopting any pattern—plain, striped, or banded—involving shades of brown, blue, white, and black. Young specimens are more commonly seen in shades of beige or light chocolate, but adults seem to have the whole range of combinations at their disposal.

So far *P. straeleni* has been found with the pattern of *L. moorii* (entirely jet black), *L.*

Perissodus straeleni seems capable of adopting almost any color pattern —plain, striped, banded, etc.— but this pattern, including a black mask, is very unusual. Photo by Glen S. Axelrod.

brichardi and other *Lamprologus* (beige), *Cyphotilapia* (banded blue and white), and *L. tretocephalus* (same pattern as *Cyphotilapia* but including also the dorsal fin).

P. straeleni breed in pairs around 25-30 meters deep, a depth at which the young *Cyphotilapia* live. Their juveniles can thus mix at an early age with the schools of their victims. Very large *P. straeleni* are only found within the schools of *Cyphotilapia* or alone and as yet have never been found near smaller species that might be frightened by their size.

This scale-ripper can thus approach a wide range of species and remain in the midst of their concentrations without being detected. When collecting *Cyphotilapia* 25 meters deep we could only sort the *P. straeleni* out when we had them in our hands by the shape of their jaws.

In many respects, i.e. morphology, breeding behaviors, and their predatory techniques, the scale-ripping cichlids of Lake Tanganyika offer a fascinating study.

BEHAVIORAL ADAPTATIONS TO PREDATION

The impact of predation on rocky coast dwellers is made clear by a number of behaviors that are directly linked to the struggle for survival, first and foremost being in the various ways fishes have developed in order to protect their spawns.

There is the awareness and caution that most display when they move about. For example there is the slanted swimming mode of *Julidochromis* when they are menaced from above, keeping one eye on the intruder and the other on the features of the rubble so that they might head directly for a hiding place. This is something that calls for special mental qualities. There is also the topographic memory of *Chalinochromis* displayed when they head unerringly toward their territory, avoiding obstacles, such as a net newly put in their path. Or the deceitful behavior of *Tropheus* when menaced from above by, let's say, a diver. Nothing in its behavior will betray that the fish has taken notice of the growing danger as it keeps browsing on a rock or zigzagging aimlessly around. Then, as we are ready to drop the net, all of a sudden the *Tropheus* dives under a rock and is gone. We discover too late that while it was zigzagging here and there in the open in a seemingly careless way, the fish had taken its bearings and was heading toward a shelter. This is how, during our first months at *Tropheus* collecting, we could catch only a few.

Deceit by predators like *P. straeleni* and perhaps also deceit by a prey like *Tropheus* are the first examples of sophisticated behavior linked to predation on the coastal biotopes of Lake Tanganyika.

Now to summarize the various aspects of predation on a rocky slope:

1. Instead of being sporadic as in open biotopes, predation is a permanent danger by day and by night.
2. Instead of using rather stereotyped long-distance chases, predators, either specialized or not, use a variety of means, such as ambush, long-distance strikes, accidental foraging, mimicry, and stealth.
3. Predation falls more heavily on the spawn and fry of rocky slopes than on any other biotope in the lake.
4. Most predation is done by individual fishes. Schooling predation in small packs has been used systematically by only one species so far discovered (*L. callipterus*). This is due to the fact that the availability of shelters and the relief of the rubble develop individualism.
5. Many species, but by far especially nestbreeding cichlids, have developed defenses against predation by means of detection organs as yet little understood or by static defenses such as spikes and plating.
6. In the same way nestbreeding and mouthbrooding cichlids have developed behaviors, some of them very unusual among fishes, directly linked to defense against predation.
7. As a whole, rock-dwelling cichlids have gone very far in developing means to protect their spawns and fry. Many of the breeding behaviors stem directly from the amount of predation on the spawns that the nestbreeding and mouthbrooding cichlids have to face.
8. Non-cichlids have as yet been poorly investigated, but the slimy mucus, apparently toxic, that *Lophiobagrus* exude when they are seized might be a harbinger of fascinating defensive adaptations among this group of fishes. As we said when we started this chapter, fear is the prevailing feeling.

267

CHART OF LATERAL LINES IN PELAGIC CICHLIDS
Number of sensory scales in each line given in % of scales in longitudinal line (from M. Poll)

Genus	Upper Lateral Line	Lower Lateral Line	Two Lateral Lines Together
Hemibates	95	45	140
Bathybates	80-100	45-50	130-150
Haplotaxodon	80-95	48-55	130-150
Perissodus	85	55	140

CHART OF LATERAL LINES IN SAND-DWELLING AND BENTHIC SPECIES
Averages per genus drawn from M. Poll and given in % of longitudinal line

Genus	Upper Lateral Line	Lower Lateral Line	Third Lateral Line (No. of Scales)
Xenotilapia	62-95	24-62	0-29
Callochromis macrops	94	54-60	–
Callochromis pleurops	73	27	–
Cardiopharynx	80	45	–
Aulonocranus	90	35	–
Ectodus	80	36	–
Trematocara	0-30	–	–
Boulengerochromis	70	50	–
Limnochromis	55-73	36-55	–

About 65% of the tubuled scales in *Tropheus* species are in the upper lateral line. The upper lateral line is very evident in this photo of a mottled copper/black color morph of *Tropheus moorii*. Photo by Pierre Brichard.

CHART OF LATERAL LINES IN ROCK-DWELLING CICHLIDS (from M. Poll)
Tubuled scales only. In % of longitudinal line

Genus	Upper Lateral Line	Lower Lateral Line	Two Lateral Lines Together
Cunningtonia	90	55	145
Cyathopharynx	95	85	180
Cyphotilapia	60	45	105
Eretmodus	75	35	110
Haplochromis benthicola	70	40	110
Lobochilotes	70	40	110
Limnotilapia	70	50	120
Ophthalmotilapia	95	80	175
Perissodus microlepis	65	60	125
Petrochromis	70	35	105
Simochromis	70	40	110
Spathodus	70	30	100
Telmatochromis sp. (dwarves)	75	55	130
T. temporalis	75	55	130
Tropheus	65	35	100
Lamprologus (x)			
(tubules only)	70	25	95
(with pored scales)	70	70	140
Lepidiolamprologus			
(tubules only)	80	35	105
(with pored scales)	80	70	150
Julidochromis/Chalinochromis			
(tubules only)	80	25	105
(incl. pored scales)	80	75	155

(x) for 38 species of *Lamprologus*

COMPARISON BETWEEN TWO SPECIES OF MIDWATER ROCKCOAST CICHLIDS
Cyathopharynx furcifer (mouthbrooder) and *Lepidiolamprologus elongatus* (nestbreeder)

Scales	*Cyathopharynx*	*Lpd. elongatus*
Longitudinal line	56	71
Upper lateral line	53	56
Tubules lower lateral line	27	27
Pores lower lateral line	19	24
Total lower lateral line	46	51
Total sensory scales	99	107

COMPARISON BETWEEN THREE SPECIES OF CICHLIDS LIVING AT GROUND LEVEL
Tropheus sp. (mouthbrooder: Kalemie), *Spathodus erythrodon* (mouthbrooder: Nyanza Lac), and *Lamprologus leleupi* (nestbreeder: North-west)

Scales	*Tropheus*	*Spathodus*	*L. leleupi*
Longitudinal Line	31	31	34
Upper lateral line	22	21	25
Tubules in lower lateral line	11	11	7
Pores in lower lateral line	–	–	13
Total in lower lateral line	11	11	20
Total all sensory scales	33	32	45

Chapter IX

Keys to the Cichlid Genera

The pattern of bars and stripes is useful in identification of the species of *Julidochromis*. This pattern is typical of *J. marlieri*. Photo by Burkhard Kahl.

1. Lower rays of pectoral fin separate from the fin; dorsal spines XV-XVI; anal spines III; gill rakers 12-14; less than 35 scales in longitudinal line; no scales on chest; body laterally compressed; copper-colored with oblique golden stripes; caudal fin rounded; maximum size 100mm .... ***Triglachromis***

Pectoral fins without any separate rays 2

2. Anal fin with 3 spines 3

Anal fin with more than 3 spines 31

3. Inner rays of pelvic fins shorter than outer rays 4

Inner rays of pelvic fins longer than outer rays; teeth conical, usually pointing forward in outer row of lower jaw; body usually elongate with slender caudal peduncle; dorsal spines XIII-XV; gill rakers 9-18; 34-43 scales in 3 lateral lines .... ***Xenotilapia***

4. Teeth of outer row more or less tricuspid 5

Teeth of outer row not tricuspid, but either bicuspid, conical, or in another shape 9

5. Dorsal spines XV-XX; teeth of outer row tricuspid (at least in part), but never conical, in thick pads; 10-16 gill rakers .... ***Petrochromis***

7 to 9 or 18-27 gill rakers; teeth in outer row eventually bicuspid, but inner teeth always tricuspid; size 350mm .... ***Sarotherodon/Tilapia***

Frontal area with a hump (higher in males) increasing with age; inner teeth conical or in part tricuspid; 6 or 7 dark blue strongly contrasting bands on a pale blue body; dorsal spines XVII-XIX; 10 to 12 gill rakers; maximum size 350mm .... ***Cyphotilapia***

Dorsal spines XII-XV; teeth tricuspid, eventually becoming conical or mixed with conical teeth 6

6. No substantial row of open-pored scales in front of tubuled scales of lower lateral line; pelvic fins with a filament reaching anal fin 8

A substantial row (10 or more) of open-pored scales in front of tubuled scales of lower lateral line; pelvic fin filaments reaching caudal fin in males and end of anal fin in females; teeth long and mobile, tricuspid or at least in part tricuspid, eventually mixed with conical teeth 7

7. Teeth mostly tricuspid with crowns curved backward, in multiple rows; 36-40 scales in longitudinal row; dorsal spines XIII; gill rakers 15-18; pelvic filaments not ended by broad spatulated tassels in males; maximum size 120mm .... ***Cunningtonia***

Teeth tricuspid (at least in part) with a crown not turned backward, in 2 to 5 rows; pelvic filaments in males always forked at tip and spatulated in brightly colored tassels.

   a) 36 to 40 scales or 62 to 74 scales in longitudinal line; teeth tricuspid or mixed with conical teeth, in 2 to 5 rows, those in front with a continuous edge or with a single row of conical teeth on the side of the jaws; silvery white, yellow, matte gray, olive green, or vermiculated dark green on pale green; snout sometimes with a proboscis; dorsal spines XII-XIV; gill rakers 17-20; maximum size 180mm .... ***Ophthalmotilapia***

   b) 48 to 64 scales in longitudinal line; teeth small and tricuspid in young fish tending to become conical in adults, in 3 to 5 rows, the outer teeth pointing forward; iridescent blue, sometimes with an orange—brown head, or entirely black with zigzagging red and green stripes; dorsal spines XII-XIII; gill rakers 15-18; maximum size 200mm .... ***Cyathopharynx***

Nota Bene: If one follows K.F.Liem (A phyletic study of the Tanganyika cichlid genera *Asprotilapia, Ectodus, Lestradea, Cunningtonia, Ophthalmochromis,* and *Ophthalmotilapia.* 1980), the distinction between *Ophthalmotilapia* and *Cyathopharynx* does not appear to be obvious.

8. Snout rounded and very protruding with a very low mouth; teeth small and tricuspid, in 2 narrow bands; body long and slender with very long and slender caudal peduncle; depth 5 times or more in body length; dorsal spines XIV; 38 scales in longitudinal line; gill rakers 15-16; maximum size about 130mm .... ***Asprotilapia***

Mouth subterminal; snout normal; teeth tricuspid, tending to become conical with age, in 2 or 3 narrow rows, inner teeth pointing backward and of same size as outer teeth; body moderately long (depth 4 times in length); eye very large; body silvery, males with a fluorescent pale lengthwise stripe in dorsal and anal fins; dorsal spines XII-XV; 36-38 scales in longitudinal line; gill-rakers 16-19; maximum size 160mm .... ***Cardiopharynx***

Nota Bene: Not to be confused with *Ectodus* or *Lestradea* .

9. 27 to 42 scales in longitudinal line 10

44 to 96 scales in logitudinal line 24

10. Teeth of outer row at least in part bicuspid 11

Teeth in outer row not bicuspid 15

11. Mouth low, snout rounded and strongly convex; lower jaw shorter than upper jaw; teeth bicuspid with a cutting edge in front, conical on the sides of the jaws, followed by a row of small tricuspid teeth; ocelli present on anal fin of males; tail never darker than body; 30-35 scales in longitudinal line; dorsal fin spines XVII-XIX (never more); maximum size 200mm ....***Simochromis***

Mouth rather terminal, if a bit low; jaws more or less equal; front teeth more or less bicuspid or in part conical; inner teeth mostly tricuspid, conical teeth in a row on the sides of the jaws; no ocelli present on anal fin of males; 32 to 37 scales in longitudinal line; dorsal spines XVIII-XX; gill-rakers 10-14; maximum size 260mm ....***Limnotilapia***

Mouth terminal; jaws more or less equal; snout not especially rounded nor very convex; less than 20 spines in dorsal fin 12

12. Lips normal 13

Lips thick and fleshy, curling back on jaws, increasingly so with age; 3 to 5 rows of compressed teeth, front ones bicuspid, inner teeth tricuspid in young fish, becoming truncate and without notch in adults; 33 to 35 large ctenoid scales in longitudinal line; dorsal spines XVII-XIX; 18 to 22 gill rakers; body striped olive green; maximum size 400mm .... ***Lobochilotes***

13. Gradual change in size between scales on chest (smaller) and those on sides 14

Scaleless patch present on chest and lower part of body; snout long and straight; jaws strong; cheeks scaleless; dorsal spines XIV-XVII; teeth unequally bicuspid, the cusps sharp; some posterior premaxillary teeth enlarged canines; inner teeth tricuspid and small, in 2 or 3 rows; weakly denticulate scales nearly cycloid, 29 to 33 in longitudinal line; body olive-green/yellow; head with distinctive black dots or short stripes; red throat and red dots on scales of sides of males; maximum size about 200mm (one species *H. horei*; see *Haplochromis horei*) .... ***Ctenochromis***

14. Anal fin in adult males with 3 to 9 ocelli (usually 3 or 4) in one or two rows; teeth cuspidate; scales always ctenoid under lateral line which has from 29 to 34 scales; dorsal spines usually XVI or fewer; caudal fin truncate or subtruncate; pharyngeal teeth all thin; sexual dimorphism strong, males more colorful ....***Haplochromis***

Same, but teeth cuspidate in young becoming in part unicuspid in adults; inner teeth mainly tricuspid and small; lateral line with 28 to 30 scales; caudal fin mostly rounded, seldom truncate; pharyngeal teeth submolariform; strong sexual dimorphism ....***Astatotilapia***

Nota Bene: See 33 for *Orthochromis* and *Astatoreochromis*

15. Head bones, including frontal, nasal, preorbital, lower jaw, and preopercle hollowed out by rows of drum-like cavities 16

Head bones not hollowed out by such cavities 17

16. 27 to 31 scales in longitudinal line; a single short, tubulated lateral line; dorsal spines VIII-XII; gill-rakers 9-25; teeth conical and small; body silvery, unpaired fins with dots, ocelli, or stripes; size 120mm .... ***Trematocara***

33-36 scales in longitudinal line; 2 lateral lines; dorsal spines XI-XIII: gill-rakers 17-20; teeth conical and small; silvery with horizontal alternating stripes of blue and yellow; maximum size 150mm .... ***Aulonocranus***

17. More than 20 spines in dorsal fin 18

Less than 20 spines in dorsal fin 19

18. Teeth with very long stem, each crown more or less spatulated with a straight edge (blade-like), in separate sets of 2 or 3 teeth off-set to the rear, those in front larger; snout and mouth broad; mouth low; dorsal spines XXII-XXV; gill rakers 10-12; beige with underparts yellow, the stripes extending to abdominal area, blue spots present on head and body; size 80mm ....***Eretmodus***

Teeth very long and conical or with an oblique straight edge, forming a more or less regular row, or regrouped in sets of 3 or 4 uneven teeth; snout and mouth pointed; head tending to develop a hump, which is substantial in one species; body beige or rosy beige, spotted by many blue spots, sometimes with horizontal dark brown stripes or two rows of brown flecks; dorsal spines XXI-XXV; gill-rakers 12-16; maximum size about 80mm ....***Spathodus***

Teeth very long, tusk-shaped, and sharp, in an uneven row, the teeth longer in front; head and snout moderately narrow; blue spots present on head and body; vertical beige stripes descending only halfway down sides; dark spot present in middle of dorsal fin; dorsal spines XXIII-XXIV; gill rakers 11-12; size 80mm .... *Tanganicodus*

19. Body not very elongate, asymmetrical with lower profile rather straight, upper profile well rounded; teeth all conical, in 3 to 5 rows, the inner teeth smaller than those in front; mouth low; caudal forked; pharyngeal teeth molar-shaped; dorsal fin spines XII-XVII; gill rakers 10-12; maximum size 180mm ....*Callochromis*

Body elongate, rather symmetrical; teeth all conical, small, in 3 to 4 rows in a single band, inner teeth smaller; mouth strongly protractile, opening in a tube-like shape; unpaired fins (eventually filamentous) obliquely striped; pharyngeal teeth conical and fine; dorsal spines XVI-XVIII; gill rakers 13-15; maximum size about 140mm .... *Reganochromis*

Body more or less symmetrical, lower profile well curved; front teeth always conical, inner teeth sometimes bicuspid 20

20. Mouth more or less horizontal, jaws equal 21

Mouth oblique and pointing upward; jaws unequal..23

21. Front teeth of lower jaw pointing forward..22

Front teeth of lower jaw vertical to the jaw; all teeth conical; pharyngeal teeth broad and molar-shaped; dorsal spines XIV-XVII; gill rakers 10-18; horizontal rows of pearly scales on a pale blue/yellow body;maximum size 150mm .... *Limnochromis*

22. All teeth conical and small, in 2-3 rows; pharyngeal teeth very thin, a crushing cushion on each side of the bone; dorsal spines XIII-XIV; 12 -14 gill rakers; body elongate and symmetrical; silvery or salmon colored with a black spot in middle of dorsal surrounded by a pale rim; maximum size 100mm................... *Ectodus*

All teeth conical and small, in 2-3 rows, separate from each other, with two very low cusps at side of main stem; inner teeth pointing backward; pharyngeal teeth thin, triangular, on a bone with a short shaft and its two apophyses in line with back of bone; dorsal fin spines XIII-XVI; 15-19 gill rakers; body elongate; plain silvery; size 140mm ....*Lestradea*

23. Lower jaw shorter, mouth low and very protractile; snout narrow and long, with very long pedicel; front teeth on lower jaw eventually pointing forward *(G. permaxillaris);* inner teeth more or less cuspidate *(G. pfefferi);* pharyngeal teeth thin and rather sparse; dorsal spines XV-XVII; caudal slightly concave; scales 32-37; size about 150mm .... *Gnathochromis*

Mouth oblique and pointing up, although both jaws are about equal; lower jaw at an acute angle with body; snout short and broad; body very short; eye very large; chest and base of pectorals scaleless; 32-34 scales; dorsal spines XIV-XV; tail rounded; opercular spot present; unpaired fins with a black edge; size 60mm .... *Tangachromis*

Nota Bene: See 25 for *Lepidochromis* and 27 for *Cyprichromis*

24. Mouth slanting very much upward, oblique or near vertical 25

Mouth terminal or subterminal and rather straight..28

Mouth low; lower profile of body nearly straight, the upper profile very curved and tapering off toward caudal peduncle; body narrow and high; 32 to 60 scales; teeth conical and small, in 4 or 5 rows; dorsal fin spines XIV-XVI; maximum size more than 300mm .... *Tylochromis*

25. Snout with preorbital sensory scale hollowed out by 6 radiating canals; snout slightly concave; body laterally compressed; teeth conical, in 2-3 rows; 44-58 scales; dorsal spines XIV-XV; caudal truncate; maximum size 150mm .... *Lepidochromis*

Snout without a sensory scale..26

26. Mouth with several rows of teeth 27

Mouth with a single row of small, conical teeth, sometimes curved back; head small; snout short, the mouth slanted upward; body laterally compressed with 64-79 scales; dorsal spines XVII-XIX; 20-29 gill rakers; maximum size 250mm.. *Haplotaxodon*

27. Upper profile very straight, the body elongate (head 4 times or more in standard length) and compressed; snout long and pointed, with a very long pedicel extending to middle of eye; mouth very protractile with small, conical teeth in 2 or 3 bands; front teeth larger; scales 34 to 71, strongly ctenoid; dorsal spines XII to XVIII; at least 20 gill rakers; size 150mm.. *Cyprichromis*

Upper profile curved and rather angular, the body short and stocky (head at most 3 times in standard length), but compressed laterally; snout short; mouth very oblique; teeth small and conical, in 2 or 3 rows (about 100 in first upper row); scales 60 to 72; dorsal spines XV-XVI; gill rakers 27-31; maximum size 280mm .... *Hemibates*

28. Two lateral lines 29

Three lateral lines, the third on caudal peduncle; body elongate; 44 to 59 ctenoid scales; caudal forked; teeth conical, in 4 to 6 rows, the front teeth large, but very small teeth in inner rows; dorsal spines XV or XVI; gill rakers 11 or 12; maximum size 260mm .... *Grammatotria*

29. A single row of very large teeth set well apart, peculiar in shape with a leaf-shaped crown set across a stout stem, curved back and with a central groove; teeth deeply embedded in thick, retractile gums; teeth eventually with two sharp lateral cusps; body laterally compressed, streamlined or not (1 species); 58 to 78 scales; dorsal spines XVIII-XX; 18-26 gill rakers; maximum size of coastal species 150mm, of pelagic species 320mm .... *Perissodus*
     More than one row of teeth 30
30. Head very large; jaws very powerful, lower protruding; main scales in part covered by layer of secondary scales; teeth in several rows, very sharp, fang-shaped, and hinged at base; 60 to 95 primary scales; dorsal fin with XIII to XVII spines; 8 to 19 gill rakers; steel gray with darker spots and vertical or horizontal stripes; up to 400mm .... *Bathybates*
     Head not especially large; no secondary network of small scales; teeth in several rows, bi- or tricuspid in juveniles, becoming conical in adult fish; scales 74-79; dorsal spines XV to XVII; 13 to 15 gill-rakers; average size 50–65cm, exceptionally to 95cm .... *Boulengerochromis*
31. Teeth bicuspid in front 32
     Teeth conical in front, some enlarged into canines; open pore scales in front of tubuled scales of lower lateral line present 34
32. Mouth terminal or subterminal; dorsal fin spines XVI-XX; head not massive or rounded 33
     Mouth very low and straight, in line with lower body profile; snout very long and curved; body stocky, high, and short; teeth of front row set in continuous cutting edge; separate conical teeth on sides of jaws; inner teeth in multiple rows forming narrow pad; dorsal spines XIX (exceptional), XX-XXI, or XXII (exceptional); anal fin with IV to VII spines; 31-32 scales; 11 or 12 gill-rakers; size 10 to 17cm .... *Tropheus*
33. Abrupt change in size between scales on chest and those on sides; chest and belly eventually bare; dorsal fin spines XVI-XX; anal fin spines III or IV; no ocelli on anal fin; ventral fins rounded, with second or third ray longest; caudal fin rounded; little sexual dimorphism; size about 120mm (fluviatile) .... *Orthochromis*
     Gradual change in size of scales between chest and area behind pectoral fin; dorsal spines XVI to XX; anal spines III to VI; anal fin with 3 to 5 rows of ocelli, involving from 6 to 20 ocelli; pharyngeal bone thick, with molar-shaped teeth; caudal fin rounded; first ventral ray longest; no sexual dimorphism; maximum size about 140mm (fluviatile or coastal swamps) .... *Astatoreochromis*

Nota Bene: Orthochromis, with 3 species out of 4 having only 3 anal spines, has to be compared with the other haplochromines in this key (#13 and #14). In *Astatoreochromis* individual specimens of one species and all specimens of another display only 3 anal spines. In this respect, caution should be exerted when identifying the haplochromines from the lake basin.

34. Inner teeth always conical 35
     Inner teeth tricuspid; 31 to 37 very ctenoid scales in longitudinal line; dorsal spines XVIII to XXII; anal with V to VIII spines; caudal fin rounded; thin oblique parallel stripes on body; maximum size 120mm .... *Telmatochromis*
35. A fleshy genital lappet protruding behind anus; snout conical and pointed; from 31 to 50 scales in longitudinal line; dorsal fin with XXI to XXIV spines; anal fin with VII to IX spines; scales present on dorsal and anal fins; caudal fin truncate, subtruncate, or slightly rounded; pattern strongly contrasted by horizontal and sometimes vertical stripes; maximum size about 130mm .... *Julidochromis*
     No fleshy genital lappet behind anus 36
36. Dorsal fin with XXII or XXIII spines; lips a bit fleshy and with thick papillae; 35-37 scales in longitudinal line; anal fin with VII to IX spines; caudal truncate, subtruncate, rounded or lyre-shaped; body elongate but stocky and cylindrical; beige, with a light blue cast behind pectorals and 2 or 3 black or brown stripes running along sides, sometimes entirely lacking or reduced to flecks; head always with black stripe running from eye to snout, another running to lip, sometimes a third on nape; maximum size about 150mm ..... *Chalinochromis*
     From XIV to XXI dorsal fin spines; anal fin with IV to XIII spines; lips devoid of thick papillae 37
37. More than 50 scales in longitudinal line; more than 50 in upper lateral line and more than 50 together in the two lateral lines; no scales on dorsal and anal fins, which are low; canine teeth strong and short; body rather cylindrical and elongate; no filaments on unpaired fins; caudal fin never lobed, but truncate or subtruncate; dorsal fin spines XVII to XXI; anal fin spines IV to VII; size at least 150mm .... *Lepidiolamprologus*
     Fewer than 50 scales in longitudinal line (except for 2 species); more often than not between 30 and 35 scales in longitudinal line; fewer than 40 scales in upper lateral line and fewer than 50 in the two lateral lines taken together; in many species scales present on dorsal and anal fins; body often laterally compressed, rather short to very short; unpaired and pelvic fins of many species often filamentous; canine teeth often slender and long; dorsal fin with XIV to XXI spines; anal fin with IV to XIII spines; caudal truncate, subtruncate, rounded, convex, crescentic, or lobed .... *Lamprologus*

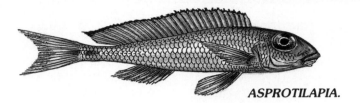

ASPROTILAPIA.

CHAPTER X.

DESCRIPTIONS OF THE CICHLID SPECIES

The descriptions of the genera and species of Lake Tanganyika cichlids are given in alphabetical order for quick reference. The main features of their ecology and feeding guilds are given in brackets. A resumé type description is given for each genus and a key to the species is provided when needed. Specific details about the ecology and various behaviors are provided whenever available.

ASPROTILAPIA Boulenger, 1901 (Rock-grazer?)

Asprotilapia is very similar in shape to the species of *Xenotilapia,* with its elongate body and very long caudal peduncle. But *Asprotilapia* can be distinguished from *Xenotilapia* by the unusual shape of its head, with its blunt, strongly curved snout and very inferior, chinless mouth.

The ethology and ecology of the single species are very poorly documented. Only females in small to large schools, sometimes several hundred fish strong, have been seen in the lake clinging to the vertical walls of massive boulders. They are seldom observed in the normal horizontal position. More often they remain still, all the bodies parallel to each other, either head down or head up or obliquely, apparently busily incubating their broods. Seldom have they been seen grazing on the rocks. We tend to believe that when we saw them on the sunlit boulders they had come up from deep water to provide increased oxygen for their broods.

The fact that females incubate their eggs in separate schools, their adoption of various body positions, and their dentition (tricuspid crowns on long and slender shafts), are features by which they can be distinguished from the sand-sifting *Xenotilapia.* One might think that somewhere in the past evolution of their ancestors the two groups evolved separately, *Asprotilapia* toward rock-grazing, *Xenotilapia* toward a life over flat sandy floors.

As rock grazers *Asprotilapia* stand apart in their morphology from most fishes in this feeding guild that have short and stocky bodies. They also have silvery scales when most rock-grazers have matte colors. Nor are they solitary or restricted to shallow waters. Their sightings have been too few and episodic for us to gain sufficient knowledge of this unusual fish.

Only one species has been found. It is apparently distributed along all the coastlines.

Brooding females of *Asprotilapia leptura* seldom are observed in the normal horizontal position. They usually hang head down on the sides of rocks. They also appear to come up from deeper water to provide increased oxygen for their broods. Photo by Pierre Brichard.

Description:

Body: very elongate with a slender caudal peduncle.

Scales: ctenoid, 38 in the longitudinal line.

Fin spines: dorsal XIV; anal III.

Fins: caudal forked.

Lateral lines: 2.

Mouth: very low, snout curved, protruding over mouth.

Dentition: small tricuspid teeth on slender shafts; in 2 rows.

Feeding type: Rock grazer.

Reproduction: mouthbrooder; 20 to 25 eggs. Synchronized gregarious spawns.

Maximum size: 120mm.

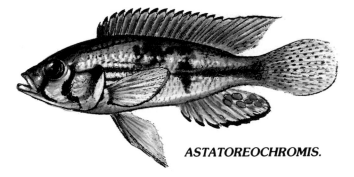

ASTATOREOCHROMIS.

ASTATOREOCHROMIS

Pellegrin, 1904 (Omnivorous swamp-dweller)

Three species are included in this genus. *A. straeleni* is found only along the northern shore of the lake. *A. alluaudi* occurs not far away in the hill country between the lake and Lake Victoria, to which basin this second species belongs.

The fishes in this genus look very much like *Astatotilapia* in shape but not in coloring, as there is little sexual dimorphism. The basic color of *A. straeleni* is orange which covers the whole body, the scales on the sides each having a crimsom dot. Several rows of yellow ocelli are present on the anal fin, which has at least 4, and sometimes more, spines.

A typical swamp-dweller, *A. straeleni* is found in the lake only at the mouths of rivers and in their associated swamps. The populations along the lake shores and inland are never very important, and one might suppose that in the swamps they are in competition with species of *Astatotilapia,* such as *A. bloyeti,* which are more abundant. A day's catch might involve only two or three dozen *Astatoreochromis straeleni,* but several hundred *Astatotilapia bloyeti.*

An excellent aquarium fish, this species is as much at ease in captivity as most fluviatile haplochromines. It is not finicky about its diet and is easy to breed.

Exported only sporadically.

ASTATOTILAPIA: See *Haplochromis.*

AULONOCRANUS Regan, 1920 (Semipelagic roamer. Drifting plankton picker)

The sole species of this genus is found in coastal and offshore waters over sandy bottoms in schools several hundred fish strong.

Males are recognized by the intensity of their typical body colors: alternating horizontal rows of blue and yellow; this is much fainter in females. In nuptial garb the males are superb, as their colors become even brighter, sometimes Prussian blue and a deep matte yellow. The unpaired fins are also deep blue and yellow; the ventral fins are bright white and filamentous. In females all fins are transparent and colorless.

By their shape *Aulonocranus* look very much like species of the *Cyathopharynx* guild.

The species is easy to recognize because of its enormous eyes and typical colors and pattern. Very common in coastal waters, it is easy to catch but rather delicate during acclimatization, although afterwards it is not a

Astatoreochromis straeleni is a typical swamp dweller, inhabiting only the mouths of rivers. Photo by Aaron Norman.

problem fish in captivity. This is an excellent fish for a large tank. It is peaceful and easy to feed, but is seldom exported because of the cost in

AULONOCRANUS.

acclimatization and freight due to its size.

Description:

Body: not very elongate and moderately deep; eyes very large; color pattern consists of alternating horizontal stripes of blue and yellow.

Scales: 33-36 in longitudinal line; 2 lateral lines.

Fin spines: dorsal XI-XIII; anal III.

Fins: ventrals filamentous; caudal forked.

Teeth: very small and conical, in 2 rows; outer teeth larger; mouth terminal.

Gill rakers: 17 to 20.

Habitat: sandy shallows.

Feeding type: omnivorous, feeding on drifting matter.

Reproduction: mouthbrooder, eggs 4mm.

Maximum size: 130-140mm; sexually mature at about 80mm.

Schools of *Aulonocranus dewindti* occur in coastal and offshore waters over sandy bottoms. Photo by Glen S. Axelrod.

BATHYBATES Boulenger, 1898 (Pelagic schooling hunter)

There are three main features by which the species of this genus can be distinguished from all other cichlids, the squamation, the dentition, and the color pattern.

Because of their pelagic roaming habits the fish are streamlined. To improve the flow of water along their body, instead of the fragmentation of the scales observed in fastmoving fishes, *Bathybates vittatus* has developed a network of very small secondary scales that surround each primary scale. The shallow groove separating each scale from the next is therefore filled in by the secondary scales and the sides of the fish are thus made smooth. Ripples caused by an uneven surface increase the drag and cut speed.

Bathybates also show perfect adaptation to their carnivorous habits by their dentition. The teeth are triangular but needle sharp. The teeth of the inner rows are hinged at their bases and lie flat against the gums, to be raised only when the fish opens its mouth to grab some prey. The whole setup reminds us of the one found in sharks. This, again, is a major departure from the usual cichlid dentition.

The color patterns of the various species involve an array of elongated patches, spots, and horizontal or vertical stripes (sometimes all of them together on the same fish), whose various combinations are typical of the species. These marks of very dark Prussian blue contrast with the steel gray of the background color, and are

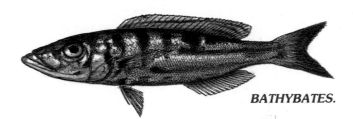

BATHYBATES.

especially strong on males. As we have said before, these patterns help them become camouflaged against their enemies or prey.

Bathybates are synchronous spawners, breeding on coastal sandy floors where their juveniles remain until they are perhaps 8cm long. These juveniles mix with schools of coastal sand-dwellers, like *Xenotilapia, Ectodus, Callochromis,* etc., on whose fry perhaps they prey. Not exported.

Description:

Body: laterally compressed and streamlined; head large, with powerful jaws.

Scales: from 60 to 95, with a secondary scale network sometimes present over at least part of the body; 2 lateral lines.

Fin spines: dorsal XIII-XVII; anal III.

Fins: caudal forked.

Teeth: fang-like and very sharp; in several rows, the inner teeth hinged at their base.

Pharyngeal bone: Y-shaped and very elongate.

Gill-rakers: from 8 to 19.

Habitat: pelagic waters.

Feeding: prey mainly on clupeids.

Reproduction: mouthbrooders; eggs about 6mm.

Maximum size: up to 420mm (one species).

BOULENGEROCHROMIS
Pellegrin, 1904 (Semipelagic predator)

This fish inspires awe in any skin diver when its pale green-yellow shape appears out of the blue haze. It also inspires emotion of a different kind because of the care with which it looks after its brood. These feelings are feelings of which hobbyists and scientists unfortunately are deprived because of the size of this fish. When 25cm long it is still very young, at 45-50cm it is middle-aged, and when fully grown it can reach a length of 95cm. Five pounders are not at all exceptional.

Along with the "Nile perches", *Boulengerochromis,* or "yellowbellies" as they are called by anglers or Nkue or Nkupi by local fishermen, is a prized game fish and renowned for the delicate taste of its flesh.

Key to the Species of *Bathybates* (after Poll, 1956)

1. Principal scales for the most part not covered by a layer of small secondary scales 2
   Principal scales for the most part covered by a layer of small secondary scales; eye oval; D.XV-XVI, 15-16; A. III, 14-16; 11-13 gill rakers on lower limb of first gill arch; 77-87 pored scales in lateral line; pattern of male consisting mainly of four longitudinal lateral bands; maximum size 420 mm ....**vittatus**

2. Body depth less than 4.6 and head length less than 3.15 in standard length; less than 16 gill rakers; interorbital space 4.7-7.85 in head; caudal peduncle 1.4-2.1 times longer than high 3
   Body depth 4.6-5.5 and length of head 3.15-3.4 in standard length; interorbital space 3.9-4.7 in head; caudal peduncle 2.2-2.5 times longer than high; eye suboval; D. XVI-XIX, 16-18; A. III, 17-19; 16-19 gill rakers; 81-96 scales in a longitudinal line; pattern typical, with transverse bands on sides; maximum size 400 mm ......
   ....**fasciatus**

3. Dorsal fin without notch; body depth 3.55-4.75 in body length 4
   Dorsal deeply notched, in two very distinct parts; body depth 3.55-3.85 in body length; interorbital space 4.7-5.0 times in head length; D. XIII-XIV, 15-17; A. III, 17-19; 10-12 gill rakers; 71-86 scales in a longitudinal line; body pattern of four longitudinal lateral stripes; maximum size 210 mm ....**minor**

4. 11-15 gill rakers on lower limb of first gill arch; caudal peduncle less than twice as long as high 5
   8-11 gill rakers on lower limb of first arch; caudal peduncle 2.0-2.1 times longer than high; D. XII-XV, 14-16; A. III, 15-16; 71-87 scales in a longitudinal line; body pattern of short vertical and horizontal stripes and dots; maximum size 270 mm..**horni**

5. Eye oval, 3.3-3.9 in head length; interorbital space 5.9-7.85 in head length 6
   Eye rounded, 3.95-4.8 in head length; interorbital space 5.25-6.0 in head; D. XV-XVI, 15-16; A. III, 16-18; 13-15 gill rakers; 75-88 scales in a longitudinal line, body pattern consisting of vertically elongated spots on anterior part of body and several horizontal stripes posteriorly; maximum size 260 mm ....**leo**

6. Body depth 4.0-4.5 in body length; head 2.75-2.95 in body length; D. XIII-XV, 15-17; A. III, 16-19; 12-15 gill rakers; 71-84 scales in a longitudinal line; body pattern consisting of horizontal rows of black dots decreasing in size from back to belly; maximum size 360 mm ....**ferox**
   Body depth 3.55-4.05 in body length; head 2.6-2.7 in body length; D. XIII-XIV, 14-16; A. III, 14-16; 11-13 gill rakers; 60-72 scales in a longitudinal line; body pattern consisting of vertical stripes anteriorly and horizontal stripes posteriorly; two large, conspicuous spots on gill cover, maximum size 250 mm ....**graueri**

The teeth of *Bathybates* sp. are triangular but needle sharp. Photo by Pierre Brichard.

The species of *Bathybates* are very streamlined, signifying their specialization as a pelagic schooling hunter. This is *B. ferox*. Photo by Pierre Brichard.

BOULENGEROCHROMIS.

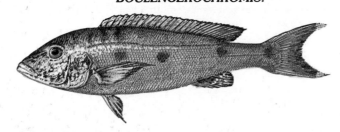

The members of our team, although we like a good fish filet as much as anybody, leave the huge beasts alone when we happen to spot them, mostly because most of the specimens seen in shallow water are either getting ready to spawn or are already looking after their brood. We just look and try to disturb them as little as we can.

Boulengerochromis is a cruising predator, rather unspecialized as far as the type of fishes it will swallow, although more often than not they prey on clupeid schools when they happen to pass by in coastal waters. These predators appear to live in small schools, probably only a few hundred strong, that roam the coastal areas down to perhaps 100m or more. There are places along the shoreline where they assemble periodically, probably to breed. These grounds can be recognized by the local concentrations of native dugouts occupied by several fishermen, all of them spending the day pulling at their lines, each with perhaps as many as 100 hooks. And this might well be the time to place an *à propos* on the wonderful skills and techniques used by native fishermen in Africa. Nowhere else are these traditional fishing skills more evident than with one of the techniques developed to catch *Boulengerochromis*.

Long ago the fishermen around the lake discovered that the Nkue were attracted to small bunches of reeds that had been ripped up by storms from the coastal swamps and were adrift in the open waters. Why? It is difficult to say, as most of the stems are too loosely packed to provide shelter or a place where other fishes on which the Nkue could prey might assemble.

As it is, the fishermen probably started to catch the Nkue around the drifting reeds with hook and line. But the whimsical drifts of the reeds did not always bring them into the areas where the Nkue are found. The fishermen also discovered that the fish scrapes against the reed stems, perhaps to get rid of parasites. So today, when you travel along the coastline, from time to time you can see in the distance a short stick rising from the water's surface. As you come by you discover that it is stuck in the middle of a short cross of balsawood, perhaps two feet long, under which a wide-meshed net not more than 4 or 5 feet square hangs down. More often than not a *Boulengerochromis* is caught by the gills in this ridiculously small trap floating in the midst of the vastness, surrounded by miles of open water. The upraised twig with a small tuft of grass or a piece of linen is there so that the fisherman who has laid the

trap will find it, and also is a way of identifying its owner to prevent theft.

The ingenuity displayed in traditional fishing in Africa, now becoming lost thanks to imported fishing gear, is a fascinating part of the reports of early explorers, where it lies scattered and forgotton.

The value of *Boulengerochromis* in fish farming cannot be overlooked. Fish farming of *Tilapia* in local communities has often been disappointing. Without proper attention they start breeding too early and their growth is stunted. Using fishes like *Boulengerochromis* or *Sarotherodon tanganicae,* which, being endemic to the lake, might adjust less easily to pond life and thus start to multiply at a later date, might perhaps provide an answer to this type of problem. Or mixing *Boulengerochromis* with *Tilapia* might, the first species weeding out the second and at the same time providing the farmers with a first class food fish. With the availability of wild caught Nkue fry, this would not be so much of a problem, if properly organized.

Obviously, the giant among the lake cichlids is not for the average aquarist and the medium-sized tank.

Description:

Body: moderately elongate and compressed. Color pale green with three dark spots on the sides; underparts yellow; blue stripes are present on the head and blue scales here and there on the sides.

Scales: 74-79 in the longitudinal line; 2 lateral lines.

Fin spines: dorsal XV-XVII; anal III.

Fins: caudal forked; pectorals long.

Teeth: small, in 3 to 5 rows; in juveniles the front teeth are bicuspid, inner rows tricuspid, all becoming conical in adult fish.

Pharyngeal teeth: bicuspid.

Gill rakers: 13-15.

Habitat: off the coast; semipelagic on soft bottoms.

Behavior: moderately gregarious.

Reproduction: substrate breeder; up to 15,000 eggs about 1mm in diameter.

Size: up to 95cm; maximum recorded weight 4.5kg.

CALLOCHROMIS Regan, 1920 (Gregarious sand-sifter)

The genus *Callochromis* includes three species and subspecies, although one of the species, *C. pleurospilus,* is now being investigated by Dr. M. Poll and might be split, rehabilitating *C. rhodostigma.* All are well suited to aquarium life, although *C. pleurospilus,* being smaller, can be at ease in a 30-gallon tank while *C. macrops macrops* and *C. macrops melanostigma* need at least a 50- or 60-gallon tank.

Fully grown *Boulengerochromis microlepis* are almost a meter in length. This is a juvenile about 15 cm long. Photo by Pierre Brichard.

In the lake all species occur between the shoreline and at least 60 meters over sand bottoms (as might be expected from their diet). The two subspecies of *C. macrops,* which are larger, can be found near rocks, *C. pleurospilus* much more seldom, but when the fish is trapped in shallow water by the afternoon surf they are often found individually against the sandstone slabs that give them protection. Normally they are gregarious and assemble in schools several hundred fish strong. They feed on the bottom and dig sand in search of the shellfish (mostly shrimps) on which they feed.

Along with *Haplochromis horei* among others (including a cichlid in lake Malawi), *C. pleurospilus* have the extraordinary habit of evading predators or seines by burying themselves head first in the sand. They can remain entirely hidden for several minutes. This behavior has not been witnessed in the two *C. macrops.*

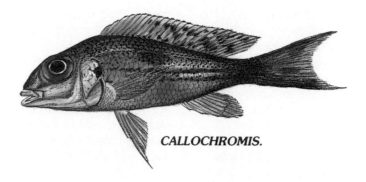

CALLOCHROMIS.

Due to the problems of catching sand-dwelling fishes without bruising them too much and its capacity to elude the seine, *C. pleurospilus* has never been a very common fish on the market, although by all means it is one of the more appealing fishes to come from Lake Tanganyika. The combination of delicate pastel colors, i.e. the shades of jade green and light blue on the sides, each scale being outlined by a crimson spot, and the creamy hues of white and pink on the fins, is rather unusual and very attractive. Other sand-sifters, like *Xenotilapia melanogenys* and *X. ochrogenys,* are more

gaudily colored and might steal the spotlight, but none is more appealing, and one might say it is even elegant. Females are less colorful than males, especially with regard to the finnage. The two subspecies of *C. macrops* represent two geographic races. The first, *C. macrops macrops,* lives in the south; the other, *C. macrops melanostigma,* in the north. A third race, sporting a yellow patch behind the gill-cover, has recently been found on the Tanzanian coastline. As yet only the northern race has been regularly exported in small numbers.

Curiously, the patterns of the two subspecies are reversed. The southern one is black with a few shimmering scales on the sides, while the northern one has an olive green, off-orange body with a few black scales here and there. The eyes are large with the top rim brown. The unpaired fins are yellow-brown, the anal fin sporting a creamy pink area.

Callochromis are known mouthbrooders. The eggs are pink, pear-shaped, and about 2mm long. Females lay about 50 eggs. Individual spawning sessions have been observed for *C. macrops* in shallow water, but never *C. pleurospilus.* It is not known if the sighting of breeding sessions by *C. macrops,* in which the male prepares the

Callochromis macrops melanostigma lives in the northern part of the lake. It is the only subspecies of this species that is regularly exported from the lake. Photo by Dr. Herbert R. Axelrod.

crater nest and females circle above (by which one might be led to think that they have no gregarious spawns on a communal site) are the rule. One would think that, given the schooling behavior of the species, the spawns by individual males were the acts of stragglers.

Callochromis in captivity can be kept in top condition quite easily and brought to breed. They relish of course shrimp and invertebrates, but will not be above eating flake foods as well.

Description:

Body: rather deep and laterally compressed, tapering off toward the caudal peduncle.

Scales: 32-37 in a longitudinal line; 2 lateral lines.

Fin spines: dorsal XII-XVII; anal III.

Fins: caudal forked.

Mouth: very low and protractile.

Teeth: small, conical, in 3-5 rows. The inner teeth smaller than those on the outer row.

Pharyngeal teeth: broadened and molar-shaped.

Gill rakers: 10-12.

Habitat: on sand floors.

Feeding: on microorganisms.

Reproduction: buccal incubation of up to 50 2mm eggs.

Maximum size: C. macrops about 150mm, *C. pleurospilus* about 100mm.

Like the other species of the genus, *Callochromis macrops melanostigma* is found over sand bottoms near rocks where they feed on invertebrates. Photo by Thierry Brichard.

Callochromis pleurospilus have the habit of evading predators by burying themselves head first in the sand. Photo by Dr. Herbert R. Axelrod.

Key to the species of *Callochromis*

1. Usually XIV or more spines in the dorsal fin; scales 32-34 in upper lateral line, 16-25 in lower.........................2
   Usually XII-XIII spines in the dorsal fin (very seldom XIV); scales 22-26 in upper lateral line, 8-10 in lower; body depth 3.33-3.6 times in standard length; pharyngeal bone with anterior blade less than ½ in dental area, teeth of pharyngeal bicuspid .... ***pleurospilus***
2. Interorbital space 6.2-6.8 in head and 2.0-3.2 in eye diameter; 2 or 3 inner rows of teeth; D. XIV-XVII; male light copper-colored, back dark olive green, black flecks on lateral scales, maximum size about 150 mm ....
   .... ***macrops melanostigma***
   Interorbital space 4.8-6.35 in head and 1.5-2.5 in eye diameter; inner teeth in a single band, not in separate rows; D. XV-XVI (never XIV or XVII); male nearly black all over, with a few silvery flecks on sides; maximum size about 150 mm .... ***macrops macrops***

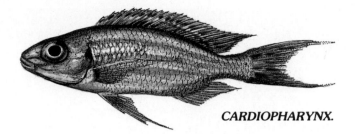

CARDIOPHARYNX.

CARDIOPHARYNX Poll, 1942 (Semipelagic microorganism feeder)

The only species in this genus *(C. schoutedeni)* is a rather common fish over sand bottoms where it roams in schools several hundred fish strong and feeds on microorganisms, such as copepods, picked out from the plankton clouds. It is occasionally exported from the northern part of the lake, but suffers from its relatively large size, more than 120mm on the average, the problems of capture, and acclimatization. It is not an unattractive species, with its silvery body and iridescent bands of pale blue along its dorsal and anal fins, but with the competition from the other very gorgeous sand-dwelling species it will never become a favorite.

Other features of its pattern are the white body underparts, the black chin and throat, and the black hue of the caudal peduncle.

Once acclimatized, keeping the fish in good health is not difficult, provided the fish is given some room to move about.

Description:

Body: Rather elongate but plump; eye very large; the mouth terminal and rather powerful; sides silvery, the caudal peduncle as well as the throat blackened, a broad iridescent light blue band on the dorsal and anal fins.

Scales: 36-38 in longitudinal line; 2 lateral lines.

Fins: dorsal XII-XV, 13-16; anal III 9-12; caudal well forked.

Teeth: tricuspid when young, becoming increasingly conical with age; in 2-3 narrow rows; the inner teeth bent backward and of the same size as the front teeth.

Pharyngeal teeth: thin and close together in a heart-shaped pattern.

Gill rakers: 16-19.

Feeding type: omnivorous on microorganisms.

Habitat: at least 3-15 meters deep, over sandy and muddy bottoms; a gregarious roamer.

Reproduction: buccal incubation; about 60 eggs approximately 4mm in diameter.

Maximum size: 150-160mm.

CHALINOCHROMIS Poll, 1974 (Rubble-dweller)

Back in 1971, inexperienced as I then was with the amazing variety of cichlids in the lake and even more unsure about their identity, it took a very unusual fish to have me exclaim, "This is a new species!"

With its beige cylindrical body and bridle-like black pattern on the head, *Chalinochromis* is of course unmistakable. In fact, it turned out not only to be a new species but a new genus as well. With luck and experience, in biotopes that have been barely scratched by explorers, one is bound to find new species. Although there is always a thrill in discovering a fish that nobody has ever laid eyes on, in such a biotope as the lake it eventually becomes routine. On average our team has found two new species a year since we came to the lake. But to find one belonging to a new genus is an entirely different matter. It is so rare

Schools of *Cardiopharynx schoutedeni* roam over sand bottoms feeding on microorganisms such as copepods picked from the plankton clouds. Photo by Pierre Brichard.

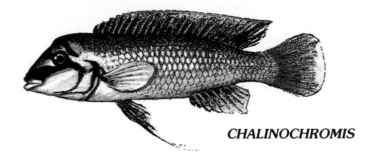

CHALINOCHROMIS

that in more than 30 years collecting I found only three such fishes.

I must be especially lucky with *Chalinochromis* for early in 1981 I found another one, this time with an oddly-shaped lyretail and three horizontal stripes. Two other races had been found in Tanzania previously.

Chalinochromis is very similar to *Julidochromis* anatomically and they are closely related. They are both quite different, however, from the species of *Lamprologus.*

Regan (1920) had characterized *Julidochromis* as having the suborbital bones replaced by teguments and more than 21 spines in the dorsal fin. According to him *Lamprologus* had the infraorbital ossified and a maximum of 21 spines in the dorsal. The two genera shared several features—scaleless cheeks, more than 3 anal spines, as well as conical teeth, of which the front ones were enlarged canines.

In fact, as it turned out, most *Lamprologus* lack the suborbital bones just as the *Julidochromis,* and both genera can have scales on their cheek. Another feature that sets the latter apart from *Lamprologus* and *Chalinochromis* is the fact that

both sexes have protruding genital papillae in *Julidochromis* and the other two genera don't. *Chalinochromis* is destinguishable from *Julidochromis* and *Lamprologus* by having very papillose taste buds on the lips.

Chalinochromis brichardi "dhoboi." Photo by H.-J. Richter.

Like *Julidochromis,* *Chalinochromis* live among the rubble in which they nest and they feed on the invertebrates hidden in the biocover growing on the rocks. They forage over the bottom a great deal and can travel in their forays more than 20 meters from the center of their territory.

They are thus more out in the open than *Julidochromis.* This is reflected in the way they raise their free-swimming fry, which graze on the rocky walls in the open and not within the recesses as *Julidochromis* fry most often do.

The fry of the two species are always banded at first, but those of *C. brichardi* progressively lose their bands, which fragment and disappear. The fry of *C. popelini,* like those of the Tanzanian race of *C. brichardi* called "bifrenatus" in the trade, keep the bands when they grow up. Those of the so-called "Dhoboi" variety retain only two rows of flecks on the side.

As an aside: Traders should avoid giving pseudoscientific latin-sounding names to fishes they import. This misuse can only add to the confusion resulting from the accelerated pace at which taxonomic revisions are nowadays published.

Chalinochromis, being very sedentary, come to learn their surroundings very well. As I already explained, they can find their way back to their nest from any part of the extensive grounds they patrol. Chased in the maze of passageways far from their nest, they will trace their way back between the rock walls in a matter of seconds. If an obstacle, such as a seine, is placed across their path they will stop, analyze the situation, find a hole under the seine, and head toward their nest. If caught again and put back far from their nest, they will arrive at the

Chalinochromis brichardi. Photo by H.-J. Richter.

285

nest area, identify the path they had taken the last time, and without fail will head toward the hole under the net.

Once a route has been memorized its features are added to its general knowledge and put to use at once among the features that had been accumulated by the fish about the layout of its surroundings. At the outskirts of its territory the fish sometimes hesitates between two routes, but the closer it comes to its central area the faster it swims. As it arrives at a crossroad it swerves right or left without stopping.

The capacity to memorize topographic features appears to be a very remarkable specialization of the substrate spawners in the lake. What is capital, in my view, is that in the nick of time and under stress they are capable of adding new information to the data already stored in their brain and deduce a new course of action. I know! Saying that a fish can have something like intelligent behavior will raise hackles on many a scientist's neck, but I stand on my facts.

Pairs are normally set up for life. Once they settle down on a nestsite they remain there year after year. I remember having found a pair of very large specimens, more than 15cm long (which is quite exceptional) under the same boulder for a period of more than 3 years—until I decided to catch them.

Breeding in captivity is very difficult to achieve, much more so even than with relocated pairs of *Julidochromis*. Even when a pair has been breeding for more than a year the relationship in captivity remains fragile. Even in 1500-

liter tanks such a pair still fight and one of the mates can be killed overnight. Altogether about 80% of the new *C. popelini* collected and paired "misfired" in this way. This is of course due to the fact that the fishes caught are separated from their mates, with whom they reached harmony when still very young. The bonds they are forced to forge afterward are never as strong as the first.

Chalinochromis are very interesting aquarium fishes because of their adaptability and good looks. They are not finicky about quarters or food and are quite hardy.

so-called "bifrenatus" form. *C. brichardi* also has two horizontal brown stripes. On *C. "dhoboi"* there are two horizontal rows of brown flecks. On *C. popelini* there are three dark brown horizontal stripes, the first along the dorsal base and extending onto the dorsal fin, the two others along the two lateral lines. The tail, which is truncate or subtruncate in *C. brichardi,* is lyre-shaped in *C. popelini* with the tips of the lobes cut off obliquely.

Scales: 36 in longitudinal line; 2 lateral lines.

Fin spines: dorsal XXII-XXIII;

Description:

Body: elongate and cylindrical; a typical pattern includes a beige background and a black bridle-like pattern on the head; sometimes a bluish hue is present on the cheek and behind the pectoral fins. On *C. brichardi* there is a black triangle at the back of the dorsal fin, but it can be missing on the

Two color forms of *Chalinochromis brichardi.* The spotted form is generally called "dhoboi." Photo by Dr. Herbert R. Axelrod.

anal VII.

Teeth: four large canines are present in front on the upper jaw, 2 on the lower; there are 4 inner rows of conical teeth.

This is a 4 cm juvenile *Chalinochromis brichardi*. The pattern changes with age. Photo by Pierre Brichard.

A young *Chalinochromis popelini*. Photo by Pierre Brichard.

Chalinochromis popelini has a distinctive lyretail that sets it apart from the "bifrenatus" form of *C. brichardi*. Photo by Pierre Brichard.

Gill rakers: only 4.
Reproduction: substrate breeder; up to 200-300 eggs; apparently not a sequential breeder like *Julidochromis;* no visible external genital papillae; no sexual dimorphism; females are a bit more plump.
Maximum size: about 150mm, usually about 120mm.

Key to the Species of
Chalinochromis:
1. Tail slightly rounded or truncate; at most 2 lateral stripes but more usually sides are plain. .... ***brichardi***
   Tail lyretail at all ages; three lateral stripes, two of them on the lateral lines .... ***popelini*, n. sp.**

The "brifrenatus" form of *C. brichardi* has a pair of lines extending the length of the body. Photo by Hans-Joachim Richter.

CUNNINGTONIA Boulenger, 1906 (Drifting plankton picker in coastal waters)

The only species of the genus, *C. longiventralis,* is easy to distinguish from the other members of the *Cyathopharynx* feeding guild by the lack of spatulate tassels at the tips of the very long ventral filaments. Contrary to the *Ophthalmotilapia,* which sport light-colored patterns, *Cunningtonia* has a dark blue, sometimes nearly black color

One of the many color morphs of *Cyathopharynx furcifer* in its natural habitat of rocky terrain. Its nest is built in sand. Photo by Glen S. Axelrod.

and seldom reaches more than 10 or 11cm.

They are found everywhere around the lake near or on rocky slopes, but build their crater nests on sand.

Because of its small size this species is probably the best suited of the guild to aquarium life. Very difficult to handle (like all the species in this group—which bruise easily), once they are acclimated they present few problems.

Description:

Body: symetrical in shape and rather plump; color pattern dark blue or blue-black with a few silvery scales on the sides; unpaired fins dull mouse-gray; pelvic filament tip white or yellow, not spatulate.

Scales: 36-41; 2 lateral lines, with a row of open-pore scales in front of tubules of lower line.

Fins: dorsal XIII, 13-15; anal III, 8-10; caudal forked.

Teeth: mobile, brush-like pads with curved tricuspid crowns in many wide bands; mouth terminal; pharyngeal teeth in a dense, velvet-like pattern.
Habitat: off the rocky slope; gregarious.
Reproduction: buccal incubator; eggs 3.5mm in diameter; crater-nest builder.
Maximum size: about 150mm, more often less than 120mm.

CYATHOPHARYNX Regan, 1920 (Drifting plankton picker; gregarious; rocky slopes)

The "blue fairy" from Lake Tanganyika is arresting in the incredible display exhibited by its iridescent blue body stationed in midwater. Although one of the most gorgeously colored cichlids in the African Rift Lakes, it loses much of its appeal in captivity when the conditions are not right. It has the added advantage of being peaceful and quiet.

Several geographical races challenge our perspicacity as to their status. Most *C.furcifer* seen in the lake are an iridescent dark blue, but there are at least four color morphs, perhaps one or two of them worthy of the status of subspecies in the parts of the lake that have already been explored.

Northern race males of *Cyathopharynx furcifer* are dark blue with a metallic sheen. This is a gregarious plankton picker. Photo by Schuphe.

The *northern race* males are dark blue with a metallic sheen. When they get excited they raise their scales ever so slightly and, in a flash, become entirely shimmering sky-blue and emerald green, including most of the forked and filamentous tail. The dorsal and anal fins are Prussian blue, striped with jet-black, the posterior part of the fins being liberally striped and dotted with pure yellow-orange. The pelvic fin tassels, of which there are two on each tip, are yellow-orange.

In the *Nyanza Lac morph* the head and body are a russet brown while the fins are rather like those of the blue morph.

The *Zambian morph* has a dark blue head with a sky blue body. The *Zambian morph at Kachese* has a jet black but matte body. The posterior part of the body and of the dorsal and anal fins are obliquely crossed by wavy bands alternating vermilion red and emerald green in metallic hues. Perhaps this is the most stunning fish in Africa. A fabulous fish!

Females of all morphs entirely lack the gorgeous colors of their mates, being pale gray and silvery with a few thin darker stripes on the body. All their fins are transparent and colorless.

Because of its size and habit of living well off the bottom, *Cyathopharynx* is not at ease in a moderately sized tank, and the males seldom display the array of colors that they display when in nuptial garb and when excited. Breeding in captivity has not been reported. They are difficult to acclimatize and it takes utmost care not to bruise them when they are captured. Added to their size, this feature increases the cost of making

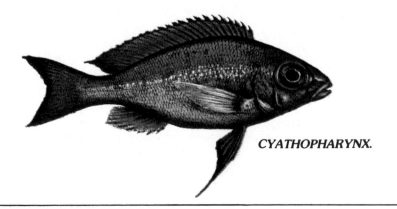

CYATHOPHARYNX.

Male *Cyathopharynx furcifer* build their crater nests on top of boulders — as close to the hovering females as possible. Photo by Schuphe.

these fishes available to the aquarists and they will never become a common sight in a tank unless aquarists agree to have them shipped when still juvenile. The wish to possess only breeding size or show specimens adds tremendously to the scarcity and costs of many Rift Lake cichlids.

We have already talked at some length about the extraordinary behavior of the males when they build their crater-nest and thus found an answer to the need to avoid the very large number of potential predators living at ground level.

Description:

Body: symetrical and rather plump, with filamentous unpaired fins; the pelvic fins of males sport a very long filament with spatulated bright yellow or orange tassels. The color generally is blue, but it can be chestnut or black.

Scales: 48-64; 2 lateral lines, the lower with a row of open pored scales in front of the tubuled scales.

Fins: dorsal XII-XIV, 13-15; anal III, 8-10; caudal forked; ventrals filamentous.

Teeth: small and tricuspid in young fish, becoming conical in adults; in 3-5 rows, the outer row pointing forward (probably used as "tweezers" to pick plankton); mouth terminal.

Pharyngeal teeth: thin and close together in a circular concave pattern.

Gill rakers: 15-18.

Habitat: lives in schools near the rocky slopes down to 20-25 meters.

Reproduction: buccal incubation; crater nest built by males most often on top of big boulders (the "penthouse" nest).

Maximum size: about 200mm.

CYPHOTILAPIA Regan, 1920 (Carnivorous rock-dweller)

A single species with two color morphs is known, one with six vertical bands, the other (in Tanzania) with seven. This is one of the most magnificent and exciting fish to have reached the aquarium world from Lake Tanganyika.

It is not rare in its native habitat (thousands of them are caught each day by native fishermen with hook and line). It is in fact a very good fish to eat, and the native dugouts are often filled to the brim with a half day's catch. The problem is to get them alive, which is very difficult because the fish lives in deep water. The big schools abound in the 30 to 40 meter layer or even deeper. There, in the twilight zone of the lake, the ghostly light blue bands of the fish seem to have a kind of luminescence of their own. The big specimens, often 350 mm long, stand side by side and move around slowly a bit off the greenish gray slope. To catch the fish at such depths is very difficult. They are quick, and dragging a 36 square meter net between two or three divers in step with the wary fish cannot be done more than three times during a dive. Afterward you have just enough strength left in you to bring the cage with the captured fish a bit higher along the slopes. This must be done slowly as the fish will very quickly show signs of decompression problems.

It is useless also, even if you get the rare opportunity, to catch more than 8-10 fish in the net at a time. They are so powerful that their concerted thrust will lift the net from the ground and they will escape. Not all the fish you thought to be *C. frontosa* will turn out to be them. In the midst of the few fish trapped you will find one or two *Plecodus straeleni* with identical blue and white bands mimicking the 'cyphos' so well that they can mix with the flock unobtrusively and, during a sudden rush, tear a few scales from the *C.*

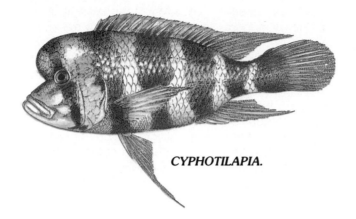

CYPHOTILAPIA.

frontosa's sides. It is impossible, without having the *Plecodus* in one's hands, to tell them apart. It is the only such case of mimicry of a species by its predator as yet reported from the lake. We will see, when dealing with *Plecodus,* that there is more to it.

Two male *Cyathopharynx furcifer* males sparring over nesting grounds as a *Cyphotilapia frontosa* passes in the foreground. Photo by Dr. Herbert R. Axelrod.

Cyphotilapia thus caught need a long time to be brought to the surface. We have never seen a *Cyphotilapia* brought to the surface alive immediately from depths in excess of 15 meters. If it is not decompressed for several hours and if it is still alive, it will certainly not survive afterward.

To overcome this problem we had to develop a new method. We brought a heavy

plastic drum down to where we planned to fish, put the captured fish in the drum, filled it with air from our diving bottles, sealed it hermetically, and brought it to the surface and our fish compound. The drum had a capacity of about 100 liters (a bit less than 30 gallons), and we could store a maximum of 6-8 middle sized fish in it. The lid had a valve, and every few hours we let some of the air out to decompress the fish. Unfortunately, the system had some drawbacks, the main one being that, after a few uses, the lid could no longer stand the two atmospheres of pressure and leaked.

We then tried metal cylinders. They were strong enough but very heavy. The trouble also was that the inner walls were rough and scraped the fish during the bumpy boat ride back to the fish

compound. Another problem was that we couldn't check on the fish; after a while in the confines of these chambers, wounds festered, the water became polluted, and fish either died or were in such poor condition that it took a month or more, if we were lucky, to get them into good condition for shipping. Under the best conditions a collecting trip, to be successful, involves three or four divers, all very experienced in catching the fish. The next day two divers must go back to the traps in which the fishes have been stored in order to raise them up a bit on the slope. This is done again the third day to finally bring them to the fish compound and start treating them in earnest. All this for an average catch of perhaps 40 'cyphos'. Even when in the best possible shape they are a bit scratched and the smallest wound is an open door for infection during acclimatization. How do we treat them? With utmost and painstaking care! It is perhaps interesting to tell, for once, a few of the secrets of fish collectors who are not in the trade exclusively to make money, but also because they like the fish.

Each 'cypho,' when out of decompression, is individually checked out of the water. The slightest scratch or bare skin (the tips of the protruding jaws are always a little scraped) is dabbed with iodine. When a fin ray has been broken it is also dabbed, although we know that the place will peel away— but the microbes are killed. The fish are then put into a

Cyphotilapia frontosa are perhaps best known for their pronounced nuchal hump developed as they grow older. Photo by Dr. Herbert R. Axelrod at the Berlin Aquarium.

Cyphotilapia frontosa command a high price because of the difficulties in retrieving them without mishap from their deep-water habitat. Photo by Burkhard Kahl.

tank individually. Whether we have 10 or 100 'cyphos,' whatever their size, each is placed in a separate tank. An unusually strong solution of copper sulfate is added to the tank, turning it a light blue. After three or more days, depending on the appearance of the fish in this bath (there is a water change every day), the fish are put into another bath with methylene blue and acriflavine. An antibiotic is put into the water and a sulfa drug added as well. This bath is changed every other day over a two-week period. The fish are fed dried food (in flakes) and tubifex worms (a 200 mm specimen will eat about a bunch of worms the size of a golf ball each day).

When they appear to be healed, which means all wounds cicatrized and covered with skin and all fins smooth and slippery with the mucus replaced on the skin, the fish are considered well enough to be shipped in a few days. This last quarantine lasts about five days.

At best, a *C. frontosa* in prime condition when caught will have to be looked after for about three weeks before being shipped abroad. It will have been handled twice daily during the first two or three days with each wound taken care of, then cured with some costly, or even very costly, drugs. Some of the antibiotics we experimented with cost $2,000 a kilogram (2.2 pounds) wholesale.

When the fish are in poor condition upon arrival they of course need more time, sometimes as long as six weeks, before they are considered in proper condition. We have often had to wait longer than this for ripped fins to grow back. And as usual there are some losses.

Very few *Cyphotilapia frontosa* taken from the lake are collected by our fishermen.

A juvenile *Cyphotilapia frontosa*. Broods are generally small, seldom more than 25, even with 20 cm adults. Photo by Hans-Joachim Richter.

In shallower water, where they are less common, they are easier to collect. Our fishermen are not allowed to use SCUBA to fish because of the hazards and latent dangers of this kind of fishing.

Not all the lake fish, fortunately, need such painstaking methods of collecting, though the most valuable fish are all given similar treatment, which explains their price. But it is well worth the effort if excessive losses are to be avoided with species like the

Description:
Body: deep and rather compressed; a hump on the head increasing with age; broad deep blue and white alternating bands.
Scales: 32-35 in a longitudinal line; 2 lateral lines.
Fins: dorsal XVII to XIX, 8-10; anal III, 7-8; ventrals with long white filament; caudal rounded; pectorals very large.
Teeth: external row conical or partly bicuspid, inner row conical or partly tricuspid
Pharyngeal teeth: all very fine, compressed, more or less bicuspid.
Mouth: inferior, protrusible and large, not especially powerful.
Gill rakers: 10-12.
Habitat: coastal, on rocks from 5-50 meters, depth increasing with age.
Feeding: omnivorous, eating in part shellfish (snails, mussels) and fish.
Reproduction: buccal incubation, about 25 eggs 5 mm in diameter are deposited.
Maximum size: in the wild 350 mm.

mouthbrooders from the rocky slopes which cannot multiply in large quantities and cannot thus be collected at random. Losses must be kept low during collecting and acclimatization. When we collected in the Congo basin these losses were less than 3% of the fish collected. We are close to this figure now with our Lake Tanganyika rock-dwellers, and recently managed to bring back 330 *Tropheus* over very bad roads with only one fish lost.

Let us get back to *Cyphotilapia frontosa.* The schools down deep number between a few hundred to sometimes as many as one thousand individuals, and they include a very large proportion of big adults. The density of the schools and the mean size of individuals dwindle when they are less deep to the point where, at a depth of about 10 meters, there are but a few young fish left, these very seldom venturing into water less than 8 meters deep. Only once during a dive (out of more than 200 dives) did I see a pair at 6 meters depth, and I still don't know what these adults were doing there.

Young fry of less than 50 mm are not seen in the open and seem to spend their lives hidden in deep recesses in the same water layer as the big adults. This is interesting because *C. frontosa,* although a mouthbrooder, seems to stay in deep water to release the fry. Most mouthbrooders go into shallow water either to release the fry or during the incubation time, probably to get better oxygenation for them. But 20 meters down is perhaps still shallow for *Cyphotilapia.*

Most mouthbrooders are near the surface when they incubate the eggs or the fry. Even some deep water fishes, such as *Lobochilotes labiatus,* which lives in the same layer as *C. frontosa,* does this. The fact that *Cyphotilapia frontosa* doesn't might indicate that the oxygen requirements of the fry, and perhaps the adults, are lower than those of most mouthbrooders. There is an advantage to this brooding method: the young fish will live in a layer where many predators don't venture and will thus be capable of surviving the terrific struggle for survival that goes on in the densely populated upper layers. It is very quiet deep down in the lake.

Broods of *C. frontosa* are small for such a huge fish. Seldom have more than 22 to 25 eggs or fry been found in mature females when they are caught, and these figures do not seem to increase much with the size of the fish. The fry are still in their mother's mouth when 25 mm long, but afterward they concentrate in deep schools.

Cyphotilapia frontosa is a ravenous eater with its large mouth capable of engulfing other fishes, but it doesn't appear to be a systematic predator. Fish 40 mm long kept in a tank with a 200 mm long *C. frontosa* were not swallowed whole but were ripped to pieces by a 'cypho' that we had kept on a diet just to see what would happen. Poll mentions that the stomach contents include fish and shellfish.

Most probably *C. frontosa* can be kept with other fishes of similar size without any trouble. It is not a bothersome fish when fed properly and is well worth the trouble.

CYPRICHROMIS

Scheuermann, 1977 (Drifting zooplankton pickers; gregarious; off a rocky slope)

This genus was separated from *Limnochromis* Regan by Scheuermann, a move that Prof. M. Poll had already suggested. It is represented by four species, of which a deep living species *(C. nigripinnis)* had never been seen alive until it was collected by our team. The first fish exported under the name *nigripinnis* were in fact *C. brieni* as Poll discovered later on, and which he identified as an undescribed species. The specific name *brieni* stands for Prof. Brien, a world famous biologist from the ULB University in Brussels.

Cyprichromis are "headstanders." Although they move about in the normal position they often remain still with their head low, and often, when they flee, dive quickly toward the deeps. This peculiar behavior appears to stem from the unusually long swim bladder. It extends beyond the first rays of the anal fin and gives unusual buoyancy to the posterior part of the body. This is especially true in two of the four species as yet identified. Mating of any of the four species has not, to my knowledge, been witnessed in the lake, but it has been in captivity. Very remarkably they breed in midwater, off the bottom, again an illustration of the trend in some mouthbrooding cichlids to avoid the crowded slopes when they mate.

Because of this gap in our understanding of the fish's behavior, we do not know whether or not they spawn at some distance from the coast and at the depth at which they

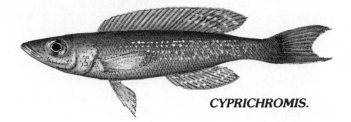

CYPRICHROMIS.

normally live, or if they come to the area in which their schools are found to mate. We suspect spawning occurs at dawn, the only time of the day when our dives have been few and far between.

Only schools in which females were already incubating eggs have been investigated. The fact that in a given school the incubated spawns often have not all reached the same stage in their development indicates that the mating sessions are more or less asynchronous, although some species appear to be even less synchronized than others in their breeding impulse.

As far as we can see the schools move about little and are more or less stratified along the slopes. The species that we found 35 meters down (C. nigripinnis) has never been seen above this level. They also appear to be restricted to towering rock pinnacles held together by calcite, landmarks around which the school swims, more often than not in mixed swarms of C. brieni and C. microlepidotus, and from which they never wander much.

They are not seen on slopes where the rubble is even and without the deep anfractuosities and caves into which the Cyprichromis release their fry at the end of the incubation period. Thus although Cyprichromis breed in midwater and are free from the need to select a site on the slope, they are still bonded to a particular type of bottom because they release their fry in rocky recesses.

If we take the other group of coastal plankton pickers belonging to the Cyathopharynx guild which release their fry not on the bottom but at the surface, we understand that the bonds of the rock-dwellers also reflect the problem of releasing their fry under the best and safest conditions.

Then, again, we can note that at first the fry of the fishes in this genus are initially cave-dwellers and afterward start to live off the slope in midwater—a drastic change in behavior indeed. Progressively the fry leave the recesses to assemble in schools near the slope and to mix afterward with the schools of adults.

The colors of all species are very bright in sexually mature males but very drab in females. Males can always be distinguished by their bright fin colors while females have colorless and transparent fins.

The color patterns of the males of the various species are as follows:

C. nigripinnis has yellow fins

Two color varieties of Cyprichromis sp. (probably leptosoma) from Zaire.

with a thin blue edge, the body being bronze with two horizontal rows of shiny sky-blue scales. *C. leptosoma* presents an unusual feature in that the males can have two color morphs in the same school, one with a bright yellow tail, the other with a dark blue tail. The colors are not interchangeable. Specimens with a yellow tail kept this color for as long as we could see. Even when preserved they had the same permanency in the color as the blue-tailed specimens. In *C. brieni* the unpaired fins are yellow-orange. *C. microlepidotus* has a dark bronze back, the underparts being lighter in color and yellow. Anal and dorsal fins are normally yellow but are swept by flashes of deep blue that appear suddenly on any part of the fin, then spread in ectoplasmic shapes to the point where they entail the whole fin. The color contracts, eventually to spread again in another shape, and eventually disappear and return again. This phenomenon, which has not been investigated as yet, appears to be linked to the moods of the fish. It might be related to dominant/submissive relationships between males.

The waving blue patterns do not appear to be brought about by the reflection of light on the fin membranes, as is most often the case with cichlids, but rather by the spread or contraction of pigments in the skin. Whether there are definite patterns in the patches does not appear probable as far as we could see; they appear at any place on the fin at random.

Juvenile *Cyprichromis* systematically are found in the

Top: *Cyprichromis brieni.* Photo by Burkhard Kahl. Middle: *Cyprichromis microlepidotus.* Photo by Schaller. Bottom: *Cyprichromis microlepidotus.* Photo by Schaller. *Cyprichromis* are headstanders, diving quickly to deeper water to avoid danger.

top reaches of the specific vertical range. They have thus been released there by the females at the end of the incubation period. As this is a phenomenon that appears to prevail among mouthbrooders in the lake, one might think they have a problem properly oxygenating their fry. They seek the best oxygenation compatible with the specific depth requirements of the species.

Having seen dozens of schools of one and then another species consistently at the same level, I had concluded that *Cyprichromis* never came higher than about 10 meters below the surface. After 10 years of diving I discovered on two separate occasions that this was not always the case, that *C. microlepidotus* and *C. leptosoma* occasionally came very close to the surface. This was an unexpected development and, although very rare, needed investigation.

On the two occasions when the fish were found perhaps two or three meters deep, they were swimming in very quiet waters. In the first case they were in a very narrow and protected cove off the Lufubu estuary; in the other they were on the leeward side of one of the Mboko Islands in the northern section of the lake, where in all types of weather they enjoy very quiet waters. Then came the discovery by Prof. M. Poll of the very large swim bladder of *Cyprichromis*

Cyprichromis microlepidotus and the other members of the genus rarely are seen shallower than 10 meters depth. Photo by Pierre Brichard.

that gives them exceptional buoyancy. One might guess that if the fish remain in deep water, it is not so much because of the availability of food or the dim light, but because they are physically unable, being more buoyant than usual, to resist the pull and tear of turbulent water, in which they lack directional control or equilibrium. In this respect one might say that they represent an opposite adaptation to the one displayed by the heavy-set

Key to the Species of *Cyprichromis* (after M. Poll)

1. Anal fin with 7-10 soft rays, dorsal with 10-14; 34-39 scales in longitudinal line; 14-16 scales around caudal peduncle..2

   Anal fin with 11-13 soft rays, dorsal with 14-18; 39-71 scales in longitudinal line; 22-27 scales around caudal peduncle..3

2. Dorsal XV-XVIII, 13-14; anal III, 9-10; eye contained 3.1 to 3.3 times in head length ....**nigripinnis**

   Dorsal XIV or XV, 13-14; anal III, 9-10; pectoral fin shorter than head; eye contained 3.7-3.9 times in head ....**brieni**

3. Dorsal XIV-XV, 14-16 (18); anal III, 11-12; scales 39-41 in longitudinal line; eye contained 3.5-3.7 times in head length ....**leptosoma**

   Dorsal XII-XIII (XIV), (15) 16-18; anal III, 12-13; 63-71 scales in longitudinal line; eye contained 3-3.5 times in head length ....**microlepidotus**

Eretmodus that are living in the surf-beaten upper layer.

Description:

Body: elongate and laterally compressed; shaped more or less like a cyprinodont (hence its name), with a deep abdominal area and a narrow and pointed snout.

Scales: 34-71, very ctenoid; 14-27 scales around the caudal peduncle; 2-4 rows on cheek; 2 lateral lines.

Fins: dorsal XII-XVIII, 10-18; anal III, 7-13; caudal forked.

Teeth: all conical; inner teeth minute in a single band that is broader in front and narrower on the sides; mouth very protractile with a very long maxillary pedicel.

Pharyngeal teeth: minute, fine, subconical; larger and forming a comblike edge along the rear of the bone, forming a sparse pattern elsewhere.

Gill rakers: 19-27 on the lower limb of the first arch.

Habitat: off rocky slopes.

Reproduction: mouthbrooder mating in midwater; about 40 eggs 3mm in diameter; fry released in recesses.

Maximum size: about 150mm.

Cyprichromis are excellent aquarium fishes and as yet the only cichlids known to remain upside down under large leafy plants. Although young fry hide in rock recesses, by the time they are 3cm long they start to appear in the open, sporting their bright yellow-orange fins, some of them remaining upside down under leaves.

Although they accept flake food, *Cyprichromis* will not be at their best unless given live shellfish. These they suck in with their long protrusible mouth. They are difficult to ship and acclimatize when caught, especially *C. microlepidotus.*

Another color variety of the *Cyprichromis* sp. from Zaire. Photo by Pierre Brichard.

ECTODUS Boulenger, 1898 (Semipelagic diatom feeder over sand)

It is often possible to catch this common but elusive species with a seine over shallow sand bottoms, provided the shallows fringe on deep water.

Ectodus is not especially beautiful but has unusually colored silvery sides, salmon in males and pale copper in females, with a large black ocellus rimmed with a white band in the dorsal fin.

Little is known about the ecology and behavior of this gregarious species, which could become a good aquarium fish. It is not difficult to keep alive. The problem again, as with all sand-dwelling species, is how to catch them in good condition.

Description:

Body: symmetrical and rather cylindrical; mouth terminal.

Scales: 35-38 in longitudinal line; 2 lateral lines.

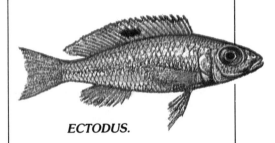

ECTODUS.

Ectodus descampsi is a semipelagic diatom feeder that occurs over coarse sandy bottoms. Photo by Pierre Brichard.

Fins: dorsal XIII-XIV, 12-15; anal III, 8-10; caudal subtruncate.

Teeth: small and conical; in 2-3 rows, the outer teeth directed outward.

Pharyngeal teeth: very thin; there is a crushing cushion on each side of the bone.

Gill rakers: 12-14.

Habitat: over coarse sandy bottoms.

Feeding: microorganisms, diatoms, algae.

Reproduction: buccal incubation; less than 25 eggs (?)

Maximum size: about 100mm.

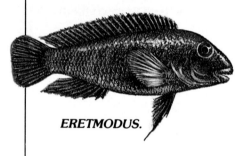

ERETMODUS.

ERETMODUS Boulenger, 1898 (Algae and zoo-biocover grazer)

Eretmodus is one of the typical surf-dwelling rock-grazers in the lake, along with *Spathodus* and *Tanganicodus.* It is represented so far by only one species *(E. cyanostictus),* which was described from specimens from the southern part of the lake. Another race exists on the Ubwari Peninsula and might eventually become a subspecies. Its range covers the coastline between Kalemie and the Ubwari. The teeth of this race, instead of being broadly spatulated as in the original species, have a quadrangular crown not much larger than the stem. Given the close relationship among the three genera one might wonder if they don't all belong to the same monophyletic group and shouldn't be put together in a single genus. The basic difference between the three genera consists of the shape of the teeth, which might simply be more variable than usual among cichlids.

They are fairly common in their habitat, which is the highly oxygenated water layer close to the shoreline and pounded by the surf, provided they can find shelter in the rubble and are able to graze. There are few if any *Eretmodus* present if the rock is not broken down into rubble, or if it is broken down too much and there are no anfractuosities left between the pebbles.

Because they are adapted to the high levels of oxygenation in the very top water layers, they are not found much deeper than 3 meters in the northern basin or 6 in the central part of the lake. But perhaps this is due to the fact that their diet might be very selective.

They live in very turbulent water and have a short, stocky, and apparently heavy body with a small swim bladder. In this respect they resemble several cichlids living in the Congo rapids, like *Steatocranus.* Like *Steatocranus* they have short but powerful pectorals, rounded tails, and a gap-toothed dentition with long tusk-like teeth with which they can rip the algal cover off the rocks.

Their negative buoyancy and short fins provide them, like *Steatocranus,* with poor swimming ability. They hop around from one piece of rock to the next and (in the lake) never rise from the bottom. When they do so in a tank it is with short jerky moves and they soon sink back to the bottom. But, like *Steatocranus,* as they are short and heavy they can resist the drag of pounding surf and not be washed away or bumped against the rocks.

When *Eretmodus* were introduced on the market they became instant favorites and have remained one of the three or four most asked-for fishes from Lake Tanganyika. Their broad snout and low thick-lipped mouth have earned them the nicknames of "schnozzle" or "tapir." These small (seldom exceeding 6-7cm) fish have sky-blue eyes and a beige body banded with light gold and brown. There are vermiculated patterns on the dorsal and anal fins and the caudal fin is rimmed with crimson red. The body has a few blue flecks here and there. They have much going for them if one considers that they are peaceful toward other fishes (if not toward their own kind), and easygoing (not too choosy) as far as food is concerned. The only requirement for their well being is to be provided with rock shelters and good oxygenation. More than other coastal cichlids they are very sensitive to oxygen levels and the first to die should the air accidentally be shut off.

Their ecology in the lake and their breeding behavior are in fact poorly documented and, as breeding in captivity has seldom occurred, little is known about these appealing fishes.

They are solitary, and move about much, but perhaps to some extent are territorial. In

One of the goby cichlids, *Eretmodus cyanostictus* lives in the turbulent, highly oxygenated shallow waters of the surf zone. Their heavy body and small swim bladder give them a negative buoyancy. Photo by Andre Roth.

The very broad snout of *Eretmodus cyanostictus* is one of the major characters of the genus. Photo by H.-J. Richter.

the surf zone, where long observation of a given fish is very difficult, mating has never been observed. The number of eggs laid by a female is around 25. They are pink and about 3mm in diameter, which is enormous for the size of the fish. When they hatch the fry are about 8mm long and when they are released by the incubating female they are about 11mm.

The race found on the Ubwari coastline is very different from specimens found elsewhere, being dark brown on the back with the underparts beige, the lower part of the bands being russet and the blue spots on the head and body missing entirely.

To distinguish *Eretmodus* from *Spathodus* and *Tanganicodus* is not difficult, even without looking at the teeth.

-if the specimen has a long and broad snout, a body barred with bands extending down to the abdominal area, it is *Eretmodus.*

-if the specimen has a narrow snout ending with a pointed mouth and a body devoid of any light colored vertical bands, it is *Spathodus;*

-if the snout is narrow, the mouth pointed, the body banded only halfway down the sides, and there is a dark spot in the middle of the dorsal, it is *Tanganicodus.*

Description:

Body: short, massive, and stocky; snout very broad and long; mouth inferior.

Scales: ctenoid, 30-32 in longitudinal line; 2 lateral lines, the lower very short.

Fins: dorsal XXII-XXV, 3-5; anal III, 7; caudal rounded.

Teeth: on a long stem with spatulated crowns set in separate groups of 2 or 3 teeth, staged from front to rear, the largest in front.

Pharyngeal teeth: subconical, of uniform size.

Gill rakers: 10-12.

Habitat: very shallow; amid rubble and large pebbles, not on smooth slabs lacking shelters; down to a maximum of 5-6m.

Reproduction: mouthbrooder, laying about 25 pear-shaped, 3mm diameter eggs.

Maximum size: 70-80mm.

The species is easily distinguished by its three lateral lines, a feature that is shared only by the sand-dwelling *Xenotilapia.* From its diet and also from this last feature we might deduce that *Grammatotria* is a fish roaming close to the bottom and not close to the surface.

But the ethology and other components of the fish are still very much a blank space in our understanding of the lake cichlids.

Description:

Body: elongate and streamlined; head big, jaws powerful, terminal; silvery.

Scales: ctenoid, 44-59 in longitudinal line; 3 lateral lines.

Fins: dorsal XV-XVI, 13-16; anal III, 9-12; caudal forked.

Teeth: conical; in 4-6 rows, outer teeth larger.

Pharyngeal teeth: very large and molariform in the center; two very large molars in the back of the bone; conical along the rear edge.

Gill rakers: 11-12 on the lower half of the first arch.

Habitat: semipelagic roamer over sand; down to at least 30m.

Reproduction: known to be a mouthbrooder; 60 to 90 eggs, 4.5mm in diameter.

Maximum size: 260mm.

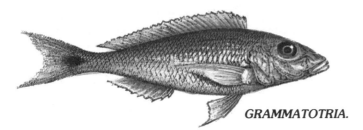

GRAMMATOTRIA.

GNATHOCHROMIS: see *Limnochromis.*

GRAMMATOTRIA Boulenger, 1899 (Semipelagic roamer, probably carnivorous; over sandy floors)

Grammatotria is very seldom seen by our team underwater, but it is often captured with seines by African fishermen. It is a powerful and fast swimmer that reaches a length of about 25 cm and whose morphology strongly suggests predatory habits. On the contrary, stomach contents point to a mollusc and diatom diet. Anyhow, the bicuspid jaw teeth and molariform pharyngeal teeth in the center of the bone point toward the bone being used as a mill to crush food, thus consistent with a diet of shellfish and snails. Were they conical and sharp, the pharyngeal teeth would have indicated a more predatory diet based on fishes.

The streamlined shape of *Grammatotria lemairei* is typical of roaming sand-dwellers. Photo by Pierre Brichard.

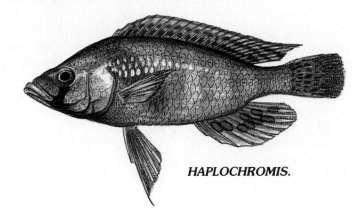

HAPLOCHROMIS.

HAPLOCHROMIS Hilgendorf, 1888 (Various habitats)

The status of the haplochromines living in and around Lake Victoria, the African rivers, and Lake Tanganyika has been reviewed recently (1979) by Greenwood. Several Tanganyikan species have been involved, but *Haplochromis benthicola* has apparently been overlooked. I am told that another revision of the same group of fishes has been undertaken, so that the future status of these fishes is still rather cloudy.

According to Greenwood, the Lake Tanganyika species should be allotted as follows to the present genera:

Haplochromis horei
   becomes *Ctenochromis horei*
Haplochromis burtoni
   becomes *Astatotilapia burtoni*
Haplochromis bloyeti
   becomes *Astatotilapia bloyeti*
Haplochromis stappersi
   becomes *Astatotilapia stappersi*
Astatoreochromis straeleni
   becoming *Astatoreochromis straeleni*
Haplochromis benthicola
   becoming *Haplochromis benthicola*
Astatotilapia paludinosa
   becoming *Astatotilapia paludinosa*
Orthochromis malagaraziensis
   becoming *Orthochromis malagaraziensis*

The fluviatile *Astatotilapia* and *Astatoreochromis* have been found only along the northern half of the lake in coastal swamps and lagoons, the southernmost being *A. stappersi,* which is found close to the Lukuga outlet. But I remember having once seen along the shore in Nkamba Bay, near a small river outlet, several fishes that looked very much like one of them. This species (which I couldn't catch) could eventually become the southernmost haplochromine of the lake basin. Many rivers in the basin are still awaiting their first exploration and I wouldn't be surprised if many species, especially in the northern part of the basin and in its eastern flank, would eventually be brought to light.

Only two species of haplochromines live in the lake proper, *H. horei* and *H. benthicola. H. horei* has been included in the genus *Ctenochromis* along with two other fluviatile species, one found in the eastern part of Tanzania, the other in the Stanley Pool rapids near Kinshasa. I must admit that I am rather reluctant to have *horei* belong to the same phyletic line as *C. polli* and feel that we still lack enough data to have these two fishes, which are so different in ecology and feeding habits, put in the same genus. The similarities between the two fishes perhaps stem from converging features. *H. horei* is a typical sand-dwelling predator along the coastline, and is never seen over rocks or in deep water. Along with the cyprinid *Varicorhinus,* it is one of the most specialized predators in its area, especially for prey such as the sand-dwelling *Callochromis* and *Xenotilapia* when they come close to the shore. Equipped with powerful jaws and a gaping mouth, *H. horei* can swallow a fish nearly half its size and is more or less a counterpart in its biotope to *Lamprologus lemairei* of the rocky habitats, at least among mouthbrooding cichlids.

Sexual dimorphism is not striking in *H. horei* but is evidenced by the red throat of the males, the many red dots of the scales on his sides, and his more vividly contrasting black stripes and dots on the head.

H. horei shares with *Callochromis pleurospilus* and a *Haplochromis* species from Lake Malawi the remarkable behavior of burying itself in the sand whenever placed in danger by a predator or by a seine. As I haven't personally been witness to this behavior I cannot say whether the technique used by *horei* is the same as the one used by *pleurospilus.*

Haplochromis pfefferi also floats back and forth between genera. Here it is very tentatively placed in *Haplochromis,* but it also is commonly put into *Limnochromis* and even into a separate genus, *Gnathochromis.*

Haplochromis burtoni (now referred to the genus *Astatotilapia* by many workers) is one of the most familiar African mouthbrooding cichlids. The egg dummy ocelli on the anal fin of the male are especially conspicuous.

The spawning behavior of *Haplochromis burtoni* is perhaps the best known of any African cichlid. Females holding eggs in the buccal cavity are especially distinctive (bottom photos). Photos by H.-J. Richter.

There are several reasons why the status of *Haplochromis benthicola* as a *Haplochromis* is very much in doubt. We can list a few of them as follows:

-there are rows of gaping cavities on the head. These are probably linked to sensory organs that have not been investigated as yet.

-the sexual dimorphism is opposite to the one displayed by haplochromines, the female being much more colorful than the male. Males are chocolate brown and have a dark blue cast on the soft parts of the dorsal and anal fins and several short rows of glittering light blue scales on the lower part of the sides. The females are bright vermilion red all over (including all the fins), with a golden background color on the sides on which scattered strongly contrasting jet black scales can be seen. Female *H. benthicola* are the most vividly colored fish in the lake.

-the fish are found at greater depths, always below 10 meters and often much deeper (hence their name), than most others.

This fish is very rare, perhaps restricted to the northern half of the lake, and little is known about its ethology. Even its status as a mouthbrooder is not sure.

Haplochromis (sometimes called *Ctenochromis*) *horei* is the only common lacustrine hap in Lake Tanganyika. Photo above by G. S. Axelrod, that below by Dr. H. R. Axelrod.

The rare *Haplochromis benthicola* is of uncertain generic status. It may belong in its own genus. Photo by Dr. H. R. Axelrod.

307

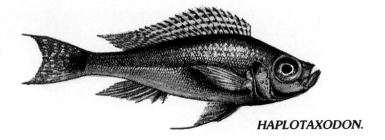

HAPLOTAXODON.

prey, which it grabs with a single row of sharp, conical, curved teeth on each jaw. The mouth is unusual for a cichlid as it is very much oblique, nearly vertical, and slanted upward with a prognate lower jaw and a very short snout. The mouth opens so high on the head that the front teeth on

HAPLOTAXODON Boulenger, 1906 (Schooling pelagic micro-zooplankton feeders)

Two species of *Haplotaxodon, H. microlepis* and *H. tricoti,* roam the open waters of the lake; perhaps we will discover more as time passes. Both are also occasionally found schooling in coastal waters along rocky coasts.

The high number of scales and the laterally compressed body unmistakably point to a fast-moving fish, although the high dorsal and anal fins, especially the filamentous dorsal, should increase the drag and slow the fish down. Probably they are not as fast as the very streamlined *Bathybates.* Our guess is that the fish relies more on a short high-speed dash to seize its

The strangely vertical mouth of *Haplotaxodon microlepis* (above and below) makes it instantly recognizable. The male (below) is much more attractive because of its bluish spots.

both jaws are level with the top rim of the eye, which is very large (equal to the preorbital area).

The background color of the body is pale green-yellow, the sides glitter a pale blue, and there are sky blue stripes on the under parts of the head and on the cheek. The dorsal and anal fins of the males have alternate horizontal bands of yellow and blue sheen. The females have more muted colors.

This is not an unpleasant fish altogether, but it is exported only sporadically. The best specimens are at least 15 to 20 cm, making air freight costs prohibitive.

The ethology of the two species and many parameters

Key to the Species of
Haplotaxodon:
1. Head length contained more than 3 times in standard length, eye less than 3 times in head length, and interorbital space more than 1.4 times in eye diameter; 3 rows of scales on cheek; more than 60 scales in upper lateral line ....*microlepis*
Head contained less than 3 times in standard length, eye more than 3 times in head length, and interorbital space less than 1.4 times in eye diameter; 2 rows of scales on cheek; less than 60 scales in upper lateral line... *tricoti*

of their ecology are still very poorly documented. They are mouthbrooders that come to coastal areas for the release of their fry, if not for mating purposes.

Description:

Body: streamlined and laterally compressed; head rather small; mouth nearly vertical and prognate; body pale green-yellow with rows of glittering scales on the sides.

Scales: ctenoid, 64-79 in longitudinal line; 2 lateral lines.

Fins: dorsal XVII-XIX, 11-13; anal III, 8-10; caudal crescentic.

Teeth: a single row of small conical teeth, curving back.

Pharyngeal teeth: conical, small, and very sharp.

Gill rakers: 20-29 on lower half of first arch.

Habitat: offshore; gregarious roamer.

Breeding: mouthbrooder.

Maximum size: up to 250mm.

HEMIBATES Regan, 1920 (Schooling pelagic predator)

An inexperienced collector might at first mistake these fish, of which there is only one species, with *Haplotaxodon*. There are simple differences to tell them apart.

-the eye of *Hemibates* is much smaller and less than the preorbital length.

-the body is squat and much more compact, while the head is larger.

-there are several rows of teeth in each jaw.

-the number of dorsal spines doesn't exceed XVI.

Again, as with so many semipelagic and pelagic species, little is known about the ethology and ecology of this species.

Description:

Body: rather short and stocky; mouth large, pointing obliquely upward, lower jaw protruding; profile of head rather angular.

Scales: 60-72, ctenoid; 2 lateral lines.

Fins: dorsal XV-XVI, 13-14; anal III, 12-13; caudal forked.

Teeth: small, conical, in 2 or 3 rows.

Pharyngeal teeth: conical and rather weak.

Gill rakers: 27-31.

Habitat: pelagic, often in front of estuaries and over mud bottoms.

Reproduction: mouthbrooder; eggs very large, about 7mm in diameter.

Maximum size: 280mm.

JULIDOCHROMIS Boulenger, 1898 (Rock-dwelling microorganism feeder; solitary; territorial)

Probably the most famous group of fishes to come from Lake Tanganyika, *Julidochromis* has become an established favorite among hobbyists. It includes 5 species so far, although in my opinion and from the specimens I collected around the lake there are already several local races, at least some of which might eventually merit specific status once we have found more refined taxonomic features by which to sort them out.

Basically, one might say that there are two dominant species of *Julidochromis*, *J. marlieri* and *J.regani,* the range of both spreading along much of the lake coastline,

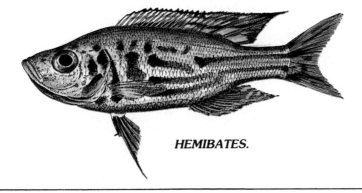

HEMIBATES.

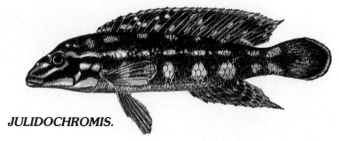

JULIDOCHROMIS.

Typical *Julidochromis regani* have a fairly distinctive pattern with the dark lines well separated by yellowish. The line below the eye is constant. Photo by B. Kahl.

both in the east as well as in the west. Both have local races that differ to some extent from the basic pattern specific to each race. To what extent these local races have diverged and whether or not they deserve subspecific status has not been elucidated.

Such are the two races of *J.regani*(?), one found along the Kapembwe escarpment near Cape Chaitika in the south and the other on the Ubwari Peninsula. Both are very dark and practically never seen in the open as they display a trend in *Julidochromis* toward cave-dwelling. The specimens seen around Cape Chaitika are so dark—they are a smoky black—that the horizontal stripes are barely visible when the fish are seen in their habitat.

Another race, this time of *J.marlieri,* is found below the shallower habitat of *J.dickfeldi* around Cape Chipimbi, also in the south. This *"marlieri"* has a much more elongate body than the usual type of *J.marlieri* found in the north and the typical color of this species but very pale. Finally, another *"marlieri"* has been found on the western shores, the color being very dark and with little white, typical of *marlieri,* but the outline of the body is much more streamlined, the head

Specimens of *Julidochromis regani* showing contrasty patterns can be very attractive. Photo by B. Kahl.

The darkest form of *Julidochromis regani* known. Photo by Pierre Brichard.

more pointed, and the snout longer, and one might think that this might well be a true subspecies of *J. marlieri.*

Two other species, *J. transcriptus* and *J. dickfeldi,* have restricted ranges only a few miles wide.

The case of *J. ornatus* was clouded in mystery until we made our long exploration along the southwestern coastline between Mpala and the Zambian border with Zaire. Until then two separate populations of *ornatus* had been found at the opposite northern and southern shores of the lake, some 650 km apart. We found *ornatus* from Lunangwa Bay to Mpala, immediately north of the range of *dickfeldi.* In this area their southernmost population is rather drab looking, but it brightens up toward Cape Zongwe. Their numbers in any of its habitats dwindle as one travels north, until they disappear around Mpala. Our exploration north of Kalemie had already shown that there

were no *ornatus* populations until one reached Uvira.

In all likelihood the progressive disappearance of *J. ornatus* appears to be linked to the presence of *Chalinochromis* species, either *popelini* or *brichardi.* As soon as either of the two latter species appears on the scene the number of *Julidochromis* one can see dwindles considerably, leaving only those little known species of *Julidochromis* that live in seclusion amid the small caves and do not appear in the open. There is thus competition between *Chalinochromis* and *Julidochromis,* one excluding the other.

There are several cases in which two species of *Julidochromis* live on the same slope, but then one apparently has the upper hand. For example there is the association between *J. regani* and *J. dickfeldi* at Kachese, the one between *dickfeldi* and the dwarf cave-dwelling species as yet undescribed at Cape Chipimbi, the one between *ornatus* and *transcriptus* near Uvira, etc. As for *marlieri* and *regani,* it would appear that whenever the two come into contact they hybridize (as at Magara) producing *regani affinis.*

Julidochromis do not occupy successive habitats; on the contrary, the species seem to alternate around the lake shores. One might wonder also why there are *dwarf* julies and *giant* julies and what the basic pattern of the body colors really is.

As far as these patterns are concerned, all *Julidochromis* species have the horizontal stripes in common, sometimes two on each side with a third

A good example of the northeastern race of *Julidochromnis regani,* perhaps representing a new species. Photo by Pierre Brichard.

on the dorsal *(J. ornatus* and *J. transcriptus),* sometimes only two on the sides and none on the dorsal (*J. dickfeldi*), and sometimes three, more or less complete on the side and one on the dorsal *(J. marlieri, J. regani regani* and *J. r. affinis).* In both groups some species have added vertical bars *(J. transcriptus, J. marlieri)* to the basic pattern. Of course it is impossible to say if these vertical bands were there first and then disappeared afterward, or the reverse. One subspecies, *J. regani affinis,* which probably should be put aside as a distinct species, in all its populations around the lake shows a trend (or is it traces?) toward vertical bars and is thus halfway between the two patterns. One species, *J. transcriptus,* has two local races, one close to the grounds of *J. ornatus* and having widely spaced black vertical bars and a pure white chin, chest, and belly; and the second, way down the coast and far from *J. ornatus* territory, is nearly all black, the vertical black bars having taken over most of the white patches (which have become mere "slits"). The chest and belly of this race are totally black. All other features of the species are common to the two races.

One would say that in the lake there is no relationship between *J. marlieri, J. regani regani,* and *J. regani affinis* when one puts together the most differentiated races of the three fishes. But it is a fact that where pure black and white *J. regani affinis* from the grounds where they breed true are in contact with *J. marlieri* populations, they crossbreed with *J. marlieri* at the border between the two flocks. This is apparent from the color patterns of some (but not many) of the *J. marlieri* in this place.

On the other hand, when one sees specimens from Tanzania, instead of the black and white color pattern of *J. regani affinis* in the north, they have the overall brown-beige color and body shape of *J. regani regani.* In the north and west *J. regani affinis* is a plump fish, like *J. marlieri,* but in the southeast it is a bit more elongate, like *J. regani regani.*

Crossbreeding in tanks between *J. regani regani* and *J. regani affinis* is no harder than between each subspecific

Julidochromis marlieri. *Julidochromis ornatus.*

Julidochromis regani, light form. *Julidochromis regani,* dark form.

Julidochromis transcriptus. *Julidochromis transcriptus.* Photos by H.-J. Richter.

race. It would be interesting to study the result of crossbreeding between the various pure, well established species. This would perhaps throw some light on the speciation that occurred in the lake and that is perhaps still underway.

Julidochromis marlieri Poll

Julidochromis marlieri is one of the most beautiful fishes to have come from Lake Tanganyika in recent years. Since its first exportation in 1971 it has reached worldwide popularity among aquarists. The fish are hardy, peaceful toward other species (in fact in the lake they are microfeeders and not carnivorous at all), and highly intelligent. It is unbelievable to see how quickly they get used to surface-feeding on dry flakes when one knows they often live in deep water and never come to the surface for feeding in the wild. They never swim in midwater but are typical rock-dwellers. They do venture into the open, but never far from the rock cover.

They breed in crevices and caves, as has already been mentioned. The behavior of the fry is not standardized. Sometimes they spread outside of the cave as soon as they are 15 mm long, and one might discover them 30 or 40 strong in broad daylight on vertical cliffs. When threatened, they do not hide in the nearby breeding cave but flee along the boulder wall. At other times the fry remain inside of the cave and apparently do not venture out until they are much larger.

We have seen that there are two methods of spawning: the large batch once a month or so and spawning in quickly repeated bursts of a few eggs at a time. We have not seen young adults have an initial big batch. Usually they start with several small ones a few days apart, then afterward might turn to large widely spaced spawns. Some adults never seem to have the large periodic broods and keep laying a few eggs every few days without ever stopping.

Pairing adults in aquaria is

tricky, especially if the fish are in narrow confines and there is only one of each sex. Pairs in the wild appear to be formed for life. Females are often larger than their consort, but it is, by far, not the rule in the lake. Very large pairs of *J. marlieri, J. regani regani,* and *J. regani affinis* have both partners of the same size.

Julidochromis regani Poll

The habitat of this species and its behavior seem to be a bit different from those of *J. marlieri. Julidochromis regani* likes to live in the open and ventures quite a lot over sand in the shallows. This is also the case with *J. marlieri,* but with a difference. Whenever *J. marlieri* is chased it will swim away, sometimes over very long distances (in excess of 20 meters), but always over rock. If surprised over sand, it will seek rufuge among the nearest rocks. The behavior of *J. regani* is quite different. It will swim toward the sand patches most of the time, even when surprised over rubble. This is not to say that *J. regani*

A rather muddily colored *Julidochromis marlieri.* Photo by Dr. Herbert R. Axelrod.

In the opinion of some, *Julidochromis marlieri* is one of the most beautiful cichlids in Lake Tanganyika. Photo by B. Kahl.

is a sand dweller, far from it; the fish has a close relationship with rocks, but this difference in behavior was worth mentioning.

Another difference is the habitat, which is much shallower, unless I am mistaken, than the one in which the biggest populations of *J. marlieri* are found. The author has never seen *J. regani* deeper than 5 meters, although it is entirely possible that there are some fish deeper. The pattern of the fish,

light beige with deep brown stripes, blends very well with the bright coastal sand floors near the tan rocks, and the fish are very hard to see whenever small particles drift in the water. The sandstone slabs (which are in very shallow water) are a pale pinkish beige when bright sunlight bathes their lightly colored biocover. The stripes on the body of *J. regani* are an excellent camouflage for the fish, even in recesses, since they break up the body shape. *J. regani* is

less visible in its habitat than *J. marlieri* because the whitish spots of *J. marlieri* literally shine in the dark.

Julidochromis regani is perhaps also less of an upside down type of *Julidochromis* than is *J. marlieri*. It is most commonly seen in the regular fish position, even in recesses, which *J. marlieri* very seldom assumes unless wandering in the open. Both fish, however, when hunted from above will flee slightly turned on one side so that they can have full

This yellow morph of *Julidochromis regani* was collected near Kiti Point. Photo by G. S. Axelrod.

An outstanding specimen of *Julidochromis marlieri*. Photo by Andre Roth.

The Kigoma race of *Julidochromis regani (affinis)* has a weak pattern. Photo by Dr. Herbert R. Axelrod.

A magnificent specimen of *Julidochromis regani* of the northern race that is now threatened with extinction by insecticides seeping from cotton fields near its very limited habitat. This race may be a separate species; investigations are now being conducted. Photo by Theirry Brichard.

vision of their enemy while they look for shelter.

This and other behavior are what make the species of *Julidochromis* such interesting pets. One cannot help but like them. Although they like dark recesses in the lake, they are not shy and learn very quickly. When a fish has been chased around several times over a few weeks, it will become more and more difficult to catch. And, as with *Tropheus* species, the memory of the hunt will stay engraved in their brains for several weeks, and they will adapt their behavior so that they might be capable of warding off further attempts at capture.

Every hobbyist knows that fishes have different ranges of intelligence. Some are dumb and never learn anything. They act according to the imprinted behavioral codes that they received from their forebearers. Other fishes show remarkable adaptability to a wide range of conditions. None show more intelligence than some of the cichlids of Lake Tanganyika. Among them, the species of *Julidochromis*, as well as *Tropheus,* are outstanding and well worth more detailed research.

Julidochromis ornatus was one of the earlier introductions of Lake Tanganyika cichlids to the aquarium trade and it is still quite popular. It is widely bred in captivity, which has led to some changes in its colors and behaviors. Because there have been few new importations of the fish, this dwarf species has generally been available only in a somewhat degenerate form (the result of long-term captive breeding) that does not always show its true colors. Photos by B. Kahl.

Julidochromis ornatus
Boulenger

J. ornatus is one of the dwarf species of *Julidochromis,* and the one with the basic horizontally striped pattern with no vertical bars. It is also the only species of *Julidochromis* with a deep yellow background that none of the other species in the genus have.

Some reports mention that fish bred in captivity from tank raised parents sometimes practice polygamy, with a male breeding alternately with two females and the three fish living peacefully together. This spawning behavior has never been noted for any species of *Julidochromis* in the wild. It is of course not at all impossible that this might happen in wild fish as well as other species of the genus. It is also possible that this apparently aberrant behavior is the result of genetic disturbances in individuals with a long ancestral background of captivity.

J. ornatus have been exported again from the lake shores for the first time since 1958. Many fish now in aquaria result from very few specimens sent to breeders years back. These fish often

Because it is easy to breed in the aquarium, *Julidochromis ornatus* has become one of the more common Tanganyika cichlids. Photo by H.-J. Richter.

A gorgeous pair of *Julidochromis ornatus*. Photo by B. Kahl.

The broken stripes of these *Julidochromis ornatus* may be due to inbreeding in captivity. Photo by B. Kahl.

show visible traces of a degenerative process under way, such as paler colors and broken stripes.

Breeding fish from wild-caught specimens is an important part of the hobby, and by selection it often results in magnificent varieties. But it is perhaps worth mentioning that past breedings of some Tanganyika cichlids in the late 50's, like *Lamprologus leleupi,* started from a very small number of breeders and resulted very quickly in the production of drab, worthless offspring. The causes of this failure were threefold: (1) too few breeders with too much interbreeding resulting in degeneration; (2) poor water quality with many trace elements or essential minerals missing in the food; (3) too intensive breeding, exhausting the breeders. With recent exports from the lake area it is probable that this situation will improve and the quality of the

tank-raised fishes will become very similar to the quality of wild fish. But it is not at all sure that this will be true with all species.

Julidochromis transcriptus Matthes

Julidochromis transcriptus is the second of the dwarf species of *Julidochromis* discovered in Lake Tanganyika. From the hobbyist's point of view, and mine, it is probably the best looking (along with the *"pure" J. regani*). The fish has everything going for it: a jet black background as well as the unpaired fins; a pure white belly in one race (called *J. "kissi"* in the aquarium trade), jet black in the second variety. The chin is pure white, and the unpaired fins are edged with a nearly fluorescent light blue. The spots on the sides, often mere slits, are white and in the *"black"* race turn a smoky black. Most probably careful breeding with selected specimens could result in a jet black fish with blue edges to the fins, a white chin, and without any other white on the body.

The trend toward melanism is sometimes apparent in *J. marlieri* when they live in very deep water in some very small areas of the rocky slopes. We have seen once, at 30 meters depth, an individual *J. marlieri* with half the body jet black. There are always specimens in this layer with a different type of white spotting. The regular type of white spotting in *J. marlieri* is not, in fact, white, but is a grayish yellow. Some specimens on the same grounds as *Lamprologus leleupi,* orange *Tropheus,* and *L. niger* have the lighter parts

A school of the dwarf *Julidochromis transcriptus* is a striking sight to behold. Photo by B. Kahl.

Julidochromis transcriptus is very similar to *Julidochromis ornatus*, differing mostly in details of the pattern and the broken nature of the stripes. Photo by Thierry Brichard.

Some dwarf *Julidochromis* are very difficult to identify. This fish has similarities to both *transcriptus* and *ornatus*. Photo by Pierre Brichard.

The strong vertical bars and the stripe below the eye help distinguish *Julidochromis marlieri*. Photo by Thierry Brichard.

of the body more yellow than in other areas of the lake. But the deep-living *J. marlieri* have the light spots rimmed with a very thin pure white line that gives these spots a "frosted" look. In reverse, *J. marlieri* living in the sunlit shadows have paler black stripes and bars, and the pale spots are larger. The intensity of the colors in these specimens doesn't change much in tanks, and these variations are permanently inbred in the fish.

Julidochromis dickfeldi
Staeck

The latest species of *Julidochromis* from the south, although belonging to the dwarf *Julidochromis* group, is a species standing well apart from all the others in the genus. One feature strikes the collector at once: the dorsal fin is much higher than in all other species of *Julidochromis* and does not have the same shape. This is the easiest species to breed.

The two *Julidochromis transcriptus* shown at middle and bottom display well the variation in the species.

Photos by B. Kahl (center) and Pierre Brichard (bottom).

Julidochromis dickfeldi is one of the more distinctive species of the genus and bears a striking resemblance in pattern to some *Lamprologus*. Photo by Dr. Herbert R. Axelrod.

All in all the whole group of *Julidochromis* species is one of the most exceptional additions to the aquarium world, providing the hobbyist with several enticing, very good looking fish with the little something extra that sets them apart from all the other African and South American cichlids. And, for a change, not one of them is a mouthbrooder.

Description:

Body: elongate and cylindrical; color pattern varying according to the species, with at least two or three horizontal stripes running along the body; sometimes with vertical bands; a characteristic pattern of black stripes on the snout and cheeks.

Scales: 31-36 in longitudinal line, 45-50 for one species (*J.ornatus*); 2 lateral lines.

Fins: dorsal XXI-XXIV, 6-7; anal VII-IX, 4-6; caudal truncate or slightly convex.

Teeth: conical; in narrow bands with strong curved canines in front; hind teeth more or less tricuspid.

Pharyngeal teeth: all small and conical.

Gill rakers: 2-5.

Habitat: always on rock, usually rubble, but one species often on sandstone slabs along sand beaches; some species down to 35 meters, perhaps more; very stationary and territorial, center of territory a crack or fissure in the rock; the extent of foraging very variable from one species to another.

Diet: lives mainly on crustaceans that they mostly pick from the zoobiocover, but more seldom from drifting matter.

Reproduction: egglayer in caves and slits; two

323

In good color, *Julidochromis dickfeldi* is a handsome fish with attractive fin colors and a delicate body pattern. Photo by J. Vierke.

spawning modes: (1) either in large spawns every 4 to 6 weeks, or (2) sequentially, in small batches of a few eggs every few days.

Maximum size: in the wild about 150mm for the two largest species (*marlieri* and *regani*), about 100mm for *dickfeldi* and *ornatus,* about 70mm for *transcriptus.*

Affinities: a) with *Lamprologus* by the similarity in morphology, the scaling on the dorsal and anal fins, the variability of the scales on the cheek, the open-pore scales in the lower lateral line, and the variability of the anal spine count. But they differ from *Lamprologus* by the elaborate and contrasting color patterns and by having more than XXI spines in the dorsal fin. They also differ by having protruding genital papillae in both sexes.

b) with *Chalinochromis* for the same reasons as with *Lamprologus,* but with the added feature that they also

Key to the Species of *Julidochromis*

1. More than 40 scales in longitudinal line; underparts yellow; a black spot at base of tail; three horizontal stripes run along body, the top stripe at dorsal base and extending onto the dorsal fin, the lowest halfway down body through lower part of eye to upper lip .... ***ornatus***
   Less than 40 scales in longitudinal line...2
2. At least 4 horizontal black stripes on body, the top at base of dorsal, the lowest, often broken, level with the pectoral; the two upper stripes merge on forehead or snout, the two lowest run through cheek and eye, and reach upper lip; eventually a fifth stripe on dorsal, which also can be plain yellow along with the tail and anal fin..***regani***

Nota bene: The horizontal stripes can be more or less enlarged and locally fused together, giving the fish a much darker pattern. This form is called *J. regani affinis.*

Two broad horizontal black bands on sides crossed by from 6 to 8 broad vertical bands, the pattern leaving between base of dorsal and belly three rows of yellow or white spots; the unpaired fins usually black with white flecks and stripes, ventral fins black, pectorals yellow ....***marlieri***

Lower part of body and head pure white; thin horizontal black stripe present under dorsal; broad stripe extends from snout to tail, the two stripes joined by vertical stripes leaving only irregular patches or mere vertical slits of pure white on sides; dorsal and caudal fins always black, with only thin white stripe near edge of fins, which are edged in black; two varieties, one with the underparts all white (var: *"kissi"*), the other nearly all black with only the chin and throat white ....***transcriptus***

Underparts white or beige; sides with network of slate blue on beige background; three horizontal black stripes, the topmost at base of dorsal; fins with bluish sheen; dorsal higher than in other species...............***dickfeldi***

have a more elaborate color pattern on the head and an identical dorsal spine count; but differing from *Chalinochromis* by having the protruding genital papillae and lacking the papillose lips.

LAMPROLOGUS Schilthuis, 1891 (Substrate spawner; various habitats but mainly rock-dwellers)

These species form the single most important group of cichlids in the lake and one which, in all likelihood, is bound to expand more than any other. Many new discoveries are to be expected once the deep sandy bottoms and the rocky slopes below 15 meters start to be explored in earnest. The genus is certainly heterogeneous and most species do not fit the criteria that have been set for them by Regan.

The polymorphism of the species in this genus strongly suggests a number of ancestral lines, but the overlap of counts and proportions prevents the systematist from identifying the lineages and sorting them out.

Lamprologus also exist in the Congo River and some of its affluents, and it is probably no accident if some of the species in the lake, which are ubiquitous in their lacustrine range, are also apparently the least specialized and the closest to the fluviatile species, such as, for example, *L.tetracanthus*. As there are now about 50 species of *Lamprologus* in the lake, some of them the result of a very long evolution, and only six in the Congo River system, one might feel that the forebearers

Julidochromis dickfeldi preparing the rocks for spawning. Photo by H.-J. Richter.

of the present stock lived in the lake basin rivers, then affluents of the Congo River.

In Lake Tanganyika most species are distributed in rocky biotopes and some of the present-day *Lamprologus* have specialized so that they can occupy the dark labyrinths, which no mouthbrooding cichlid apparently did. They manage to survive in the midst of a dominant non-cichlid group of specialized species. To do so they had to develop exceptional qualities, which several species indeed hint at (*L. furcifer, L. prochilus,* and *L. niger*) but that as yet have not been investigated.

Other species have developed a morphology that makes them fit for survival in the open, in full view of the predators around. Such are the armored *L. moorii, L. compressiceps, L. calvus, L. brichardi,* etc.

The genus, in fact, displays a variety of morphological and anatomical adaptations that point to a very long lineage from a common ancestor. If one accepts that such an ancestor lived in the Malagarazi River before the river was captured by the sinking lake basin, then one has to believe that this ancestor must have lived perhaps more than 20 million years ago.

Facing page: Jaw-locking behavior in *Julidochromis dickfeldi* is typical of many *Lamprologus*-like cichlids from Lake Tanganyika. Photo by H.-J. Richter.

Although not very variable, *Julidochromis dickfeldi* does have a "blue" morph (above) in which the pelvic and anal fins are largely bright blue compared to a normal specimen (below). Photo above by Pierre Brichard, that below by Dr. P. A. Lewis.

LAMPROLOGUS.

If they came to the lake at a later time, this would mean that during its history the lake could have had the same outlet as today and, due to the mountain-building activity in the area, that the lake has risen in line with the gain in altitude of its outlet. Unfortunately, the Lualaba (upper Congo) and the Lukuga Rivers have not been explored as yet. Nobody knows if there are other *Lamprologus* species in the Lualaba and lower Lukuga.

The main body of *Lamprologus* species has several interesting features:

The *"average" Lamprologus* has a moderately long and compressed body, from 17 to 18 dorsal fin spines and from 5 to 7 anal fin spines, 32 to 38 ctenoid scales in a longitudinal line, and two lateral lines, of which the lower has a row of open pored scales in front of the tubules. The cheek is usually scaleless, but sometimes it is covered with several rows of scales. Sometimes scales grow on the cheek only in old, individual fish *(L. savoryi).* The sensory neuromasts on the head, whether on the suborbital, maxillary, gill plate, or on the nape of the neck, are well developed and display an amazing variety of shapes.

The teeth are always conical, but some of the teeth in front on both jaws are enlarged into a row of canines, the central ones being smaller than those at the ends of the row.

This is the picture of the average *Lamprologus,* but there are many fishes that disagree with this portrait. We find species with as few as 4 anal spines while others have as many as 13; dorsal spines range from 14 to 21. The scales are more or less ctenoid, some with raised denticles that even grow on the tail. The dorsal and anal fins can be elevated from a normally curved body outline or from a groove made of raised scales. The open pores of the lower lateral line, which are normally single for each scale, can be double, triple, or even quadruple, etc. Each of these features by its variability provides us with a clue about the ecology of the fish, but some might also be brought about by convergent evolution rather than be significant of a phyletic relationship. Let us say that *Lamprologus* appears singularly diversified, that the characters that Regan used to separate this genus from *Julidochromis* and *Telmatochromis* have been found inadequate, and that a redefinition of the genus is needed. As yet the various lines, with the sole exception of the cruising predators, are hard to separate because of

The predacious, roaming *Lamprologus*-like cichlids are now assigned to the genus *Lepidiolamprologus.* This pair of *Lpd. elongatus* is guarding a spawn. Photo by D. Scheuermann.

the overlap of proportions and counts. With the increased pace at which new species are now being added to the *Lamprologus* roster, we might hope to find more clues in the near future, and before attempting a revision we might have to wait until we have more insight into the genus.

The sandy floors between 20 and 100 meters of depth and the slopes along rocky coasts below 25 meters should provide us with many new fishes in a short time. One might conservatively estimate that in the long run a total of more than 100 species will be involved in this group of fishes.

The following list of species includes the latest data available on the various species.

This "Magara" male differs from typical *Lamprologus brevis* in the bright orange head spot and the filamentous tail corners. Photo by H.-J. Richter.

Lamprologus brevis
Boulenger

(Shell-dweller?)
One of the smallest species, reaching a maximum of 4cm, *L. brevis* has become increasingly available to hobbyists in recent years. It has a very squat and angular body that is laterally compressed and a short head with a very oblique mouth. It is often mistaken for *L. ocellatus,* from which it can be distinguished by (a) the profile of the snout, which is straight or convex in *brevis* and concave in *ocellatus* (due to its high-set eye); (b) the mauve cheek, the golden stripes, and the white-edged fins of *brevis.*

Barely had the confusion between these two species ended than new discoveries in the north again raised questions. First the discovery of the species so-called "magarae" that I at first identified as a potentially new species, of which I am now less sure. This fish reaches a size of about 7cm. The tail is subtruncate as in *brevis* but there are two short filaments about 5mm long on each corner of the fin. Males have a bright orange spot on each temple. It is difficult to imagine that sexually mature males of *L. brevis* when 4cm long could lack this mark on the head and have the corners of the caudal rounded, and still be the same species as "magarae". Still, all taxonomic counts and measurements overlap for the two fishes. Moreover, the habitat of "magarae" and *L. brevis* have not been found to be identical. "Magarae" lives in deeper water, below 30 meters, at the foot of rocky slopes, but over sand; it occurs in very large schools that can be more than a thousand strong. It was first discovered on the northwestern coast (Uvira) at a depth of only 8-10 m, as a small school of what were probably stragglers. *L. brevis* appears to be a schooling species that is known to live in *Neothauma* snail shells and also lives in shallow waters.

Another as yet unidentified species looking much like *L. brevis,* being as small as that species but lacking the typical pattern, has also been found in the north quite recently.

All are excellent aquarium fishes but should be kept in their own tank without the larger species with which they cannot compete for food. Small shelters should be provided. If *Neothauma* shells are not available, Burgundy snail shells provide a perfect substitute. Given the gregarious habits of these pygmy cichlids it would be best to keep several together, with a higher ratio of females to prevent fights between the males.

Microorganism feeders, they should preferably be given live foods such as daphnia, cyclops, or *artemia*. Fry are raised on *artemia* nauplii. When adult, the fishes can be fed white worms or blood worms as well.

Lamprologus brichardi Poll
(Zooplankton picker)

One of the most famous fish to come from the lake, along with *Julidochromis* and *Tropheus, L. brichardi* was introduced to the market in 1971 and, although perhaps one of the easiest cichlids to breed in captivity, it is still a favorite as a wild-caught fish. Because of its qualities this very certainly could be the fish by which a hobbyist can be introduced to the world of African lake cichlids. Very hardy and peaceful, it adds the sleek elegance of its shape to any tank and also provides the thrill of seeing it multiply.

It is a micro-feeder, like so many other rock-dwelling *Lamprologus,* feeding partly on the zoobiocover but mostly on the *"bugs"* hopping about as they drift as part of the plankton clouds. They thus rise from the bottom, but not much more than about 50cm, and remain stationary. Once a school of these fishes has been located it will be found month after month and year after year on the same stretch of slope. They compensate for the weakness of their defensive means by being very prolific, and pool their resources together to defend their broods, which are raised in what has been called a "nursery". Even half-grown youngsters from previous spawns participate in the defense of the following generations. The schools are

Lamprologus brevis "Magara" is a typical shell-dweller using *Neothauma* shells. Photos by H.-J. Richter.

331

In males of typical *Lamprologus brevis* (above and below) the corners of the tail are rounded. Photo above by Thierry Brichard, that below by E. C. Taylor.

Lamprologus brichardi currently is one of the most popular fishes from Lake Tanganyika. Photo by H.-J. Richter.

static, allowing the sexually mature adults to find a nest site on the ground under the umbrella of the main flock, which they join later on. The largest fry are thus integrated into the school without any problem.

The habitat of the species, at least in the northern part of the lake, extends from 3 meters below the surface to 25 or 30m deep, depending on the slope, but not deeper. Because of decompression problems when they are captured below 4-5 meters, only the top part of the school is subjected to fishing. Contrary to pelagic schools, the parts of the school of *L. brichardi* swimming deeper are not affected by the commotion above, so that one realizes that far from being a school it is simply an aggregation of individual fish affected solely by what is happening in their immediate neighborhood.

Intrusion by enemies, be they man or a predator, doesn't dislocate the school, as it is so huge and static, but provokes local retreats or feeble attempts to flee (for which the fish are poorly equipped), changing the layout of the flock in ectoplasmic shapes. Such a school can be 100,000 fish strong in the northeastern part of the lake, but so far nowhere else have schools of *L. brichardi* so large been discovered. More often than not they involve less than 1,000 fish.

In the lake *L. brichardi* displays a total lack of aggressiveness and I have never seen a hard-pressed parent make anything more than a timid maneuver to repel an intruder. Their lack of aggressiveness in the lake is such that other *Lamprologus,*

such as *L. sexfasciatus,* have been seen tolerating them in the midst of a swarm of fry barely 4mm long. Reports on the aggressiveness of *L. brichardi* in captivity might be due to the fact that being single instead of in a group, they have to display more energy to save their brood, or perhaps also to the fact that to survive with other fishes in cramped quarters they had to secure a territory and hold it. In the lake they would leave the bottom and join again with the main flock above the ground. It would not be the first time that strong territorialism has developed in a tank among fishes that are not territorial in the wild.

Because of the static nature of the flocks several races of *L. brichardi* have developed in the lake. The Burundi morph does not have exactly the same pattern as the one found

The elegant finnage and subtle colors of *Lamprologus brichardi* are related to its status as a schooling fish in moderately deep water. Although often called aggressive in aquaria, they are totally nonaggressive in the lake. Photo above by Pierre Brichard, that below by H. J. Richter.

This is a relatively new albino mutation of *Lamprologus brichardi* developed by commercial breeders. Photo by Dr. H. Grier.

just across the lake south of Uvira. Nor is the Ubwari morph similar to the one found in the southern part of the lake. For this reason *L. brichardi* could be considered as just another local race of *L. pulcher* and put into its synonymy.

Lamprologus buescheri
Staeck

This is a new species discovered by Mr. Büscher during a trip to the southern shores of the lake in August, 1982, and in February of 1984 by my son Thierry on the Ubwari peninsula.

Although, of course, this species is difficult to bring to the surface alive, Mr. Buscher managed to bring a trio back to Switzerland and my son brought five home, among which were some youngsters. One might hope to see this valuable species, not exceeding 6-7cm in length, soon become available to aquarists.

It is unmistakable with its horizontal stripes, crescentic caudal fin, and a shape reminiscent of *L. furcifer*. It lives off the bottom but not in midwater, and rather deep, between 15 meters and more than 25 meters. This is one of the new species that hint at a strong layering of nestbreeding cichlids on a rocky coast.

Lamprologus callipterus
Boulenger
(Scavenger-predator)

This is a poor aquarium fish but a fascinating species because of the variety of its behaviors. It has a voracious appetite and is a perennial roamer, a species that, because of this last feature, puts to lie at least in part my theory that substrate-spawning cichlids are territorial and only mouthbrooders are roamers. Remaining still only when it is getting ready to attack a prey, *L. callipterus* moves about in small packs, never quietly but

at a frantic pace, following the leader in its moves, stopping and digging the sand ravenously when it so does, or grazing on the biocover. Before the last one has started feeding the head of the column has started moving again. It is always on the lookout for any unusual behavior that might betray that a fish is in trouble, or simply watching its spawn.

By the way they assemble around a prey and alternate their attacks, they look like a pack of wolves. Even the huge *Boulengerochromis* parents cannot prevent the destruction of their brood once 20 or 30 *L. callipterus* are pressing their attacks. Bold when they are together, but very timid when they happen to be alone or in pairs or trios they beat a hasty retreat as soon as their enemy meets their threat.

It took me seven years to discover how and where this very common fish breeds, and it was not without some

Lamprologus brichardi from Ubwari.

Lamprologus pulcher from Chaitika.

Lamprologus buescheri.

Lamprologus modestus.

Lamprologus sp.

Lamprologus sp.

Lamprologus calvus.

Lamprologus compressiceps. Photos by Pierre Brichard.

The elegant *Lamprologus brichardi*.

The yellow morph of *Lamprologus brichardi* known as the "Daffodil" form.

Lamprologus buescheri. Photos by H.-J. Richter.

Lamprologus callipterus is an opportunistic feeder that roams the lake in "wolf packs." Photo by Tetsu Sato,

measure of amazement that I discovered them guarding their fry in a pile of *Neothauma* shells along with *L. multifasciatus.* As soon as I approached the pile, I could see juvenile *L. callipterus,* already 4cm long, dive head first into the shells and remain motionless, probably believing they were hidden although most of their body protruded grotesquely from the shell. A dozen or so adult *L. callipterus* had retreated and could be seen in the distance. Later on I observed parents guarding their much smaller fry in other piles of shells.

Thus it is not only the dwarf shell-dwelling cichlids that make use of the snail shells as a shelter, but much larger species like *L. callipterus,* which reaches 15cm, as well, and it is in this light that one has to look at *L. ornatipinnis,* whose captures have often been made by a dragnet along with the shells. *L. ornatipinnis* does not live in the shells, for which it is much too large, but simply spawns in the shells.

Another form, called *L. reticulatus,* was described by Boulenger, but was put into synonymy by Poll. The specimens that were collected by Lestrade at Nyanza Lac, although very similar to *L. callipterus,* have different head proportions and should be considered as a geminate and valid species.

Lamprologus calvus Poll (Microorganism feeder on zoobiocover)

When you catch an unmistakable fish like *L. compressiceps* you don't have to check the identification keys to know what it is. But I was surprised at the end of 1975, during my first visits to the southern waters, to see two different looking *L. compressiceps* on the Cape Chaitika coast. One was yellow, like the one I had caught in the north, while the other was steel gray with many pearly white dots on the body and unpaired fins. I had put them aside without further ado as possible examples of sexual dimorphism.

Two years later, as I was sorting my specimens out, I put two equal sized fish in the same pail and gasped, *"But they are different fish!"* As seen from above the yellow one was plumper, bulkier, and had a shorter and broader snout. Seen from the side it was also stockier and deeper bodied. I brought enough material to Prof. Poll to have him accept the fact that indeed there were two *"compressiceps".*

In the course of his investigations he discovered that *L. compressiceps* had scales on the head down to the interorbital space while the other had scales that stopped short of the head and was therefore called *calvus* or baldy in latin. It could not have

337

L. caudopunctatus Poll
(Midwater microorganism feeder)

One of the species that I classified as unknown when I first saw it back in 1976 and later identified and described by M. Poll, was *L. caudopunctatus,* so-called because of the fine checkerboard pattern on its tail. It is not found in the northern half of the lake to my knowledge, which is odd because the fish moves about quite a bit.

Moderately gregarious and seldom standing still, schools of this fish swim about one meter from the bottom, over rock, although one might think from its color pattern and anatomy that it also lives over sand. Spawning has not been witnessed and nothing much is known about this species, i.e. whether it has communal spawning grounds or mates in separate pairs, which, from the behavior of the school, appears dubious. Not growing much over 6cm, most often a bit less, it is not at all lacking in appeal with its sky-blue eyes

been a local race of *L. compressiceps* as the two lived together. Another morph, the all-black *compressiceps,* he said could eventually be a third species but he was not sure.

My late discovery of *L. calvus,* linked to the cohabitation of *Tropheus duboisi* and *T. moorii moorii* at Bemba, which had been a challenge ever since the two species had been found together, started a train of thought that lead eventually to the theory of the fluidity in the past of the coastlines.

L. calvus is the most attractive fish of the two, but it also is probably less hardy in a tank and less common. Highly specialized, this slow-moving, high strung fish needs live foods and shelters to be kept at its best. Very sensitive to the environment, too, it needs crystal-clear water to be happy. Picking its food very slowly and with utmost precision from the tiny cracks it is investigating, this fish cannot compete when there are very gluttonous fishes around. It has been bred in captivity, which is an achievement not so many hobbyists can boast about.

Although it makes a poor aquarium fish, *Lamprologus callipterus* has some of the most interesting behavior of all the Lake fishes. Photo by Dr. Herbert R. Axelrod.

The Kigoma Yellow form of *Lamprologus calvus.* Photo by Pierre Brichard.

Lamprologus compressiceps from Tanzania.

Lamprologus calvus from Kachese.

Lamprologus compressiceps "Red Tail" from Tanzania.

Lamprologus falcicula.

Lamprologus furcifer Orange.

Lamprologus leleupi melas.

Lamprologus niger.

Lamprologus leleupi. Photos by Pierre Brichard.

The rather plain *Lamprologus caudopunctatus* feeds readily on flake food. Photo by Pierre Brichard.

and yellow dorsal and anal fins.

It has no counterpart in the northern basin.

This species is easy to keep alive even on flake food. It could become a welcome addition to a tank devoted to the smaller Tanganyika cichlids if it could be exported at low cost. Breeding in captivity has not been reported but should be easy. It might breed in shells.

Lamprologus christyi
Trewavas & Poll
(Unknown ecology and ethology)

Mistaken by me in my earlier book for *L. cunningtoni,* this species, as far as I know, has never been exported from the lake and has not even been rediscovered in recent times. It stands out from the main body of conspecifics by several features, such as a relatively high number of scales in the longitudinal line that are very denticulate ctenoid and a strongly forked and lobed tail. It doesn't belong to the group of cruising predators that also have a higher than average number of scales in the longitudinal line.

Nothing is known about the color pattern, the ecology, or the behavior of this elusive species. It appears to belong to the same line as *L. furcifer* and *L. buscheri.*

L. compressiceps Boulenger
(Rock-dwelling zoobiocover pecker)

This is one of the most specialized *Lamprologus* species, being strictly infeudated to a given type of rocky habitat, namely rubble with very few patches of sand. This is borne out by the fact that *L. compressiceps* is not found on isolated rock outposts that it could have reached by crossing sand barriers that other rock-dwelling fishes could pass. This is also supported by the fact that it is often absent from areas where the rocks are smooth and support only a sparse biocover or where the biocover is covered by silt. In fact, *L. compressiceps* is among the first to disappear when the conditions are not up to its requirements.

Although common in its habitat, it is never found in large quantities, and when an area has been fished for this species it takes several

months, perhaps a year, before they make their comeback. This is probably due to a low fertility rate. The spawns, very difficult to discover in the recesses, never appear to involve more than a few dozen eggs.

The morphology of this species stands out from the main line of *Lamprologus* by two features:

1. Feeding as it does on crustaceans hidden among the narrow cracks and tiny gullies of the rubble surface, this fish has a very narrow and pointed mouth that can be inserted between the anfractuosities of the rock walls. The jaws can also be extended forward. To do this and still be capable of retracting them back in line with the head, they are guided in their movement by two long bones, the premaxillary pedicels, which glide in a groove on the snout. Fishes that have strongly protractile jaws *(Cyprichromis* and *L. prochilus* for example) always have a very long premaxillary pedicel. *L. compressiceps,* like all fishes, often "yawn", the reason for which is unknown. Sometimes, as they extend their jaws the pedicel pops out of its groove at the end of its course and the mouth cannot retract anymore. The fish is then doomed unless quick action is taken, i.e. by pulling gently at the upper lip bringing the pedicel back in line with the groove and sliding it back into its setting. Otherwise the fish will suffocate. *L. compressiceps* are especially prone to this kind of accident when they are frightened.

The teeth of *L. compressiceps,* which in front are long and curved, are used as "tweezers" to pry the prey loose from its niche.

Another adaptation of *L. compressiceps* with respect to its diet is the mobility of the eyeballs in their sockets. Cichlids as a rule can roll their eyeballs to some extent in their sockets like many other fishes (some, like the large characoids in Africa can do much less as their eyes are embedded in a transparent skin). More than other *Lamprologus* species, and along with another zoobiocover pecker, *L. fasciatus,* and of course *L. calvus, L. compressiceps* can roll its eyeballs nearly to a 90° angle with their normal plane, and thus "squint" along its narrow snout at very close quarters into the tiny cracks in which the crustaceans are hiding. The fish are thus often seen slowly scrutinizing the surface of the rocks before they peck with utmost precision only once at something that to us is invisible.

Thus, the specialization to this type of feeding has brought in several fishes identical adaptations—a very narrow and pointed mouth, very protractile jaws, long tweezer-shaped teeth, and eyes that can focus at short range across the bridge of the snout.

We might add that the body has to remain steady during the time the rock is explored, and that the fish has to adopt unusual positions in order to get into line with the tiny shelter opening. *L. compressiceps* are thus usually seen head down or on their side looking at the rocks. As keeping immobile is not readily achieved in turbulent waters, the species is not found close to the surface or the shore when the going gets bumpy. Also, among all the species of *Lamprologus, L. compressiceps* offers the biggest surface to lateral pressure and would tend to be

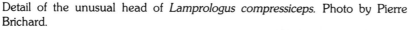

Lamprologus compressiceps can be a rather rather drab fish. Photo by W. Hoppe.

swept away and bumped against rocks.

Thus they prefer quiet waters and their highest density, once adult, is found below 10 meters. Younger fish, though, are found in quiet coves in water about 1 meter deep.

2. This species, along with *L. calvus,* is one of the more armored fishes of the rocky habitats. The dorsal and anal spines, which are exceptionally long and strong, are above the average number found in other *Lamprologus*—21 spines in the dorsal fin and up to 13 in the anal fin. Moreover, strongly denticulate ctenoid scales cover the sides and even extend along the spines of the dorsal and anal fins. They are also found way down the tail membrane. The fish, by raising its scales ever so slightly, can present to an enemy a bristling

Detail of the unusual head of *Lamprologus compressiceps.* Photo by Pierre Brichard.

Lamprologus fasciatus (swimming).

Lamprologus meeli.

Lamprologus modestus variety.

Lamprologus sp. "Walter". Photos by H.-J. Richter.

A hump-headed *Lamprologus furcifer.* Photo by J. Shortreed.

Breeding has not to my knowledge been observed in the lake nor has it been achieved in captivity and little is known about such behavior.

Lamprologus furcifer
Boulenger
(Cave-dweller; name meaning: "carrying a fork").

A very secretive and intriguing cave-dweller, *L. furcifer* is one of the favorites of the aquarium world and one of the least adapted to life in captivity of all of the coastal cichlids from the lake.

Very much different from the common *Lamprologus* mold, it is one of the very few species that have adapted to life in darkness and deep caves. This adaptation is manifest by several features including a very large and globular eye and a fragmentation of the squamation resulting in a higher than average number of very ctenoid and denticulate scales (the denticles bristling out from the scales' plane). The scales extend onto the dorsal and anal fins and cover most of the tail, on which the scales, although smaller, are also very much denticulate. The lateral lines are well developed, the lower one unusually so, and sensory openings gape on the head. All these are features that point to a highly sensitive detection system. Another cave-dwelling *Lamprologus, L. prochilus,* shares these features, so that we might think that adaptation to life in deep recesses involved for those cichlids an evolution toward reduction of the size of the scales parallel to the increase in the size of the denticles and of the various detection organs. But contrary to *L. furcifer,* which is slender, *L. prochilus* is stocky

In shape and color *Lamprologus furcifer* is quite distinctive. Photo by G. S. Axelrod.

345

Lamprologus furcifer is a cave-dweller living exclusively under boulders. Photo by J. Shortreed.

be a major adaptation for a cichlid.

L. furcifer has, along with *Julidochromis,* the uncanny ability to remain upside down, on its side or head-down, for any length of time. Very seldom does the fish orient normally.

A boulder usually harbors several specimens constituting a harem, one male mating with several females. This is not unheard of among *Lamprologus* and has already been documented 25 years ago for *L. werneri* and *L. congoensis,* two fluviatile species. A male *L. furcifer* can have as many as 5 or 6 females, each in a different slit in the rock.

Once they have become free swimming, the fry spread out from each nest on the walls of the caves, which become communal foraging grounds for all the fry. By that time they are slate gray and very hard to distinguish against the dark rock. They are very individualistic at a very early age and when two fry meet head-on they already display a formidable temper. They face each other at close quarters with their mouth wide open in a very threatening posture. The aggressiveness between adults is such that wild caught specimens cannot be kept together without very heavy losses. They don't usually lose time shredding the other fishes' fins, but bite at their sides and with their strong and sharp canines tear out whole pieces of flesh. Against other species they do not display this aggression.

There are usually well over a hundred fry creeping on the roof and walls of the caves used by *L. furcifer,* and

and apparently belongs to another line in the genus.

On the other hand, several cave-dwelling cichlids, of which *L. obscurus* provides a good example, have not developed smaller and more numerous scales. On the contrary, they have fewer than average and their lateral lines do not appear to have been exceptionally developed.

There are thus evolutionary trends among the *Lamprologus* living amid the rubble that appear to contradict each other, and we will need the discovery of more species from this special type of habitat to identify the adaptations they underwent.

Dark purple in color, *L. furcifer* is difficult to make out in the caves as it blends so well with the background. Very often they are detected only by the shiny white canines that are visible in the dark when the mouth is shut, or by the orange cast of their big eyes.

They live exclusively under big boulders. When they appear on the outside walls, which is often, they remain close to the vertical walls in the shaded areas. They never leave the protection of the boulders or caves to venture out into the open, at least by daylight, and I have never seen one feed in the wild. It might be possible that *L. furcifer* has a nocturnal nyctemeral cycle of activity. Were this supposition proven to be correct, in itself it would

progressively they spread on the outside walls until they depart individually and start looking around for a cave of their own. By the time they are 4cm long they have already settled down.

Apparently *L. furcifer* found around the lake look very much alike and one cannot tell the difference at first glance between a specimen from Burundi and another from Cape Chaitika. But around Cape Chipimbi, on the western side of the Bay of Cameron, a few have been found to have the posterior half of the dorsal, the entire anal and caudal fins, and the caudal peduncle a matte yellow-orange. Whether they constitute a subspecies or simply mutant specimens has not been determined.

Keeping *L. furcifer* alive in a tank is not too difficult, but having them in top condition so that they might spawn is another matter. As in the wild they do not move about in the open in strong light, and remain hidden from sight. Deprived of the large boulders that they are used to patrolling, they are restricted to a small corner of the tank and don't move much, unless their neighbors are small enough not to present a threat.

Because they do not venture into the open they often miss most of the food and start wasting away. They are not opposed to earthworms and feeding them will help them remain in tiptop shape.

This juvenile of *Lamprologus furcifer* bears little resemblance to the purplish adults. Photo by J. Shortreed.

Lamprologus hecqui
Boulenger

(Shell-dweller)

This species was named after Capt. Hecq, a Belgian officer stationed at Kalemie before WWI. Another of the dwarf species, *L. hecqui* has never been collected in recent times and probably is simply a sand-dwelling *Lamprologus* spawning in *Neothauma* shells. It otherwise roams the empty sandy bottoms in schools, which its relatively small size (80mm maximum) strongly suggests. From its stomach contents it is thought to eat shrimp and, despite its small size, also baby snails. Its breeding behavior is entirely unknown.

Recently hatched fry of *Lamprologus leleupi*. Photo by H.-J. Richter.

has a higher vertebral count. The body is thus definitely longer. Also, the orange body doesn't lose any of its brightness in captivity. This subspecies has been found only along a rather short stretch of coast in Tanzania.

The two *L. leleupi* (*l. leleupi* and *l. longior*) are the only cichlids as yet found in the lake sporting a bright matte yellow. There is a dwarf *Petrochromis* around Cape Chipimbi that has the same color, but unfortunately as soon as it is removed from the lake the fish turns a transparent copper. I also remember having captured one day in the south a bright yellow *Telmatochromis caninus* which obviously was a color mutant, but next morning before I could get a good picture of the fish the color had lost all its sparkle.

Lamprologus leloupi Poll

This fish was named after Leloup, one of the scientists attached to the Belgian expedition of 1946. It is a species described from a single specimen caught by M. Poll. From its morphology one would believe that it belongs to the same line as *L. caudopunctatus.* The ecology and ethology are unknown.

Lamprologus lemairei Boulenger

(Predator)

This species was named after Ct. Lemaire, a Belgian officer and fish collector stationed on the lake shores before WWI.

Again, one of the most voracious and predatory species of the genus and not

An adult yellow *Lamprologus leleupi*. Photo by H.-J. Richter.

Lamprologus leleupi with a batch of hatching eggs. Photo by H.-J. Richter.

suited for the aquarium. It has a very big head with an enormous mouth allowing it to swallow fish nearly its own size. They are not roamers and are ill-equipped for speed with their short and stocky body. They are very often seen near rocks, sometimes lying on the floor in the open, motionless. They are highly solitary, and pairs are seldom seen. The female lays its eggs in rock recesses, and the fry are taken care of by the female only.

The diet is, of course, carnivorous and composed exclusively of fish. The coiled intestine attains 50% of the standard length.

Lamprologus meeli Poll
(Sand-dweller)
L. meeli was named after van Meel. It has never been collected nor seen underwater in the northern part of the lake.

Lamprologus lemairei. Photo by Tetsu Sato, *Midori Shobo.*

It is one of the few species of *Lamprologus* with more than 40 scales in a longitudinal line and is also one of the dwarfs—the maximum size being 67 mm. The diet is microcarnivorous (insect larvae and copepods), which is reflected by the length of the intestines (60% of the standard length).

Nothing is known about the behavior. Probably a shell breeder.

Lamprologus modestus Boulenger and **L. mondabu** Boulenger

(Omnivorous and ubiquitous)

There are few differences between these two species, which explains why Poll at first put them into synonymy. He recently reversed this decision. *L. mondabu* (a native name) prevails in the northern basin and *L. modestus* in the southern. The first species has a concave tail, the second a rounded one. A third species, called "staecki" by me prematurely in my earlier book, with a lemon yellow stripe across the pectorals, has since been considered by Poll to be *L. modestus,* of which I am not so sure. One of the most common and ubiquitous fish in the lake, it is seen over sand as well as on rock, but usually near the surface and near the shore. The normal color is pale beige, with the edges of the unpaired fins a pale blue.

Their nests have never been discovered among rocks, although they are easily identified when the fish live in the sand plains. They dig a deep tunnel propped against a big shell or a stone in the open even when nearby rock outcrops could help them in the building of their shaft. Both parents watch over the fry.

The central pharyngeal teeth are molar-shaped, which agrees with stomach content inventories that include snail shells. The intestines reach 70% of the standard length. The teeth of the two species, along with those of *L. petricola,* belonging to the same line, are peculiar. The central canines are very long and strong and bunched together like those of a rabbit, perhaps to scoop sand.

Lamprologus leleupi. Photo by Dr. H. Grier.

A faded *Lamprologus meeli.* Photo by H.-J. Richter.

Lamprologus moorii Boulenger

This species was named after Moore, the first explorer who sent several fish collections to Boulenger from the lake. It has not been discovered in the northern part of the lake and is restricted to the southern shores. This species has several unusual features for a *Lamprologus,*

Lamprologus meeli is a very poorly known dwarf cichlid. Its relationships with *L. hecqui* and *L. boulengeri* are uncertain.

Lamprologus multifasciatus. *Lamprologus ocellatus.* Photos by H.-J. Richter.

Lamprologus mondabu. Photo by Dr. Herbert R. Axelrod.

Lamprologus moorii in black breeding color. Photo by Dr. Herbert R. Axelrod.

A young or non-breeding *Lamprologus moorii.* Photo by Pierre Brichard.

including the very high number of canines in both jaws (the upper in excess of 10) and a very long intestine (200% of the standard length). This should indicate a very different diet from the one normal to all *Lamprologus,* i.e. flesh in every form. This feature seems borne out by the stomach contents of *L. moorii,* which consist of algae and ostracods, but the tremendous array of sharp, curved, hook-like canines, interlocking when the jaws are closed, might point toward a flesh-ripping diet. This fish is one of the best looking fish in the wild, with flowing fins, jet-black body, and a sky-blue edge to the tail. But aquarists report it has a very nasty temper.

Lamprologus multifasciatus
Boulenger

(Shell-dweller)
This is perhaps the cutest and most attractive dwarf *Lamprologus,* with its hazelnut colored background and about 10 golden stripes (that give it its name) on the body. It lives in *Neothauma* shell piles at the foot of the rocky slopes.

An adult at 25mm, *L. multifasciatus* feeds on tiny invertebrates in the wild and adapts quite well to dried food in captivity. It is always found in rather deep water where the shell piles are not rolled about by the surf. It is thus caught more by accident whenever one discovers one of these piles. *L. multifasciatus* is a perfect fish for the small tank, along with those species that have already been mentioned as suitable for new cichlid amateurs.

Lamprologus mustax Poll
(Rubble-dweller)

The name *mustax* means "moustache," referring in this case to the dark brown stripes rising from the sides of the mouth.

A typical rubble-dweller like so many *Lamprologus* species (*niger, leleupi,* etc.), *L. mustax* is found only on the southern shores of the lake, especially in Cameron Bay. It was mistaken for *L. leleupi* at first by the first collectors in the area—which it definitely is not—because of the yellow color, which is typical of the species. It is chestnut brown on the back with creamy white underparts, a white chin and cheek, and all the fins bright yellow.

The high and plump body of females and the more slender one of males is the only noticeable sexual dimorphism as is the case in many species of substrate spawners. It is not rare in its preferred habitat around Cape Chipimbi. Like the other species in this group, the fish has been exported more or less sporadically and has been bred in captivity. Conspecific aggression is rather high as with all rubble-dwellers as they usually are very territorial.

The ethology of this species is as yet poorly documented. It breeds in rubble.

Lamprologus niger Poll
(Rubble-dweller)

Apparently *L. niger* belongs to the same species group as *L. leleupi.* It occurs on the northwestern shores of the lake and is not found, to my knowledge, in the southern half of the lake nor in Burundi. It appears to occupy the same ecological niche as *L.leleupi,*

A white-chinned *Lamprologus mustax.* Photo by Pierre Brichard.

Lamprologus niger is certainly not a colorful species. Photos by Dr. Herbert R. Axelrod.

Lamprologus lemairei juvenile.

Lamprologus mondabu.

Lamprologus mustax female.

Lamprologus mustax male.

Lamprologus multifasciatus.

Lamprologus niger.

Lamprologus ocellatus.

Lamprologus ocellatus with parasite. Photos by Pierre Brichard.

Lamprologus mustax is one of the more interesting recent discoveries. Photo by Yasuyuki Kurasawa, *Midori Shobo.*

which might explain why *L. leleupi* does not extend much along the coastline, especially on the Ubwari Peninsula where they are replaced by a strong population of *niger.*

L. niger is very abundant in their habitat, which is the piled-up rubble, the only species found along with them and in about equal numbers being *L. savoryi.* In these areas several other rubble-dwellers do not appear in strength. These are *L. brichardi, L. schreyeni,* and perhaps *L. compressiceps,* certainly not *Julidochromis,* of which there are very few indeed.

Together, the two species, *L. niger* and *L. savoryi,* appear to occupy most of the available territories. *L. niger* is well

equipped to have the upper hand, being well armor-plated, the scales very ctenoid with strong denticles extending onto the unpaired fins, having a high number of spines in the dorsal (20) and anal (8-9) fins, well developed lateral lines, and rows of very large sensory cavities on the head.

The average size is 6-7cm, seldom more, but can reach to a bit less than 9cm. The color is, in adults, a very dark brown, nearly black (hence the name *niger),* with all unpaired fins displaying a typical checkerboard pattern. The eyes are dark red and the pectorals a rich reddish brown. All this adds up to a rather attractive fish.

Juveniles have such a different pattern that at first it was hard to recognize them as *L. niger.* They are entirely red-

copper or an orange-vermilion-red without any checkerboard pattern on the unpaired fins until nearly 4cm long. Then the colors darken and the crisscross pattern appears on the fins.

Under the eye of *L. niger,* in the infraorbital area, a fleshy crescentic pad of a paler hue follows the outline of the eye socket. This pad is found in several other rock-dwelling *Lamprologus* such as *L. brichardi, L. leleupi, L. leleupi longior, L. mustax, L. savoryi,* etc. Whether or not this crescent-shaped pad bears a relationship to the disposition of the infraorbital ring of tubular bones in this series of species has not been investigated. This feature might perhaps be of phyletic value.

L. niger could become an

Lamprologus obscurus is a cave-dweller that seldom leaves the shelter of the rubble. In many ways it is similar to *L. niger*. Photo by Pierre Brichard.

excellent aquarium fish were it to be exported, which is doubtful because of the remote areas where it is being found. Exporters will not go out of their way to make costly collecting trips for fishes whose value is not exceptional.

Lamprologus obscurus Poll
(Cave-dweller)

L. obscurus is a species that I was fortunate to discover on the coastline south of Cape Chipimbi in Zambia after previous explorers had missed the fish. No wonder! It remains hidden in the rubble and never comes into view in the open. Thus it is a typical so-called cave-dweller. This is reflected by its color, a deep chestnut lightened by a few scattered off-white scales.

A typical *Lamprologus* with a very long premaxillary pedicel and angular chin, it might be the southern counterpart of *L. niger.*

Not large, 6cm appears to be its maximum size. It could

well become another attractive rock-dwelling nestbreeder, were it found in an area from which exports of inexpensive fishes are not hindered by high collecting and transportation costs.

Lamprologus ocellatus
(Steindachner)

(Shell-dweller)

L. ocellatus is one of the dwarf cichlids. It is of little value to the hobbyist because of its lackluster colors and because it looks much like *L. brevis* (but *brevis* is much better looking). It can be distinguished from *brevis* at once by the concave outline of the snout, which appears slightly caved-in. This is caused by the eyes, which are set high in the head and whose top rim bulges above the snout. The colors are not unattractive but altogether a bit drab. The underparts are silvery, a bluish or greenish cast occurs behind the pectorals, it has an olive-beige back, and the fins are poorly colored.

It lives in *Neothauma* shells or at least breeds in them, and has been found only in small schools and often mixed with *L. brevis,* with which perhaps the species competes for possession of the shell piles. It is a microfeeder on invertebrates like so many *Lamprologus* species.

The shell-dwelling *Lamprologus* have been noted as having a much reduced lower lateral line. It often appears to be missing entirely and no tubuled scales are to be seen. In fact, it would be likely that the tubules did not grow on the neuromast cavities which remained as open pores. The lower lateral line of *Lamprologus* as a genus, and especially of the shell-dwelling species, turns out to be much more developed than was first thought.

L. ocellatus is found on appropriate grounds all around the lake and as yet no geographic races of any interest have been found. It is exported sporadically and has been successfully spawned in captivity. Not choosy about its food, but better off with live invertebrates, it is easy to maintain in top condition. This is another good fish to try one's hand at as an introduction to the African cichlid tank.

Lamprologus ornatipinnis
Poll

L. ornatipinnis is definitely not a shell-dweller because of its size (8-9cm). It doesn't fit into a shell, but very probably, like so many other so-called "shell-dwellers" and *L. callipterus,* it is a shell spawner. It is always found swimming off the sand bottom

Spawning in *Lamprologus ocellatus*. The female (top right) enters the empty shell, lays her eggs, and the male (left and bottom) fertilizes them. These fish have fangs (large canine teeth) that are curved and very sharp. Photos by H.-J. Richter.

and sometimes over rock, in whose recesses it perhaps might spawn as well.

I have never as yet seen *L. ornatipinnis* in the lake except singly or as pairs or trios, and never in schools. They might, however, be gregarious for all we know and the fish captured or seen might have been stragglers. One of the species roaming the sunken sand plains, *L. ornatipinnis* perhaps commonly stays deep, which explains why the fish is seldom exported.

Although its body is rather a plain greenish beige, the fish is unmistakable because of the striking pattern of its unpaired fins (*ornatipinnis* means ornate fins), which are all alternately striped in black and white.

From what we could see it has a very peaceful temper, and its lack of aggression or territorialism would make this fish another valuable addition to a tank. Collected only by sheer luck during far in between trips, the fish will never become common.

Lamprologus petricola Poll

A yellow morph of *L. moorii* coming from near Cape Kabeyeye was, in my earlier book, thought to have been *L. petricola*. The real fish was only discovered and brought back recently by my son Thierry during a deep dive.

This is one of the most elusive species of the lake, of which only very few have been sighted on the northwestern coast of the basin north of the Ubwari and so far always in rather deep water.

It is a pale beige in color. The head is unmistakable, as the snout is really long, oblique, and straight, and the mouth is small and in line with the lower profile of the body.

A very low mouth is common among the rock-grazing mouthbrooders such as *Tropheus, Simochromis,* and *Eretmodus,* but *L. petricola* is the only substrate breeder as yet found with such a morphological evolution. The mouth being rather narrow, one would tend to believe that they are not rock grazers. The stomach contents, investigated by Poll, consisted of insect larvae, which points toward a rock-pecking diet. Strangely, all the specimens that he collected from the western coast were captured in shallow water. The front teeth, like those of *modestus,* are very long and strong.

The pattern of the species looks much like that of *L. modestus,* but the fins have a peculiar pattern consisting of creamy white spots and short stripes on a transparent background.

The breeding behavior of this funny-looking

Lamprologus has not been clarified, all specimens seen in the wild having been alone at the time. The name *petricola* means "living in the rocks."

Lamprologus prochilus
Bailey & Stewart

I believe this to be one of the most specialized species among nest-breeding cichlids. But although we can guess at what the specializations and adaptations are we still do not know how they work and why. Everything about this fish appears to be very unusual.

This is the only fish I know of that when seen from above is symmetrical, the head being very pointed and tapering off in the same way as the caudal peduncle. From the side the fish looks a bit like *L. compressiceps,* with a deep body and a big head with powerful jaws forming a very oblique mouth. The very denticulate but small scales, as in many rubble-dwellers, extend along the rays of the dorsal and anal fins and well over the tail. The lateral lines are extensive and the neuromasts buried in the infraorbital area open up into a row of open pores.

The other sensory cavities on the jaws and the snout are also very large. In this respect *L. prochilus* has followed the same adaptations as *L. furcifer*—strong static defenses including 20 or 21 dorsal and 7 or 8 anal spines, and what should add up to excellent detection systems. This appears to be a case of convergent evolution of various lines of *Lamprologus* living close to or within the rubble.

Probably the two species do not have the same ecological

The seldom-seen *Lamprologus ornatipinnis.* Photo by Pierre Brichard.

Lamprologus tetracanthus, female.

Lamprologus tetracanthus, male.

Lamprologus savoryi.

Lamprologus sexfasciatus.

Lepidiolamprologus pleuromaculatus.

Lamprologus prochilus.

Lamprologus ocellatus.

Lamprologus meeli. Photos by Dr. H. R. Axelrod, Pierre Brichard, and H.-J. Richter.

niche. Given the morphology of the head, one is not surprised to find that *L. furcifer* is a predator, feeding on small fishes living in the caves, nor would one be surprised to learn that, despite its frightful jaws, *L. prochilus* is, like *L. compressiceps,* living on invertebrates. The long premaxillary pedicel means of course that the mouth can open forward much moreso than in species having a shorter pedicel. But we have seen in *L. compressiceps* that the long pedicel fully extended is the only buttress keeping the fully extended jaws in line with the cranium. The bone has a propensity to jump out of its groove at the end of its slide leaving the mouth ajar. Thus a very long pedicel on a narrow snout is a delicate mechanism that cannot bear the stresses involved in the violent twisting efforts of a large prey animal trying to get free. Fishes like *L. compressiceps, L. prochilus,* or *Cyprichromis* spp., which all have a long pedicel and can project their jaws well in front of the head, can only capture defenseless prey like small invertebrates.

The color pattern of *L. prochilus* is unusual as well, but well adapted to the dim light penetrating the deep recesses. A pink mauve is the best approximation one finds to describe the color, with a few irregular darker bands on the sides. I once managed to capture a breeding pair in which pink prevailed, which made for a very unusual fish to look at. Juveniles appear to be still with the parents when 3cm long, and they are a brighter rosy mauve. One cannot say that, given the squat body of the species, they were any better looking for this fact than

their parents!

The fish does not appear to be very prolific, and at most perhaps 20 fry 2cm long were found together, which probably means that the spawn did not involve more than 50 eggs.

The habitat is definitely under the rubble. Never has a single specimen been captured that had been seen first in the open, or even poking its nose out from under a stone. Thus the fish is very secretive and one of the most typical cave-dwellers in the lake.

Perhaps it will turn out that *L. compressiceps, L. calvus,* and *L. prochilus* stem from a single ancestor, the first two species adapting to life in the open, the last to life in the deep recesses. But we will have to wait for an investigation on the anatomy of these species and many others before we can reach any conclusions.

What is remarkable about the cave-dwelling cichlids in the lake is that only nest breeding species took to the recesses. As yet no mouthbrooding cichlid has been found in the lake, nor for that matter a really specialized cave-dweller in Lake Malawi, to have undergone major adaptations toward life in cave-like recesses. As a result, none of the various mouthbrooding cichlid lines in the lake have developed anatomical or physiological adaptations to permanent obscurity, confined space, and promiscuity with a host of non-cichlid predators.

Perhaps the nestbreeders succeeded in this respect because they are more sedentary and they could thereby take a gradual step toward life in the dark. Rock-

dwelling mouthbrooders like *Tropheus, Simochromis,* and *Petrochromis* more or less enter the recesses and might stay within them for a while, but sooner or later they come out again and wander about in the open. Their occupation of an anfractuosity is therefore not permanent.

The coloration of cave-dwelling fishes, be they cichlids or not, is usually very dark, in varieties of brown, blue, purple, and black. The colors are matte, never silvery. This feature is shared by catfishes and *Mastacembelus.* There are very few pale fishes in the deep rubble, which is the opposite of what we see in the open where dark colored fishes are very few. *L. moorii* along with the very large *L. compressiceps* are the sole exceptions I know of. In some cases this rule is so strict that with fishes that display a variety of patterns in the same habitat, one can identify the place on the slope where they live by their coloration. For example there are the melanistic varieties of *Julidochromis.*

L. prochilus, on the other hand, has a very soft and plump body, contrary to *L. compressiceps* which has a hard and lean body. As a result the fish is bruised at the slightest mishandling and the slightest scratch brings its downfall, as infections spread quickly and cannot be stopped.

Difficult to find, difficult to handle and acclimatize, and difficult to keep alive and in proper condition, *L. prochilus* will probably never become a common sight in a tank. Yet it probably will remain one of the most remarkable species to have been found in the lake.

The name *prochilus* means

"with the strong chin" and comes from the Greek pro-cheilos, alluding to the long premaxillary pedicel typical of the species.

Lamprologus pulcher (Poll) (Zoobiocover and drifting invertebrates)

This species was at first considered a subspecies of *L. savoryi* but was set apart as a distinct species by Poll. Aside from the preserved specimens, little was known about the ecology of the fish, and no specimens had been observed, much less caught, underwater until I discovered it in the south.

The fish is known to be a rock-dweller and is probably microphagous like *L. brichardi,* to which it seems to be closely related.

Lamprologus savoryi Poll (Rubble-dweller)

Not at all a bad looking fish, *L. savoryi* has a very nasty temper that makes it a poor aquarium fish unless kept with larger fishes. This is not too difficult because the maximum size of *L. savoryi* doesn't exceed 80 mm by much. The color is dark blue with very dark blue bands. The fins are elongated into filaments, not as long, however, as those of *L. brichardi.*

The fry and the adults hide in the rock rubble and are never seen in the open. The fish spawn in deep, dark recesses, a few fry being hatched in each spawn. We never saw more than a dozen fry, which is extremely low for egglayers and for a species of *Lamprologus.* The fry are watched over by the parents until they are quite large, at least 30 mm long. At that size

the young *L. savoryi* can easily be mistaken for young *Cyphotilapia,* with the lighter bands, and the head and the lips being a pale blue, contrasting with the deep blue bands.

It is one of the most secretive fish in the lake because of its hiding places and is seldom captured although not at all uncommon in its habitat. The diet, according to examination of the stomach contents, included Diptera, copepods, and vegetable matter (quite unusual for the genus). The coiled intestine reaches 70% of the standard length, which

would indicate a rather omnivorous diet. The southern race, instead of being bluish pink, is smoke-colored and shows no bands in the wild.

This species was named after Mr. Savory, District Commissioner in Kigoma in 1946.

Lamprologus schreyeni Poll (Rock-dweller)

Lamprologus schreyeni is one of the dwarf rock-dwelling species of *Lamprologus* living deep inside rock crevices. It is never seen in the open where its very small size would make it an immediate victim of the

Under water many of the cichlids appear as dark fishes with little obvious pattern or distinguishing features. Photo of *Lamprologus* sp. by G. S. Axelrod.

Lamprologus savoryi has a very nasty temperament. Photo by Dr. Warren E. Burgess.

363

Lamprologus brichardi var. *Lamprologus brichardi* var.

Lamprologus brichardi var. *Lamprologus brichardi* var.

Lepidiolamprologus cunningtoni. *Lamprologus savoryi.*

Lamprologus moorii. Photos by H. J. Mayland, D.
Lamprologus moorii. Schaller, and R. Stawikowski.

predators of the "outside" world.

The species belongs to a group of small rock-dwellers that spend all of their time in hiding, being discovered only by chance. Their capture can only be made during underwater SCUBA explorations made at leisure. When new species of *Lamprologus* are discovered in Lake Tanganyika most probably shy rock-dwellers will see their numbers increase substantially.

This species, although quite small, has a high number of dorsal fin spines (XVII-XIX) and a low number of anal spines (V). The pharyngeal teeth include two enlarged molar-shaped teeth in the center rear that would tend to indicate a diet including, perhaps, small snails or clams. This is not at all out of line with their habitat—among rocks with a rich biocover.

Breeding has never been researched. The number of fry seems to be about 20, perhaps more; they are watched over by both parents. The nests appear to be very narrow slits in the rocks, out of reach of any other fish except young mastacembelid eels or *Telmatochromis bifrenatus.*

L. schreyeni has not been found outside of the northern basin, and in this basin outside of the eastern coast. It thus seems to have a very restricted range. This is shared by another typical cave-dweller, *L. prochilus.* One might conclude that with strong bonds to a very special type of habitat they have little leeway in colonizing new territories and biotopes.

L. schreyeni is named after A. Schreyen, who collaborated as a member of the Fishes of Burundi team in the discovery of this species.

Lamprologus sexfasciatus
Trewavas & Poll

(Rock-dweller, in the open.)

This species is without doubt one of the most spectacular fishes in Lake Tanganyika, with its rich deep blue bands alternating with pale blue, the iridescent green and electric blue spot on the opercle and the striking pale blue edges to the unpaired fins. The species is very close to *L. tretocephalus,* but it is very easy to tell them apart.

Both species are found on the same type of substratum (rocks), but they are never found together. None have been discovered on the northeastern coast of the lake, although they are rather abundant on the eastern coast from the Tanzania-Burundi border toward the south. On the northwestern coast they are common but not abundant. This is one of the striking examples of the variety of coastal fish populations around the lake.

L. sexfasciatus is distinguished from its close relatives by the fact that it has 6 dark bands (of which 4 start along the dorsal length), whereas *L. tretocephalus* only has five (of which three start along the dorsal length). The two species have peculiar flat molar teeth on the center and rear of the pharyngeal bone, which has a different shape in *L. tretocephalus*—and an unusual one at that.

Both species live over the rubble, not in it. The size of *L. sexfasciatus,* according to specimens in collections, is said not to exceed 90 mm, but I caught some that were 12 cm long.

The diet is said to include snails, among them *Neothauma.* The shell of this snail is very hard and it is thus in line with this diet that the pharyngeal bone and teeth are so strong. The coiled intestine reaches nearly 90% of the standard length, which is quite long.

A geographic race has been found in 1983 in the southern part of the Tanzanian coastline. The back and sides of this race are a rich yellow, the other parts pale blue, the bands appearing to be less contrasting than in the more common variety. It doesn't appear from the investigation conducted by Prof. Poll that the new variety deserves even subspecific status.

Lamprologus signatus Poll
(Shell-dweller?)

This species belongs to the same group (and probably lineage) as *L. ocellatus,* with which it has many affinities.

A very low number of gill rakers (4-5), of dorsal (XV) and anal (V) spines, and a naked chest, belly, and nape are the features of this dwarf *Lamprologus.* On each jaw there are two outsized fang-shaped canines. The eye is very big and the interorbital space very narrow (7.25-10 times in head length). The habitat is reported to be deep water on soft, muddy bottoms. The fish feed on small shrimp. The maximum size of collected specimens was 53 mm, at which size the fish were mature. Nothing else is known of this species other than it belongs to the shell-dweller group.

Although *Lamprologus savoryi* closely resembles *Lamprologus brichardi,* it is an extremely nasty fish. Photo by Pierre Brichard.

Lamprologus sexfasciatus is one of the more spectacular cichlids in the Lake. Photo by Shuichi Iwai, *Midori Shobo.*

Lamprologus stappersi
Pellegrin

This fish was described by Pellegrin from only a single specimen collected in or near the Sambala River (?) near Pala in Tanzania by the French fathers that were at that time in the area. This would then be the only *Lamprologus* as yet found in a river of the lake basin. Most probably it is a lacustrine species that ventures into river mouths. It has never been seen or collected since its original capture. By general features it appears to belong also to the shell-dweller group of fishes.

This fish was named after Dr. Stappers, a South African who collected the fish and who died in East Africa during WW I.

Lamprologus tetracanthus
Boulenger

(Ubiquitous omnivore)

This is one of the least specialized forms of *Lamprologus,* found mainly on sand all around the lake without noticeable local races.

This species has general primitive features (among which is the lowest anal spine count of all *Lamprologus)* and might look much like some of their ancestors did.

Lamprologus tetracanthus is one of the best looking sand-dwelling species of *Lamprologus,* with its alternating rows of pearly and beige scales, four or five vertical bands, and long ventral and anal fins, the latter spotted. The caudal and dorsal fins are also spotted and have a white or yellow band at the edge. Each ray of the dorsal fin is tipped either with bright red or black. Unfortunately, *L. tetracanthus* grows to a maximum size of nearly 200 mm and of course is not to be kept with other fishes of a size it might swallow. Two molars at the back of the pharyngeal bone, stomach content inventories including mainly snails and fish, and the coiled intestine reaching 90% of the standard length would lead us to believe that the species is not one of the typical predators of the lake.

Breeding has been observed in the open on sand flats in a crater-nest dug from the bottom. The fry number several hundred and are closely watched by the parents. We don't remember having seen the fish over rock.

Although too large for the average aquarium, *Lamprologus tetracanthus* is a stunning fish. The white spine tips may indicate that this is a female. Photo by Mitsuyoshi Tatematsu, *Midori Shobo.*

Haplochromis benthicola, female.

Lamprologus sexfasciatus, yellow Tanzania race.

Lepidiolamprologus attenuatus, juvenile.

Lepidiolamprologus elongatus.

Lepidiolamprologus profundicola.

Lepidiolamprologus elongatus, juvenile.

Ophthalmotilapia nasutus, Burundi.

Ophthalmotilapia nasutus. Photos by Pierre Brichard.

Lamprologus toae Poll
(Rock-dweller)

In more than 10 years of collecting neither I nor any member of my team ever saw *L. toae* on the northeastern coast of the lake, although it has been fished south of Burundi and its border with Tanzania, down to the Malagarazi. On the west coast it is a common fish down to Kalemie. It does not exist in the southernmost part of the lake. Thus it is one of the species providing a clue about the separation of the lake lengthwise into two distinct entities.

It also provides a clue, along with *Tropheus, Simochromis, Petrochromis,* and *L. leleupi,* to the way the colonization of the northern basin occurred, and to the successive appearance of several ecological barriers, in this case the one built by the Ruzzizi River delta, the coastline between Nyanza Lac and Rumonge, the ford between the tip of the Ubwari Peninsula and the eastern coast, etc.

The species is a rock-dweller, living in the open. It is rather small (maximum size 100mm). The fish doesn't look like a typical *Lamprologus* at first. It is difficult to say, but there is something different in the appearance of the fish which is not at all unattractive, with its very big eye, the fins well erect, and the scales in a horizontally contrasting pattern.

It has a very special dentition shared with *L. moorii.* The canines are not very sturdy, nor are they very much differentiated from the other teeth or from each other. The mouth is also straight in front instead of oval as in most species of *Lamprologus.*

The red tips of the dorsal fin spine membranes may be an indication that this *Lamprologus tetracanthus* is a male. Photo by Pierre Brichard.

A subadult *Lamprologus toae* about 70 mm long. Photo by Pierre Brichard.

The coiled intestine reaches 60% of the standard length, and the diet is reported to be mainly aquatic insect larvae.

When *L. toae* is looking after its fry, numbering according to our count about 60 when they become free-swimming and come into the open from their birthplace among the rocks, both parents take good care of their brood. One of them is often seen sailing after an intruder at full speed. They appear in the northern part of the lake to occupy, more or less, the ecological niche of *L. moorii*. Like *moorii* (but a little less) they like to rise above the rubble. *L. toae* also doesn't display the formidable temper of its southern counterpart. Strangely, the two species are as yet the only ones found in the lake that answer, to some extent, the definition of *Lamprologus* by Regan. They have the infraorbital chain of bones, at least in part. It is entirely possible that they both belong to a distinct line of the genus. Were this to be the case one would think that some of their ancestors turned south of what was to become the Lukuga area sand mound, while the other (*L. toae*) developed to the north of this tremendous ecological barrier.

Lamprologus tretocephalus. Photo by B. Carissimi.

L. toae is differentiated from all other *Lamprologus* species by having an open sensory infraorbital canal along with the remnants of the infraorbital bones.

This fish was named after N'Toa Bay where it was first found.

Lamprologus tretocephalus Boulenger (Rock-dweller)

This is again one of the species that, like *L. toae,* are found in the northern basin only on the west coast, the Ubwari, and then south to the Malagarazi delta and the Lukuga outlet, but not beyond.

Much has been said already about this species in the discussion of *L. sexfasciatus.* The fish is said to reach about 140 mm and has a diet consisting of shells and aquatic insect larvae. The coiled intestine is only 30% of the standard length in contrast to *L. sexfasciatus* where it reaches 90%. The species are certainly differentiated by their basic diet, although *L. tretocephalus* is also reported to eat some clams; in the lake most molluscs of this type usually have thin and brittle shells. This is perhaps as good a reason as any to explain the huge difference in intestine length.

L. tretocephalus is of course one of the best Lake Tanganyika additions to the aquarium fish inventory. It is peaceful toward other fishes, although fights might occur between them when too crowded in small tanks.

Lamprologus wauthioni Poll (Shell-dweller)

L. wauthioni is one of the shell-dwellers, or at least shell-spawners, and a dwarf species. The concave snout and large eye present some similarity with *L. ocellatus* but the fish can be easily distinguished by the fact that *L. ocellauts* has from 7 to 9 anal spines and *L. wauthioni* always has 6. There are a few dark patches on the sides and all unpaired fins seem to have a number of ocelli. But the color pattern has in fact not been documented, as few captures of this species have occurred since it was first discovered in 1946 in the central part of the lake. It might belong to the same group as *L. caudopunctatus* (along with *L. leloupi* and *L. meeli).*

Definition of the Genus:
-all scales ctenoid;
-all teeth conical, in several rows, the front teeth of the outer row always including at least 2 enlarged canines on the sides of each jaw and a set of from 2 to 10 intermediate frontal teeth;
-cheeks naked and tegumentous, scaleless, or covered with a field of scales in rows or not;
-anal fin with from 4 to 13 spines and dorsal with 14 to 21 spines;
-inferior tubulated lateral line always preceded by a row of open pore scales, eventually complete.

One of the most favored and spectacular of the Lake Tanganyika *Lamprologus* is *Lamprologus tretocephalus*. The brilliantly colored adults (above) are very peaceful. Even the juveniles (bottom) are attractive. Photos by Shuichi Iwai and Yasuyuki Kurasawa, *Midori Shobo*.

371

Key to the Species of *Lamprologus* and *Lepidiolamprologus*
(This key is for field identification only; for taxonomically more certain identification please refer to original descriptions.)

A. More than 45 scales in longitudinal line and more than 45 scales in upper lateral line (*Lepidiolamprologus*). (See also B and C)
  1. 47 to 60 scales in long. line; dorsal spines XVII-XIX,9-11; anal fin V-VII,7-9; gill rakers 9-12; lat. lines: 45/17 ....*pleuromaculatus*
  2. 63-72 scales in long. line; dorsal fin XVII-XVIII,11; anal fin VIII-IX,8-9; lat. lines 54/22; cheek covered with multiple rows of scales separated from the naked area by a raised fold of skin ....
  ....*profundicola*
  3. 61-71 scales in long. line; D.XX-XXI,10-12; A.IV-V,7-9; lat. lines 50/22; 6-8 gill rakers ...*cunningtoni*
  4. Eye large, 2.2 times in head depth, top rim level with snout profile; D.XVII-XX,9-12; A.IV-VII,7-9; long. line 66-73; lat. lines 55/20; 13-17 gill rakers ....*attenuatus*
  5. Eye small, about 4 times in head depth, top rim not level with and below snout profile; 66-74 scales in long. line; lat. lines 55/27; D.XVII-XVIII,10-12; A.V,7-9; 11 to 14 gill rakers; body with 4 or 5 whitish spots on sides and small pearly thin strips all over body and fins ....*elongatus*
  6. 66-73 scales in long. line; lat. lines 56—26; D.XVIII) 10-11; A.V 8-9; gill-rakers 10; body smoky gray with 3 or 4 wavy black horizontal stripes .... *kendalli*

B. More than 40 scales in longitudinal line and less than 40 in upper lateral line (*Lamprologus,* part). (See also A and C)
  1. Dorsal and anal fins in part covered with scales; tail lobed.
    a. Dorsal spines XIX-XX; anal spines IV-V; 50 to 60 scales in long. line.................................*christyi*
    b. Dorsal spines XX; anal spines V-VII; 46 to 54 scales in longitudinal line ....*furcifer*
    c. Dorsal spines XIX-XX; anal spines VI; 41-45 scales in long. line ....*buescheri*
  2. Dorsal and anal fins not covered with scales; tail not lobed.
    a. Dorsal spines XVII-XIX; anal spines VI-VII; lower lateral line atrophied; 48-54 scales in long. line; maximum size 60mm; ....*hecqui*
    b. Dorsal spines XVII-XIX; anal spines VI-VII; lower lateral line not atrophied (more than 20% of long. line); 42-50 scales in long. line; size 67mm.. *meeli*
  3. Dorsal and anal fins scaled; tail not lobed.
    a. Dorsal spines XX-XXI, anal spines VII-VIII; 48-57 scales in long. line; maxillary pedicel very long; body seen from above entirely symmetrical from snout to tail; maximum size about 120mm ....
    .... *prochilus*

C. Less than 40 scales in longitudinal line (*Lamprologus,* part). (See also A and B)
  1. Not more than 15 spines in dorsal fin.
    a. Anal spines V; 35 scales in longitudinal line; lower lateral line atrophied; body striped vertically; dwarf species..*signatus*
    b. Anal spines VI; 30 to 33 scales in longitudinal line; body plain, with black spot in dorsal fin; many fang-shaped frontal teeth; dwarf species ....*kungweensis*
  2. More than 15 spines in dorsal fin; V-VI anal fin spines.
    a. Four median anterior teeth bunched together between the canines and as long as the canines; all frontal teeth long, strong, with blunt crowns.

    (1). Mouth terminal; 33-37 scales in long. line; tail rounded .... *modestus*
    (2). Mouth terminal; 33-36 scales in long. line; tail straight or concave ....*mondabu*
    (3). Mouth very low, level with lower profile of body; snout straight and oblique; 33-36 scales in longitudinal line .... *petricola*
    b. Median anterior teeth not bunched together, but spread between the canines at more or less regular intervals.
    (1) Lower lateral line not atrophied (tubulated scales in excess of 20% of scales in longitudinal line).
      # Dorsal, anal, and caudal fins not covered with scales; caudal rounded or truncated.
        @. Always IV anal spines; dorsal fin XVIII-XX,29-32; 36-40 scales in long. line; body with several horizontal rows of pearly scales ....*tetracanthus*
        @@. Anal fin spines V-VII; body mottled.
          —Head about 2.5 times in standard length; 33-37 scales in long. line; body olive yellow with darker patches ....*lemairei*
          —Head more than 3 times in standard length; 35-37 scales in long. line; body olive yellow with darker patches ....*callipterus*
        @@@. Anal fin spines V-VII; dorsal fin XVII-XVIII,25-27; silvery rows of pearly scales on sides.
          —34 scales in long. line; eyes blue, body pale yellow, tail pale gray; maximum size 60 mm ..
          ....*caudopunctatus*

Lamprologus brichardi.

Lamprologus brichardi.

Lamprologus brevis "Magara".

Lamprologus brevis.

Lamprologus callipterus.

Lpeidiolamprologus attenuatus, nuptial colors.

Lamprologus compressiceps.

Lamprologus caudopunctatus. Photos by Dr. Herbert R. Axelrod and H.-J. Richter.

–33 scales in long. line; tail typical with margin black, then a pure white band rimmed with two black stripes; maximum size about 60 mm .... *leloupi*

@@@. Anal fin spinesV-VII; dorsal fin XVII-XIX,27-29; body with alternate bands of dark blue and pale blue or white.

–33-35 scales in longitudinal line; 6-8 gill rakers; 3 black stripes in dorsal fin *tretocephalus*

–34-37 scales in longitudinal line; 11-13 gill rakers; 4 black stripes in dorsal fin .... *sexfasciatus*

##. Dorsal, anal, and caudal fins more or less covered with scales, at least along their bases.

@. Caudal rounded or truncated, without filaments.

&. Maxillary pedicel very long.

%. Body laterally compressed, deep; IX-XIII anal spines.

–Depth of body 40% of standard length; nape covered with scales; anal spines IX-XII; dorsal spines XX-XXI; 31-34 scales in long. line; max. size 120mm ..... *compressiceps*

–Depth of body about 35% of standard length; nape not covered with scales; anal spines X-XIII; dorsal spines XX-XXI; 32-34 scales in long. line; max. size 120 mm .... *calvus*

%%. Body elongate and cylindrical; X anal spines; darker triangular bands on back and sides .... *fasciatus*

%%%. Body normally shaped; VI-VIII anal spines; dark colored with clear patches .... *obscurus*

&&. Maxillary pedicel short or medium size.

%. 36-38 scales in long. line; 6-8 teeth in front of each jaw, the middle ones shorter; pectoral fins orange, eyes red; juveniles red; maximum size approximately 80 mm........ .... *niger*

%%. 31-35 scales in longitudinal line.

¢. 8-12 anterior teeth that are long, sharp, hooked, and nearly as long as canine teeth.

–Anal spines VII-IX; dorsal XIX-XXI,28-30; 33-35 scales in long. line; body black; maximum size 110 m...*moorii*

–Anal spines V-VI; dorsal spines XVII; 29-33 scales in long. line; body mottled with silvery scales... *toae*

¢¢. 6-8 anterior teeth, the median much smaller than canines; body yellow, beige, or brown.

–Anal spines VI-IX; dorsal spines XIX-XX; 33-34 scales in long. line; body bright yellow ....*leleupi leleupi*

–Anal spines VI; dorsal spines XIX-XX; 33-34 scales in long. line; body chestnut brown...*leleupi melas*

–Anal spines V-VI; dorsal spines XIX; 33 scales in long. line; body bright orange-yellow; more elongate than other subspecies ....*leleupi longior*

–Anal spines V-VI; dorsal spines XIX; 32-35 scales in long. line; lower part of head whitish; body hazelnut brown; unpaired fins yellow ....*mustax*

@@. Caudal fin always lyretail; anal spines V-VI; dorsal spines XVIII-XX.

&. More than 15 gill rakers; body depth 25-28% of standard length........... *gracilis,* sp. nov.

&&. Not more than 14 gill rakers; body depth 29-35% of standard length.

%. Only 8½ scales in lower transverse line; only 7 gill rakers .... *olivaceous,* sp. nov.

%%. 9½ scales in lower transverse line; 8-10 gill rakers .... *pulcher*

%%%. 10½-13½ scales in transverse lower line.

–8-9 gill rakers; anal spines V-VI; dorsal spines XVIII-XIX; 33-36 scales in long. line; lat.

l. scales $\frac{18\text{-}26}{6\text{-}10}$, body purple black, all fins black, a crescentric yellow stripe at the caudal margin, bloodred opercular spot; preopercle and opercle very scaly ....

.... *splendens,* sp. nov.

–9-10 gill rakers; anal spines VI; dorsal spines XIX-XX; 33-36 scales in long. line; lat. l.

scales $\frac{19\text{-}23}{6\text{-}7}$ transverse lower line scales10½ to 11½; no opercular spot; body rather

plump, plain pale purple, eye very blue, filaments of lyretail tipped with white; only the

opercle scaly; few scales on upaired fins .... *crassus,* sp. nov.

–7-14 gill rakers; anal spines V-VI; dorsal spines XVIII-XIX; 34-36 scales in long. line;

lat. l. scales $\frac{18\text{-}22}{8\text{-}11}$, transverse lower line scales 10½ to 12½; no scales on opercle or

preopercle; unpaired fins very scaly; opercular spot and stripes present; body beige with

yellow spots on the scales; unpaired fins all filamentous, the filaments long and white ..

.... *brichardi*

Lamprologus leloupi.

Lamprologus leleupi, Lunangwa.

Lamprologus gracilis.

Lamprologus olivaceus.

Lamprologus brevis "Magara".

Lamprologus compressiceps, orange.

Julidochromis ornatus, Kapampa.

Julidochromis sp., Kapampa. Photos by Pierre Brichard.

—8-9 gill rakers; anal spines V-VI; dorsal spines XIX-XX; 31-34 scales in long. line; lat. line scales $\frac{19\text{-}25}{4\text{-}9}$; transverse lower line scales 10½; no scales on preopercle or opercle, unpaired fins very scaly; no opercular spots or stripes; body plain olive green-beige; unpaired fins plain, with lower part of anal and caudal fins dark, filaments not white .... ***falcicula*** sp. nov.

Nota Bene: *L. olivaceous* and *L. crassus* are sympatric at Luhanga; *L. gracilis* and *L. brichardi* are sympatric at Matanza; *L. brichardi* and *L. falcicula* are sympatric on the Burundi coast.

Nota Bene 2: Lower transverse line scales are counted in oblique forward row upward starting with scale closest to first anal spine.

@@@. Caudal fin truncated, outermost rays filamentous; anal spines V-VII; dorsal spines XVIII-XX; body banded blue or purple/white, or entirely smoky black; 32-35 scales in long. line; tail straight, with outer rays extending in short, black filament; cheek covered with several rows of scales in old specimens; juveniles always strongly banded in light and dark blue; 70 mm .... ***savoryi***

(2). Lower lateral line atrophied (less than 20% of long. line)

—Anal spines V-VII; dorsal fin XV-XVII (mode XVI), 23 to 25; 32-36 scales in long. line; unpaired fins with many thin contrasting black stripes; max. size about 10 cm .... ***ornatipinnis***

—Anal spines V; dorsal spines XVII-XVIII; 32-32 scales in longitudinal line; body dark brown; maximum size about 45 mm .... ***schreyeni***

—Anal fin count VI-VIII,24-26; 29-31 scales in longitudinal line; many vertical stripes; size 25 mm .... ***multifasciatus***

—Anal spines VII-IX; dorsal fin XVI-XVIII,23-25; 25-30 scales in longitudinal line; olive, abdomen silvery, with green sheen on sides and light blue around pectorals; maximum size about 45 mm .... ***ocellatus***

—Anal spines VI-IX; dorsal fin XVIII-XIX,25-28; 30-34 scales in longitudinal line; beige with thin vertical golden stripes and mauve sheen on cheek; tail truncate; maximum size about 45 mm .... ***brevis***

—Similar to *brevis* but tail truncate with two short filaments on outer rays and with orange spot on temporal area and the unstriped sides of body are spattered with blue dots; maximum size about 45 mm .... ***brevis*** "Magara"

—Anal spines IX; dorsal fin XVII,25; 31 scales in longitudinal line; color not known; about 60 mm .... ***stappersi***

—Anal spines V; dorsal spines XVII-XVIII; 30-31 scales in longitudinal line .... ***wauthioni***

376

Lepidiolamprologus kendalli.   *Lepidiolamprologus kendalli.*

Lepidiolamprologus near *kendalli.*   *Lamprologus buescheri.*

Lamprologus near *moorii.*   *Lamprologus fasciatus.*

Lamprologus modestus. Photos by H. J. Mayland and
Lamprologus sp. ("Walter").   D. Schaller.

LEPIDOCHROMIS Poll, 1981
(Semi-benthic)

Neither of the two species in the genus has been seen alive in its habitat and nothing has been reported about the ecology or the ethology of these unusual cichlids. They are remarkable because of the large sensory scale they are endowed with on each side of the snout. The scale is grooved by six radiating canals, but how it works has not been determined as yet.

Description:
Body: short and deep; laterally compressed; mouth very oblique; eye very large.

Scales: from 44 to 55 in longitudinal line; 2 lateral lines, both well developed; one pored scale on each side of snout.

Fins: dorsal fin XIV-XV,8-10; anal fin III,7-8; caudal fin truncate.

Teeth: conical, in 3 rows in upper jaw and 2 rows on lower jaw.

Pharyngeal teeth: conical, on triangular bone.

Gill rakers: 16-17.

Habitat: open waters; semipelagic, rather deep.

Reproduction: undocumented.

Maximum size: about 140mm.

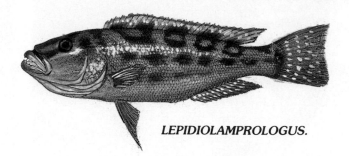

LEPIDIOLAMPROLOGUS.

LEPIDIOLAMPROLOGUS
Pellegrin, 1903

One group of *Lamprologus* species stands well apart from the other species of the genus (see key to *Lamprologus* and *Lepidiolamprologus* above). As their ecology and behavior are in line with predatory and roaming habits, and different from those of the other species, they are dealt with as a separate genus, *Lepidiolamprologus.*

Six species are involved in this group, three of which are ubiquitous and found everywhere around the lake (*L. elongatus, L. cunningtoni,* and *L. profundicola*), two are sibling (geminate) species, one of them living in the south (*L. attenuatus*) and the other in the north *(L. pleuromaculatus),* and the sixth, a species closely related to *elongatus, L.*

kendalli, living only in the south.

The cruising *Lepidiolamprologus* are not small fishes, and I remember having once seen a salmon-sized *L. profundicola* that was well above the 45cm reported as the maximum size for this species. Most specimens of *Lepidiolamprologus* exceed 15cm. Their size and speed allow them to roam freely, well above the bottom. Thus, when they have to flee, which is seldom, it is into the open and they do not try to hide in the rubble.

They thus represent among the *Lamprologus* relatives a trend toward life in open waters, and perhaps they could have become pelagic with the anatomical features they developed, were it not for two behavioral characteristics. First: they are solitary and we have seen that all open water fishes have to bunch together if they want to survive. Second: they are nestbreeders, which means that at least 4 or 5 times a year, if not more, they are immobilized on a nestsite for several weeks in a row by the need to protect their brood.

The six species do not share exactly the same ecological features, as might be expected:

L. elongatus, L. kendalli, and *L. profundicola,* the three most

---

Key to the Species of
Lepidochromis:
1. From 44 to 49 scales in longitudinal line..........................***christyi***
   From 51 to 55 scales in longitudinal line.....................***bellcrossi***

Lepidochromis Poll, 1981, has been found to be preoccupied. It has been replaced with the name *Greenwoodochromis* Poll, 1983.

Lepidiolamprologus elongatus.

Lepidiolamprologus cunningtoni.

Lamprologus furcifer.

Lamprologus sp.

Lamprologus leleupi.

Lamprologus leleupi.

Lamprologus meeli.

Lamprologus lemairei. Photos by Dr. Herbert R. Axelrod and H.-J. Richter.

Lamprologus niger.

Lamprologus obscurus.

Lamprologus moorii, yellow form.

Lamprologus moorii, black breeding form.

Lamprologus modestus.

Lamprologus mustax.

Lamprologus mondabu.

Lamprologus mondabu. Photos by Pierre Brichard.

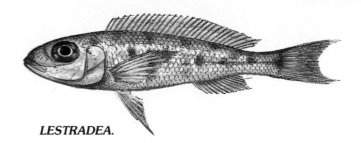

LESTRADEA.

closely-linked species, are rock-dwellers and do not roam the wide open sandy bottoms. *L. profundicola* lives deeper than the other two and is seldom found near the surface. All three breed in rubble, but the fry do not remain hidden for long, and as soon as they become free-swimming they emerge from the hole in which they were born. By that time they are, I would say, between 3 and 4 mm long. The spawn of *L. profundicola* might involve more than a thousand eggs, while the other two species lay only several hundreds.

L. cunningtoni, L. attenuatus, and *L. pleuromaculatus* prefer sand and dig large crater-shaped nests, even when there are rocks offering good shelters nearby, a fact that shows that the choice of a nest site can be very revealing of a species' ecology.

Not that these fishes are against protecting their nest with a prop. They will choose to dig their crater under a sunken branch or a waterlogged trunk whenever there is one available. One might thus say that these species avoid the rocks, and in fact they are very seldom seen hovering above the rubble.

L. cunningtoni as a rule prefers to swim close to the sand bottom. The other two species prefer to hover well above the ground, especially *L. attenuatus,* which, of all the *Lepidiolamprologus,* is the one met the farthest away from the slope, roaming over water 20 meters deep close to the surface. Their fry, often in small schools, are commonly seen among fry or juveniles from the *Cyathopharynx* guild, along with the fry of pelagic

scale-rippers. On the other hand, the fry of *L. elongatus* are known to fall prey to the juvenile coastal scale-rippers, like *Perissodus microlepis.*

Of all the *Lepidiolamprologus, L. attenuatus* is the most gorgeously colored, with alternating horizontal stripes of yellow and vermilion red, suffused with a few sky-blue metallic scales. As a whole, however, this genus is not noteworthy for its bright colors, the other species being mainly gray or smoke or a muted blue in shade.

With their roaming habits and the large size they reach, these fishes need a large tank. Although preying on other fishes, like so many carnivorous and voracious species they are much less bothersome than might be expected. They will not attack other fishes that are too large. When they cannot swallow prey whole they leave it alone. In fact, the most bothersome fishes are those that nibble, graze, or rip chunks of flesh from their neighbors.

L. elongatus is the most common of the six species in the lake as well as in tanks, the others having been exported only sporadically at best.

LESTRADEA Poll, 1943

The two subspecies of *L. perspicax* are semipelagic roamers known to feed mainly on microorganisms. Their teeth are tricuspid but with an elongate and rather sharp central crown, the inner teeth being bent backward. The exact use of this type of dentition is not clear.

Streamlined and silvery, *Lestradea* are obviously built for speed. The head and back being plain gray, they are not very attractive and so far have been exported only sporadically from the northern part of the lake.

Description:

Body: elongate, with a very large eye.

Scales: 37 to 40 in longitudinal line; 2 lateral lines, the upper complete.

Fins: dorsal fin XIII-XVI,13-16; anal fin III,9-11; caudal forked.

Teeth: 2 or 3 rows of conical teeth, with two low cusps on the stem; the inner teeth bent backward and all of the same size.

Pharyngeal teeth: thin, close together in a triangular pattern.

Gill rakers: 15-19.

Habitat: shallow, over sandy bottoms.

Feeding: microorganisms.

Reproduction: buccal incubation; more than 20 eggs about 2mm in diameter.

Size: about 140mm.

Lamprologus mustax. R. Stawikowski. *Lamprologus mustax*. H. Linke.

Lamprologus mustax. H. J. Mayland. *Lepidiolamprologus elongatus*. H. Linke.

Lepidiolamprologus attenuatus. R. Stawikowski. *Lamprologus mondabu*. R. Stawikowski.

Lamprologus leleupi longior. H. Linke. *Lamprologus leleupi*. B. Kahl.

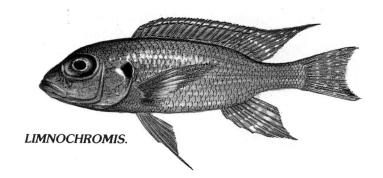

LIMNOCHROMIS.

LIMNOCHROMIS Regan, 1920

Of the four species currently assigned to *Limnochromis*, three *(abeelei, auritus, staneri)* certainly belong here, while *permaxillaris* has been assigned to the separate genus *Gnathochromis* (sometimes also including *Haplochromis pfefferi)*. Only *L. auritus* has been seen and exported recently. The ecology and behavior of the other two species have not been documented.

L. auritus has a rather short, plump, cylindrical body featuring horizontal rows of light-reflecting scales, alternating with rows of blue or cream-colored scales. The unpaired fins are translucent with arrays of flecks in the same colors. The pelvic fins are white and a bit filamentous. The body has four faint broad vertical steel-gray bands extending halfway down the sides.

The species is noteworthy among the lake mouthbrooders by the fact that both parents participate in the incubation of the brood, which implies that the mates are bonded in a monogamous relationship lasting at least as long as the fry are incubated.

An easygoing and adaptable fish of long standing with aquarists, *L. auritus* is rather easy to breed and fairly

common, although its supply doesn't meet the demand.

Collected with huge seines dragged over hundreds of meters of sand floor to the shore from depths of 10 meters and more, the fish, along with the other sand-dwelling species, are in such poor condition that few survive. Fished for food by the local fishermen, who do not want to risk losing most of their catch were they to bring in their nets slowly and more cautiously, sand-dwelling fishes are very costly to collect for the hobby and, due to their very high losses, uneconomical to handle.

Selective fishing by divers is impossible, because sand-dwellers are concentrated in hard-to-find local

concentrations in hazy waters in which one could spend days trying to locate one.

All this explains why sand-dwelling fishes, although some are among the most gorgeous fishes in Africa, are so seldom exported, and then only from the northern basin and often at less than cost.

Because of the exceptional properties of the lake water and also because of the fact that the lake is isothermic, Tanganyika lake cichlids are not easy to acclimatize and ship abroad. Most sand-dwellers are more delicate than most other cichlids. They cannot stand a sudden drop in temperature, even a few degrees, without danger to their well-being. Coming from well aerated and clean water, they cannot stand pollution. When the temperature drops below 20°C (68°F) during their shipment, the fishes will go through a kind of thermal shock when unpacked, from which they will recover only by a painstaking process of gradual mixing of their packing water with the one into which they are going to be stored. This is why the Tanganyika

Limnochromis auritus is the only member of the genus commonly imported. Photo by Dr. Herbert R. Axelrod.

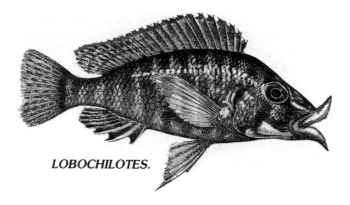

LOBOCHILOTES.

LOBOCHILOTES Boulenger, 1915

This large fish, one of the largest cichlids in the lake (or anywhere), has very peculiar lips. Very fleshy when the fish are young, they start to thicken more when the fish grows older, becoming more and more like a kind of rubbery tubing all around the jaws. When adult, the fish has a set of very thick lips curling back on the outside of the mouth. These lips are in no way immobile, and the fish can control them to some extent. Seeing the fish eat in the lake, one gains the impression that the lips might be prehensile and thus very useful in helping the fish dig shells from the sand bottoms in which they are embedded, and perhaps even more to seize the crabs that are a regular part of the fish's diet. The pharyngeal teeth of the species are very thick and flattened and used as crushers where the shells are broken down into pieces, the body being swallowed and the shell fragments spit out.

There are places in the lake on rocky slopes where the bottom is littered with a vast quantity of freshly broken bivalves. It is highly probable that *Lobochilotes,* the only mollusc-eating fish of such size in the lake living on these same grounds, is responsible for these big meals. What is noteworthy is the fact that these broken shells are always concentrated on huge flat boulders, protruding like a table from the nearby sand slopes. They appear thus to have been brought there deliberately.

Lobochilotes when young is a striking fish in any tank because of a number of deep green vertical stripes on the body as well as a mottled array of dots and small stripes of deep green and orange–brown. For this reason the fish has gained a lasting popularity among hobbyists. Feeding has presented few problems nor does breeding present problems provided one is ready to accommodate the fish in large quarters.

This fish again illustrates the trend of buccal incubators to come up with their fry from the deep layers to the upper, well oxygenated, layer. It is then that the diver can see the big adults looking over their hundreds of youngsters in very shallow water. The juvenile fish will spend their first year in this habitat, individually and not in a school like those of *L. dardennei.* Later on they will move down the slopes, to 40 meters or more, apparently to wander in the semidarkness.

Description:

Body: powerful, with a big head, the profile from the first dorsal ray to the lips rather straight, tapering off toward the tail.

Color: green, 12 or 13 thin stripes on the sides, becoming faint with age.

Scales: ctenoid; 33-35 in long. line; 2 lateral lines.

Fins: dorsal XVII-XIX,10-11;

The large size and distinctive vetically banded pattern make *Lobochilotes labiatus* a readily recognizable species. Photo by Dr. Herbert R. Axelrod.

The Sud du Lac (southern race) form of *Lobochilotes labiatus* has the vertical barring less distinct. Photo by Pierre Brichard.

anal III,7-8; caudal truncate; pectorals large.

Teeth: all rounded and truncated, outer teeth bicuspid in young.

Pharyngeal teeth: conical, rounded, but molariform at the back of the bone.

Mouth: terminal, lips fleshy when young, becoming more so with age, very thick and curling on maxillary; perhaps prehensile.

Gill rakers: 18-22.

Habitat: always seen close to rocks, but suspected to roam the sunken sand plains in search of bivalve molluscs, crabs, and other invertebrates.

Reproduction: buccal incubator laying several hundred eggs.

Maximum size: about 400mm, making it the third largest cichlid in the lake (after *Boulengerochromis* and *Lamprologus profundicola*).

OPHTHALMOTILAPIA

Pellegrin, 1904

The genus *Ophthalmochromis* is now considered a junior synonym of *Ophthalmotilapia,* the latter taxon being enlarged to accommodate *O. nasutus,* the two subspecies of *O. ventralis,* and *O. boops.*

This genus, along with *Cyathopharynx* and *Cunningtonia,* form a guild of mostly plankton pickers, whose habitat is the water column above the bottom along rocky slopes. It now includes six species.

There is a variety of tooth shapes, even within a species, so that one day probably *Ophthalmotilapia, Cunningtonia,* and *Cyathopharynx* might be united into a single genus, the last having priority.

In every species sexual dimorphism is strong. The pelvic fins of males are so long that they reach to the tail and

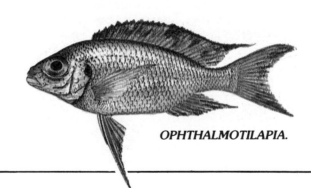

OPHTHALMOTILAPIA.

387

eventually end in colorful spatulated tassels *(Ophthalmotilapia* and *Cyathopharynx)* or a point *(Cunningtonia).*

The diversity of patterns displayed by the local races is flabbergasting and might hide the presence of several more species. Thus it would not be surprising if in the future several local races will be considered worthy of subspecific or even specific status.

Without attempting to classify all the races that have been discovered, but not often photographed in their habitat, here is a list of the most noteworthy color morphs in each species.

O. nasutus:
 Magara race: matte gray, the sides paler and with a bit of yellow; thin dark gray stripes on the sides; all unpaired fins opaque gray, sometimes with a pale blue sheen.
 West coast northern basin race: more yellow on the body; the ventral tassels are pure white instead of canary yellow.
 Chipimbi southern race: vermiculated dark green on olive green; tassels lemon yellow.

O. ventralis ventralis:
northern subspecies; pale matte blue overall.
O. ventralis heterodontus:
southern subspecies; very outstanding sterling silvery white overall.
 Herringbone pattern (found on the central west coast): a ghostly white color overall, with dark gray patches on the caudal peduncle, abdominal area, and back near the profile of the body, giving the impression that these areas have been chewed on by another fish and are missing. A very unusual pattern indeed!
Only the northern race is regularly exported. Care as with *Cyathopharynx.*

Description:
Body: not elongate, with rounded profile; fins long, flowing, the pelvic fins with very long filaments reaching to the tail in males, to the end of anal fin in females.
Scales: 36 to 74; 2 lateral lines, the lower anteriorly with a row of open-pore scales extending the line of tubuled scales (as in *Lamprologus).*
Fins: dorsal XII-XIV,9-16; anal II-III,8-11; caudal well forked, often a bit

filamentous; pelvics split at the tip, each tip ending with a spatulated flat colorful tassel.
Teeth: tricuspid, then mixed with conical teeth or entirely conical; in 2 to 5 rows; teeth of inner rows sometimes spatulated; outer teeth larger and close to the rim of the jaw in a nearly continuous edge.
Pharyngeal teeth: in a densely packed cushion of small conical teeth.
Mouth: terminal, with lower jaw shorter. In one species *(O. nasutus)* a fleshy protruding proboscis extends from the snout over the upper lip.
Gill rakers: 16-20.
Habitat: from 2 to 10 meters deep on average, well off the slope.
Feeding: on microorganisms drifting by in plankton clouds.
Reproduction: buccal incubator; perhaps 20-25 eggs 4mm in diameter; release of fry close to shore and near the surface; crater nest on sand, most often between the rocks; gregarious, but the spawns of the school are asynchronous.
Maximum size: O.nasutus 180mm; others 120mm.

Key to the Species of *Ophthalmotilapia*:
1. 39 scales in longitudinal line; 36 in upper lateral line; snout narrow and with a proboscis; interorbital space convex or straight; spatulated teeth in inner rows..............***nasutus***
   33-37 scales in longitudinal line; 33 in upper lateral line; interorbital space concave (supraorbital bones raised); conical teeth in inner rows...............................***ventralis***
   62-74 scales in longitudinal line...***boops***

Male *Ophthalmotilapia ventralis*.   School of *Ophthalmotilapia ventralis*.

Male *Ophthalmotilapia ventralis* with the head in high contrast to body coloration.   Jaw-locking in males of *O. ventralis*.

Male combat in *Ophthalmotilapia ventralis*.

Fertilization of the eggs in *Ophthalmotilapia ventralis*. Photos by Pierre Brichard.

Mating in *Ophthalmotilapia ventralis* occurs near rocks. The female in the top photo has mouthed the end of the male's ventrals. Photos by Pierre Brichard.

has two of these tassels, the four of them are several times the size of the eggs that are going to be laid by the female.

Note: The tassels are a bright golden yellow in the Burundi *nasuta*; in other races they can be bright white or deep orange. The eggs of this species are a dull yellow, very different in all races in shape, size, brightness, and color from the tassels.

6. The female entered and took the tassels into her mouth for a short while, after which the male left the nest but remained nearby, checking on what she was doing. The female started depositing one or two eggs at a time and rather leisurely took them into her mouth. This was repeated several times during which time the male came and hovered alongside the nest rim, not coming any closer to her. One gets the feeling from his behavior that he is spraying the whole area with sperm.

7. After about a half dozen eggs were laid by the female, the male entered in front and in line with her and the procedure described in #4 and #5 was repeated, as long as she had eggs to lay.

Altogether it is only a few times that the female sucks the ventral fin tassels of the male during the spawning session. Most of the eggs are laid and taken into the mouth without the male being there. Sometimes a small batch is laid when, just before, she missed the tassels spread out in front of her, but anyhow took a mouthful of water from the area just behind the tassels. As a result of what is happening during the breeding session of a pair of *O. nasutus* one cannot decide whether the tassels of the male, although at times mouthed by the female, *are mistaken by her as her own eggs.* The male is not there as she is laying and picking up her eggs. The first contact she has with the tassels occurs *before* she has laid a single egg.

There is thus every reason to believe that the tassels are not egg dummies. What is their use then? Tentatively one might think that they play a double role in the mating procedure:

1. As a part of the male's nuptial garb. Perhaps a very important one as the males of the *Burundi nasutus* are a very dark and matte blue all over, the only other colors being the orange of the pectoral fins. The very bright yellow tassels at the tips of the ventrals are thus the most visible color patch of the male, even when the fish is "at rest" and not breeding.

2. As a signal from the female to the male (who has his tail toward her and doesn't see what she is doing) that she is ready and needs sperm for the eggs that she is about to lay.

There appears to be an exchange of signals between the two partners:

a) When the male bends his

two ventral fins so that they stretch along the same side of his body (right side as far as I could ascertain). This signal perhaps tells the female that he is ready to release sperm.

b) When the female sucks the tassels telling him that he can proceed and that she is ready to lay her eggs.

The bright yellow or salmon

Portait of a male *Ophthalmotilapia ventralis.* Photo by Pierre Brichard.

colored but very thin tips of *Cunningtonia*'s long ventral fins cannot be mistaken by the female for one of her eggs but can act as a stimulus and signal just as efficient as the one we have described in detail.

Let us keep in mind the elaborate spawning behavior of other Tanganyika cichlids such as *Simochromis* and *Petrochromis.* In these genera the rearmost ocellus goes as far as mimicking *in relief* an egg although it isn't more than a lure. Let us remember as well the fact that many (but not all) races of *Tropheus* males have anal fin spots so small that they cannot possibly be confused by a female for one of her eggs.

Why doesn't a male *Tropheus* (one of the most efficient and sophisticated mouthbrooding cichlids) have

an anal ocellus, inflatable or not, like those of *Simochromis* (a genus to which, by all means, it is closely related) if these ocelli still really act as egg-dummies and improve the efficiency of the spawn of these fishes? Many mouthbrooding cichlids in the lake are devoid of any spot, dot, or ocellus on their lower fins.

The conclusion that we can tentatively reach after all the facts have been put together is that as yet *not a single endemic cichlid female from Lake Tanganyika has been found to mistake the lower fin's ocelli or spots for her eggs.* Fertilization of the eggs appears to be triggered by visual signals, physical contact, and, foremost, the olfactory sense of the female in search for the sperm in the nest.

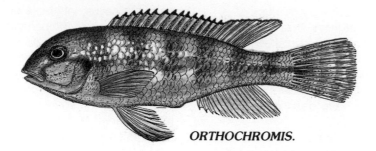

ORTHOCHROMIS.

ORTHOCHROMIS Greenwood, 1954 (Mountain torrent dweller)

Orthochromis is a haplochromine that has been reported from the headwaters of several lake tributaries, especially from the Malagarazi River, the Congo basin, and the Cunene River in Angola.

The species from the Malagarazi *(O. malagaraziensis)* displays the usual adaptations of haplochromines living in fast moving waters, namely rounded fins, a lack of scales in the thoracic area, and the second or third ray of the ventral fins being the longest. Sexual dimorphism is rather weak, the males lacking the egg-dummy ocelli on the anal fin. There are XVI-XX dorsal fin spines and III anal fin spines.

It has not been collected nor exported recently.

PERISSODUS Boulenger, 1898 (Scale-rippers)

This genus now includes the seven species that in my earlier book had been distributed among the genera *Perissodus, Plecodus,* and *Xenochromis.* It consists now of a group of scale-ripping fishes of which five (*eccentricus, elaviae, paradoxus, multidentatus,* and *hecqui*) are pelagic and schooling and two *(straeleni* and *microlepis)* are coastal over rocks. Except for *P. straeleni,* all have streamlined bodies. *P. straeleni* is squat and has a deep and laterally compressed body.

These fishes stand apart from the mainstream cichlids by their dentition, basically composed of teeth with a leaf-shaped crown that is curved backward and planted across a long stem deeply buried in the gums. The shape is reminiscent of a pickax. Each crown has a cusp on each side that hooks onto the edge of a victim's scale. Pressure is applied by the powerful jaws and the scale pops out of its seating.

The ethology of the pelagic roamers is very poorly documented. Nothing else is known other than they are gregarious mouthbrooders coming to the shoreline to release their fry in coastal waters. There the juveniles stay for a while with the fry of midwater coastal species and start preying upon them. It would be very difficult and certainly very dangerous for these young fish to start at once roaming in open waters in search of suitable prey.

The Sud du Lac form of *Perissodus straeleni* mimics both *Lamprologus sexfasciatus* and *Cyphotilapia frontosa.* Photo by G. S. Axelrod.

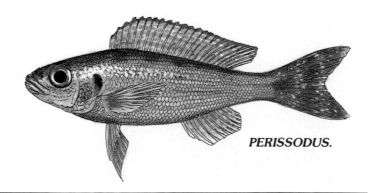

PERISSODUS.

The breeding behavior of the two coastal species is, as we have said, very unusual among the lake mouthbrooders. Buccal incubation appears to be used only as a last resort and the fry from the beginning of their free-swimming state spend much more time foraging for themselves than in their parents' mouth. This is due probably to the many small eggs laid by the female, each with only a small yolk sac, and the monogamy of the mates. *P. straeleni* in this respect stands very close to the nestbreeders and apart from the other species, as from our personal observations it would appear that the pairs are permanent, breed repeatedly, and raise their offspring in the same nest.

It is also a very peculiar fish with its capacity to mimic the color patterns of several other fishes.

Description:

Body: usually elongate, with one exception, *P. straeleni,* which has a high, squat body; mouth powerful and terminal.

Scales: from 58 to 78 cycloid scales in longitudinal line; 2 lateral lines.

Fins: dorsal XIII-XX,11-15; anal III,9-13.

Teeth: a single row of very large, stocky teeth featuring a leaf-shaped crown with a central groove set across a solid stem; the teeth are buried in the retractile gums.

Pharyngeal teeth: small, pointed, and few; sometimes buried in a wrinkled skin.

Gill rakers: 18-26 (*hecqui* between 47 and 57).

Habitat: two species are coastal, the others are pelagic.

Perissodus microlepis. Photo by Pierre Brichard.

Perissodus paradoxus. Photo by Dr. Herbert R. Axelrod.

Variant of *Perissodus straeleni* with split black bands. Photo by Pierre Brichard.

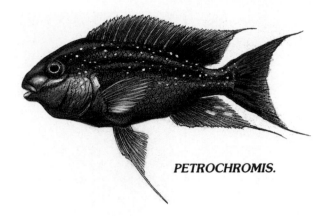

PETROCHROMIS.

Reproduction: up to 300 eggs, one of the largest spawns recorded for a lake mouthbrooder; the two coastal species form pairs with both parents looking after their brood.

Size: P. multidentatus is probably not more than 120mm, *P. straeleni* not more than 160mm, *P. microlepis* not more than 120mm; the others up to 320mm.

PETROCHROMIS Boulenger, 1898 (Rock-grazers)

This is one of the main genera of typical rock-grazers in the lake with their brush-like pads of tricuspid teeth on a long stem hinged at their base. There appears to be a case of convergent evolution with *Petrotilapia* from Lake Malawi and *Sarotherodon tanganicae* from this lake.

Six species are currently known, five of them with a wide range around the lake and one *(P. trewavasae)* probably restricted to the shoreline between Luhangwa and Sumbu on the west coast.

Several local populations are present in the lake, the one around Chipimbi in Cameron Bay being entirely orange-yellow and probably remains very small, another, called "ephippium" (meaning "with a saddle") in the trade is a new subspecies.

The "ephippium" looks very much like a *P. trewavasae* but lacks the long fin banners so typical of that species. "Ephippium" has a wide range, encompassing the southern coast around Cape Chaitika as well as the entire west coast up to the northern basin. The best local population has the males in nuptial garb chocolate brown on the sides and

Outline of the species of *Perissodus:*

1. 58-61 scales in longitudinal line, $\frac{44\text{-}46}{24\text{-}32}$ scales in the two lateral lines; body depth less than 3 times in length; dorsal spines XVIII; basically chestnut with a blue sheen in several bands; maximum size 160mm ....***straeleni***

2. 57-65 scales in longitudinal line, $\frac{46\text{-}57}{23\text{-}37}$ scales in the two lateral lines; body depth more than 3 times in length; dorsal spines XVI-XVII; teeth $\frac{22\text{-}45}{19\text{-}40}$ on the two jaws; body plain with two black spots at base of soft dorsal; feeds mainly, if not exclusively, on copepods; maximum size more than 300 mm ....***hecqui***

3. 69-73 scales in longitudinal line; $\frac{56\text{-}63}{34\text{-}46}$ scales in the two lateral lines; body depth more than 3 times in length (as with all following species); dorsal spines XVI-XVII; teeth $\frac{33\text{-}40}{30\text{-}35}$ on the two jaws; tail striped; maximum size 120 mm .... ***multidentatus***

4. 68-78 scales in longitudinal line; $\frac{57\text{-}68}{39\text{-}46}$ scales in the two lateral lines; dorsal spines XVIII-XX; teeth 18-20/12-14 on the two jaws; a black spot on caudal peduncle; maximum size 290 mm .... ***paradoxus***

5. 64-67 scales in longitudinal line; teeth $\frac{54\text{-}58}{36\text{-}38}$ on the two jaws; maximum size 320 mm ....***eleviae***

6. 65-70 scales in longitudinal line; $\frac{45\text{-}51}{22\text{-}32}$ scales in the two lateral lines; dorsal spines XVIII; teeth 18-26/16-18 on the two jaws; maximum size 120 mm ....***microlepis***

7. Same as *paradoxus* but with teeth slanted sideways to the left or right in the jaws .... ***eccentricus***

Petrochromis fasciolatus, female.

Petrochromis polyodon, southern race.

Telmatochromis vittatus.

Reganochromis calliurus.

Telmatochromis bifrenatus variety.

Telmatochromis dhonti.

Photos by Pierre and Thierry
Brichard and H.-J. Richter.

Triglachromis otostigma.

Telmatochromis temporalis.

abdomen while the back and all the unpaired fins are entirely an opaque and bright yellow.

The southern variety of "ephippium" has only a saddle-shaped yellow rectangle running along the base of the dorsal fin. All unpaired fins are halfway between chocolate brown and a dirty yellow. Like *P. trewavasae,* "ephippium" have small light colored specks on the body, and when excited the males also have the lower part of the cheek a pale white/pink (in the lake only). As *trewavasae* and "ephippium" have separate ranges, it is perhaps best to think of them as subspecies deserving closer investigation.

Several well-known species, on the other hand, have developed local races as well, and the variety of mood patterns has sometimes made it very difficult to identify a specimen at first sight. There are, however, only minor differences between the garb of a *P. fasciolatus* from the north and one from Cameron Bay.

P. polyodon displays much more variability and there are several well-known varieties. There is the new "red-tail" race coming from the northern basin in which males sport a light blue unstriped pattern and the soft parts of the dorsal, anal, and caudal fins being entirely a bright vermilion red. Females are striped.

The best known variety that is regularly exported from the northern basin is the mottled blue-orange-yellow.

Another famous

Petrochromis has been introduced lately called, rightly, in the trade the "Nyanza Lac" race, which is the place in which it was originally found. The body in the best population is a light beige suffused with light blue. The head and dorsal and anal fins are of a deeper blue, the pectorals being orange or salmon, the tail having a checkered blue pattern. This species doesn't reach the size of *P. polyodon* or *P. fasciolatus,* perhaps only to 160-170mm at most.

The Nyanza Lac blue *Petrochromis,* not found north of this place on the eastern shore, again points toward an ancient ford between the Ubwari and the Nyanza Lac area, as it is found along with the green morph of *Tropheus* on the two sides of the Ubwari

Petrochromis polyodon among the rubble. Photo by Pierre Brichard.

and up north on the western shore. It might perhaps be a twin species of *P. famula.*

Identifying a *Petrochromis* during an exploration of new grounds is often tricky and one has to be cautious before talking abut a new species. According to their sex and moods (and much more than *Tropheus*) local populations of *Petrochromis* display a variety of color patterns.

From the aquarist's point of view *Petrochromis* are not for the inexperienced hobbyist and, as a whole, might be considered as poor aquarium fishes.

There are several sound reasons for this situation. First, they are large and bulky fishes, most often on the move. As such they require a large tank. Then, they are mostly vegetarian and need to be fed very often. If not, they are bound to start grazing on the aquatic plants—which they will tear to shreds. After they have dispatched the plants they will start nibbling at the other fishes with disastrous results. They will start wasting away in a few weeks unless given regular meals. It is seldom that I have seen a well-fed *Petrochromis,* especially *P. trewavasae,* in a hobbyist's tank.

Being bulky, they will also be expensive depending on the air freight involved. As the turnaround time in the pet dealer's shop is slow for this type of fish and he is bound to have losses, he will compensate by taking a higher than normal profit margin, which is understandable. Finally, the fish are aggressive among themselves and pairs are difficult to bring to spawning condition. Thus

Petrochromis polyodon. Photo by Dr. Herbert R. Axelrod.

hobbyists have trouble recouping their initial investment with the sale of the spawns, which are small. This is why *Petrochromis* are expensive and rather scarce.

The mating behavior of *Petrochromis* is poorly documented. This is due to the fact that in the wild, although mating preliminaries are often conducted in the open above the bottom, the partners are discreet when it comes to the actual spawning session and retreat to the rocks.

As the fry grow inside of the female's mouth, she goes more and more often to the rubble where she remains in hiding, probably getting her

The form of *Petrochromis polyodon* found in the northern part of the Lake. Photo by Pierre Brichard.

brood familiarized with their future surroundings. As buccal incubation in the lake rock-grazers lasts well over a month, it means that four out of five females are hidden, out of sight, in the rubble at all times.

When fish collectors are catching fishes, it is obviously the males that they mainly catch because they can be caught out in the open while the females remain out of reach. This explains why dealers and aquarists alike complain all too often that shipments from the lake include too few females.

Were the collectors only to ship paired fish and discard the excess males, they would lose at least two thirds of their catch (often very small to start with) and prices would really become prohibitively expensive. The whole economics of fish-collecting in the African Rift lakes would be upset and many species would stop being available.

Description:

Body: deep, massive, and plump; head and snout broad.

Scales: 32-36 in longitudinal line; 2 lateral lines.

Fins: dorsal spines XVII-XX; anal spines III; tail truncate.

Teeth: tricuspid on very long stems and hinged at their base; the crown is curved backward; very close together in two separate brush-like pads on each jaw; mouth very low and broad, with thick fleshy lips, the lower jaw shorter. The fish curls its lips back when it feeds, applying the tooth pads to the rocks, closing and opening its jaws several times vigorously to scrape the rock clean of its epiphytes.

The thick lip pads and multiple teeth of *Petrochromis polyodon* are very distinct in this photo by Dr. Herbert R. Axelrod.

Gill rakers: 10-16.

Habitat: always over rock; some species tolerate murky water.

Feeding: rock biocover scraper.

Reproduction: buccal incubation; approximately 50 eggs 4-5mm in diameter.

Maximum size: variable according to the species; between 100 and about 200mm.

Key to the Species of Petrochromis
with the description of a new subspecies.
(After Matthès, Yamaoka, and Brichard)

Snout strongly covex; mouth extending to vertical through anterior eye rim; caudal fin well covered with scales; chin profile concave ....*macrognathus*

5. Caudal fin slightly lobed, outer rays longer, not spotted .... *polyodon*
Caudal fin truncated, not lobed, but with outer rays shorter than central rays, spotted..........*famula*

6. Dorsal, anal, and caudal fins exceedingly filamentous, dorsal and anal filaments often extending beyond posterior edge of caudal; body and all fins uniformly velvety dark chocolate with a few scattered pearly scales on the sides forming or not faint rows, less on females; males with 2 to 4 anal ocelli, females without or with one, two, or three small orange dots; range restricted to area around Cape Chipimbi and northward inside of Kapampa Bay, nowhere else ....*trewavasae trewavasae*
Dorsal, anal, and caudal fins moderately filamentous, dorsal and anal filaments extending to caudal but never beyond; body olive-brown to yellow-orange, fins lighter in color; a broad saddle-like rectangular light-colored patch under dorsal fin, contrasting with darker body color. This striking color pattern is found among all the populations that have been identified from Mpulungu to Cameron Bay in the south and from Moba to Uvira on the west coast. No populations have as yet been described from the eastern coasts. Doesn't exist in Burundi waters ....*trewavasae ephippium,* new subspecies

1. Jaws equal or upper jaw protruding........................2
Lower jaw protruding ....*fasciolatus*

2. Tail not crescentic, but truncate, subtruncate, or slightly rounded; snout straight or curved, with a hump only in the occipital area 3
Tail always crescentic; all unpaired fins with long filaments; cephalic hump on interorbital and frontal areas..6

3. The two jaws equal or upper only slightly protruding; snout and forehead strongly curved; tricuspid teeth not in a dense pad ....*P. orthognathus*
Upper jaw notably extending in front of lower; tricuspid teeth in dense pads...........................4

4. Snout straight from mouth to frontal area; mouth not reaching vertical through anterior eye rim; tail scaleless; profile straight...........................5

Discussion: Although more than a hundred specimens of each subspecies were examined no significant morphometric differences were found between the two allopatric subspecies aside from the spectacularly long fin filaments of *P. trewavasae trewavasae.* Not a single population of *P. trewavasae ephippium* was found on 700 km of east coast that has been thoroughly explored. More important, in our opinion, is the fact that the color pattern typical of the second subspecies is clearly evident in all populations over the very large range of this subspecies. This is unusual among the various *Petrochromis,* but for two notable exceptions: *P. fasciolatus* and *P. orthognathus* which also display a good homogeneity of the color pattern of their various populations. The diversity of patterns of the many populations of *P. polyodon* makes their identification often difficult in view of the fact the meristic characters of the various species so far often identified often overlap.

In *Petrochromis fasciolatus* the lower jaw protrudes. Photo by Pierre Brichard.

Notice the protruding upper jaw in this *Petrochromis* sp. Photo by G. S. Axelrod.

An unidentified *Petrochromis* from the south of the Lake. Photo by Pierre Brichard.

REGANOCHROMIS Whitley, 1929 (Drifting plankton picker)

Christened by Regan *Leptochromis,* it was pointed out by Dr. Warren E. Burgess that Whitley had replaced that name with *Reganochromis,* as *Leptochromis* had already been used for a marine fish.

Two species make up this genus, of which only one has been sporadically exported. The other is known only from preserved specimens.

This fish is seldom collected because of the average depth at which it lives over huge sandy bottoms. Accidental captures of a few stragglers do not hint at a gregarious behavior or abundance. But the fish are schooling and not especially rare, they are simply mostly out of reach. Forty meters appears to be the average depth at which they live. A diver has a scant 20 minutes to try to locate a school, while at the same time dragging a 20-30 meter long seine along, deploying his net, pushing the fish in, removing them, and storing them in a trap. All this after he has sighted the fish, of course. It is a very exhausting and unrewarding job, and still the fish will have to be brought back to the surface over the following days. Not one dive out of 10 is successful. Even when you spot a school of 500 sand-dwelling fish, you are lucky to get 100 in the net. This is why species living deeper than 20-25 meters should just as well be forgotten by hobbyists.

The diet of *Reganochromis* consists mainly of the largest shrimps available in the lake, which are about 25mm long, but they also eat fry of other fishes such as *Lates* (Nile

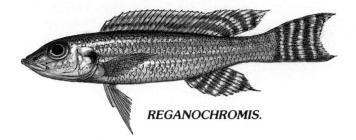

REGANOCHROMIS.

perches).

The mouth is very protractile and about 7-8mm across. It opens into a kind of tube-shaped extension, and the fish are capable of sucking their prey in. Thus *Reganochromis* will not bother fishes that they cannot capture by this method. Were their prey too strong it could, by its violent twisting moves when grabbed, strain the delicate mechanism of the jaws.

If *Reganochromis* could only be bred they would become excellent aquarium fishes. Their shape is elegant and there are rows of mother-of-pearl scales on the sides. The unpaired fins are endowed with strongly contrasting yellow and black stripes. They are hardy fishes, which is not often the case with sand-dwellers, and not choosy about food in captivity. Copepods, daphnia, etc. would bring them into top form.

Description:
Body: very elongate and rather cylindrical, with horizontal rows of very brilliant scales; unpaired fins striped.
Scales: from 37 to 59 in longitudinal line; 2 lateral lines.
Fins: dorsal spines XIV-XVIII, anal spines III; caudal forked; pelvics with outer rays longer than inner rays.
Teeth: small, conical, in 3 to 4 rows in a single band, the teeth in the outer row larger (sand-sifter?).

Pharyngeal teeth: subconical and very thin.
Mouth: rather large and terminal; very protractile and in the shape of a tube.
Gill rakers: 13-15.
Habitat: rather deep, but coastal over sand bottoms; at least to 60m.
Feeding: omnivorous, but mainly on shrimp.
Reproduction: no buccal incubation reported, but probable, although ripe females had up to 200 eggs, 2mm in diameter, in their ovaries.
Maximum Size: about 150mm.

Key to the Species of *Reganochromis:*

1. First 4 or 5 dorsal spines with long filaments; all spines with shortened membranes rising only halfway up the spines; dorsal spines (XIV)-XV-(XVI); anal spines III; 51-59 scales in longitudinal line; $\frac{37\text{-}45}{27\text{-}44}$ scales in 2 lateral lines; 26-29 scales around caudal peduncle; no scales under eye; chin black; lower jaw protruding ....**centropomoides**

Dorsal membranes reaching close to tip of spines; dorsal spines XVI-XVIII; anal spines III; 37-39 scales in longitudinal line; $\frac{27\text{-}29}{9\text{-}19}$ scales in 2 lateral lines; 18-20 scales around caudal peduncle; 2-5 scale rows under eye; chin white ....**calliurus**

The slender form and protruding jaws of *Reganochromis calliurus* are very distinctive. Photo by Pierre Brichard.

403

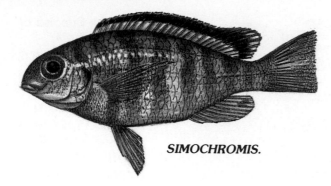

SIMOCHROMIS.

SIMOCHROMIS Boulenger, 1898 and
PSEUDOSIMOCHROMIS Nelissen, 1977

Six species have been recorded for *Simochromis,* but one of these was placed by Nelissen into a new genus in 1977 *(Pseudosimochromis curvifrons).* They obviously all belong to the same line and one might discuss the questionability of splitting the six species into two different genera.

Both are also close to *Tropheus* and probably share a common ancestor, but in this case the anatomical and ecological differences are such that the two subsequent lineages that developed in the lake are quite distant. *Simochromis* are close to the mainstream cichlids by having only and always 3 anal fin spines; *Tropheus* always have a minimum of 4 and often more, 5 or 6 being most frequent among all the local races and species.

Simochromis also normally have 19 or fewer dorsal spines, while *Tropheus* most often are endowed with 20 or 21 spines, sometimes even 22, and it is a rare specimen that has only 19. There are usually much fewer than 40 teeth in the front row of the upper jaw

of *Simochromis,* which as a rule has a rather narrow mouth, and more than 40 in *Tropheus* of similar age. *T. duboisi,* though, has fewer than 40 teeth. All *Simochromis* have anal fin ocelli, of which one (the rearmost) bulges with the fin movements. *Tropheus* do not have ocelli on the anal fin, only occasionally small flecks or spots too small to mimic one of their eggs. The *Simochromis* mouth is oval or rounded while that of *Tropheus* is straight. One might also add that as a rule the tail of *Tropheus* is darker in color or at least of the same shade as the body while in *Simochromis* it is paler, most often colorless.

Although they inhabit the same rocky coast, the habitats are much shallower for *Simochromis. Simochromis* seldom go deeper than 10 meters, and often much less for many species, while the vertical range of *Tropheus* everywhere around the lake is much deeper. Only one species of *Simochromis*, as yet undescribed, is found around a depth of 10 meters or a bit deeper. Thus the parameters of their ecological niche overlap, but some of them are not quite the same. One might add that *Simochromis* are often found in places where

Tropheus are not, such as rocks covered with silt or even mud (*S. diagramma* and *S. babaulti),* which shows that they are less choosy about the quality of the biocover they graze on.

From a taxonomic point of view the case of the pharyngeal apophysis of the two genera is interesting. This apophysis in *Simochromis* is said to be of the *Haplochromis* type, but in *Tropheus* it belongs to an intermediate state between the apophysis of *Tilapia* and that of *Haplochromis.* As the apophysis has been extensively used for the phyletic classification of many African cichlids, one might wonder of how much value it can be for some of the lake lineages when one knows that, despite all the other data pointing to a common ancestry, the apophyses of *Simochromis* and *Tropheus* would tend to place them well apart. The same could be said about the similarity between *Limnotilapia* and *Simochromis.*

Several species of *Simochromis* are more or less frequently exported from the lake:

Simochromis babaulti Pellegrin

This is one of the smallest and hardiest species along with *S. marginatum* and *S.*

PSEUDOSIMOCHROMIS.

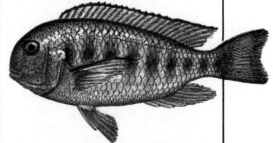

Simochromis babaulti is very attractive and hardy as well. Photo by Dr. Herbert R. Axelrod.

pleurospilus. It is identifiable by the 8 broad alternating steel blue and lemon yellow bands on the sides and a short black band on the first spines of the dorsal fin. Its maximum size is about 10 cm. It is very intolerant of any other rock-grazer, less so of the other species, and very territorial and aggressive toward its conspecifics. It cleans off a patch of ground between rubble to breed. Apparently *S. babaulti* does not use the bulging anal ocellus to attract females. The female most often bumps the male in the abdominal area, as well as in the anal fin or the caudal peduncle.

Simochromis marginatum
Poll

This is a species very similar to *babaulti,* but from which it can easily be distinguished by its having a black stripe running the entire length of the dorsal fin along the margin. The maximum size is around 10cm. It is not found on the northeastern coast.

Simochromis pleurospilus
Nelissen

This is the most attractive species of *Simochromis,* with 6 or 7 horizontal rows of crimson dots on the sides. It is found only along the southern coast, where it outnumbers *S. babaulti.* It has the same habitat and same size as *S. babaulti.*

Simochromis diagramma
(Günther)

S. diagramma is the largest of this group of species and is identified by the 10 to 13 oblique stripes on the sides and a broad head with vermiculated stripes. The sides are steel-blue, the back olive-green, and the underparts pale yellow. Only one anal ocellus is present on the southern populations, but two or three in northern populations with always one ocellus bulging.

Pseudosimochromis curvifrons (Poll)

This is a funny looking fish with a deep body and head, both very much compressed laterally, but especially the head. The unusually narrow head has a small, very low mouth at the end of a long, nearly vertical snout. The maximum is about 13cm. It is jade green in the northern part of the lake and blue in the southern part. The southern race is often found with large irregular patches of deep blue, which might be brought about by a skin disease.

Simochromis margaritae
Axelrod & Harrison

This species was found in Kigoma harbor by Glen Axelrod but has not been

Note the odd head shape of *Pseudosimochromis curvifrons*. Photo by Pierre Brichard.

The status of the genus *Limnotilapia* is currently being debated. Although some authorities merge it with *Simochromis*, it is retained as distinct here. A school of juvenile *Limnotilapia dardennei* swims over a sandstone slab (above); an adult is pictured below. Photo above by G. S. Axelrod, that below by Dr. Herbert R. Axelrod.

exported since. It looks much like *S. marginatum*, having a black stripe along the entire margin of the dorsal fin.

Less gaudy in their colors than *Tropheus* and with much fewer local geographical races, *Simochromis* are not among the favorites of cichlid amateurs. But several species probably will have to be added soon to their present roster. A strongly contrasted pattern has been seen on a specimen along the Ngoma escarpment south of the Ubwari. It displayed a beige body with a number of thin wavy vertical chocolate stripes on the sides.

On the other hand, the Nyanza Lac race, which has a lemon yellow tail with the two outer rays black, has not been described as yet.

Description:

Body: stocky and rather broad with a blunt head and strongly convex snout; mouth very low.

Scales: 30 to 35 in longitudinal line; 2 lateral lines.

Fins: dorsal spines XVII-XIX; anal spines III; caudal truncate or slightly forked.

Teeth: one row of bicuspid teeth in front, conical teeth on the sides of the jaws; up to 6 rows of tiny tricuspid teeth behind.

Pharyngeal teeth: more or less bicuspid.

Mouth: low, with lower jaw shorter.

Gill Rakers: 5 to 13.

Habitat: rubble in shallow water, most often not exceeding 5 meters in depth and never exceeding 10 m.

Feeding: feed on biocover.

Reproduction: buccal incubation of about 40 eggs 6mm in diameter.

Maximum size: diagramma about 200mm; others 110-140mm.

Key to the Species of
Simochromis (including
Pseudosimochromis):

1. Interorbital width less than 25% of head length; 30-34 teeth; body with 8 alternating steel blue and lemon yellow bands; short stripe on anterior part of dorsal .... ***babaulti***

   Interorbital width more than 25% of head length 2

2. Postocular length at least 45% of head length; 26-28 teeth; body plain blue or green .... ***curvifrons***

   Postocular length less than 45% of head length 3

3. 9-12 oblique stripes on sides; from 38-46 teeth in front row of upper jaw .... ***diagramma***

   At most 8 vertical stripes on sides..4

4. Depth of caudal peduncle 86% of its length; 30-38 teeth in front row of upper jaw; rows of crimson dots on sides ....

   .... ***pleurospilus***

   Depth of caudal peduncle 110-120% of its length; 36 to 38 teeth in front row of upper jaw .... ***margaretae***

   Depth of caudal peduncle 90% of its length; 44 to 55 teeth in upper front row .... ***marginatus***

Nota Bene: Both *S. margaretae* and *S. marginatus* have a black stripe along the entire margin of the dorsal fin.

Simochromis diagramma has characteristic oblique vertical bands. Photo by Dr. Herbert R. Axelrod.

Note the dark marginal band on the dorsal fin of *Simochromis marginatus.* Photo by G. S. Axelrod.

The bright red spots of *Simochromis pleurospilus* make it a desirable species. Photo by Pierre Brichard.

407

An unidentified *Simochromis* (possibly *pleurospilus*) from the southern end of the Lake. Photo by Pierre Brichard.

The unstriped body and bright blue spots identify *Spathodus erythrodon*.

SPATHODUS Boulenger, 1900 (Rock-grazer)

This genus, composed of two species, is very close to *Eretmodus* and *Tanganicodus* and shares the same type of habitat, the shoreline strewn with pebbles and small rubble, where it darts from rock to rock between crashing waves. Even more than the two other genera, *Spathodus erythrodon* lives in very shallow water. We have never discovered any specimen deeper than one meter. Most are seen in less than a foot of water. *S. erythrodon* in the lake is a little jewel; the body is a rosy pink with mauve, blue, or green dots on the head and body depending on the light. It is unfortunate that the fish loses some, but not much, of its basic rosy body color in a tank to become a beige-brown. Although known to reach the same maximum size as *Eretmodus,* most *Spathodus erythrodon* are much smaller, seldom exceeding 50 mm in total length.

They are peaceful companions in a tank and are not afraid to wander about in the open, as they are used to living in open sunlight in the wild. They are not choosy about their diet and will thrive on dried flakes although, as rock pickers, brine shrimp, frozen or alive, would be a welcome addition for their well being.

Spathodus marlieri Poll has been discovered and exported lately from the southernmost part of the Burundi coast in the northern part of the lake near the Tanzanian border.

The behavior of this species is quite different from that of *S. erythrodon. S. marlieri* can be seen swimming over the rock

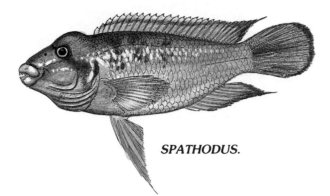

SPATHODUS.

the northern habitat, or a different ecological niche, or competition from *Eretmodus*.

Description:

Body: stocky and short; no vertical bands; two or three rows of blue dots on the body *(S. erythrodon)* or none at all *(S. marlieri);* head of *S. marlieri* with a hump extending to between the eyes.

Scales: 30 or 31 in a longitudinal line; 2 lateral lines, not complete.

bottom for long distances, alone or in pairs, instead of jumping from place to place between the pebbles as *S. erythrodon* does. *S. marlieri* is larger and averages between 60 and 80 mm in length. In the lake the basic color is nearly black with a few green spots on the head and none at all on the body. The temper of the fish is awful toward its own kind and not very good toward other fishes. As such it will never become one of the aquarium favorites.

Although geographical races have not as yet been discovered among the *Spathodus*, it is remarkable how, on the northeastern coast, their populations are restricted to a few bays instead of being spread out along the pebble coastline (as in *Eretmodus*). Along the first 40 km from north to south on the east coast only one bay shelters *Spathodus erythrodon,* and not a single specimen has been found in the other bays and coves, although *Eretmodus* has been discovered in large numbers along the entire coastline. The fact that not a single *S. marlieri* has ever been found along the same stretch might be explained by the long sand beaches separating the southernmost rocky coast of Burundi, where it lives, from

Aquarium specimens of *Spathodus erythrodon* unfortunately lose much of their bright coloration. Photo by Dr. Herbert R. Axelrod.

The dull colors and strange head hump of *Spathodus marlieri* do not make it an especially attractive species. Photo by Pierre Brichard.

The presence of vertical stripes only on the lower half of the body (usually) and a black spot in the dorsal fin are good identifying characters for *Tanganicodus irsacae.* Photo by W. Staeck.

Mouth: rather narrow, but less so than in *Spathodus.*

Gill rakers: 11 or 12 on the lower half of the first gill arch.

Habitat: very close to the shoreline in shallow water, over rocks in the rubble.

Feeding: on rock biocover.

Reproduction: buccal incubation.

Maximum size: about 70 mm.

Affinities: with *Eretmodus,* from which it differs by the following: 1) tusk-like teeth in a single row, 2) body striped only on the lower half, with 2 rows of blue dots on the upper half and a black spot on the dorsal, 3) mouth narrow and not broadened terminally, and 4) head without the long snout so typical of *Eretmodus.* With *Spathodus,* from which it differs by the following: 1) teeth not arranged in sets and not blunt, and 2) color pattern having half stripes on the body and the black spot which *Spathodus* lacks.

TELMATOCHROMIS Boulenger, 1898 (Rocky habitats)

The genus *Telmatochromis* is very close to *Lamprologus* and *Julidochromis,* two other egg-laying groups of cichlids from the lake, and like them have several powerful canines on each jaw and a relatively high number of anal spines. Three of the five species appear to have developed a strict bond toward a certain type of rocky habitat in pure lake water, whereas the other species, *T. dhonti* (formerly *caninus)* and *T. temporalis,* have been discovered over many types of rock substrates even in very muddy water and are, in fact, among the most common and ubiquitous fishes of the lake.

Telmatochromis temporalis and *Telmatochromis dhonti* were among the first fishes to be exported from the lake a few years back and both are now established as regular aquarium cichlids. Both species have been bred without much problem. As in their native habitat, the fishes are not light shy and swim in the open. Because of their behavior in the lake, where they might prey upon small fishes, they should not be trusted with much smaller fishes in the aquarium.

Telmatochromis bifrenatus, *T. vittatus,* and *T. burgeoni* are so different from these other two species that it is with reluctance that one accepts their grouping under the same generic name. They live in rather deep water for such small species, specimens having been found as deep as 20 meters, although they seem to prefer a range between 5 and 10 m. Although most of the time specimens are discovered wandering in the open close to the rocks or over coarse sand between rock patches, they never assemble in schools, even small ones. The fish, although often very close to neighbors, lead solitary lives. When in danger they will seek shelter between the piled-up rubble or in the tiniest hole wide enough for them to enter.

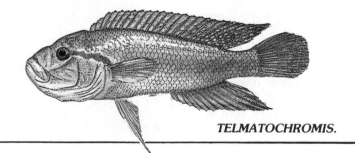

TELMATOCHROMIS.

It is also in these miniature caves that they will mate and rear their young. In the danger-strewn world they live in, amid countless enemies, the fry don't wander into the open until they are capable of a fast escape. Not before they are at least 15 mm long can one discover them along with mature fish.

As dwarf aquarium fish, it is difficult to imagine a better pair than *T. bifrenatus* and *T. vittatus.* With a maximum size of less than 100 mm, their elegant and slender shapes not unlike that of *Julidochromis,* and the ease with which they adapt to aquarium life, not hiding but swimming in the open and readily accepting most types of commercial food, one might say that they have everything going for them. It is unfortunate that after a few spawnings the original striking color pattern, with its two longitudinal black stripes on a very pale beige background, tends to become dull and much less appealing.

Of the three dwarf species, probably the most colorful is *T. burgeoni,* which I have collected only once. This species is missing both from the northern and southern shores. Seen from some distance in the lake it looks very much like *Julidochromis ornatus.*

None of the *Telmatochromis* are difficult to keep and they all breed in captivity. *T. bifrenatus* is, foremost among them, an ideal beginner's fish.

Description:

Body: elongate to very elongate, with the occipital crest becoming a slight hump; diagonal and/or horizontal stripes on the body; three species, *T.*

Telmatochromis bifrenatus is a dwarf species. Photo by B. Kahl.

Telmatochromis dhonti was formerly called *T. caninus,* a synonym. Photo by H.-J. Richter.

Key to the Species of
Telmatochromis:

1. Body depth included less than 4 times and head included less than 3.35 times in standard length 2

Body depth included more than 4 times and head length included more than 3.35 times in standard length 4

2. Teeth of inner rows all tricuspid 3

Teeth of inner rows increasingly conical with their distance from outer edge of jaw; 5-10/4-10 canine teeth; oblique stripes on sides, more visible on hind part of body; maximum size 120 mm.. ***dhonti***

3. 7-20/7-21 canines in front of jaws; without oblique stripes; maximum size 100 mm .... ***temporalis***

6-6 canines in front of jaws; 2 long horizontal stripes; maximum size 51 mm... ***burgeoni***

4. Body depth included 4.3 to 4.4 times and head length included 3.7 to 4.0 times in standard length; eye included 4.55 to 4.65 times in head length; 12-15/12-13 canines in front of jaw; 2 black horizontal stripes along sides of body, upper at base of dorsal fin; maximum size 86 mm........................... ***vittatus***

Body depth included 5.3 to 5.7 times and head length included 3.35 to 3.65 times in standard length; eye 3.9 to 4.0 times in head length; 8-10/6-8 canines in front of jaws; body with 2 horizontal stripes; maximum size 53 mm(?) .... ***bifrenatus***

bifrenatus, *T. burgeoni*, and *T. vittatus*, are very slender fish.

Scales: 31 to 37 in a longitudinal line, very denticulate.

Lateral lines: 2, not complete.

Fins: dorsal XVIII to XXII,6 to 10; anal V to VIII,5 to 8; ventral long; caudal rounded; all unpaired fins usually tipped by a filament.

Teeth: outer teeth always conical, often including from 6 to 8 canines; inner rows made up of small tricuspid teeth.

Pharyngeal teeth: conical or subconical.

Gill rakers: 3 to 9 (*T. vittatus* and *T. bifrenatus* have 3-4).

Habitat: always rock; although *T. vittatus* and *T. bifrenatus* are always in very clear water, the other three might be found on rocks in muddy water.

Feeding: microorganisms, omnivorous.

Reproduction: nest-breeders.

Maximum size in the wild: T. bifrenatus and *T. burgeoni* reach 60 mm; *T. vittatus* grows to about 90 mm; the other species of *Telmatochromis* are about 120 mm.

Telmatochromis temporalis (above) differs from *dhonti* (right) by lacking the golden ear stripe and the thin oblique bands on the sides. Photos by Aaron Norman and Dr. H. R. Axelrod.

Telmatochromis vittatus.   Telmatochromis dhonti.

Telmatochromis temporalis. These are the most often-seen species of the genus. Photos by H.-J. Richter.

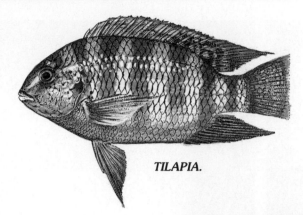

TILAPIA.

Color patterns of *Tilapia rendalli* change with behavior. Top to bottom: Female sexual display; parents guarding fry; fright coloration. Photos by J. Voss.

TILAPIA Smith, 1840/*SAROTHERODON* Rüppell, 1854

S. tanganicae is the only species of *Sarotherodon/Tilapia* found on coastal bottoms far from river estuaries. It is also the only *Sarotherodon/Tilapia* commonly found over sand and rock and not associated systematically with mud bottoms in the lake. *S. karomo* is found in the Malagarazi River delta but is more abundant upstream. *T. rendalli* has a range in the southern part of the lake basin but has been transplanted for fish-farming purposes into the Ruzizi River valley. This fact explains its presence along the northern shores of Lake Tanganyika but not in the lake proper. It has not been found, as yet, on the western or eastern shores.

S. nilotica is represented along the northern shores by the Lake Edward and Lake Kivu subspecies, *S. nilotica eduardiana,* but has not been discovered in the southern part of the lake.

Another species of *Sarotherodon,* first believed to be *S. nilotica,* was identified from a single specimen in the Malagarazi River delta on the central eastern coast. Other specimens of the species were

Tilapia rendalli guarding fry. Photo by J. Voss.

A young *Tilapia tanganicae*. Photo by G. S. Axelrod.

discovered around the mouth of the Lukuga River outlet where it arrived from the Upemba depression in Zaire. This species is called *S. upembae* Thys van den Audenaerde, 1964, and is not actually a part of the lake fauna.

Of the *Sarotherodon* and *Tilapia* species identified in Lake Tanganyika and its basin, only *S. tanganicae* has been able to colonize the whole length of the lake and can be considered as a true lacustrine species. Many adult schools have been discovered in the south along rocky shores and were seen grazing on the

Spawning in *Tilapia karomo*; the male has red in the dorsal fin. Photo by H.-J. Richter.

Detail of the head of *Tilapia karomo*.

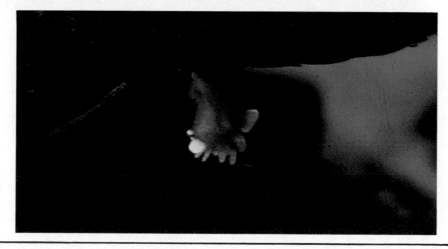

Detail of the anal papilla of *Tilapia karomo*. Photos by H.-J. Richter.

Tilapia rendalli. Note the "tilapia spot" in the dorsal fin. Photo by J. Voss.

rocks in typical *Tropheus* or *Petrochromis* style. This is reflected in their dentition, the teeth being assembled into brush-like pads very similar to those found in *Petrochromis*. No wonder, then, when the first specimens of *S. tanganicae* were collected, they were considered as being *Petrochromis*. Fluviatile *Tilapia/Sarotherodon* feeding on aquatic plants have very different teeth, i.e. assembled into rows, those in the first row forming a continuous cutting edge.

T. nilotica, found in the Ruzizi delta, is probably a latecomer to the lake basin, which it assumedly reached only when the Kivu basin was captured by the lake, probably not more than 10,000 years ago. One of the few true Nilotic fauna species in the lake basin (along with a few haplochromines), this species indicates only a recent contact of Lake Tanganyika with this river system, and indirectly at that, via one of its headwaters, Lake Victoria, whose fauna is mostly distinct from the one found in the Nile.

S. tanganicae is one of the most gorgeous fishes in the lake with a rainbow of colors, specifically a sky blue-jade green body, each scale being enhanced by a crimson dot. The unpaired fins are vermiculated in shades of bright yellow, orange, metallic green and blue, with crimson around the edges. Unfortunately, only medium-sized specimens are so beautifully attired, which means that the fish need to be a minimum of 20-25 cm before starting to show their colors off. They reach about 35cm, at which time they are truly magnificent, but weigh up to 1.5 kg.

All by themselves in a large tank they are a sight to behold, which is just as well because they do not tolerate many intruders in their realm. Only fair-sized nest-breeders or the biggest mouthbrooding cichlids will be capable of holding their own.

As for plants, what has been said about *Petrochromis* holds true for *S. tanganicae.* They have a ravenous appetite, and those plants that have escaped their attention because they are too coarse will be uprooted unless they are anchored under a big stone.

S. tanganicae has, perhaps, a bright future as a food fish, were the fishfarmers to realize that they have here a fish that could be made to spawn in ponds, but not as early as most riverine *Tilapia* are used to. This explains why, all too often, attempts to farm them in Africa have led to very disappointing results, such as a production of dwarves.

Coming from the lake, *S. tanganicae* could perhaps have a slowed-down spawning cycle in a pond and grow to full size.

S. karomo, the shovel-snouted *Tilapia* from the Malagarazi, is a very beautiful fish also, with its Prussian blue back and sides and orange underparts, but to my knowledge it has not been exported frequently. The teeth of this species stand halfway between the brush-like tooth pads of *S. tanganicae* and the cutting edges of *T. nilotica.*

Key to the Species of
Sarotherodon/Tilapia:
The subdivision of *Tilapia* into *Sarotherodon, Oreochromis,* and *Tilapia* according to the presence or absence of buccal incubation, etc. has not been taken into account for taxonomic identification.

1. Gill rakers 18 to 27 on lower part of first gill arch; outer teeth bicuspid or tricuspid 2
· Gill rakers 7 to 9 on lower part of first gill arch; outer teeth bicuspid and larger than the inner teeth; maximum size 330 mm; non-endemic... ***Tilapia rendalli***
2. Teeth in wide strips tricuspid only (even in front row), or with a few bicuspid teeth 3
· Teeth bicuspid in the front row, tricuspid in the rear, in narrow strips; gill rakers 21 to 27 on lower part of first gill arch; pharyngeal teeth in a felt-like pad with a sharp angle in front; maximum size 500 mm; non-endemic ....***Sarotherodon nilotica eduardiana***
3. Very broad felt-like dental pad; pharyngeal teeth surface also like felt but pad not pointed in front; gill rakers 18-20 on lower part of first gill arch; snout 2.0-2.8 times longer than eye; maximum size 320 mm; endemic to the Malagarazi River, not in the lake proper beyond the river delta.......................... ....***Sarotherodon karomo***
· Felt-like dental pad not as broad; felt-like surface of pharyngeal teeth heart-shaped and with a point in front; gill rakers 23-27; snout only 1.1 to 2.4 times longer than eye; maximum size 440 mm; endemic to the lake proper ....***Sarotherodon tanganicae***

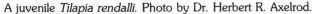

A juvenile *Tilapia rendalli.* Photo by Dr. Herbert R. Axelrod.

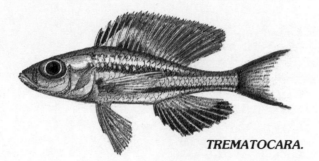

TREMATOCARA.

TREMATOCARA Boulenger, 1899 (Benthic diatom and microorganism feeders)

No *Trematocara* that I know of has been collected since Professor Poll participated in the 1946-47 Belgian Mission to the lake. We therefore have to rely on his reports about this fascinating group of deep-living cichlids. None of the eight species as yet recorded has been collected in less than 30 meters depth, and some have been fished at the bottom of the oxygen-bearing layer. These fishes thus present the best case of acclimatization among cichlids to poor oxygen levels in the lake.

What we know about these elusive fishes is very little, as nobody has as yet been diving deep enough to encounter a school of any one of the species. Collections made by Poll reveal that they are gregarious and mouthbrooders. In a few cases females were caught still with a few eggs in their mouth, but this was a stroke of luck. Due to the tremendous decrease of pressure as the catch was hauled in from so deep, all should have lost their spawn. An additional clue is provided by the size of the eggs found in ovaries. Even in species less than 6cm in length they reach 2mm in diameter. As we have seen, no nestbreeding cichlid in the lake has eggs much more than 1mm in diameter.

We can only guess at their preferred ecological conditions. They live in permanent dimness and probably are light-shy. Otherwise, we wouldn't see them rise at night from their deepest levels to one closer to the surface. This nyctemeral pattern of vertical migrations might follow that of the zooplankton.

All *Trematocara* have very large eyes, which helps them spot their prey in whatever light penetrates into their dark realm. It seeps through the water layer from the surface, sometimes as much as 200 meters away. In this respect we have an entirely opposite physiological answer to lack of light than the one we observe in riverine fishes. In rivers, deep-living fishes in mud-laden waters where vertical penetration of light is not more than 30-50 cm, as in the Congo River in Kinshasa, tend to lose their sight. Cases of microphthalmy, or atrophy of the eyes, are many and are found among *Mastacembelus* eels and catfishes, and lead to the loss of the organ. There are several blind fishes in African rivers, among them in the Congo basin alone a catfish eel, a mastacembelid eel, a barb, and even the only blind cichlid as yet known in the world *Lamprologus lethops.* Why then not in the lake, where cichlids respond to the dimness by boosting the efficiency of their eyes in order to pick up the slightest glow reflected from the bottom ?

What has prevented riverine fishes from responding to dim light or near-total darkness by boosting the efficiency and number of light-sensitive cells in their eyes as the lake cichlids did? We might find an answer in the quality of the lake water. To start with it is crystal clear, untainted by color pigments from decaying vegetation releasing humic and tannic acids, as happens in tropical rivers, and without the fine mud particles that so quickly hamper the penetration of light below the surface.

In the lake, given the crystal clearness of the water, boosting the light-receptive capacity of the eye leads to better vision. In a river, laden as it is with fine silt and already tainted by the dissolved pigments, increasing the eye efficiency is pointless. The murkiness and color of the water will curtail the range and perception.

In the lake we have several other examples of cichlids having adapted their eyes to peculiar conditions. Cave-dwellers like *Lamprologus furcifer* or *L. prochilus* have unusually large eyes; *L. compressiceps, L. calvus,* and *L. fasciatus* have eyes that rotate very much in their sockets so that they can squint at close quarters.

Trematocara stand apart from all other cichlids in the lake by several features, some of them quite unusual:
-*Trematocara* have atrophied

lateral lines. None has a lower lateral line and in all of them the upper lateral line is reduced to very few scales. Contrary to *Lamprologus* and affiliated genera, the missing tubuled scales are not replaced by open-pore neuromasts. The lines are simply missing. Prof. Poll says that the scales are caducous, which means that they fall off very easily and then grow back. This is a very unusual condition for cichlids and one that in the lake is shared only by clupeids. Clupeids also lack lateral lines and one might be concerned about the efficiency of the lateral sensory organs, depending so much as they are on the good setting of the supporting scales. If the scales were not permanently set, the sensory organs had to disappear. Why we find converging evolution in clupeids and *Trematocara* and why the scales are deciduous, we don't know.

-All species of *Trematocara* are endowed with big cavities in their facial bones. These cavities, covered with a thin skin, extend in continuous rows on the lower jaw, the preopercle, around the eye, on the snout, and in the postocular and nuchal areas. The extension of the rows varies with each species. In some the rows extend only onto the jaw, the snout, the preopercle, and under the eye.

Given the lack of lateral line organs one would be led to believe that these drum-like cavities are sensory organs enabling the fish to perceive what is happening in front of him and on his sides. Perhaps these cavities are amplifying the clicking sounds of microorganisms as they hop by like so many finely tuned microphones? Pending a study of these organs one can only guess.

Trematocara have one last distinctive feature—their teeth. Unlike those of other cichlids, theirs look very much like those of clupeids. They are small, conical, very fine, and thickly packed into a band running near the jaw edge. It is a dentition that is well suited to its purpose—grabbing tiny animalcules. Some of the species have an enormous and gaping mouth, opening much like the lake clupeids' mouth.

There is thus not only one convergence but several between these cichlids and the lake clupeids.

Most species are very colorful in their own way, with strongly contrasted patterns featuring jet black stripes along the body in various arrays. Some have black spots ocellated with white on the dorsal fin. The body sides are silvery or endowed with sky-blue glistening scales. One species can be distinguished by the lack of all these features, being plain silvery (*T.macrostoma*). These strongly contrasting patterns are typical of males because, as a rule, females are much less colorful. Thus sexual dimorphism is strong.

The sizes range from 45mm to 150mm (one species) thus they all could become favorites among cichlid amateurs because of their unusual patterns if they could only be made available, which is very doubtful.

Description:

Body: moderately elongate.

Scales: 27 to 31 cycloid scales in a longitudinal line.

Lateral line: only one, the upper, which is reduced to a few tubes.

Fins: dorsal VII-XII,10 to 13; anal III,8 to 11 (rays longer than spines); ventrals long; caudal forked; pectorals large.

Teeth: very minute; conical, in a single narrow band.

Pharyngeal teeth: subconical, conical, bicuspid; in some species the central posterior teeth are rather molar-like.

Mouth: terminal; *the bones of the head* (nasal, frontal, pre- and suborbitals, preopercle, and mandible) *with rows of excavations,* separated by narrow bridges *and covered by a thin skin.*

Gill rakers: 9-25 on the lower half of the first gill arch.

Habitat: deep, along the coastline, sometimes in 200 meters. All the species seem to tend toward a nyctemeral vertical migration, ascending at night into shallower waters.

Feeding: microorganisms.

Reproduction: nothing is known of their behavior except for the fact that captures commonly involve many specimens.

Maximum size in the wild: depending on species, from 50 mm to 150 mm.

423

Key to the Species of
Trematocara:

1. 15 to 25 gill rakers on lower part of first gill arch; 3 to 9 sensory organs on pre- and suborbital bones; caudal peduncle 1.35-2 times longer than high.............................2

   9 to 12 gill rakers on lower part of first gill arch; 8 (seldom 9) sensory organs on pre- and suborbital bones; jaws equal and lips rather thick; dorsal fin edged with black stripe along its entire length; maximum size 100 mm .... ***marginatum***

2. Body depth included 3.0-4.2 times and head length included 2.5 to 3.0 times in standard length; 3 to 9 sensory organs on pre- and suborbital bones (except in *T. variabile,* which has only 5 sensory organs but on the other hand has 15-17 gill rakers) 3

   Body depth included 3.7-4.1 times and head length included 2.75-3.0 times in standard length; 5 hypertrophied sensory organs; lower jaw protruding; a typical black spot on chin, more visible in female; 17 to 21 gill rakers on lower part of first gill arch; maximum size 115 mm .... ***nigrifrons***

3. Minimum of 4 sensory organs on pre- and suborbital bone 4

   Only 3 sensory organs; caudal peduncle very narrow, twice as long as high; body depth included 3.8 to 4.2 times in standard length; an elongate black spot at front of spiny dorsal; maximum size 68 mm .... ***kufferathi***

4. A full row of 8 or 9 hypertrophied sensory organs... 5

   An incomplete row of 4 to 7 hypertrophied sensory organs 6

5. Body depth included 3.0 to 3.25 times in body length; eye included 2.5-3.0 times in head length; maxillary not reaching to middle of eye; teeth not set on external part of lips; dorsal fin with median spot; maximum size 150 mm .... ***unimaculatum***

   Body depth included 4.1 to 4.2 times in body length; eye included 3.5 times in head length; maxillary reaching middle of eye; teeth set in part on external part of lips; dorsal fin without anterior black spot; maximum size 45 mm .... ***macrostoma***

6. Only 4 hypertrophied sensory organs on pre- and suborbital bones..................................7

   5 to 7 hypertrophied sensory organs on the pre- and suborbital bones; body depth included 3.4 to 4.1 times in body length; 15 to 17 gill rakers on lower part of first gill arch; black spot on anterior part of dorsal fin; maximum size 87 mm .... ***variabile***

7. Eye oval, included 2.65 to 2.95 times in head length; 15 to 20 gill rakers; dorsal edged with black band in anterior part of fin; maximum size 75 mm................. ***stigmaticum***

   Eye round and very large, included 2.3 to 2.65 times in head length; 21 to 25 gill rakers; dorsal with broad black band at mid-height; maximum size 67 mm .... ***caparti***

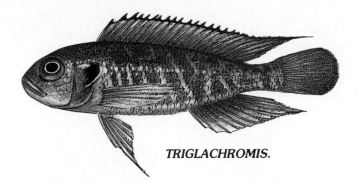

TRIGLACHROMIS.

TRIGLACHROMIS Poll & Thys van den Audenaerde, 1974

It was quite by accident that in 1972 the author, while experimenting with new packing tranquilizers for this fish, discovered the very unusual feature that led Pr. Thys van den Audenaerde to separate *T. otostigma* from *Limnochromis*.

T. otostimga has the lowest rays of the pectoral fin separate from the other rays of the fin and not bound together by a membrane. These independent rays can be bent down by the fish, and it is possible, although it hasn't been proved, that they might act as feelers when the fish, in its very muddy habitat, has very poor visibility. Unless I am mistaken, it is the only case of separate pectoral rays reported from African freshwater fishes.

One of the staple fish to come from Lake Tanganyika to the aquarium world, *T. otostigma* has never been a problem as far as acclimatization is concerned. It is more surprising that, at least to the author's knowledge, there have been no reports of breeding the fish in captivity.

T. otostigma is again one of those unusual Lake Tanganyika cichlids with much less color than many of the better known varieties from Lake Malawi, but with body patterns not often found anywhere else. The appeal of the fish depends on its metallic copper background with contrasting oblique golden stripes; this pattern is strongly enhanced by the white paired fins and the black edge on the tail.

Much smaller than most Rift Valley cichlids, *Triglachromis otostigma* is a good community tank fish provided, as with all cichlids, that its companions are of comparable size.

Description:
Body: elongate.
Scales: absent from in front of the ventral fins as well as around the base of the pectorals; 35 to 37 on a longitudinal line; oblique, alternate rows of pearly scales on the body.
Lateral lines: 2, the upper complete.
Fins: dorsal spines XV or XVI, each tipped with a soft filament, and 8 to 10 rays; anal III,7 or 8; first rays of ventrals with a filament; pectorals with *lower rays free from the fin,* the next rays partially welded into the fin; caudal rounded, not forked as in *Limnochromis*.
Teeth: conical, in 2 or 3 rows, the outer row well separated from the next row and with much larger and rather horizontal teeth.
Pharyngeal bone: very small rather bicuspid teeth.
Mouth: low, horizontal and broad.
Gill rakers: 12 to 14.
Habitat: typical mud-dweller, coastal, from shallow to rather deep water.
Feeding in the wild: diatoms and microorganisms, but also omnivorous.
Reproduction: the breeding mode is unknown; buccal incubation has not been reported.
Maximum size in the wild: not in excess of 100 mm.

Triglachromis otostigma is a popular aquarium species. Photo by H. Arment.

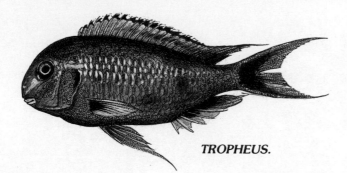

TROPHEUS.

TROPHEUS Boulenger, 1898 (Rock-grazer)

The various species, subspecies, and local varieties have already been covered elsewhere in this book and need not be discussed at length here, but aspects of their ecology and ethology remain to be considered.

The habitat of *Tropheus* is solid rock, either anchored and part of the hard core of the coast, interlocking rubble, or even sandstone slabs lining a sand beach. Loose pebbles rolling about in the surf and silted-over habitats are avoided. The vertical range depends on the depth at which the epiphytic algae keep on growing and varies from one species to another and from one part of the lake to another. The distribution of each species around the lake had depended on their bonds toward rocks, but it also depended on the maximum expansion the rocky coast it lived on reached in the past.

The density of a population of *Tropheus* depends on the availability of food and shelter, especially the latter for the fry. The ratio of adults and fry is thus very variable.

The average and maximum sizes reached by individuals in a population might reflect the scarcity of abundant or overabundant food, such as at Nyanza Lac, where the local

Tropheus brichardi never exceed 10cm in total length, whereas in other places they commonly exceed 14cm.

The social behavior of *Tropheus* is based upon individualism and mobility. They do not form schools, although it might happen that they merge into loose concentrations when their density is high.

Territorialism on a slope has been deduced from short observations in the lake and from many reports on captive fish, but the permanent occupation of a piece of ground similar to that of nestbreeding rock-dwellers has not been witnessed and very probably will not.

Given the psychic adaptability of *Tropheus* and the narrow confines of a tank, it is normal that these extremely individualistic fishes secure privacy in a tank by staking out a place where they claim ownership. Unless their quarters are very roomy, these fish will display such an aggressiveness that after a few days only one is left. This occurs mainly between males, but females have also been reported, when on their own, to fight with each other. The aggressiveness drops when several dozen fish are kept together; the instinct to fight for room appears to be blunted by the inanity of fighting off too

many opponents.

There are degrees in the aggressiveness and territoriality between the various species and even races. So far the most aggressive species appears to be *T. brichardi* from Nyanza Lac. Very aggressive also is *T. duboisi*. If one *duboisi* and one *T. moorii moorii* are put together in a large tank, it occurs much more frequently that the *T. moorii* will be killed, and sometimes other fish as well. Among the *T. moorii moorii* the least aggressive toward its own appears to be the Burundi "red" race.

Breeding *Tropheus* has been achieved quite regularly and up to now it does not appear that too many attempts at producing hybrid color morphs have been successful, at least in Europe.

About 15 races are bred by the fish-collectors in Burundi because the supplies of wild-caught specimens from small or hard-to-reach populations would never equal the demand. Aside from hybridizing, these collectors try to produce the purest and best fry from first rate breeders carefully selected from the best wild fishes collected. The fish are raised in ponds filled with lake water and given natural foods. Thus the quality and pure lineage of the fry cannot be placed in doubt.

The ratio of males versus females for breeding purposes is important; too many males and the females soon become exhausted by too many spawns. In the lake ripe females spawn. Then, as time passes during their incubation, they retreat more frequently to the rubble. They can do so as well when they need a

Tropheus sp. Kanyosha. H. J. Mayland.   *Tropheus* sp. R. Stawikowski.

Tropheus moorii kasabae, Mpulungu. R. Stawikowski.   *Tropheus* sp. "Murago". R. Stawikowski.

Tropheus sp. "Kirch-fleck". D. Schaller.   *Simochromis diagramma*. B. Kahl.

Petrochromis fasciolatus, Tanzania. R. Stawikowski.   *Limnochromis auritus*. H. J. Mayland.

Tropheus duboisi showing the typical blue head and white midbody band. Photo by G. S. Axelrod.

moments rest between two spawns. In captivity they cannot find a place away from the males so that perhaps tank-raised fry could have been born from physically exhausted females. Given the very high price paid by professional breeders for their stock, it is only normal that they try to recoup their investment as quickly as they can.

Description:

Body: deep in front, tapering off toward the tail; head blunt and broad; body stocky and broad.

Scales: denticulate, 31 to 32 in a longitudinal line.

Lateral lines: 2, neither complete.

Fins: dorsal XX or XXI, 5 or 6; anal IV to VI, 5 to 7; ventrals with short filament; caudal truncated, slightly forked; pectorals long.

Teeth: an outer row of bicuspid teeth have a continuous cutting edge; the sides of the premaxillary bone have strong, curved conical teeth.

Pharyngeal teeth: more or less bicuspid, thin, and compressed.

Mouth: very low, very straight, and transverse.

Gill rakers: 11 or 12 on lower half of first gill arch.

Habitat: restricted to rocky slopes with a heavy biocover; to 15 m deep.

The typical blue eye of this *Tropheus brichardi* is very visible. Photo by H.-J. Richter.

Some *Tropheus* varieties. Left column, top to bottom: *duboisi*, wide yellow band; *moorii* fed on algae; *moorii* green with orange dorsal. Right column, top to bottom: *duboisi*, broad band; *moorii* "Moliro"; *moorii*, Bemba orange.

Feeding: typical rock-grazer.
Reproduction: buccal incubation; about a dozen eggs (maximum number of eggs in a wild caught specimen 14), 7 mm in diameter.
Maximum size in the wild: T. moorii: 165 mm; *T. duboisi:* about 120 mm. *T. brichardi:* 140 mm.
Affinities: with *Simochromis,* from which it differs by having: 1) more than 3 anal spines; 2) never less than 20 spines in the dorsal fin; and 3) a deeper habitat, never in muddy water.

Whitish spots are typical of *Tropheus duboisi,* but they weaken with age. Ph by Dr. Herbert R. Axelrod.

Key to the Species of
***Tropheus* (after M. Nelissen
and G. Axelrod)**

1a. Anal spines IV; mouth length 1⁄3 of mouth width, which is a bit greater than interorbital width; preorbital length less than eye diameter; cheek depth more than 3⁄2 of eye diameter; body brown, green, or gray, sometimes with broad stripes in adults; juveniles striped...***annectens***

b. Anal spines IV; mouth length 1⁄2 of mouth width; mouth width a bit greater than interorbital width; cheek depth more than 3⁄2 of eye diameter; body with several broad bands; dorsal and anal fins filamentous, tail crescentic.............................***polli***

c. Anal spines V; mouth a bit larger than interorbital width; mouth length 1⁄3 of mouth width; preorbital length smaller than eye diameter; cheek depth more than 3⁄2 of eye diameter; snout longer than eye diameter; posterior part of body, including tail, black; tail truncate or slightly forked .... ***moorii moorii***

d. Anal spines V-VI; mouth length 1⁄2 of mouth width; mouth rounded with rounded lips; body black; head also black but as if covered with blue skin; white or yellow/white band from dorsal fin to abdomen; juveniles black, spotted with white.............***duboisi***

e. Anal spines VI; snout length 3⁄2 of eye diameter, which is only 1⁄2 of cheek depth; body with broad lighter colored patch on side in abdominal, pectoral, or caudal peduncle area; sometimes with about 10 thin vertical light colored stripes; juveniles striped; otherwise as in *moorii moorii* .... ***moorii kasabae***

f. Anal spines VI-VIII; preorbital length less than eye; mouth width more than 3⁄2 of interorbital width; cheek depth less than 3⁄2 of eye diameter; color in shades of green, chocolate, or yellow, most often broadly banded .... ***brichardi***

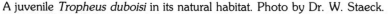

A juvenile *Tropheus duboisi* in its natural habitat. Photo by Dr. W. Staeck.

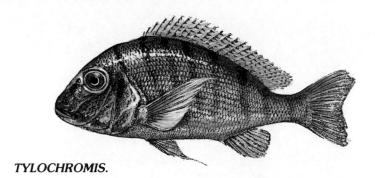

TYLOCHROMIS.

TYLOCHROMIS Regan, 1920 (Swamp-dweller)

Only one species is endemic to the lake, but five more exist in the Congo basin. The single species has not been caught recently but was well documented by Poll who caught many around Kalemie. The juveniles are coastal and gregarious, but adult fish wander over the sandy bottoms by themselves. They can do this because they reach 330mm in length and few predators will attack a fish that size.

T. polylepis is not as gorgeously attired as its counterpart from the Stanley Pool near Kinshasa, which I occasionally collected. I found it to be one of the most beautiful, if not the most beautiful, fish in the river. The head and anterior part of the body are a flamboyant orange, the posterior part green-blue, and all fins are a display of bright yellow, blue, and crimson vermiculated stripes. The species from the lake is unfortunately less colorful.

As with *Limnotilapia dardennei,* juvenile fish less than 20cm do not display any of the colors of the adults, so that it would take a very patient hobbyist to buy such a fish, were it to become available, and wait until it develops its colors.

Description:
Body: deep and narrow, laterally compressed, the head large and deep, the body tapering off strongly toward the caudal peduncle.
Scales: 32-60 in a longitudinal line.
Lateral lines: 2, both complete.
Fins: dorsal XIV-XVI,13-15, anal III,7-9.

XENOTILAPIA Boulenger, 1898 (Sand-dwellers; gregarious)

Of all the species of fishes Lake Tanganyika shelters, none are more elusive for a frustrated fish collector than the 13 species of *Xenotilapia.* This is not because the fish are rare. Many species along the northern shores are caught with seines by professional fishermen in very large quantities, one might even say by the tens of thousands each day.

The problem with this kind of seine fishing is that the fish never survive the catch because the seines are of a peculiar type. They are very high and are laid very far from shore in water ten or even fifteen meters deep, where they sink and are then hauled back to shore with ropes hundreds of meters long by a score of fishermen. Rolled between the mesh, bruised or crushed by pieces of wood and pebbles, and finally dragged into shallow water over coarse sand, the fish are damaged beyond rescue. Worse still, at the depth at which they were captured by the net it is impossible for them to reach the shallow water near the beach without a strong case of the bends (decompression sickness), with an inflated bladder and the intestine protruding from the mouth or the anal orifice.

For years we have brought the best looking fish by the hundreds from the fishermen, but to no avail; losses were too high. When we were very lucky we saved one percent of the fish thus collected.

We tried to see what could be done. But over these vast expanses of sand floors with a slight slope, the turbidity of the water was such that no one could see more than a few feet. The colors of *Xenotilapia* blend wonderfully well with their biotope. The only way to see any was to put your eyes at sand level and try to see the shape of the fish against the lighter background of the

XENOTILAPIA.

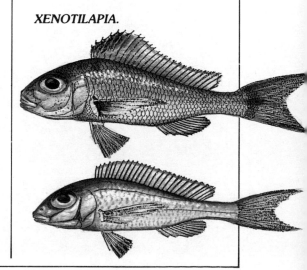

Xenotilapia longispinis longispinis. Photo by G. S. Axelrod.

Xenotilapia flavipinnis. Photo by Dr. Herbert R. Axelrod.

Xenotilapia ochrogenys. Photo by Pierre Brichard.

water. Even so, the problem was to find a sizable school, to get the other divers that were looking around on their own back to the school, to stretch the net into the proper direction, and then try to chase the fish into the net. As I said before (more than once) this is not the easiest way to catch fishes.

Being in a state of shock after their frantic maneuvers to avoid capture, *Xenotilapia* miss the delicate adjustment of the swim bladder pressure during the time they are hauled up to the boat, a climb that they would be capable of in normal circumstances. So, again, they die.

One species is occasionally seen over sand patches near the rocky slopes. So it happens that when on a collecting trip for other fishes, we may happen upon one of their schools. If we have the right type of fishing gear along, we of course try to catch the whole bunch, numbering usually between 500 and one thousand fish, so that for a few weeks this *Xenotilapia,* which we called *flavipinnis* pending its description by M. Poll, might be found by hobbyists at their dealers.

Another *Xenotilapia, X. sima* is also found in very small bands, at most a few dozen strong, on the rocky slopes. Unfortunately, this unmistakable fish, with its short and heavy head and large oval yellow eyes, lacks the gorgeous colors of the other species.

God knows they can be beautiful! Nothing in the cichlid world can beat a *X. ochrogenys* or *X. melanogenys* male in full nuptial regalia. The head, blunt and short in the first fish but streamlined in the second species, is a rainbow of iridescent orange, Prussian blue, emerald green, and mauve, while the body is basically fawn colored with horizontal stripes of dull yellow and shimmering hues of pink, blue, and crimson. The unpaired fins sport broad yellow-orange bands and the caudal peduncle shimmers with iridescent dark and pale blue. Words cannot express the amazing beauty of these two species.

Other species, which have not yet been identified and might be new, display several rows of mother-of-pearl scales, a very large and blunt head (much like *X. ochrogenys)* and wavy patterns of yellow-orange on the dorsal and anal fins; or there is a bright golden yellow patch behind the pectoral fins, with the unpaired fins also being striped with yellow-orange.

Thus all *Xenotilapia* could easily become the darlings of the aquaristic world, if they could only be brought in alive. All *Xenotilapia,* once used to tank life, will thrive on common types of food and will present no problem to the average hobbyist. They need some space since they are roamers, but they will not bother any fish unless it is really a very small one. Breeding them in captivity has as yet not been reported.

In the wild it appears that all females in the same school are ripe at the same time and mate, incubate, and probably release their fry on the same spot at the same time. The fry are thus formed into a communal school of assorted sizes as are the adult schools. Needless to say, *Xenotilapia* are very gregarious and

Sand-dwelling *Xenotilapia* blend well with the bottom and may be almost invisible. These are *Xenotilapia ochrogenys*. Photo by G. S. Axelrod.

These *Xenotilapia longispinis* are hovering over the top of a flat boulder rather than hiding in the sand. Photo by G. S. Axelrod.

Xenotilapia melanogenys. Photo by Aaron Norman.

Xenotilapia sima. Photo by Pierre Brichard.

Xenotilapia ochrogenys in partial breeding color. Photo by Thierry Brichard.

respond to all the group stimuli of very gregarious fishes. In a very recent observation of a spawning school, we followed one female who repeatedly went down into several male's nests, spawned, got her eggs fertilized, and swam away. The eggs of one female were thus fertilized by different males.

We have only once succeeded in discovering a huge school of the fish mixed with another of *X. ochrogenys* in shallow water over favorable grounds for catching them, but as the fish were busy nesting and spawning on this site, we preferred just to enjoy the sight. Once again, with these deep-living fishes, often deeper than 60 meters, it is apparent that they come up the slopes to mate in shallow water and it is there that the schools stay after mating has taken place. The females can then incubate the eggs and fry in their mouths in the upper, oxygen-rich layer. Once the fry have grown and have been released, these buccal incubators will migrate back to the deep layers where they spend their lives.

Description:

Body: elongate, with a large, deep head, body tapering off from the nape to the caudal peduncle, which is narrow.

Scales: 34 to 43 in a longitudinal line.

Lateral lines: 3, the upper complete, the other two not.

Fins: dorsal XIII-XV,11 to 19; anal III,7 to 18; ventrals *with outer rays always equal to or shorter than inner rays;* caudal forked.

Teeth: very small, conical; in 2 or 3 rows; outer teeth of the lower jaw pointing forward (typical of sand-sifters).

Key to the Species of
***Xenotilapia:* (from M. Poll)**

1. Teeth set in a minimum of 2 rows, the teeth from lower front row pointing more or less in horizontal plane; pharyngeal teeth rather molar-shaped, at least in central part of bone; interorbital space 4.8-8.25 in head length 2

    Teeth very thin and miniscule, in one row at edge of each jaw, the row on lower jaw not pointing in a well defined direction; pharyngeal teeth all thin; interorbital space included 3.8 to 4.6 times in head length; maximum size 80 mm .... ***tenuidentata***

2. Body depth included less than 5 times in standard length; slope of head always rather convex; anal fin with fewer than 15 soft rays 3

    Body depth included 5.0 to 5.9 times in standard length; slope of head straight or barely convex; 13 to 18 gill rakers on lower part of first gill arch; D. XIII-XV, 16-19; A. III, 15-18; maximum size 150 mm...***melanogenys***

3. Body depth included 4.25 to 4.85 times in standard length; ventral fins with inner ray longer than outer ray; maximum size about 110 mm...4

    Body depth included 3.3 to 4.1 times in standard length; ventral fins with inner ray equal to or longer than outer ray; maximum size in excess of 125 mm..............................5

4. Gill rakers 9-12 on lower part of first gill arch; caudal peduncle 1.65 to 2.0 times longer than high; eye 2.5-3.75 times in head length; maximum size 110 mm ....
    ....***ochrogenys ochrogenys***

    Gill rakers 11-14 on lower part of first gill arch; caudal peduncle 1.9 to 2.5 times longer than high; eye 2.3 to 3.1 times in head length; maximum size 103 mm ....
    ....***ochrogenys bathyphilus***

5. Gill rakers 13-18 (seldom 13)...6

    Gill rakers 9-14 (seldom 14)...9

6. Dorsal fin with spines longer in middle (in the male) 7

    Dorsal fin with spines increasing in length from first to last..................................8

7. Longest dorsal spine shorter than half head length; inner ray of ventral fins longer than outer ray; no spots on anterior part of dorsal fin; maximum size 163 mm ....
    ....***longispinis longispinis***

    Longest dorsal spine equal to or longer than half head length; inner ray of ventral fin about equal to outer ray; dorsal fin with several small black spots in front; maximum size 173 mm.. ***longispinis burtoni***

8. Gill rakers 13-15 on lower part of first gill arch; A. III,8-10 (usually 9); caudal peduncle 1.9-2.1 times longer than high; ventral fins with inner ray longer than outer ray; maximum size 130 mm .... ***nigrolabiata***

    Gill rakers 15-17 on lower part of first gill arch; A. III,7 to 9 (usually 8); caudal peduncle 1.65-1.9 times longer than high; ventral fins with inner ray equal to or barely longer than outer ray; maximum size 125 mm .... ***ornatipinnis***

9. Lower jaw not protruding 10

    Lower jaw very protruding; 11-14 gill rakers on lower part of first gill arch; D. XIV-XV,11 to 14; A. III,10-12; ventral fins with inner ray longer than outer ray; maximum size 164 mm... ***sima***

10. Dorsal fin with more than 12 soft rays 11

    Dorsal fin with XIV or XV spines and 10 to 12 rays; A. III,7-9; 9 to 11 gill rakers; ventral fins with equal rays; male with striped caudal fin; maximum size 156 mm ....
    ....***caudafasciata***

11. (triplet). 11 to 14 (usually 12) gill rakers; D. XIII to XV,13-15; A. III,10-11; ventral fin with inner ray longer than outer ray; maximum size 153 mm...........................***boulengeri***

    13 gill rakers; D. XIV,14; A. III,10; ventral fins with inner and outer rays equal; maximum size 104 mm..***lestradei***

    12-13 gill rakers; D. XII-XIII,13-14; A. III,8 or 9; ventral fins with inner ray shorter than outer ray; body depth 3.3 to 3.6 in S.L.; large black spot in middle of dorsal; dorsal fin with spines number eight to XI longer than other spines ....***spilopterus***

Detail of the head of *Xenotilapia flavipinnis*. Photo by Pierre Brichard.

Xenotilapia spilopterus. Photo by G. S. Axelrod.

Xenotilapia flavipinnis. Photo by Pierre Brichard.

Pharyngeal teeth: in two papillose cushions located on the sides of the bone.

Mouth: low, straight, and small, but protractile.

Gill rakers: 9-18 on the lower half of the first gill arch.

Habitat: not deep, *always over sand,* even on sand patches between rubble on rock slopes.

Feeding: copepods, small shrimps.

Reproduction: buccal incubation involving up to about 50 eggs, size 3 mm.

Maximum size: about 150 mm, but many species much less.

Xenotilapia (Enantiopus) melanogenys

These very specialized sand-dwelling cichlids breed in an "arena." The whole school, or a substantial part of it, swims into shallow water, picking a site on sandy bottoms where all sexually mature fish will start breeding in crater nests individually built by the males. The typical nest is between 35 and 40 cm across, the smallest being about 20 cm across. These nests, with a flat, shallow floor and slightly raised rim, are dug so close to each other that the rim of one is usually also part of the next nest's rim. Seen from above the interlocking nests leave little spare room between them, their array looking very much like a mosaic of big and small saucers thrown at random.

Late arriving males have only two options open to them. 1) Squeeze a much smaller nest in between the large ones — which they often do. 2) Build a regular size nest at the outskirts of the arena.

As yet it is not possible to say if the efficiency, i.e., the number of females a male will mate with, will be better in a small nest in the middle of the main arena or in a large nest lying out of the hub of the main spawning activity. Perhaps it makes no difference!

About 70-100 cm above the bottom the main flock, consisting of immature males and unripe or already mouthbrooding females, swims about. Skimming the nest rims, males and females that are ready to spawn crisscross the arena, the former in search of a site on which to build their nests, the latter looking for a nest with a male to spawn with. Although nests lie very close to each other, there is very little aggression displayed by the males defending their crater. They very seldom, if ever, trespass on another's territory. As long as no female has entered the nest the owner will confine his animosity toward intruders only to the extent of the full display of his nuptial colors and the spreading of his gill-covers. On the other hand, as soon as a female arrives at the nest the male will savagely repel any intruder to the point of interrupting an actual mating session that may be in full swing.

It is not difficult to tell a male in full spawning readiness from one that isn't, as the former displays an extravagantly colored garb. Without entering into needless detail, one can describe the mating garb as such: the entire snout (not only its tip as is normal) becomes jet black up to the interorbital space. This blackened area is followed on the occiput and nape of the neck in front of the dorsal fin by a broad iridescent turquoise green patch. The sides of the body are shimmering mauve, pink, and yellow. Ventral and anal fins become jet black, except for their bases which become pale blue. The normal colors of these fins involve a pattern of orange stripes and zigzagging dots. The first half of the dorsal fin has a row of 4 to 5 black dots, the remainder of the fin being bright yellow. The flamboyant colors of the mating garb of the males appear only as they start working on their nests and increase in intensity to reach their climax when mating occurs.

The activity commences as follows: A female enters the nest and comes to rest on the floor quite motionless. The male goes into a frenzy and with jerking moves starts to circle around her with his body leaning on its side. All his fins are outstretched and his chin protrudes as if he, himself, were mouthbrooding (which is not the case), as if it were a reminder of what all this is about. For several minutes she will remain apparently oblivious to the male's frenetic antics. Then, tentatively, she will make a half-hearted move sideways, which is the signal to the male that she is ready. This signal triggers the ultimate frenzy of the male. He gets very close and all of a sudden the female starts gyrating with him, her mouth close to the genital opening of her mate. Suddenly, after the merry-go-round has been going on for some time, one or two eggs are expelled and fall to the ground. The male fertilizes the spawn as he passes over the eggs and the female snatches them up as she completes the circle. The eggs are thus laid, fertilized, and secured in the female's mouth in a fraction of a second. It takes a long time and several gyrations before another egg is laid.

It often occurs that, for no apparent reason, a female after having entered a nest shows no interest in mating, to the point of feeding herself by picking up a mouthful of sand. She then departs and swims toward the main flock, crossing on her way several other nests with their eager males. It might be that she doesn't feel at this time the urgency of getting rid of her eggs. It also happens that after having had several eggs fertilized by one male she goes to another nest to get a new batch of eggs fertilized by another male as the eggs in her ovaries enter the genital duct. It is the female who decides whether or not the conditions are right for her to lay her eggs.

The mating session of a *Xenotilapia* school lasts several days.

A few features of interest remain to be told about the reproduction of these fish: 1) We haven't seen any male incubating, mouthbrooding being the exclusive duty of the females. 2) The school doesn't remain in very shallow water after the mating session has ended and the arena is abandoned. 3) Mating among *Xenotilapia* is a community affair, but stragglers or fish that didn't reach sexual maturity at the time the school mating occurred remain in coastal waters to mate individually or in small groups. It is in those cases that I could observe the star-shaped crater-nests to which I alluded in my earlier work in which a male erects small elongate mounds,

Xenotilapia spilopterus. G. S. Axelrod.

numbering between 8 and 11, with a small central depression. It thus appears that there are two types of nest built by *melanogenys*.

One has to be cautious when assessing the behavior of the lake cichlids. It is more sophisticated and adaptable and much less stereotyped than is common among other fishes, even other cichlids.

Xenotilapia flavipinnis. Dr. H. R. Axelrod.

Xenotilapia sima. Dr. H. R. Axelrod.

Xenotilapia ochrogenys. Pierre Brichard.

Xenotilapia near *flavipinnis.* Pierre Brichard.

Pierre Brichard at work. Dr. H. R. Axelrod.

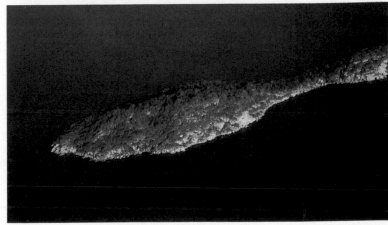

Rocky spit in Lake Tanganyika. Dr. H. R. Axelrod.

CHAPTER XI
THE NON-CICHLID FISHES

Eighteen families include around 43 genera that in turn contain about 110 species of fishes that are not cichlids but live in the lake basin. Most live in the lake affluents. The endemism of fluviatile non-cichlids is much weaker than that displayed among the lacustrine species.

The origin of the fluviatile and lacustrine species can be traced back to the major river systems passing in the neighborhood of the lake. First and foremost is the Congo River, from which most of the non-cichlids in the lake area are derived, then comes the Nile and perhaps even the Zambezi Rivers. Being nested in a fracture of the Earth's crust where orogenesis was strongly at work, many headwaters could be successively captured by one and then another major river in the area.

This explains why the fluviatile species are often less endemic than their lake counterparts. Then, also, riverine fishes entering another river didn't feel as much pressure to adapt and speciate in order to survive in their new habitat as the fishes entering the lake. The endemism of non-cichlids in the lake is not as spectacular as the one displayed by the lacustrine cichlids, but it is nevertheless remarkable.

Not all the fishes that entered the lake felt the pressure to undergo major adaptations. Probably the genetic drift of these species was low, or their anatomical and physiological features were already suited to this change of habitat. Those fishes coming from the lake now are exactly like or very similar to the specimens caught elsewhere in Africa. Not surprisingly they are widespread on the continent and occupy a variety of biotopes.

Before starting a survey of non-cichlids in the lake basin one should keep in mind two facts:

a) None of the major streams, not even the Malagarazi and the Ruzizi, much less the smaller tributaries, have been fully explored. Only the affluents near the Lukuga, the mouth of the Malagarazi and the mouth of the Ruzizi have been explored. This is due to the fact that the rivers flow over very difficult ground and cannot be explored by boat because of falls and rapids, nor by any other means, as they lie off the beaten path. Only the upper reaches of the Malagarazi can be reached by road in Burundi. The Mutambala, Lugufu, and Lufubu run in deep gorges, the first in a virgin mountain forest, the two others in semi-desert areas. Thus many species remain to be discovered in the rivers.

b) The habitats provided by the lake affluents should be split between those high up in the mountains, with cool or cold fast-running waters, and those meandering through coastal plains where currents slow down and the water has time to become heated. The first are rocky habitats, the second are swampy.

The distribution of non-cichlids in the lake affluents throws some light on their ecological niche, of which practically nothing is known.

Small cyprinids (such as some *Barbus*), kneriids,

Leptoglanis, and amphiliids are restricted to the cool or cold waters of mountain torrents. The water temperature can be as low as 10°C and often is colder. The fishes are small, probably because this type of habitat has only limited amounts of food available, mainly insect larvae, and large fishes would be swept away by the very strong floods that in the rainy season flush the torrents. Only small species find shelter and enough food. The populations of these fishes are not very dense.

As the torrents eventually find their way to a coastal plain, before emptying into the lake the rheophilic species are replaced by other fishes such as typical swamp-dwellers. These include catfishes such as *Bagrus, Chrysichthys, Malapterurus,* mochokids, and cyprinids, like *Barilius, Labeo, Varicorhinus,* and the larger species of *Barbus.* Very specialized swamp-dwellers appear, like the "prehistoric" *Protopterus* and *Polypterus,* living fossils of past ages, as well as cyprinodonts, anabantoids, *Clarias,* mormyrids, characoids, and not a few cichlids associated with swamps.

When they entered the lake non-cichlid species as a rule, with few exceptions, roamed and settled everywhere around the coastlines. Non-cichlids are much less localized in the lake than coastal cichlids. *Malapterurus, Auchenoglanis,* the spiny eels, and *Synodontis,* to name a few, are found everywhere along the shores.

Only a few of the families that entered the lake display a trend toward rapid speciation. A noteworthy example of this process is *Lates,* to which belongs the famed Nile perch. Throughout Africa only one species, *Lates niloticus,* is found in all of the major African rivers. Other species appeared here and there (whose fossil remains have been discovered), but only a single one has thus survived in most areas. However, in Lake Tanganyika there are four species of *Lates* cohabiting— *L. niloticus* isn't one of them! *Synodontis* and *Mastacembelus* are two other genera that have done well in the lake and multiplied into a number of endemic species.

The families of non-cichlids living in the lake basin are poorly documented as far as their behaviors are concerned. First, this is because many adapt very poorly to captivity. The second and probably foremost reason is that they are often ungainly and lack the appeal of smaller and more attractive fishes like the cichlids, and few aquarists bother to keep and study them. Non-cichlids of the lake area are mostly neglected by aquarists and are thus not available from imports for scientific institutions as well. Which shows the positive impact of the hobby on research.

Our survey will necessarily be fragmentary. Keys to genera and species have been developed with the view of using them in the field. They are based whenever possible only on external features. In case of doubt one would be wise to go back to the original description.

Family LEPIDOSIRENIDAE

Protopterus aethiopicus is the sole lungfish found in the lake. It is easy to identify because the dorsal fin starts midway in the distance between the occipital crest and a point above the anus.

P. aethiopicus from various localities in Africa are basically gray, more or less marbled, or dotted with tiny black stripes. Those from the lake display a gray background with faint darker blotches. So far most specimens have been found in the Ruzizi delta where they are common. Lungfish are known to endure a long period of desiccation during the several months of the dry season.

When marshes dry up, the fish coil and wrap themselves up with their large tail. They then start exuding a gelatinous slime that hardens into an airtight cocoon except for a small opening in front of the mouth. Breathing is slowed down so that the loss of moisture is reduced. At the start of the rainy season the tunnel in which the fish lives in suspended animation becomes moistened and the hardened cocoon melts away and is discarded as the fish becomes free-swimming once again and quickly recovers its full volume.

The teeth of *Protopterus* are very special. A strong bony ridge links two lateral canines on each jaw. As the very powerful jaws close the canines fit into cavities in the opposite jaw and the two ridges come into contact and slide alongside each other with a guillotine effect. The bite of a *Protopterus* always inflicts severe wounds and a large specimen can in a single bite cut off several fingers.

The giant *P. aethiopicus* from Lake Edward, reaching a length in excess of two meters,

The head of a lungfish has a distinctive appearance because of the shape of the mouth and the large sensory pores and canals. This is *Protopterus annectens*. Photo by Dr. D. Terver, Nancy Aquarium, France.

Lungfishes are among the oddest fishes on earth. The strangely reduced and almost filamentous pectoral and ventral (pelvic) fins are delicate and probably serve in part as sense organs. Photos of *Protopterus annectens* by Dr. Herbert R. Axelrod.

Protopterus annectens. Note the obvious canals and the gill remnants on the fish above.

can cause frightful amputations, more so because they don't let go. Manipulating lungfish is thus always dangerous as one does not expect lightning-fast action by such a sluggish-looking fish. When seized they have the uncanny capacity to slip backward and then, instead of trying to flee as any other sensible fish would do, they rush forward, attack, and bite. Even simply scratching a lungfish on the sides will cause them to turn swiftly and strike. They are carnivorous and will not hesitate to snap at a fish passing by, but they are also fruit-eaters and especially relish the fruits of the *Thalamus* palm-tree and from the palm-oil tree *(Elaeis)*. In captivity they eat fish, chopped meat, worms, and even flakes. They are one of the very few fishes to eat tadpoles and frogs.

Protopterus are strange animals—no wonder they are involved in the local folklore. It so happened several times that when I was roaming the virgin forest of Central Zaire, local fishermen told me that the lungfishes could climb trees. This of course I dismissed out of hand as another fisherman's tale. But one day I had a Catholic Father tell me the same story and he had been a witness to the fact. As it turned out his explanations were plausible. In the central part of the virgin forest in Zaire the trees grow in water, the whole area being flooded permanently. The palm-oil trees *(Elaeis)* thus stand in waist-deep water in the forest swamps, and decaying fronds remain to rot on the trunk. Many trees are not very high and their fruit

bunches are only one or two meters above water level. Thus lungfish, attracted by the fallen nuts that they find around, can slither up the tree by means of the footholds they get among the stumps of fronds and reach the fruits. So the story of the tree-climbing lungfish stands a good chance of being true after all!

Family POLYPTERIDAE

Ganoid scales set in oblique rows and paired fins set on peduncles are features that indicate very ancient fishes. The dorsal fin of polypterids is composed of a set of individual rectangular bifid spikes, each with its own membranous lappet, that fold down into a groove of the thick skin.

Two species live in the lake basin swamps, both having originated from the Congo River basin. *P. endlicheri congicus* reaches 100cm in length and, although a typical swamp-dweller, I saw one large specimen on the rocky slopes of Cape Chaitika.

The other, *P. ornatipinnis*, was found by Poll in the estuary of the Malagarazi. Never found anywhere else around the lake, this fish, along with *Tetraodon mbu*, provides a clue to the fact that a long time ago this river ran toward the upper course of the Congo River, the Lualaba. *P. ornatipinnis* in the delta are not as brightly colored as they are in the Kinshasa rapids. All *P. ornatipinnis* in the tropical fish trade come from the Kinshasa area. Reaching 60cm and being fundamentally predators, Polypteridae are not suitable aquarium fishes, except for their basic hardiness and longevity in captivity.

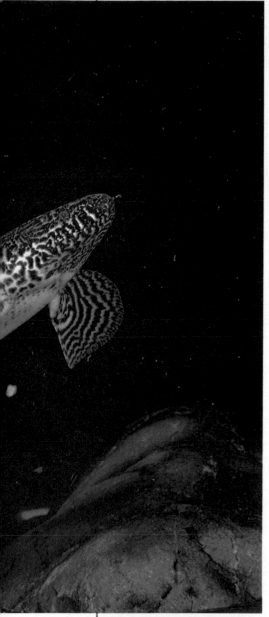

Most of the bichirs are coarse, dull fishes, but *Polypterus ornatipinnis* is an exception to this rule. Both the fins and the body are relatively attractive. This species has bred in captivity.

Key to the *Polypterus* of the Lake Tanganyika Basin

1. Snout much longer than lower jaw; eye included 8.5 times in head; X or XI dorsal spines, first spine well behind pectoral fin; 62 or 63 scales in a longitudinal line, 24 to 25 between occiput and first dorsal spine; Malagarazi River .... *ornatipinnis*

Snout much shorter than lower jaw; eye included 8 to 14 times in head; XII-XV (more often XIII or XIV) dorsal spines, first very close to pectoral fin; 55 to 59 scales in a longitudinal line, 11 to 15 between occiput and first dorsal spine; river estuaries .... *endlicheri congicus*

Family CLUPEIDAE

One cannot say that clupeids would be good aquarium fishes as they are next to impossible to keep alive for more than a few days in captivity and are even hard to take out of the water without seeing them die immediately.

Two anatomical features set them apart from characoids with which, superficially, they could be confused:

a) Their scales are deciduous, which means that they fall off at the slightest contact. The clupeids in the lake lack lateral lines and a dorsal adipose fin.

b) The scales on both sides of the fish meet in the ventral area without overlapping, but are raised in a zigzagging crest or keel.

Two genera live in the lake:

(1) *Limnothrissa,* in which teeth are set not only in the jaws but on the tongue and palate as well. The maxillary is as wide in front as it is at the back of the jaw.

(2) *Stolothrissa,* in which the teeth are restricted to the jaws and the maxillary bone is spatulated, i.e. wider in front than in back.

Called Ndakala in the north and in Zaire, Ndagaa in Tanzania, and Kapenta in Zambia, clupeids are a very important source of animal protein around the lake and as far away as Kinshasa and Lusaka. Since the 1960's they have been successfully transplanted into Lakes Kivu and Kariba.

Although commerically fished by thousands of tons, remarkably little is known about their ethology. Spawning sessions have never been witnessed. One does know that their spawnings are cyclical and during which months they happen, based on the young fry discovered afterward. Surveys report years of prodigious numbers of these fishes but as yet nobody can tell if the abundance is localized or entails the whole lake. Fishing is done by trawlers using huge nets several hundred meters long and a hundred high during moonless nights. Powerful beams of light are directed at the surface that progressively attract the fishes and their predators. When the concentration of clupeids reaches a given density the

This rather dull brown bichir, *Polypterus* sp., is still quite small, but it will soon outgrow most standard tanks. Photo courtesy *Midori Shobo*.

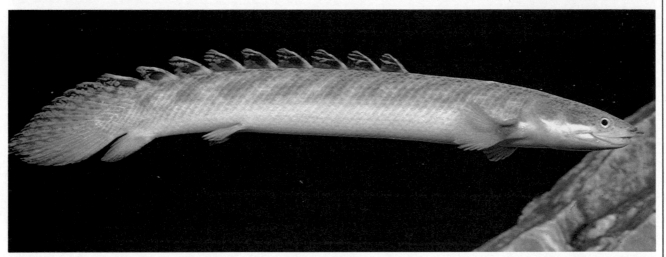

Polypterus sp., possibly *P. lowei* or *P. senegalus*. Photo courtesy *Midori Shobo*.

Albino (more appropriately xanthic) bichirs are rare but are always popular and in demand. This specimen was imported from Stanley Pool. Photo by E. C. Taylor.

Polypterus endlicheri congicus is often called the Tanganyika fossil fish, and admittedly it does look like the popular conception of a "living fossil." These fish are nasty and almost impossible to kill, but they do make interesting conversation pieces. Photos by G. S. Axelrod.

Top: *Malapterurus electricus*. E. Balon. Center: *Heterobranchus longifilis*. E. Balon. Bottom: Close-up of the mouth of the electric catfish, *Malapterurus electricus*. Photo by K. Paysan.

net is drawn shut and hauled up. Sometimes it is so heavy with fishes that some have to be released before the net can be brought on board without risk of the craft capsizing.

Clupeids are so abundant that 16 species of predatory fishes have developed in the lake just to prey on these silvery fishes. They are always under a frightful stress, very mobile and jumpy, and always on the move. In captivity they need to be hyperoxygenated and need room to maneuver. None, to my knowledge, has survived more than a couple of days in captivity.

Family MORMYRIDAE

This is probably the most typical family of African freshwater fishes, their main habitat being the Congo basin. This is not reflected in the lake area, which harbors only a few species of which only one lives in the lake itself *(Hippopotamyrus discorhynchus)*.

Mormyrids, as we have seen, are poorly adapted to the lake ecology, mainly because their eyesight is so poor, but probably also because they don't find the type of food they are used to in the lake. Only *Mormyrops* are predatory. They are used to and adapted to murky waters and mud and many feed quasi-exclusively on mosquito larvae and especially the so-called bloodworms. Hence the development of a long trunk on the "elephant-nosed" mormyrids and the fleshy tactile bumps on the chin of several species with which they poke into the mud. There are no mosquitos on the lake, and were one so rash and stupid as to come to the lake

Limnothrissa miodon is one of the two endemic genera and species of herring found in Lake Tanganyika. Both are of great commercial importance. Photo by E. Balon.

water to spawn, it wouldn't last long, nor would the larvae stand a chance of developing. Mormyrids could perhaps live on the zooplankton organisms if they had good eyesight.

It is a very rare occasion when one discovers a mormyrid in the lake. Only *Hippopotamyrus discorhynchus* was sighted twice, each time in a shadowy recess on a rocky coast (Cape Magara).

Mormyrids have a strong appeal for hobbyists, especially the long-snouted species. But they are basically poor aquarium fishes with a bad record of longevity. Their requirements cannot be met in the narrow confines of a tank.

Key to the Genera of Mormyrids in the Lake Tanganyika Basin (from Taverne)

1. Very long dorsal fin .... *Mormyrus*
   Short dorsal fin 2
2. Mouth in a tube-like extension 3
   Mouth not at tip of a tube-like extension............... 4
3. With chin barbel at end of tube.............................
   .... *Campylomormyrus*
   Without chin barbel at tip of tube..........................
   .... *Mormyrops (Oxymormyrus)*
4. Chin bulge fleshy and globulous ....*Marcusenius*
   No fleshy chin bulge...5
5. Caudal fin with very short lobes; body very
   elongate; 10-36/10-36 bicuspid teeth on each
   jaw; teeth of palate and tongue reduced..........
   .... *Mormyrops*
   Caudal fin with elongated lobes; body rather short;
   never more than 10 teeth on lower jaw; well
   developed teeth on palate and tongue...........6
6. Large *bony* chin made up of recurved lower jaw
   bones; upper jaw protruding over lower jaw .....
   .... *Hippopotamyrus*
   No bony chin.. *Pollimyrus*

Top: *Mormyrops deliciosus.* Center: *Mormyrops longirostris.* Photos by E. Balon. Bottom: *Marcusenius* sp. Photo by E. C. Taylor.

Mormyrus sp., possibly *rume*. Photo by E. C. Taylor.

Marcusenius sp. Photo by Dr. Herbert R. Axelrod.

Family KNERIIDAE

Kneria and *Parakneria* are very specialized fishes appearing in the mountain torrents east of the Lualaba (upper part of the Congo River) and spreading toward the lake.

They show interesting adaptations to life in fast-moving waters. Like most rheophilic fishes, be they cichlids, cyprinids, or catfishes, fishes living in fast-moving waters often have an elongate and flat body, which helps them by offering less resistance to the flow of water. *Kneria* belong to this type. They have adapted also in another field—breeding techniques.

In torrents it is often difficult for a fish to remain at a given place, much more so for a pair busy spawning. To solve the problem males have developed suction disks on either side of the head so that when the time comes he can keep his mate alongside and fertilize her spawn.

They are small fishes, 5-6cm long at most, used to cool (if not cold) strongly aerated waters. The body is often more or less marbled and the unpaired fins are enhanced by a few brighter spots that can be yellow, orange, or even vermilion.

None has been exported because it is difficult to reach the headwaters of their habitat. Their diet, undocumented as yet, should consist of microorganisms and possibly include the aquatic larvae of insects.

Only one species, *K. wittei,* has been found in the lake area. But there might be more in the streams cascading toward the lake, most of which have never been explored.

Characters of the Family (according to Poll):

-Mouth, deprived of teeth and of barbels, very low and protractile, located under the snout. A circular suction-disk is present on the gill cover, followed by a serrated ridge in the postopercular area on males only. Tail lobed. Scales very minute.

Characters of *K. wittei:*

-Body depth 4.6 times in body length; head length 5 times in standard length; snout 3.25 times in head length; eye 3.75 times in head length; interorbital space 2.6 times in eye diameter. Dorsal fin III,7i, originating in second half of standard length of the body; anal fin III,7; ventrals I,7; pectorals 15; caudal lobed. Scales 78 in long. line.

Families ALESTIDAE and CITHARINIDAE

There are very few characoids in the lake basin and even fewer in the lake itself, although the Congo River is one of the richest habitats of characoids in Africa.

Four genera with a total of only seven species live in the lake basin. Four species have been discovered in the lake, two being tigerfishes (*Hydrocynus),* the two others belonging to the group of very large river characins (*Alestes).* None is endemic to the lake.

Hydrocynus (Family Alestidae)

The so-called tigerfishes live more in the large river estuaries than in the lake, although in the southern parts they are a common sight in front of the swamps and river mouths.

They are in Africa the counterpart of the dreaded South America *piranha* and their reputation as fish predators is well deserved. Fortunately, they are less bloodthirsty and even the giant of the two species, *H. goliath,* does not often (and then only by accident) attack man or a mammal. *H. vittatus* reaches one meter and 10kgs, *H. goliath* 2 meters, the largest specimen ever caught weighing 85 kgs. Worldwide there isn't a single freshwater predator as powerful and frightening as a full sized *H. goliath,* because it is endowed with a set of teeth unmatched by any other fish, except perhaps by a large barracuda or shark.

Their teeth are triangular and up to 3cm long, needle sharp, and coated with a very hard enamel. The side of each tooth is honed to a razorlike edge. When the powerful and hard jaws snap shut the huge teeth enter cavities in the opposite jaw and the lateral edges of the teeth come into contact. Thus a bite first lacerates the tissues, then the razor-sharp edges cut them like so many guillotines. As the mouth is enormous and crescentic, whole chunks of meat can be ripped off in a single stroke. Attacks on fishermen or swimmers are very few, but in one case I saw a man whose whole calf had been bitten off.

Fortunately, when *H. goliath* attacks man it is always by mistake in murky waters. They are attracted by splashing and rush toward the commotion. In one personal contact with *H. goliath,* I had been water-skiing and fell into the water. As I was floating face-up waiting for the boat to pick me

The large tigerfish, *Hydrocynus goliath*, is a common but feared resident of Lake Tanganyika. At 2 meters in length, it is one of the very largest predatory freshwater fishes. Photo by Shuichi Iwai, *Midori Shobo*.

up, I suddenly felt a tremendous pressure on my side and a huge *H. goliath* burst out of the water at full speed, sailed over me, and disappeared with a splash on the other side.

Often *H. goliath* move in pairs and if one of the mates get hooked on a line it often happens that the second fish attacks its mate. In one such case only the anterior part of the victim was brought on board. It was still 10kg, and the whole fish was estimated to have been about 1.5m long and should have weighted about 20kg. Thus the missing

piece probably measured more than 70cm and weighed 8 or 9 kg. It had been cut transversally, which means that the offender had swallowed whole, tail first, a length of 70cm, weighing nearly 10kg before closing its jaws.

These few examples illustrate the power displayed by medium-sized tigerfish. No wonder, then, in the clear waters of the lake, when one of us found himself under the close scrutiny of a tigerfish during a dive, he didn't feel quite at ease and wished he were out of the water. Moreso

because when the fish breathe they have the nasty habit of displaying their fangs.

H. vittatus, the smallest of the two, can be distinguished by its pattern—a golden yellow with several black horizontal stripes *(H. goliath* is a plain steel blue). The smaller species is much more common in the southern part of the lake than the other species, but we have never seen one in the northern part. They are usually seen in small bands and quite often mixed in a school of *Tilapia tanganicae,* on which they might prey selectively.

The requirements of fast-

Key to the Species of *Hydrocynus*:

1. Body longitudinally striped with several rows of black-spotted scales; A. III,10-13; 8 to 10 gill rakers; size to 1 m .... ***vittatus***
   Body without horizontal rows of black-spotted scales; dorsal as well as lower caudal lobe yellow or red; A. III,13 to 16; 10 to 12 gill rakers; size to 2 m .... ***goliath***

moving fishes such as tigerfishes regarding a high level of oxygenation are obvious. They are always on the move, fast growing, voracious, and dangerous to handle (I have been bitten more than once by small juveniles). They are among of the most difficult characoids to keep alive in captivity. They stand at the antipodes of what a hobbyist should wish in his tank. Nothing is of course known about their breeding habits.

Alestes (Family Alestidae)

Two species of *Alestes,* and very large ones at that, roam the coastal waters of the lake, *Alestes rhodopleura* and *A. macrophthalmus.* They reach in excess of 30cm and are far from being common. I have seen only a few isolated specimens in my   years on the lake. In rivers they are much more common, rove in schools, and might be considered as omnivorous predators, along with *Barilius* and *Varicorhinus.*

Micralestes and *Bryconaethiops* (Family Alestidae)

These two genera of typical fluviatile characoids have only a single species each along the western shores of the lake, where they live in affluents of the Lukuga outlet. Of the two species, *B. boulengeri* is the most attractive with a dark blue back and silvery sides marked by a huge black spot followed by a broad black stripe. The dorsal fin sports long filaments.

Unfortunately, I remember the fish as a very difficult characin to acclimate to captivity and most died in my hands. *M. stormsi* is one of the featureless species of alestid, with a silvery body faintly marked with a golden horizontal stripe. *Micralestes* are among the hardiest of African characoids.

Family Citharinidae

Very poorly represented in East Africa, citharinids have in Central Africa one preferred habitat—the central virgin forest of Zaire, where they display a wide variety. Two *Distichodus* have secured a foothold around the mouth of the Lukuga, and we have seen an adult specimen of *D. sexfasciatus* on the rocky coast of Cape Magara, which indicates that they are spreading in the lake.

Alestes longipinnis, one of the many confusing species of African tetras. Photo by B. Kahl.

Key to the Genera of Alestidae from the Lake Tanganyika Basin (from M. Poll):

1. Pluricuspid teeth in 2 or 3 rows on each jaw.........2
   Unicuspid teeth, sharp and very large, in a single row on each jaw...........................**Hydrocynus**
2. Teeth in 2 rows on upper jaw; first dorsal ray above or behind ventral fins 3
   Teeth in 3 rows on upper jaw; first dorsal ray in front of ventral fins .... **Bryconaethiops**
3. Upper jaw teeth of inner row conical; lower jaw with only 2 inner teeth; lateral line complete ....**Micralestes**
   Upper jaw teeth of inner row obliquely truncated or molar shaped; only two inner teeth on lower jaw; lateral line complete; large size as adult ... .**Alestes**

Key to the Species of Alestes:

1. Dorsal fin II,8; anal fin III,12-16; scales in a longitudinal line 23-29......................................2
   Dorsal fin II,8; anal fin II,16-20; scales in a longitudinal line 39-45; no peduncle spot.......... ....**macrophthalmus**
2. First rays of dorsal fin above ventrals; 23-29 scales; peduncle spot present; body deep ....**imberi**
   First ray of dorsal fin well behind base of ventrals; 28-29 scales; peduncle spot present; snout to dorsal fin profile straight ....**rhodopleura**

Micralestes acutidens. Photo by Dr. Herbert R. Axelrod.

Distichodus fasciolatus. Photo by Dr. Herbert R. Axelrod.

Brycinus chaperi. Photo by Dr. Herbert R. Axelrod.

Distichodus, although omnivores, are mainly vegetarian fishes. No wonder they are seen more in swamps and aquatic plants beds. *Citharinus gibbosus* is a mud-sifter and as such is hard to keep alive in a tank for any length of time.

Distichodus sexfasciatus has become one of the most sought-after fish for aquarium purposes. Amateurs apparently don't know that this fast–growing fish reaches in excess of 1.5m and a weight of 25 kg.

455

Distichodus fasciolatus. Photo by J. Greenwald.

Distichodus sexfasciatus.

Distichodus sexfasciatus. Photo by H.-J. Richter.

Key to the Genera of Citharinidae of the Lake Tanganyika Basin

1. Scales cycloid; body very deep and laterally compressed, maximum twice as long as high .. ..*Citharinus*

   Scales ctenoid; body more than twice as long as high, moderately compressed.......*Distichodus*

Key to the Species of *Distichodus*:

1. Eye small, 5.7 times in head length.......................2

   Eye large, 3.2 times in head length; body long, 2.7 to 3.0 times longer than high; snout short, included three times in head length; teeth in two rows on each jaw, 28 in outer row of upper jaw, 24 in outer row of lower jaw; 11 anal rays, 18 dorsal rays; 70 scales in a longitudinal line; origin of dorsal fin in front of ventral fins; large spots all over the sides; size about 350 mm..... ....*maculatus*

2. Body very deep, only 1.7 to 2.2 times in standard length; head large, 4.0 to 4.3 times in S.L.; 16 to 18 teeth in outer row of upper jaw; 60 to 68 scales in a longitudinal line; about 6 vertical black bands on yellow-orange body; all fins red (juvenile); the pale background darkens with age as do the fins; maximum size in excess of 1000 mm..*sexfasciatus*

   Body elongate, depth 2.5 times in S.L.; head small, 4.8 times in body length; 2 rows of teeth; upper jaw with 29 teeth, lower with 25; 68 to 78 scales in a longitudinal line; 18 to 20 thin vertical bands on a dark green body; maximum size about 350 mm.........................*fasciolatus*

The genus *Distichodus*. Left to right, top to bottom: *Distichodus* sp.; *Distichodus decemmaculatus; Distichodus affinis; Distichodus maculatus; Distichodus fasciolatus; Distichodus lusosso.* Photos by B. Kahl.

A young *Citharinus gibbosus*. Rough handling easily strips the scales off these fish. Photo by Pierre Brichard.

Family CYPRINIDAE

Found everywhere in Africa, but less common in virgin forests, the favorite habitat of the cyprinids appears to be savannah rivers and swamps and fast-moving waters of mountain streams. There are relatively few of them in the lake basin, although they amount to 36 species, or one third of all the non-cichlid fishes in the basin, or nearly two thirds of all fishes living in the rivers. This illustrates how much cyprinids dominate the rivers in East Africa.

One should split them into two main groups:

(1) The dwarf species living high up in mountain torrents. Many would be fit for aquarium life were they just a bit more colorful. Among them I would point to *B. minchini* with its yellow-green body with a few dark spots scattered on the sides. The unpaired fins are pale yellow in females, but orange or vermilion in males. Males have a conspicuous orange-red spot on the gill cover.

Rheophilic barbs move about a great deal in small schools in their cool waters. One can see them darting here and there over the pebbles of their habitat. Despite their small size, barbs are known to be predatory and destroy the spawns and fry of the fishes they live with.

(2) The larger species of *Barbus, Labeo,* and *Varicorhinus* live in the warmer waters of slow-moving streams and in coastal lagoons. They are omnivores but also predatory, and as they grow as big as a carp they can wreak havoc on fry and spawns of other fishes.

Barilius are the equivalent of a trout in the streams where they live. They are very swift and rush toward anything that drops into the water. This is how they catch many insects. A large *Barilius* has a striped body in many iridescent shades and is a magnificent fish. Unfortunately, although I have kept a few in my tanks, they are unfit for aquarium life. They are predatory, have high requirements as to their oxygen level, and need a lot of room.

Varicorhinus are the only large barbs living in coastal waters everywhere around the lake shorelines. They are one of the main predators close to the shore, and one can see their small "gangs" moving swiftly in the surf.

Engraulicypris, on the contrary, have taken to open waters.

**Key to the Genera of
Cyprinidae of the Lake
Tanganyika Basin:**

1. Dorsal fin entirely in front of anal fin; anal fin with maximum of 7 branched rays............2
   Dorsal fin at least partly extending beyond origin of anal fin, which has a minimum of 10
      branched rays..4
2. Mouth terminal or inferior, with more or less developed lips 3
   Mouth inferior, without lips; lower jaw with a cutting edge in a horny sheath....................
   ....*Varicorhinus*
3. Mouth inferior, with very well developed sucking lips and with an inner cutting edge
      covered in a horny sheath; first rays of dorsal fin always well in front of first rays of
      ventral fins .... *Labeo*
   Mouth terminal or inferior; lips relatively large, not shaped as a sucking organ, with or
      without a cutting edge; first rays of dorsal fin somewhat in front or somewhat behind
      the first ventral rays... *Barbus*
4. First ray of dorsal fin in front of first ray of anal fin.. *Barilius*
   Dorsal fin entirely above anal fin ....*Engraulicypris*

Close-up of the head of a barb, *Barbus*, showing the barbels. The number and size of barbels is quite variable in these hard to identify fishes.

Key to the Species of
***Barbus* of the Lake**
Tanganyika Basin:

1. Scales longitudinally striated..2
   Scales radially striated..7
2. Number of scales in lower lateral line not more than 40 3
   Number of scales in lower lateral line between 44 and 47; 5 or 6 scales between lateral line and base of ventral fins; 16 scales around caudal peduncle; single barbel on each side of mouth; maximum size 100 mm..***tropidolepis***
3. Not more than 3 scales between lateral line and base of ventrals................................ 4
   3 1/2 to 4 scales between lateral line and base of ventrals; 34 to 39 scales on lateral line; dorsal well forward; 2 long barbels on each side of mouth; maximum size 360 mm...***urundensis***
4. Maximum of 32 scales in lateral line; less than 3 scales between lateral line and base of ventral fins; anal fin doesn't reach tail; first dorsal ray soft and thin; 2 barbels......5
   34 to 35 scales in lateral line; 3 scales between lateral line and base of ventrals; anal fin reaching tail; first dorsal ray thick and bony; 2 long barbels at least equal to eye, which is very small (5.0 to 5.5 times in head length); maximum size 500 mm; affluents .... ***altianalis***
5. Eye included 4.0 to 4.25 times in head length..6
   Eye very small, 5.7 times in head length and 2.16 in snout length; lips thick, with median lobe; 2 barbels on each side; first dorsal ray soft and smooth, shorter than head length; 26 to 32 scales in a longitudinal line; a dark stripe on upper and lower edges of tail; maximum size 27 mm; Congo affluents, lake, and Lukuga River.....................
   .... ***caudovittatus***
6. Body depth 4.0 in standard length; head length 3.5-4.0 in standard length; interorbital space 3.0-3.3 in head length; eye 4.0 in head length; 2 short barbels on each side; D. IV,10; last single dorsal ray thin, soft, and smooth; maximum size 85 mm; Lukuga River and its affluents .... ***euchilus***
   Body depth 3.5 in standard length; head length 4.0-4.5 in standard length; interorbital space 2.6 in head length; eye 4.0-4.25 in head length; 2 short barbels on each side; D. III,10; last single ray of dorsal bony at its base, segmented at its tip, soft, and smooth; maximum size 180 mm; Lukuga River and its affluents........................***pojeri***
7. Lateral line complete..9
   Lateral line not complete, reduced to at most its anterior half..........................8
8. Maximum of 3 scales in lateral line (which might be totally missing); last single dorsal ray short, strong, bony, and serrated, a bit more than half head length; 21 to 23 scales in a longitudinal line; 10 scales around caudal peduncle; one barbel on each side; maximum size 48 mm; affluents of Malagarazi River..............***aphantogramma***
   4-13 scales in lateral line; 12 scales around caudal peduncle; 2 barbels on each side; maximum size 57 mm; high affluents of the Malagarazi River; other features very similar to *aphantogramma*...***oligogrammus***
9. Not more than 32 scales in lateral line...10
   34-39 scales in lateral line; 6½ - 7½/5½ - 6½ scales in transverse line; 16 to 18 scales around caudal peduncle; first single dorsal ray bony, serrated, as long as head; 2 barbels on each side, second as long as eye; maximum size 120 mm; Lukuga, East and South Africa...***paludinosus***
10. 22 or 23 scales in lateral line; base of ventrals entirely under dorsal fin; maximum size 140 mm; Central Africa, Lukuga River..***nicholsi***
    24 to 32 scales in lateral line; base of ventral fins, at least in part, in front of first dorsal rays..11
11. At least 4½/3½ scales in a transverse line; 2 scales between lateral line and base of ventral fin; 10 scales around caudal peduncle 12
    3½/2½ scales in transverse line; 1½ scales between lateral line and base of ventral fin;

460

8 scales around caudal peduncle; 24 very narrow and very high scales in lateral line; dorsal III,7-8; last single ray bony, thin, smooth, and equal to head length; 2 short barbels; black spot behind opercle; scales edged in black; maximum size 110 mm; western affluents of the lake ....*lufukiensis*

12. Last single ray of dorsal strong, bony, and serrated...13
Last single ray of dorsal thin, soft, segmented, and smooth..16

13. At least 24 scales in lateral line and 4½/4½ in transverse line.......................................14
23 scales in lateral line, 4½/3½ in transverse line; eye, snout, and interorbital space approximately 3 times in head length; dorsal III,8; base of ventral fins under first dorsal rays; caudal peduncle 1.5 times longer than high; 12 scales around caudal peduncle; only 2 scales between lateral line and base of ventrals; maximum size 70mm; Congo River basin and affluents of the lake ....*miolepis*

14. Base of ventral fins partly in front of dorsal 15
Base of ventrals well in front of dorsal fin; caudal peduncle 1.5 times as long as high, surrounded by 11 to 14 scales; 2 barbels about size of eye; dorsal III,7, placed midway between eye and tail; 24-28 scales in lateral line, 4½/4½ in transverse line; 2 or 2½ scales between lateral line and base of ventrals; black stripe from eye to tail; East and South Africa and lake affluents; size 90 mm ....*eutaenia*

15 (quadruplet). Caudal peduncle 1.5 to 2.0 times longer than high; 28 to 30 scales in lateral line; 4½/4½ - 5½ scales in transverse line; 2 or 2½ scales between lateral line and base of ventrals; 12 scales around caudal peduncle; 2 barbels, second very long; 3 black spots on body, second spot sometimes double, the three spots sometimes connected by black line; Lakes Edward and Kivu, and Upper Malagarazi; maximum size 115 mm...*pellegrini*

Caudal peduncle 1.5 to 1.6 times longer than high; 26 to 30 scales in lateral line; 4½ - 5½/4½ - 5½ scales in transverse line; 2½ to 3 scales between lateral line and base of ventrals; 12 scales around caudal peduncle; 2 barbels, second very long; a thin black stripe along body ending in a small black spot on caudal peduncle; Lakes Edward and Kivu, and various affluents of the lake; maximum size 120 mm . *serrifer*

Caudal peduncle relatively long (1.5 to 1.6 times longer than high) surrounded by 12 scales; 25-26 scales in lateral line, 4½/4½ in transverse line; 2 to 2½ scales between lateral line and base of pelvic fins; 2 barbels; dorsal III,7, midway in the body length; silver streak along body; maximum size 85 mm; Kilimandjaro, Masai Steppe, and Malagarazi...*kerstenii*

Caudal peduncle long (2 times longer than high), surrounded by 12 scales; 25-26 scales in lateral line, 5½/6½ in transverse line; 2½ scales between lateral line and pelvic fins; 2 barbels; dorsal III,7; several black spots in a line on the sides; opercular spot orange or red in males; unpaired fins yellow, orange, or red; maximum size 90 mm; area between Lakes Victoria and Tanganyika................................*minchini*

16. Dorsal III, 8 branched rays 17
Dorsal III, 7 branched rays; caudal peduncle 1.5-2.0 times longer than high; 25 to 30 scales in lateral line; 4½/4½ in transverse line; 2½ or 3 scales between lateral line and base of ventrals; 12 scales around caudal peduncle; rounded black spot at base of caudal fin...*urostigma*

17. Dorsal well in first half of body length; caudal peduncle twice as long as high, with 12 scales around peduncle; 27 to 32 scales in lateral line; 4½ - 5½/4½ scales in transverse line; 2½ to 3 scales between lateral line and base of ventrals; 2 barbels, second long; 4 to 7 black spots in a line above lateral line; sometimes a black stripe on dorsal ridge in front of dorsal fin; maximum size 66 mm; East Africa and lake affluents ....*lineomaculatus*

Dorsal halfway between eye and tail; caudal peduncle 1.6 to 2.0 times longer than high, with 12 scales around peduncle; 27 to 29 scales in lateral line; 4½/4½ in transverse line; 2½ scales between lateral line and base of ventrals; black stripe meeting lateral line on caudal peduncle; maximum size 90 mm; various lake affluents . *taeniopleura*

Barbus sp.

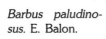

Barbus paludinosus. E. Balon.

Barbus eutaenia. G. Marcuse.

462

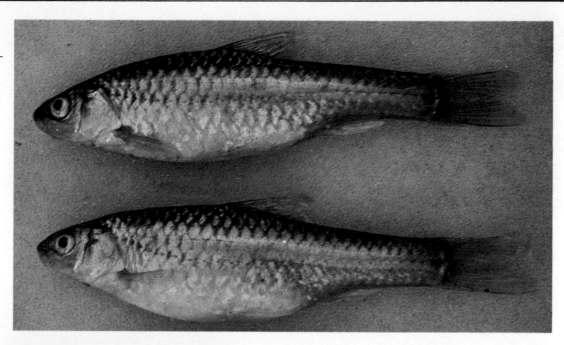

Barbus lineomaculatus. E. Balon.

Barbus multilineatus. Dr. H. R. Axelrod.

Labeo weeksi.

Barbus nicholsi. Dr. H. R. Axelrod.

A juvenile specimen of *Labeo cylindricus*. Photo by H. Hansen.

Key to the Species of *Labeo* from Lake Tanganyika:

The number of scales of the Tanganyika basin labeos and their dorsal and anal finnage are very similar. Noteworthy are the shape of the caudal peduncle and, for some species, the shape of the body, the inner surface of the lips, the placement of the eye in the head, and even the shape of the dorsal fin.

1. Transverse folds on inner side of lips 2
   No transverse folds on inner side of lips; caudal peduncle twice as high as long; dorsal fin halfway between snout and tail; D. III,11, with a convex shape; anal fin reaching the caudal fin; fins red, red dots on each lateral scale; maximum size 780 mm; Congo basin, Lukuga outlet .... ***lineatus***
2. 16-18 scales around the caudal peduncle 3
   12-14 scales around the caudal peduncle; D. III,10-11 with upper edge concave; maximum size 400 mm ....***dhonti***
3. D. III or IV,8-10; anal fin III,5 4
   D. III,12-13; anal fin II,5; dorsal fin with upper edge very convex; anal fin very long; caudal fin well notched; 37 scales in lateral line; 5½/6½ in transverse line; 4 scales between lateral line and base of ventrals; 16 scales around caudal peduncle; maximum size 700 mm; Congo and Lukuga Rivers .... ***velifer***
4. Caudal peduncle as high as long...5
   Caudal peduncle 1.3 to 1.5 times longer than high; body depth 4.0 to 4.6 in standard length; head 4.7 times in standard length; eye median; 38-39 scales in lateral line; 5½-6½/6½-7½ scales in transverse line; 3 or 4 scales between lateral line and base of ventrals; 16 to 18 scales around caudal peduncle; maximum size 400 mm; overall color olive green; East Africa lake affluents and in the lake proper...........***cylindricus***
5. Eye small, entirely lateral and toward rear of head, included at least 7 times in head length; caudal peduncle 1.5 times higher than long; snout long and very protruding, 1.5 times in head length; D. III,11, with upper edge convex; 3 scales between lateral line and base of ventrals; color green with red dots, fins red; maximum size 680 mm; Lukuga River only .... ***kibimbi***
   Eye superolateral, less than 7 times in head length; caudal peduncle as high as long; D. III-IV, 9-10, with upper edge straight or a bit concave; 4 scales between lateral line and base of ventrals; maximum size 235 mm; Lake Rukwa and upper Malagarazi affluents .... ***fuelleborni***

Key to the Species of
***Varicorhinus* of the Lake**
Basin:
1. Number of scales in lateral line fewer than 40...2
   Number of scales in lateral line more than 40 3
2. Dorsal fin IV,10; anal fin III,5; 34 to 36 scales in lateral line; 5½/5½ scales in transverse line; 3 scales between lateral line and base of ventrals; maximum size 210 mm........ ....***ruandae***
   Dorsal III,10; anal II,5; 30 or 31 scales in lateral line; 4½/4½ scales in transverse line; 2½ scales between lateral line and base of ventrals; maximum size 310 mm ....***stappersi***
3. Dorsal III,10; anal III,5; lateral line with 46 scales; 8½/8½ scales in transverse line; 5½ between lateral line and base of ventrals; maximum size 270 mm ....***leleupanus***
   Dorsal IV,8-10; anal III,5; 64 to 70 scales in lateral line; 13½ - 14½/14½ - 15½ scales in transverse line; 9 or 10 between lateral line and base of ventrals; maximum size 550 mm...***tanganicae***

Key to the Species of
***Barilius* from the Lake**
Tanganyika Basin:
1. More than 50 scales in lateral line 2
   Less than 50 scales in lateral line 3
2. 82 scales in lateral line; 13½/7½ scales in transverse line; 4 scales between lateral line and base of ventrals, 26 around caudal peduncle; snout very pointed, 1.5 times in head length; mouth reaching rear of eye; 16 to 17 vertical bands on body; anal fin III,17; maximum size 260 mm...***tanganicae***
   54 to 61 scales in lateral line; 9½-11½/6½-7½ scales in transverse line; 2½-3½ scales between lateral line and base of ventrals; 18 to 20 scales around caudal peduncle; 6 to 14 vertical bands on body; anal fin III-IV, 12-15; maximum size 210 mm..***moorii***
3. Maximum of 45 scales in lateral line..4
   46 to 48 scales in lateral line; 8½/4½ in transverse line; 2 or 2½ scales between lateral line and base of ventrals; 16 to 18 around caudal peduncle; dorsal III, 9; anal III, 13 to 15; mouth reaching posterior ⅔ of eye; dorsal fin clear, red edge to tail; maximum size 175 mm..***salmolucius***
4. Caudal peduncle with 14 scales; 3 scales between lateral line and base of ventrals; 42 to 45 in lateral line; 7½/4½ scales in transverse line; dorsal fin II, 9-10; anal III-IV, 11-16; eye 4.0 to 5.75 times in head length; 8 or 9 bands on body; dorsal blackish; maximum size 190 mm; affluents of Upper Malagarazi ....***neavii***
   Caudal peduncle with 15 or 16 scales; 1½ to 2 scales between lateral line and base of ventrals; 8½/4½-5½ scales in transverse line; 41 to 44 scales in lateral line; dorsal III, 9; anal III, 13; eye 3.5 to 4.5 times in head; 8 to 13 bands on body; rear edge of dorsal black; origin Congo River, only found in Lukuga outlet; maximum size 130 mm ....***ubangensis***

Barilius christyi. Photo by Dr. J. Gery.

Barilius ubangensis. Photo by Dr. J. Gery.

Barilius barna.

Juvenile *Barilius*, possibly *B. moorii*. Photo by Dr. Herbert R. Axelrod.

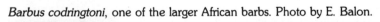

Barbus codringtoni, one of the larger African barbs. Photo by E. Balon.

Key to the Species of
Engraulicypris **from the Lake**
Tanganyika Basin:

Body depth 4.5 to 5.5 times in standard length; anal fin III, 16 to 18; transverse scales 6½/2½; first rays of anal fin just in front of first rays of dorsal fin; origin Congo River, affluents of the Lukuga outlet; maximum size 70 mm....................................*congicus*

Body depth included 5.5 to 6.5 times in standard length; anal fin III, 18 to 20; transverse line with 7½/1½ scales; origin of anal well in front of first dorsal rays; maximum size 75 mm ....*minutus*

Family BAGRIDAE

The number of species of this family to be found in the lake and its basin is rather small, but they present a very interesting set of adaptations to their various habitats, be they in mountain streams or in the lake.

Bagrids in the lake include giant species such as found in *Auchenoglanis,* of which I have seen specimens reaching 70kg, or *Chrysichthys grandis,* of which thirty years ago I once saw a specimen more than 2 meters long, 60cm across, and weighing 190kg. But there are also dwarf species not exceeding 7 or 8 cm, such as in the genus *Lophiobagrus.* Several genera in the lake have developed endemic species and some of the genera are even endemic to the lake, which shows the amount of adaptations they went through.

Some species still have to return to rivers in order to breed thus giving us a clue that their adaptation to the lake ecology has not been perfect.

In rivers bagrid fishes are usually found over mud and sand. It thus came as a surprise when we found nearly all of them living in deep recesses on the rocky slopes of the lake. Not that they refrain from foraging at night or even by daytime in the open over the huge sand floors underlining the rocks, when they live deep, but they return to their territory afterward.

Lophiobagrus Poll

Lophiobagrus is a peculiar fish not exceeding 8cm, although most specimens captured are between 5 and 6cm. Like all bagrids it is endowed with a defense in the form of a strong stubby spine in the dorsal fin; the first ray of the pectoral fin is also a bony "spear". With most catfishes, these spines are coated with a skin that, when the spine is stuck in the enemy's flesh, brings about a very painful burning sensation, and sometimes in man a violent allergic reaction.

Lophiobagrus might have improved its defense system over the other catfishes: whenever one is handled it exudes a viscous transparent slime with toxic properties that kills the fish it is stored with. A small *Lophiobagrus* that I put into a jar with a formaldehyde solution exuded so much slime that it turned the solution into a gelatinous, elastic ball. *L. cyclurus* is not an attractive fish, with its short and stubby body, and being pale purple on the back, pale pink on the sides, and yellowish underneath.

Phyllonemus Boulenger

The two *Phyllonemus* species, on the other hand, are among the most appealing catfishes I have seen anywhere, with their dark purple body, square flat head, large black eyes, and enormous "whiskers". These, in fact, are tactile filaments, sprouting from their snout, corners of the mouth, chin, and even thorax. When the fish moves about the longest "whiskers" jut forward, then bend back in graceful arches. With them *Phyllonemus,* which are seldom more than one cm across, can detect anything that lies or drifts in a swath 6 or 7 cm across. One species has even developed spatulated tips at the end of its pair of main barbels, which illustrates the advantage these fish possess with their formidable array of sensors.

Phyllonemus are not exceedingly rare on a rocky slope, but they hide in the recesses and come out only at night, when they forage. For this reason it is only very seldom that one is caught. Their life is in fact poorly documented and, typically, never has their breeding been observed nor a nest been found.

Along with *Synodontis,* the two species of *Phyllonemus* are natural and valuable

Phyllonemus typus.

Phyllonemus typus.

Bagrus ubangensis.

Lophiobagrus cyclurus.

Lophiobagrus cyclurus.

**Key to the Genera of
Bagridae of the Lake
Tanganyika Basin:**

1. Nasal barbels present...2
   Nasal barbels absent 4
2. Dorsal rays 8-11; adipose long... ***Bagrus***
   Dorsal rays 5-6; adipose short..3
3. Caudal rounded; maximum size about 100 mm; lips dark viewed from below ....
   .... ***Lophiobagrus***
   Caudal notched or forked; maximum size 200-770 mm; lips usually light viewed from
   below .... ***Chrysichthys***
4. Palatine teeth present; maxillary barbels often widened at anterior end by membrane,
   very long.. ***Phyllonemus***
   Palatine teeth absent; maxillary barbels normal..5
5. Anterior nostrils dorsal; size small, 45 mm maximum...................................... ***Leptoglanis***
   Anterior nostril on upper lip; size large, to over 1000 mm ....***Auchenoglanis***

**Table for the Identification of
the Species of *Chrysichthys*
in the Lake Basin:**

Head 1.3 to 1.7 times longer than wide; gill rakers 17-20; interorbital space 0.33 to 1.0 in
eye; upper and lower lips seen from underneath pale .... ***sianenna***

Head 1.1 to 1.5 times longer than wide; gill rakers 13-16; interorbital space 0.75 to 1.8
times wider than eye; upper and lower lips seen from underneath pale. ***brachynema***

Head 1.1 to 1.2 times as long as wide; 10 to 12 gill rakers; interorbital space 2.1 to 2.8
times wider than eye; upper and lower lips seen from underneath pigmented ....
....***grandis***

Head 1.4 to 1.6 times as long as wide; 9 to 10 + 1 gill rakers; interorbital space 1.3 times
wider than eye; upper and lower lips seen from underneath pale ....***stappersii***

Head 1.2 to 1.5 times longer than broad; 8 to 10, sometimes 11, gill rakers; interorbital
space 1.0 to 1.8 times wider than eye; upper and lower lips seen from underneath
pale... ***graueri***

Head 1.0 to 1.25 times longer than wide; 5 to 8 gill rakers; interorbital 1.5 to 2.0 times
wider than eye; upper and lower lips seen from underneath pale ....***platycephalus***

**Key to the Species of
Phyllonemus:**

1. Maxillary barbel very long, 1.2 to 1.5 times longer than head, ending in broad black
   feather-like membrane; head 1.6 to 1.75 times longer than broad; snout 1.3 to 1.4
   times broader than long; maximum size 100 mm .... ***typus***
   Maxillary barbel very long, 1.4 to 1.65 times head length, without a feather-shaped
   membrane; head 1.5 to 1.75 times longer than broad, the sides appearing concave;
   snout 1.4 to 1.55 times broader than long; maximum size 100 mm.............***filinemus***

Chrysichthys brevibarbis.   *Chrysichthys ornatus.*

Chrysichthys laticeps.   *Chrysichthys sianenna.*

Auchenoglanis occidentalis.   *Auchenoglanis occidentalis.*

Auchenoglanis occidentalis.   *Auchenoglanis biscutatus.*

complements to a tank devoted to Tanganyika cichlids, as they are scavengers and will take care of leftovers. But one should remember that all catfishes, small as they may be, relish the spawns and fry of other fishes and in this respect should not be trusted.

Chrysichthys Bleeker

The third species of bagrid that might find its way into a tank is *Chrysichthys sianenna,* one of the smallest species in the genus as it doesn't grow to more than 20cm, although captures usually involve specimens between 10 and 15cm. It is one of the few *Chrysichthys* species in the lake to have specialized toward a life on the sandy bottoms. This is reflected in its coloration, metallic copper on top and sides, pure white underneath. Its very large eyes are implanted high on its head so that the fish enjoys an excellent vision upward as well as at ground level. *Chrysichthys* usually have short and massive bodies, but this species is elongate. Another unusual feature, which is enough to identify it at once as *C. sianenna,* is the head. Instead of being rounded, it has a typical triangular shape with a pointed snout and a rather small mouth for a *Chrysichthys.*

Often captured in large numbers in beach seines, the fish seldom survive the ordeal and it takes time and good care to bring a few back to health, which explains why this good-looking catfish is seldom seen in a tank. Once acclimatized it makes a good pet, is rather hardy, and will not bother a fish that it cannot swallow. Shrimp, fish meat, or even flakes will keep it in good health.

Leptoglanis Boulenger

Leptoglanis are small catfishes that don't live in the lake but in torrents high up in the mountains. They are very peculiar with an elongate body and the low setting of the pectoral fins, which are well developed as are the pelvic fins. With their large paired fins moved by powerful muscles they can secure a four-point grip on plant stems of the bottom. If *Leptoglanis* lose their hold they are such poor swimmers they will be swept away by the fast-moving waters they live in. They then might drift until they can get another hold on the substrate.

Their ecological niche is rather constraining and they live poorly in a tank.

Leptoglanis brevis Boulenger

Body depth 5.5 to 5.8 times in standard length; head length 4.5 to 5.0 times in standard length; head flat, barely longer than broad (1.0 to 1.15 times); snout well rounded, 1.2 to 1.3 times longer than postorbital area; 1.7 to 2.0 times in head length; eye medium, 3.75 to 5.0 times in head length; maxillary barbel broad and as long as head; 6-7 gill rakers; D. I, 6, with a short but strong spine, much closer to the head than to the tail and well in front of ventrals; adipose fin long and low; anal fin 10-11; caudal concave; maximum size 45 mm; mountain torrents.

Auchenoglanis Günther
Auchenoglanis occidentalis (Valenciennes)

Body depth 3.6 to 4.8 times in standard length; head length 2.8 to 3.8 times in standard length; head width 1.1 to 1.7 times in head length; snout a little more than two times longer than postorbital area; occipital bony plate in contact with interneural shield; premaxillary teeth in two groups, each twice as broad as long; mandibular teeth in two small and well separated groups; maxillary barbel 1.6 to 2.7 times, outer mandibular barbel 1.1 to 2.0 times, inner mandibular barbel 2.8 to 4.5 times in head length; dorsal fin I,7; A. III-IV,7-8; adipose fin height 4 to 6 times in fin length; marbled or spotted gray, with a few black spots arranged in rows; in excess of 1000 mm.

Family MOCHOKIDAE

There are close to 150 species of mochokid catfishes in Africa, of which more than 100 belong to the genus *Synodontis.* Seven species of *Synodontis* have colonized the lake and I have personally seen another one in the southern part of the lake that, unfortunately, I couldn't catch. Quite a few more might still be found in some of the major rivers.

Six of the seven species recorded so far are endemic to the lake, and all of them have a typical pattern, the main feature of which consists of a white band bordering all the fins. Some of them have villous skins. Contrary to riverine *Synodontis,* which have smooth skins, theirs is rough and ribbed by microscopic folds, whose usefulness has not been

Chrysichthys brachynema.   *Chrysichthys stappersi.*

Gnathobagrus depressus.   *Chrysichthys ornatus.*

Chrysichthys ornatus.   *Chrysichthys walkeri.*

Leptoglanis rotundiceps.   *Leptoglanis dorae.*

discovered as yet. One species, *S. granulosus,* received its name because of this very feature, but *S. petricola* and *S. polli* to some extent also have villous skins.

It has long since been a common opinion that *Synodontis* are nocturnal, i.e. active by night and rest in the daytime. I had often caught *Synodontis* on hook and line during my days in Central Africa and wondered about the fact that most appeared to be diurnal fishes. I was thus not utterly amazed during my first dives in the lake to discover that the *Synodontis* I saw were not asleep during the day but very active—milling about, foraging, and even courting. It all depended on the amount of light reaching the area we were exploring. In strong light they hid under the rocks; deeper down they were out of their holes. As it turned out *Synodontis* are not nocturnal but light-shy. In the twilight of the 40 meter depth one can occasionally see a school of *S. multipunctatus,* several hundred strong, swimming around.

Contrary to the consensus, they are not infeudated to mud floors either, at least systematically, as many species are found exclusively on rocky slopes. It all depends on the species considered. In the Congo River some species also live in swamps and others are found exclusively in the rapids (*S. acanthomias, S. longirostris, S. brichardi,* and *S. ornatipinnis,* to name but a few) and have become adapted to these habitats.

Most *Synodontis* are omnivores and feed on whatever they encounter during their foraging. However, some species in the lake have become specialized and eat only snails. They rip the opercle from the shell and then nibble at the mollusc. The problem with *Synodontis* having a specialized diet is that they will not change their habits when they are captured and given another type of food. They will starve to death. Only with patience can one teach them to take anything else. When the change of diet has been accomplished the problems are not over, as we discovered with *S. multipunctatus,* one of the malacophagous species. The bones of the fish become brittle, and should it happen that the spines on the pectoral fins become broken when they are released from the nets, instead of healing and growing back they keep on breaking until finally they disintegrate. With proper treatment, including the addition of calcium to their diet, the process eventually stops and the fins recover their shape.

S. multipunctatus is one of the best-looking species in the whole of Africa, with a golden beige background, pure white underparts, and dotted liberally with round black spots. Only the pelvic and anal fins are creamy white, the other fins being black with a broad ivory white edge. Juveniles are the most beautiful specimens, with a golden sheen and a few contrasting black marks on the body. The number of spots increases with age and the color turns dark beige in very large specimens, so that they are less striking when they reach 25cm (which is the maximum size) than when they were 10 cm long.

Of the seven species, only *S. nigromaculatus,* with a pale gray color, yellow ventrally, lives in swamps and is the only non-endemic species, having been recorded from other areas to the south of the lake.

S. granulosus is considered by some amateurs as the most beautiful *Synodontis,* with its dark blue or dark purple color devoid of any marks and all fins, except the pelvic and anal fins, which are white, being jet black with a broad white band. Juveniles are a sight to behold, being jet black all over and all the fins being entirely ivory white. *S. granulosus* are very rare and hard to catch. Only a dozen or so are caught each year and as they fetch very high prices each capture, always by chance, is considered as the event of the dive. They are unfortunately difficult to acclimatize and rather delicate at first. Most captured specimens are infected by *Argulus* parasites, which might account for their scarcity in the lake.

S. petricola and *S. polli* are less rare and caught more often. They are difficult for a novice to tell apart. Both have a dark vermiculated pattern on a brown, beige, or blue-gray background. Perhaps a hundred specimens of the two species are caught in a year. They are caught when they are found foraging in the open. Catching a *Synodontis* when it is hiding in its deep recess is a very difficult task.

Capturing *S. multipunctatus,* on the other hand, is much easier as most are caught in the huge beach seines used by teams of fishermen. *S. multipunctatus* are thus caught every day, but fishermen hate having them in their nets as their barbed

Synodontis petricola.   *Synodontis petricola.*

Synodontis polli.   **Synodontis polli.*

Synodontis polli.   *Hemisynodontis membranaceus.*

Brachysynodontis batensoda.   *Brachysynodontis batensoda.*

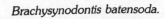

*S. polli Gosse, 1982, is a new name for *S. eurystomus* Matthes, 1959,
preoccupied.

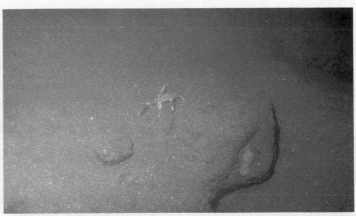

Synodontis multipunctatus. *Synodontis multipunctatus.*

Synodontis multipunctatus. *Synodontis petricola.*

Synodontis polli. *Synodontis marmoratus.*

Synodontis greshoffi. *Synodontis nigrita.*

Synodontis brichardi was named in honor of Pierre Brichard.

Synodontis granulosus. *Synodontis nigromaculatus.*

Synodontis nigromaculatus. *Synodontis polli.*

Synodontis multipunctatus.

Synodontis budgetti. *Synodontis haugi.*

Synodontis frontosus.

Key to the Species of
***Chiloglanis* of the Lake**
Tanganyika Basin (from M.
Poll):

1. Body height more than 5.5 times in standard length; eye included 7.5 to 9.0 times in head length; 12 to 14 mandibular teeth in a bunch; base of adipose fin twice in distance between this fin and dorsal fin ....*lukugae*

Body height less than 5.5 times in standard length; eye included 6.5 times in head length; 10 mandibular teeth set in a transverse row; base of adipose fin only 1.65 in distance between this fin and dorsal ....*pojeri*

Key to the Species of
***Synodontis* from the Lake**
Basin (from Matthes):

1. Mandibular teeth 27-55 3

    Mandibular teeth 14-25 2

2. Eye large and protruding, on side of head, 2.55 to 5.25 times in head length; maxillary barbel 0.75 to 1.5 times as long as head; maximum size 60 mm ......*multipunctatus*

    Eye small, supralateral, 4.8 to 9.3 times in head length; maxillary barbel 0.55 to 0.75 times as long as head; maximum size 360 mm ....*dhonti*

3. Adipose fin high, maxillary barbel 0.9 to 1.35 times as long as head; top of head and occipital-nuchal shield rough and not covered by skin; adult size in excess of 200 mm 5

    Adipose fin low; maxillary barbel short, 0.5 to 0.8 times as long as head; top of head and occipital-nuchal shield covered by skin; adult size less than 200 mm 4

4. 40 to 55 mandibular teeth; snout 1.5 to 1.9 times longer than postorbital area; body depth 3.6 to 4.0 times in standard length; maximum size 150 mm ....*polli*

    27 to 40 mandibular teeth; snout 1.35 to 1.6 times longer than postorbital area; body depth 4.0 to 5.2 times in standard length; maximum size 150 mm ....*petricola*

5. (triplet). 35 to 50 mandibular teeth; humeral bone narrow, pointed, and keeled, 2.35 to 2.85 times longer than high; snout 1.0 to 1.4 times longer than postorbital area; maxillary barbel without a membrane; body gray to steel blue, without spots; maximum size 430 mm .... *granulosus*

    27 to 40 mandibular teeth; humeral bone broadly triangular and pointed, 1.2 to 2.4 times longer than high; snout 1.25 to 1.45 times longer than postorbital area; maxillary barbel with a narrow membrane at its base; caudal peduncle 0.87 to 1.15 times longer than high; body gray with countless small spots, even on fins; all fins without white edge present on all other *Synodontis* from the lake; maximum size 320 mm ....*nigromaculatus*

    39 to 48 mandibular teeth; humeral bone broad, triangular, and blunt, 1.7 to 2.4 times longer than high; snout 1.4 to 1.85 times longer than postorbital area; maxillary barbel with a rather broad membrane; caudal peduncle 1.25 to 1.48 times longer than high; maximum size 600 mm ....*lacustricolis*

Synodontis eupterus.

Synodontis robertsi.

Synodontis longirostris.

Synodontis longirostris.

Synodontis longirostris.

Synodontis nigromaculatus.

Synodontis njassae.

Synodontis nigromaculatus.

Synodontis petricola. *Synodontis multipunctatus.*

Synodontis longirostris. *Synodontis* sp.

Synodontis sp. *Synodontis* sp.

Synodontis longirostris. *Synodontis* sp.

spines get stuck in the mesh and take a long time to be dislodged. As they are not eaten by fishermen they are just considered as a big nuisance and treated accordingly. Few survive the scratches and breaking of the spines. Some can be nursed back to health during a process that can take as long as two months of assiduous treatment, which explains the high cost and relative scarcity of *S. multipunctatus* on the market.

In captivity, *Synodontis* from the lake are hardy and endowed with a good longevity. They will not be finicky about their diet and are excellent scavengers. They contribute to the good condition of the other fishes living with them as they take care of many leftovers that might pollute the water. They will not bother other fishes, but should not be trusted with a brood (no fish for that matter could be so trusted).

They will come out from their recesses at meal time and wander all around the tank in search of food. One common complaint that I often hear is that they remain hidden. Being light-shy, it is only natural that they refrain from going around in the harsh light of a well-lit tank. Give them subdued light in part of the tank and they will come out.

Ichthyophthirius is a common parasite of *Synodontis* in captivity, especially of the lake species. The latter don't suffer from ich in the lake, where it is unknown. They were thus unable to build defenses by inborn immunity. *Synodontis* as a rule are very sensitive to ich because the parasites

embed themselves in the mucous coating of the bare skin of the fish and are thus well protected against drugs, like copper sulfate, that are used to kill the parasites. On the other hand, most *Synodontis* do not tolerate the drugs at the dosage needed to kill the parasites and die from copper poisoning. Thus one has to use copper sulfate sparingly at 1⁄3 or 1⁄4 the normal dilution so as not to kill the fish. It should also be used over a much longer period. Progressively the disease will disappear without trouble for the catfishes. Fortunately, fishes that have suffered from a strong attack of ich have developed a temporary immunity against another outbreak, and progressively will build natural defenses.

Chiloglanis Peters

It is a rather spectacular adaptation through which the species from this genus have gone to be able to live in their habitat—fast-flowing waters. Their mouth is merely a long and narrow slit in the middle of a huge sucking disk. In front of the mouth two large pads concentrate the teeth, which are used to graze vegetal biocover from the rocks. This unusual setup allows the fish to feed while remaining stuck to the rock without risk of being swept away by the currents. The suction created by the disk is so powerful that one cannot pry the fish loose by brute force without inflicting mortal wounds. Sucking disks are an anatomical adaptation very common among rheophilic fishes everywhere in the world. In Africa a few

species that are endowed with sucking disks or a vacuum apparatus are kneriids, cyprinids, and mochokids.

Chiloglanis are so specialized that they cannot feed in any other fashion. Thus in a tank, after having cut a swath through the green algae carpeting the rocks and the plate glass, they soon start to become emaciated and eventually die in a few weeks time. Basically they are strong and hardy, provided that they get well aerated water. It is simply a fact that in the confines of a tank, epilithic algae don't grow as fast as the fish grazes.

Most species are olive green, providing them with excellent camouflage, but a few have mottled or marbled patterns. In the lake basin they are restricted to mountain torrents and live in cool water with a low pH.

Family AMPHILIIDAE

The "walking catfishes," as I called them when I caught my first specimens in the Congo basin, live in fast moving waters of the main streams and their affluents. This means that they need high oxygen levels and have undergone adaptations to this habitat.

The main one, similar to the one displayed by other rheophilic fishes, is reflected by the position and development of their pectoral and pelvic fins. The two pairs of fins are set far apart, the pectorals well in front and low on the body, the pelvic fins well behind and horizontal instead of more vertical. This setup provides the fish with a

Synodontis multipunctatus.   *Synodontis multipunctatus.*

Synodontis petricola.   *Synodontis polli.*

Synodontis acanthomias.   *Synodontis nigromaculatus.*

Synodontis nigromaculatus.   *Synodontis robbianus.*

four-point support on the bottom, and as the fins are spread out and large, they help the fish resist the pull and tear of the water flowing by. In this task the fish gets an additional grip with its mouth, which is rubbery and very low, as befits a grazer, and by the fact that the first ray of its paired fins is composed of thick, rubbery, rough skin, providing a better hold on the slippery surface of the rocks and reducing the risk of bruises due to the battering of the current should the fish lose its grip. Elastic skin surface underneath the body is a feature shared by many rheophilic fishes. We find similar adaptations in cichlids (*Orthochromis, Steatocranus, Teleogramma*), cyprinids (*Garra*), and various other families of catfishes.

The "walking" *Amphilius* got its nickname when I got my first specimens and observed that they could, out of water, move about in a very peculiar way. Alternately supported on the pectoral and pelvic fins on one side, then the other, they manage to wriggle forward at an impressive pace. No doubt that it is in this fashion that in their native turbulent waters they manage to creep forward on the slippery surface of the rocks and perhaps even find their way out of the water up the small cascades that break the flow of mountain torrents.

Walking catfishes are not large, usually well under 10cm, and, contrary to *Chiloglanis*, make excellent aquarium fishes, although their coloration is nothing to get excited about, being olive green or steel gray with rosy underparts.

Key to the Identification of the Genera of Clariidae of the Lake Tanganyika Basin (from M. Poll)

1. A rayed dorsal fin and an adipose fin present 2
   Only the rayed dorsal fin present; no adipose fin 3
2. Adipose fin large; sides of head protected by bony shields..... .... *Heterobranchus*
   Adipose fin short; sides of head not protected by bony shields; cranial roof shield reduced .... *Dinotopterus*
3. Head, at least in part, protected on sides by bony shields; cranial arch roof-shaped; head devoid of longitudinal median groove.. *Clarias*
   Sides of head not protected by bony shields; cranial arch not roof-shaped; strong longitudinal median groove.................. .... *Tanganikallabes*

Family CLARIIDAE

This large family is distributed over the whole of Africa and even includes a few marine species. They also live in Southeast Asia.

There are only seven species in the lake basin belonging to four genera, one of which is endemic to the lake (*Tanganicallabes*) and a second being shared with Lake Malawi (*Dinotopterus*).

To colonize the lake clariids often had to adapt and lose important features, such as the need to go periodically to the surface for a gulp of atmospheric air. The endemic genus *Tanganicallabes* also displays a reduction of the cranial skeletal structure (several bones missing); this feature is shared by *Dinotopterus*.

Among the lake dwellers *Heterobranchus longifilis* is the only one reaching a respectable size. Specimens 20kg or more are not rare. *Clarias* are often seen as ugly fishes: they are gray, slippery, look vicious, and have long "whiskers". *H. longifilis*, with a slender shape and pale yellow-gray with white underparts, manages to be less repelling than the other species. I got this feeling perhaps a long time ago because it was a baby *H. longifilis* that became my first pet. I got it in the drying-up grasses of a flooded savannah at the start of the dry season. It was then a mere 3cm long. Six months later it was 10cm long and after a year it had grown to more than 20cm. Five years after its birth Hippolyte, as I called him, was close to one meter long and probably could have been bigger if I had always fed him properly. But as it was he accepted only live fish—a dead one he wouldn't touch—and raw hamburger meat, of which when he was full size he gulped a pound a day in two or three bites. His appetite was incredible. So was his resiliency.

Amphilius grandis.   *Amphilius grandis.*

Phractura near *intermedia.*   *Amphilius longirostris.*

Amphilius platychir.   *Phractura ansorgii.*

Phractura ansorgii.   *Phractura ansorgii,* fry.

 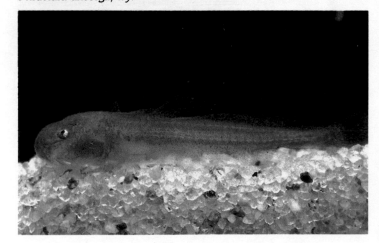

It happened twice that during one of my safari trips I left him home for a few days under the care of a friend. He jumped out of his drum and fell into the garden one night and was left to wriggle on the ground during the rest of the night, then in bright sun the next day. I found him next evening 20 yards away, covered with sand, his back and sides sunburned, the flesh on his belly scraped bare. I cleaned him as well as I could, daubed him entirely with mercurochrome, and put him back into clean water. He recuperated so fast that in little more than three weeks there were no traces left of his ordeal.

This short story helps explain how *Clarias* trapped in drying-out rivers of East Africa manage to survive the dry season in overheated mudpools. Needless to say they are probably the most hardy of all African fishes and are long-lived. *H. longifilis* could perhaps outlive its owner.

Tanganicallabes and *Dinotopterus* are rock-dwelling clariids living deep in the recesses and discovered only when they are flushed out of their hiding places with tranquilizers. They don't exceed 30cm and have a long, slender eel-shaped body. Like other clariids living in the rocks of the Kinshasa rapids the two lacustrine species have a rounded head, most clariids having an oval cranium.

The two species are also a dark chestnut color.

Key to the Species of *Clarias* from the Lake Tanganyika Basin
1. Length of head less than 4 times in standard length; ventral fins equally distant from snout and tail .... *mossambicus*
   Length of head more than 4 times in standard length; ventrals closer to snout than to tail 2
2. Dorsal and anal fins joining with caudal base .... *theodorae*
   Dorsal and anal fins separate from caudal base ...*3*
3. First ray of dorsal fin distant from head by half head length .... *ornatus*
   First ray of dorsal fin distant from head by head length ....*liocephalus*

Family MALAPTERURIDAE

The famous African electric catfish has apparently colonized the lake for a very long time, as it is found in all habitats and along the whole length of the lake. On rocky slopes, and even more on rocky outcrops isolated in the midst of huge sand plains when they offer appropriate shelters, they are quite common.

The author knows of such an isolated place in a shallow area where the concentration of electric cats and their average size reach an incredible level. In this area, one might say that there is one electric catfish at least 300 mm long (sometimes one meter long) every two or three meters along the rocky shelters. Squeezing a hand into these holes often results in a very jolting experience—but not dangerous for the diver who stays calm. The electric jolts appear to be less powerful than with specimens of an identical size in the Congo River. It is not at all impossible that the lake electric catfishes living in very conductive water (because of the high mineral salt content) need less power, with the result that direct contact provokes a less powerful electrical impulse. In the Congo River, on the contrary, the jolts are extremely powerful, and direct contact might be dangerous when wading chest-deep in the river. This might be due to the fact that the Congo *Malapterurus* needs more power to achieve the same protective or hunting jolt in poorly mineralized water. Anyway, it would be interesting to look into this problem and find an answer to the question.

What is rather difficult to explain is the low number of *Malapterurus* fry found in the places where the concentration of the large adults is highest. It appears from this observation that either they have special spawning grounds where they migrate for the spawn or that the losses are very high. Never, among the hundreds of

Clarias angolensis.   *Clarias angolensis.*

Channallabes apus.

 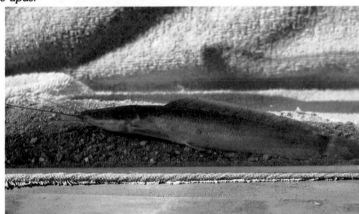

Gymnallabes typus.   *Clarias gariepinus.*

Gymnallabes typus.   *Gymnallabes typus.*

This large specimen of the electric catfish, *Malapterurus electricus,* lacks the typical black and white banded area at the base of the caudal peduncle. The spotting on the body is also heavier than in most specimens. Photo by E. Balon.

Head (above) and body (below) views of a fairly typical small electric catfish. Photo above by Dr. Herbert R. Axelrod, that below by E. Balon.

Malapterurus discovered by the author and his team during five years of underwater fishing, was there one batch of babies discovered. Even when they are about 30 mm long, the baby electric catfishes are living on their own among the rubble.

Family CYPRINODONTIDAE

Eastern Africa is the poorest part of the continent for many species of fish mainly because it is poor in large and small river basins. In the case of the cyprinodonts, this holds especially true because the family seems to have proliferated much more in dense forest areas with high acidity and low mineralization of the water. The only three genera that are represented in East Africa are *Nothobranchius,* of which none have been found in the lake basin, *Aplocheilichthys* and the endemic lake genus and species *Lamprichthys tanganicanus.*

Aplocheilichthys, represented by *A. pumilus,* lives in swamps and lagoons near the lake edge and is not found in the lake itself but all along the lake perimeter. The eggs seem to have a fantastic resistance to drought and to outside factors, if one might judge from the author's personal experience with these fish. Having to put wild caught plants in an outside artificially made breeding pool, the author's team collected *Ceratophyllum* in nearby swamps. To disinfect them from any snail or insect egg, the plants were put into a strong solution of copper sulfate, knowing that most of the plants would rot away but that some stems would sprout anew and start growing. The plants remained in the solution for several hours and then were put into a freshly dug pool lined with new polyethylene sheeting. Fifteen days later there were young *Aplocheilichthys pumilus* fry swimming in the pool. The resistance of the eggs to unfavorable outside conditions might very well explain the very large area in which this species is found.

These tiny *Aplocheilichthys* are hardy and have fewer requirements as to their water conditions and diet than any other cyprinodont. They thrive in neutral or slightly acid water and are also at ease in a DH of 5 German degrees or more. They will take dried food without trouble and retain their original bronze-green metallic sheen. They are not a bother to the other fish. What's more, they are very active, always on the move, and for the beginner have the added charm of easy spawning.

Although not especially colorful, *Aplocheilichthys pumilius* is an attractive and very hardy species that does well in aquaria. Photo by Dr. Herbert R. Axelrod.

If kept in small schools of four or five females and one or two males, the African lampeye *Aplocheilichthys pumilius* can put on quite a show in a properly lighted aquarium. The sides and eyes are very iridescent. Photo by Pierre Brichard.

489

Lamprichthys tanganicanus is quite another matter. The only endemic cyprinodont in the lake, it is the giant of the family, reaching about 150 mm (a little less for the less colorful female). Far from being a swamp or an acid brook inhabitant like the other genera, it roams at will far from the shoreline, often in large schools, but sometimes all by itself. As such it is one of the fishes collected with the clupeids by the traditional native collecting methods or by the big commercial trawlers.

They are sometimes a nuisance for the diver when they follow him in his search for fishes. Always in search of food and inquisitive, they follow the diver everywhere, and whenever small fry try to escape the man's hands or nets and rush into the open, the *Lamprichthys* set upon it and the diver's victim is at once done away with.

At other times when two divers haul a big net to catch some of the bigger lake fishes, hundreds of *Lamprichthys* follow the divers and get entangled in the nets. It's a nuisance to remove them afterward from the net, and there is a considerable loss of time.

The spawning of *Lamprichthys* in the lake has been observed very often. In typical roamer fashion a pair meets in midwater, both partners descend to the rock substrate on the lake slopes, find at random a small corner (for example where two flat slabs join) and, swimming very slowly, lay and at the same time fertilize the eggs. The female is closest to the rock, the male a little behind (his mouth about at the female's ventrals) and in the open water both lean on their sides a little. The eggs and the male's sperm are totally clear and invisible. The author or members of his team have never seen a *Lamprichthys* egg, as it is invisible. The shape of the egg could only be guessed at in the lake by its apparently different refraction than the surrounding water. This phenomenon, or the smell of the eggs, might attract the predators living on or around the nearby rocks, since they rush to the spawning site as soon as the *Lamprichthys* descend to spawn. They lay in wait a few centimeters away and, as soon as the cyprinodonts depart, jump on the place where the eggs have been laid, apparently not missing many of them. Thus not many of the eggs stand a chance to survive, laid as they are on bare slabs amid countless predators.

But some do survive, enough of them to insure the survival of many young *Lamprichthys.* Two facts are noteworthy: first, this pelagic fish lays its eggs on unprotected grounds against great odds and has fully transparent eggs—which is the best protection it could afford its spawns under the circumstances; second, several rock-dwelling cichlids, especially *Telmatochromis dhonti* and *T. temporalis,* know what it means when they see two *Lamprichthys* descending along the rock slabs. They stop at once what they are doing at the time, come near, and several of them stand watch and wait until the spawning is finished. As soon as the *Lamprichthys* start to leave, they rush toward the eggs. The pairs are not a foot away before the eggs have already been gobbled up. Many times we have seen this scenario happen with a growing feeling of the apparent waste. But then, there are so many *Lamprichthys* around that it doesn't matter, and it feeds the other fishes.

The *Lamprichthys* fry, as might be expected from schooling pelagic fishes, bunch together apparently as soon as they are born, and it is not rare to swim across a school several hundreds or even thousands strong. Like their parents, the young *Lamprichthys* wander around endlessly, falling prey, like the clupeids, to the big roaming predators.

To catch and acclimatize these cyprinodonts is quite a problem, in fact just the reverse of the problem with cyprinodonts from the acid forest brooks. *Lamprichthys* is hard to acclimatize to another type of water when the mineral salts it is used to are missing. The fish develop sores and fungus on the body and die in a matter of days. Once acclimatized, they are one of the most gorgeous freshwater fishes ever. The body is an iridescent aquamarine, sparkling with each move of the fish. Even the tail flashes subtle green-blue because of the presence of minute scales well into the rayed fin. The male's body is still more enhanced by half a dozen horizontal rows of bright aquamarine spots, the long anal fin, the dorsal fin loosely hanging over the side of the body and both fins being an overall pale orange color. The fish is incredibly beautiful—there is no better word.

Spawning behavior in the Tanganyika lampeye, *Lamprichthys tanganicanus*. The male is more colorful than the female and has higher fins. The eggs are virtually invisible. Photos by H.-J. Richter.

Lamprichthys can be kept in a tank provided much attention is given to a few rules. The DH should be around 10° to 14°; the temperature should be about the lake average (26-26.5°C); a little salt should be added to the water (perhaps even magnesium sulfate); and lots of room should be provided as the rather large fish are moving around most of the time. Feeding is easy. The fish will rush toward any type of food dropped on the surface of the water and eat it. But, as it is endowed with a ravenous appetite, it should be fed several times a day to compensate for an apparently rapid metabolism.

Family ANABANTIDAE

The only anabantid in the lake area is *Ctenopoma muriei,* one of the more elongate forms of the genus. It is apparent that the many *Ctenopoma* from the Congo basin virgin forests and savannahs, all of them swamp or slow brooks dwellers and rather sluggish, could not climb the fast currents of the Lukuga outlet toward a habitat which is not to their liking. *Ctenopoma* are basically soft and acid water fish. The lake doesn't offer them this type of water. One might thus say that the only species of *Ctenopoma* to live in the lake basin is not a recent invader but was one of the few species that have spread eastward from the main grounds of the genus in central West Africa.

C. muriei, which is not one of the most striking species of the genus, is a dull fish with a brown body and a few black specks scattered at random. It is a typical swamp inhabitant

Key to the Genera of Cyprinodontidae of the Lake Tanganyika Basin:
1. Dorsal fin with 10 to 12 rays; anal fin with 14 to 16 ....**Aplocheilichthys**
Dorsal fin with 13 to 16 rays; anal fin with 24 to 30 ....**Lamprichthys**

The genus *Lamprichthys* includes, at the moment, only one valid species, *L. tanganicanus.* At one time another species, *L. curtianalis* David, 1936, had been described. This was later placed in synonymy with *L. tanganicanus.* The author believes that there are probably two different species of *Lamprichthys* in the lake, one with a long anal fin, the other with a much shorter one. For this reason, until the matter is settled the key to the identification of both species of *Lamprichthys* is given.

Key to the Species of *Lamprichthys* of the Lake Tanganyika Basin:
1. Anal fin 2.6 to 3.2 in standard length; posterior rays of anal fin longer than anterior rays; eye 3.0 to 3.6 times in head ....*tanganicanus*
Anal fin 3.5 to 4 in standard length; anterior rays of anal fin longer than posterior rays; eye 2.75 to 3.0 in head length.............................*curtianalis*

Ctenopoma muriei is the only anabantid recorded from the vicinity of Lake Tanganyika. Photo by Dr. Herbert R. Axelrod.

and has never been collected, or even seen, in the lake itself.

Family MASTACEMBELIDAE

Eleven of the twelve species of spiny eels living in the lake dwell on rocky slopes and only two might be considered as mainly sand-dwelling species. The occupation of the anfractuosities of the rocky habitats by so many different spiny eels probably has had far-reaching effects on the occupation of the bottom by cichlids.

It is probably because of mastacembelids that one can assess the poor occupation of the recesses by cichlids. It is thus of interest to dwell at some length on the various species of spiny eels and see which adaptations they eventually went through to occupy in such strength their rocky habitat, especially because spiny eels amount to 25% of the non-cichlid fishes in the lake (13 species out of 55).

Sand-dwelling spiny eels number only two species, *M. cunningtoni* and *M. ophidium* (although the latter is also occasionally found on rocky slopes). Only two species, *M. frenatus* and *M. cunningtoni*, breed in rivers. All the others reproduce in the lake, which shows to what extent they have adapted to the lake ecology. Speciation had thus been very much at work when the eels came into the lake, and in strong contrast to the situation in Lake Malawi where only one species lives.

The spiny eels of Lake Tanganyika also display an astonishing variety of shapes and color patterns. As a rule they are much more diversified than the spiny eels found in most of the other African habitats.

Some of the spiny eels, although found all around the lake, are more common in one part of the shorelines. *M. moorii* is very common in the northern part of the lake and rare in the southern part while *M. frenatus* (the only spiny eel also found regularly in rivers) is the most common spiny eel in the southern part of the lake and exceedingly rare in the northern part. But most other spiny eels appear to be distributed more or less evenly around the lake, even the most reclusive and rock-infeudated species.

Basically, one should divide the spiny eels into two groups, the main one involving the majority of species reaching at least 30cm in length, the second involving three dwarf species. The dwarf species, *M. platysoma, M. micropectus,* and *M. tanganicae,* are all rock-dwellers. Among the large spiny eels, the biggest is *M. moorii,* growing to more than 70cm long and a weight of 1,000gr. It lives exclusively on rock, but wanders into the open to forage by daytime. It is not seen foraging at night. It remains in its hole most of the time, like a moray eel, and will attack fishes that pass by. *M. frenatus,* with a pale beige background color and regularly spaced chevrons on its back, ventures out more than *M. moorii,* and will slither out of its den to attack fishes in trouble several yards away. It is one of the main scavengers on a rocky coast. Several individuals will attack a trap and try to grab fishes through the mesh.

The density of *M. moorii* and *M. frenatus* can be phenomenal, with at least one specimen per 100 sq.m. of bottom being average. In Ruziba the density of *M. moorii* (large adults) was one specimen every two or three meters of sandstone slabs. Adults of the two species can swallow an adult *Tropheus* or *Cyathopharynx. M. frenatus* appears also to have a diurnal cycle of activity.

M. albomaculatus and *M. flavidus* are very seldom seen and should be considered as very rare. The first is chestnut brown with faint pale spots, becoming white toward the tail.

The second species is plain pale gray or pale yellow speckled with countless tiny black dots. Nothing is known about their predatory behavior or cycle of activity.

M. ellipsifer, M. zebratus, and *M. plagiostoma,* which do not grow much in excess of 30cm, belong to a line of spiny eels in which the coloration, white underneath, very pale beige on the sides, and darker on the back, is decorated with a set of stripes, bands, or ovals that are generally brown or chestnut. Juveniles of one of the species that were not identified because of their status, were found to be alternately banded in black and white and were very attractive indeed. All three of these spiny eels are nocturnal.

M. cunningtoni is especially plentiful around the Ruzizi delta. It is a typical sand-dweller, which is reflected by its color pattern, which for a spiny eel is rather unusual. It is plain copper on the back, becoming lighter on the sides and pure white underneath; the tail tip is jet black and lined with white. Not at all unattractive.

M. ophidium is one of the spiny eels in the lake with a very peculiar shape. The body is thin and slender, the head large, short, fleshy, and with a very short snout, and the eyes are set very much forward. At first sight *M. ophidium* looks very much like a viper, and the first reaction of anybody upon seeing one is to think, "Caution! A snake!". *M. ophidium* reaches 60cm and lives mainly on sand, where its slender pale gray or pale beige body with faint markings provides excellent camouflage. It is the only spiny eel in the

lake from which concentrations of thousands of young fry a few cm long have been observed in quite bays during some months of the year. It appears that the spiny eels might migrate and have synchronous spawns, but as yet this observation applies only to *M. ophidium* and not to the other species.

It is well known that sand-dwelling fluviatile species of spiny eels bury themselves in the sand to lay in ambush waiting for prey to pass by, or to do so as a protection against predators. The sand-dwelling species in the lake have not been reported to have the same behavior, but one would not be surprised to find that they, too, bury themselves in the sand.

The three dwarf species are very interesting fishes indeed. *M. tanganicae* is probably the most common species of spiny eel along a rocky shore, with its main habitat very close to the shore itself. Very few are found deeper than 5 meters. Slender, but not exceeding 20cm, its cylindrical yellow, olivaceous, or green body mottled with countless little dark specks is generally spotted when one turns the rubble, under which they hide, over. Too small to swallow a fish as small as an adult *Spathodus erythrodon, M. tanganicae* lives mainly on crustaceans, insect larvae, and fish fry. In the surfbeaten zone of the shoreline they are the only predators the fry from rock-grazers have to worry about, provided they stay in the rubble.

M. tanganicae find their way into tanks devoted to Tanganyikan fauna and have become favorites of hobbyists.

Mastacembelus frenatus.

Mastacembelus moorii.

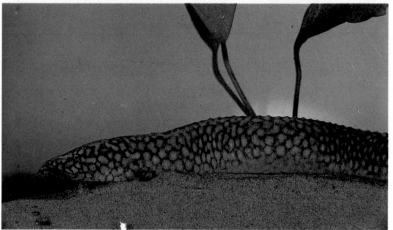

Mastacembelus moorii.

Mastacembelus ophidium.

Mastacembelus ellipsifer.

Mastacembelus tanganicae.

495

Once acclimatization has been accomplished and provided they have shelters in which to hide, the small spiny eels feel at ease in captivity. They can be kept in top form with chunks of fish meat or shellfish, or better yet with live freshwater crustaceans.

M. platysoma has the same requirements and would have become a much sought after pet, because it is a very cute fish, were it not for its extreme scarcity. It is practically unknown as only very few specimens are caught, and those quite by accident. It is rare in its habitat, which extends from the top layer on a rocky slope to at least 30 meters deep.

From time to time one discovers a *M. platysoma* winding its tortuous way among the rocks or even swimming in the open just above the ground. It is unmistakable because of its peculiar shape: the body is very short and very deep for a spiny eel. This is even accented because the dorsal and anal fins are well developed and of the same background color as the body, which is dark chestnut, so that swimming with the two fins erected the eel appears to be even shorter than it really is. The snout is very long and pointed and the eyes are large and black and rimmed with gold. The chestnut color of the body is covered with black vermiculations. *M. platysoma* has the uncanny habit of coiling itself on the bottom and sometimes to even swim in the shape of a coiled "S," in which the spiny eel is helped by the development of its unpaired fins. This is why when he discovered the eel Dr. Matthes

Key to the Species of *Mastacembelus* of the Lake Tanganyika Basin (from H. Matthes):

I. Anus more or less halfway between snout and tail

A. Strong preorbital and preopercular spines; D. XXVI-XXIX; 26-28 scales in a transverse line; pectoral fin 2.25 to 3.3 in head; body plain copper, belly white .... *cunningtoni*

B. Head spines in young, but tiny or absent (hidden under skin) in adults; D. XXV-XXIX; 32-35 transverse scales; pectoral 3.3 to 4.7 in head .. ..*moorii*

C. Head spines absent (seldom a small preopercular spine in young)

    1) D. XXIX-XXXIV; pectoral fin 6.5-14.75 times in head; body plain purple ....*micropectus*

    2) D. XXX-XXXV; pectoral fin 2.7-4.75 in head; rostral appendix 1.0 to 1.9 times longer than eye; body dark and marbled .... *frenatus*

    3) D. XXXIV-XXXV; pectoral fin 3.5 to 4.3 in head; rostral appendix 0.6 to 1.05 times longer than eye; body pale yellowish with darker spots...................................... *flavidus*

II. Anus much closer to tail than to snout

A. Preorbital spine present

    1) D. XXIII-XXV; body 6.6 to 7.5 times longer than high; flat, plain colored body .... *platysoma*

    2) D. XXIX-XXXII; body 10.5 to 11.5 times longer than high; crescent-shaped stripes present... *ellipsifer*

B. No preorbital spine

    1) D.XXIII-XXV; mouth low; body 8.9-10.5 times longer than high; body with dark stripes on pale brown background..*zebratus*

    2) D. XXX-XXXII; mouth low; body 10.6-12.1 times longer than high; saddle-like bars on back, body pale beige .... *plagiostomus*

    3) D. XXXIII-XXXVI; mouth terminal; head 5.75 to 6.9 times in standard length; 25 to 30 transverse scales; countless whitish dots on a brown body........................ *albomaculatus*

    4) D. XXXVI-XLII; head 7.0-8.8 in standard length; 17-21 transverse scales; many vermiculated stripes on a thin, light brown body.. *tanganicae*

III. Anus closer to snout than to tail

D. XXII (young)—XXXIII; body pale pale beige; unmistakable viper-like head with eyes very much in front of head........................ *ophidium*

nicknamed *M. platysoma* the "seahorse" eel, a name the fish fully deserves. Its diet in the lake consists mainly of microorganisms and probably fish fry. It is not especially hard to keep in captivity, provided the tank is well aerated and free from organic pollution. None has been caught that exceeded 15cm.

M. micropectus, the third dwarf species, has evolved along entirely different lines than the seahorse eel. The worm-like body is extremely long. The pectoral fins have suffered such an atrophy that they are reduced to mere stubs and have lost most of their usefulness; the dorsal and anal fins are much reduced in height. The swimming ability of *M. micropectus* is thus very poor, contrary to *M. platysoma,* and the fish probably is more used to crawling around the rocks in the obscurity of its dwelling deep in the labyrinths, than to swimming.

Its color pattern reflects the habitat—pale purple with pinkish underparts. The depigmentation of the skin reminds one of the similar color of several spiny eels living under the sandstone slabs of the Kinshasa rapids. Like them, *M. micropectis* also has very small eyes (another contrast with *M. platysoma* which enjoys large eyes) set in a short, fleshy head and a very short snout.

Apparently *M. micropectis* has gone a long way from the standard *Mastacembelus* anatomy to adapt to a very constraining ecological niche. Because of these specializations the fish does very poorly in captivity and usually dies after a few days.

The spiny eels behavior is poorly documented, which is unfortunate as they play such an important role on the rocky slopes. Their impact, especially on cichlids, cannot be stressed too much. Their density of occupation of the rock recesses can be phenomenal (one would not be far off the mark when thinking that there are on average 30 to 50 spiny eels, all species and sizes put together, per 100 sq. meters of substrate).

Juveniles of most species and of course two out of the three dwarf species should be welcome in a tank, to which they would add variety and life. But, and this is especially true in the large species, a spiny eel should not be trusted with a fish it can swallow.

Family CENTROPOMIDAE

The *Lates,* or Nile perches as they are called, are represented in most of central Africa, aside from Lake Tanganyika, by the species *Lates niloticus.* This common fish is found in lakes, large rivers, and streams from the Nile to the Niger and often reaches enormous proportions. The few other known *Lates* species are restricted to large lakes. Fossil *Lates* remains have been discoverd, especially around Lake Rudolph, of which some belong to now extinct species. It is thus rather queer that Lake Tanganyika, which lacks *L. niloticus,* has been the birthplace of as many as four other species, one of which is a dwarf form. It is probable that the five lake species have evolved from one or several forms of now extinct fishes and that *Lates niloticus* never

entered the lake basin area before the lake was formed. One argument, in addition to the others, is to think that the lake area never was part of the Nile basin. Although the Congo River harbors *L. niloticus,* it is possible that the fish couldn't or wouldn't climb the successive steps from the Lukuga outlet cascading down from the lake toward the Congo River.

The Nile perch is one of the giants among the African fish fauna; many specimens have been caught with a weight well in excess of 150 kg, and the record is over 200 kg. Not one of the four Tanganyika *Lates* reaches such a tremendous size, but they are very important as food fishes for the people living on the lake shores and even farther inland. The lake perches live in schools, prey on the clupeids roaming in the lake and make up a very significant portion, if not the most important part, of all the non-clupeid commercial catch.

They are caught at night when the powerful lamps used for fishing have attracted the clupeids toward the nets. All the predators, *Lates, Bathybates, Haplotaxodon, Hemibates,* etc., follow this growing concentration of their prey and gorge themselves to the limit. While the splashing fight goes on they are getting trapped in the net. This is how the predators of the lake are caught (along with their victims). The catch is so important that at least two countries bordering the lake, Burundi and Zambia, are now planning to export to foreign markets frozen or smoked lake perches.

Needless to say,

A juvenile *Lates mariae*. Photo by Dr. Herbert R. Axelrod.

Lates microlepis with a close-up of its head. Photos by G. S. Axelrod.

centropomids are not aquarium fishes, but research into the possibility of their being introduced into foreign waters might prove valuable. All over Africa, *Lates* flesh is considered as the very best and fetches very high prices.

Lates apparently breed close to the shores, although spawning has never been observed, since many young *Lates,* often in schools, are discovered along the rocky shoreline hidden in small patches of reed stems. The size of the fry increases with depth and it is not rare to find several dozen fish 25 to 30 cm in a small school 10 meters or a little more deep. The density of the schools and the size of the fish might vary considerably, but it appears probable from the data available that the schools are made up from specimens of a rather uniform size.

Family TETRAODONTIDAE

Pufferfishes are represented in the lake by a single species of Congolese origin— *Tetraodon mbu.* The species is localized in the Malagarazi River and is found nowhere else around the lake.

As there are no *Tetraodon* in Lake Victoria and no *T. mbu* in the Nile, one is led to believe that the fish came in from the upper Congo. As it is not capable of swimming against strong currents, one has to admit that it reached the upper reaches of the Malagarazi before the river was cut off from the Congo by the lake then in the process of being built, and that the lower course of this river was then lying in lowlands without any rapids or falls. Had the fish entered the lake basin quite recently it would have had to climb falls and rapids and would be found around the Lukuga River outlet. *Tetraodon mbu* is thus a species several million years old.

T. *mbu* is the largest of the African freshwater puffers. I have seen one about one meter long and 30cm across with a weight of about 10kg. It is of course seldom that one discovers such a huge *T. mbu*, but just as difficult to catch a specimen less than 10cm long, the average size being between 15 and 30 cm. It is the largest of the four Congo puffers, the other three being *T. miurus, T. schoutedeni,* and *T. duboisi.* Not one of these species should be put into a tank along with other fishes. They are carnivorous and predatory and although they feed on molluscs (oysters, mussels, snails, and bivalves) they will attack any fish.

They came by their name (meaning "four toothed")

because of their parrot-shaped dentition, consisting of two teeth fused together on each jaw. With these teeth they are capable of breaking open a mollusc shell so that they can gobble up the flesh. The crushing strength of their jaws is enormous, as I could see when I tested a 25cm long *T. mbu* with a piece of hardened leather 5mm thick. One bite was enough to punch a hole in

The really large species of Nile perches, such as *Lates niloticus* (above) and *Lates calcarifer* (below) do not occur in Lake Tanganyika. Photo above by Dr. Herbert R. Axelrod, that below by Ken Lucas, Steinhart Aquarium.

the leather! Thus pufferfishes have to be handled with care— a 30cm puffer can crush a finger. As for their predatory habits, I learned about them the hard way at the start of my professional collecting career. I had stored 10,000 *Phenacogrammus interruptus* and about 50 *T. schoutedeni* 4 to 6cm long in a large basin. Every day I had to remove two or three hundred badly mutilated tetra corpses. I thought they had been attacked by the puffers after their death, thinking that the slow-moving puffers couldn't catch a swift tetra even if they tried very hard. I realized after a couple of days that many of the *Phenacogrammus* were moving about with here and there a piece missing from their fins, and then started to pay a bit more attendion to what was going on.

The balloon-shaped puffers could approach their prey even when facing them without awakening their suspicions because puffers can move so quietly and without the sudden jerks of power displayed by other predators. As soon as they were at close range the puffer bit and a piece of tetra was gone. More often than not they attacked the eyes, which explained why so many tetras had been beheaded. The damage inflicted by the puffers was so widespread because they never bit a fish twice. They always went for another.

Little is known about the breeding habits of African pufferfishes aside from the fact that they breed seasonally.

To my knowledge they have few enemies, which might be due to the fact that their bile is toxic and lethal even at only a small dosage. They should thus thrive and proliferate— which they do not. All puffers are rather common in their river habitat, only *T. duboisi* being exceedingly rare (I caught about 12 in my 20 years on the Congo River). Perhaps *T. duboisi* is a natural hybrid between *T. miurus* and *T. schoutedeni* ?

Although being common in the river, puffers represent only a very small fraction of the fauna. This might be due to sporadic epidemics which, every few years, kill them by the thousands during the dry season. The fact that *T. mbu* could find plenty of food in the lake in the form of the many species of bivalves, mussels, and snails, but hasn't colonized the lake, might be attributed to the properties of the lake water, which might not be fit for the puffers.

CONCLUSIONS

If we divide the various families of non-cichlid fishes according to their habitats we discover a few interesting facts. Cyprinids are the dominant family everywhere except in the lake itself, but especially in small brooks and

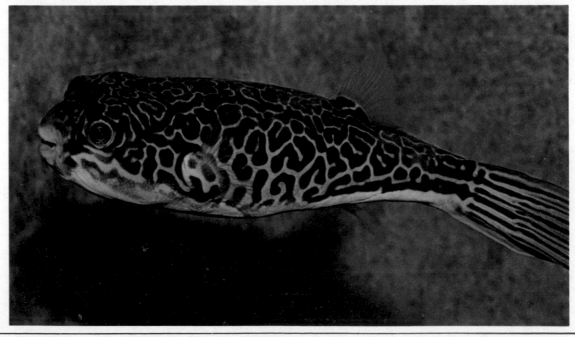

The brightly colored puffer *Tetraodon mbu* is quite rare near Lake Tanganyika and seldom collected. Photo by J. Vierke.

At a length of about one meter and a weight of 10 kilograms, *Tetraodon mbu* is one of the larger freshwater fishes in Africa. Photo by E. C. Taylor.

mountain torrents. In these habitats the large species are missing, probably because food is scarce and that which is there cannot sustain large fishes. Lower on the slopes water heats up, rivers become more important, floods are more progressive, plankton becomes more abundant, and the density of fishes as well as their variety increases.

The big rivers, aside from housing the large cyprinids, are also the realm of bagrids, lungfishes *(Protopterus),* as well as characoids and mormyrids. In the lake mochokids and *Mastacembelus* together involve 20 species out of 55 non-cichlids, but the biomass of non-cichlid families is heavily weighted in favor of the centropomids, the so-called Nile perches, and the clupeids. Both of these fishes have

There is a possibility that *Tetraodon duboisi* (below) is a natural hybrid between *Tetraodon mbu* and *T. schoutedeni* (above). Photos by Pierre Brichard.

Key to the Identification of the Species of *Lates* of the Lake Tanganyika Basin (after Poll)

1. Body depth included 4.5-5.0 times in standard length .... *stappersi*

   Body depth included less than 4 times in standard length 2

2. Anal fin truncated; caudal fin crescent-shaped or at least concave; eye small, 3.6-6.4 (adult) in head; interorbital space 5.5-7.5 in head; gill-rakers 12-15 .... *microlepis*

   Anal fin rounded; caudal fin rounded or straight; interorbital space 7-10 in head; not more than 10 gill rakers 3

3. Eye very large, 3.1-4.8 in head; D. VII I,11-13; in specimens over 25 cm the 8th spine is between the two dorsal fins; 95-120 scales in lateral line; caudal rounded or straight; 9 or 10 gill rakers...*mariae*

   Eye medium, 2.2-6.8 (adult) times in head; D. VII I,12 or 13; 88-99 scales in lateral line; caudal rounded; 5-8 gill rakers ...... *angustifrons*

colonized the pelagic waters. In coastal waters the catfishes and especially the spiny mastacembelid eels, have made life very difficult for the rock-dwelling cichlids, although it is as yet difficult to assess their impact on cichlid behavior. As we gain understanding of the ethology of non-cichlid fishes in the lake, we will also better understand their life in the murky waters of Central Africa.

An adult *Lates microlepis*, a commercially valuable species. Photo by Dr. Herbert R. Axelrod.

The stunning reddish mutation of *Lates mariae*. Photo by E. Balon.

502

N	Genus and Species	Orig. Descr.	Date	Habitat
	1. CICHLIDS:			
1	Asprotilapia * leptura *	Boulenger	1901	rock
2	Astatoreochromis straeleni *	(Poll)	1944	swamps
	vanderhorsti *	(Greenwood)	1954	rivers
4	Astatotilapia bloyeti	(Sauvage)	1883	swamps
	burtoni	(Günther)	1893	swamps
	paludinosa *	Greenwood	1980	river
	stappersi *	(Poll)	1943	swamps
1	Aulonocranus * dewindti *	(Boulenger)	1899	sand
7	Bathybates * fasciatus *	Boulenger	1901	Pelagic
	ferox *	Boulenger	1898	pelagic
	graueri *	Steindachner	1911	pelagic
	horni *	Steindachner	1911	pelagic
	leo *	Poll	1956	pelagic
	minor *	Boulenger	1906	pelagic
	vittatus *	Boulenger	1914	pelagic
1	Boulengerochromis * microlepis *	(Boulenger)	1899	sand
2	Callochromis * macrops macrops *	(Boulenger)	1898	sand
	macrops melanostigma *	(Boulenger)	1906	sand
	pleurospilus *	(Boulenger)	1906	sand
1	Cardiopharynx schoutedeni *	Poll	1942	sand
2	Chalinochromis * brichardi *	Poll	1974	rock
	popelini *	Brichard	MS	rock
1	Cunningtonia * longiventralis *	Boulenger	1906	sand/rock
1	Cyathopharynx * furcifer *	(Boulenger)	1898	sand/rock
1	Cyphotilapia * frontosa *	(Boulenger)	1906	rock
4	Cyprichromis * brieni *	Poll	1982	rock
	leptosoma *	(Boulenger)	1898	rock
	microlepidotus *	(Poll)	1956	rock
	nigripinnis *	(Boulenger)	1901	rock
1	Ectodus * descampsi *	Boulenger	1898	sand
1	Eretmodus * cyanostictus *	Boulenger	1898	rock
1	Grammatotria * lemairei *	Boulenger	1899	sand
3	Haplochromis benthicola *	Matthes	1962	rock
	horei *	(Günther)	1893	sand
	pfefferi *	(Boulenger)	1898	ubiquit.
2	Haplotaxodon * microlepis *	Boulenger	1906	pelagic
	tricoti *	Poll	1948	pelagic
1	Hemibates * stenosoma *	(Boulenger)	1901	pelagic
5	Julidochromis * dickfeldi *	Staeck	1975	rock
	marlieri *	Poll	1956	rock
	ornatus *	Boulenger	1898	rock

* denotes an endemic genus or species.

regani *	Poll	1942	rock
transcriptus *	Matthes	1959	rock
44 *Lamprologus brevis* *	Boulenger	1899	rock
brichardi *	Poll	1974	rock
buescheri *	Staeck	1982	rock
callipterus *	Boulenger	1906	ubiquitous
calvus *	Poll	1978	rock
caudopunctatus *	Poll	1978	rock
christyi *	Trewavas & Poll	1952	rock
compressiceps *	Boulenger	1898	rock
crassus *	Brichard	MS	rock
falcicula *	Brichard	MS	rock
fasciatus *	Boulenger	1898	rock
furcifer *	Boulenger	1898	rock
gracilis *	Brichard	MS	rock
hecqui *	Boulenger	1899	sand
kungweensis *	Poll	1956	?/shells
leleupi leleupi *	Poll	1956	rock
leleupi longior *	Staeck	1980	rock
leleupi melas *	Matthes	1959	rock
leloupi *	Poll	1948	?
lemairei *	Boulenger	1899	rock
meeli *	Poll	1948	?
modestus *	Boulenger	1898	Ubiquitous
mondabu *	Boulenger	1906	Ubiquitous
moorii *	Boulenger	1898	rock
multifasciatus *	Boulenger	1906	sand/shell
mustax *	Poll	1978	rock
niger *	Poll	1956	rock
obscurus *	Poll	1978	rock
ocellatus *	(Steindachner)	1909	sand/shell
olivaceous *	Brichard	MS	rock
ornatipinnis *	Poll	1949	rock/shell
petricola *	Poll	1949	rock
prochilus *	Bailey & Stewart	1977	rock
pulcher *	Poll	1949	rock
savoryi *	Poll	1949	rock
schreyeni *	Poll	1974	rock
sexfasciatus *	Trewavas & Poll	1952	rock
signatus *	Poll	1952	?
splendens *	Brichard	MS	rock
stappersi *	Pellegrin	1927	river
tetracanthus *	Boulenger	1899	sand

	toae *	Poll	1949	rock
	tretocephalus *	Boulenger	1899	rock
	wauthioni *	Poll	1949	?
6	*Lepidiolamprologus* * *attenuatus* *	(Steindachner)	1909	ubiquitous
	cunningtoni *	(Boulenger)	1906	ubiquitous
	elongatus *	(Boulenger)	1898	ubiquitous/rock
	kendalli *	(Poll)	1977	ubiquitous/rock
	pleuromaculatus *	(Trewavas & Poll)	1952	ubiquitous
	profundicola *	(Poll)	1949	rock
2	*Lepidochromis* * *christyi* *	(Trewavas)	1953	bottoms
	bellcrossi *	(Poll)	1976	bottoms
1	*Lestradea* * *perspicax perspicax* *	Poll	1943	sand
	perspicax stappersi *	Poll	1943	sand
4	*Limnochromis* * *abeelei* *	Poll	1949	bottom
	auritus *	(Boulenger)	1901	bottom
	permaxillaris *	(David)	1936	bottoms
	staneri *	Poll	1949	bottom
3	*Limnotilapia* * *dardennei* *	(Boulenger)	1899	ubiquitous
	loocki *	Poll	1949	?
	trematocephala *	(Boulenger)	1901	?
1	*Lobochilotes* * *labiatus* *	Boulenger	1898	rock
3	*Ophthalmotilapia* * *boops* *	(Boulenger)	1901	sand/rock
	nasutus *	(Poll & Matthes)	1962	rock
	ventralis ventralis *	(Boulenger)	1898	rock
	v. heterodontus *	(Poll & Matthes)	1962	rock
1	*Orthochromis malagaraziensis* *	(David)	1937	river
7	*Perissodus* * *eccentricus* *	Liem & Stewart	1976	pelagic
	elaviae *	(Poll)	1949	pelagic
	hecqui *	(Boulenger)	1899	pelagic
	microlepis *	Boulenger	1898	rock
	multidentatus *	(Poll)	1952	pelagic
	paradoxus *	(Boulenger)	1898	pelagic
	straeleni *	(Poll)	1948	rock
7	*Petrochromis* * *ephippium* *	Brichard	MS	rock
	famula *	Matthes & Trewavas	1960	rock
	fasciolatus *	Boulenger	1914	rock
	macrognathus *	Yamaoka	1983	rock
	orthognathus *	Matthes	1959	rock
	polyodon *	Boulenger	1898	rock
	trewavasae *	Poll	1948	rock
1	*Pseudosimochromis* * *curvifrons* *	(Poll)	1942	rock
2	*Reganochromis* * *calliurum* *	(Boulenger)	1901	sand

	centropomoides *	Bailey & Stewart	1976	?
5	Simochromis * babaulti *	Pellegrin	1927	rock
	diagramma *	(Günther)	1893	rock
	margaretae *	Axelrod & Harrison	1978	rock
	marginatus *	Poll	1956	rock
	pleurospilus *	Nelissen	1978	rock
2	Spathodus * erythrodon *	Boulenger	1900	rock
	marlieri *	Poll	1950	rock
1	Tangachromis * dhanisi *	(Poll)	1949	bottom
1	Tanganicodus * irsacae *	Poll	1950	rock
5	Telmatochromis * bifrenatus *	Myers	1946	rock
	burgeoni *	Poll	1942	rock
	dhonti *	(Boulenger)	1919	rock
	temporalis *	Boulenger	1898	rock
	vittatus *	Boulenger	1898	rock
4	Tilapia karomo *	Poll	1948	river
	nilotica	(Linné)	1758	swamp
	rendalli	Boulenger	1896	swamp
	tanganicae *	Günther	1893	ubiquitous
8	Trematocara * caparti *	Poll	1948	bottoms
	kufferathi *	Poll	1948	bottoms
	macrostoma *	Poll	1952	bottoms
	marginatum *	Boulenger	1899	bottoms
	nigrifrons *	Boulenger	1906	bottoms
	stigmaticum *	Poll	1943	bottoms
	unimaculatum *	Boulenger	1901	bottoms
	variabile *	Poll	1952	bottoms
1	Triglachromis * otostigma *	(Regan)	1920	mud bottom
5	Tropheus * annectens *	Boulenger	1902	rock
	brichardi *	Nelissen & Thys	1975	rock
	duboisi *	Marlier	1959	rock
	moorii kasabae *	Nelissen & Thys	1977	rock
	moorii moorii *	(Boulenger)	1898	rock
	polli *	G. Axelrod	1977	rock
1	Tylochromis polylepis *	(Boulenger)	1900	swamp
10	Xenotilapia * boulengeri *	(Poll)	1942	sand
	caudafasciata *	Poll	1951	sand
	flavipinnis *	Poll	1985	sand
	longispinis burtoni *	Poll	1951	sand
	longispinis longispinis *	Poll	1951	sand
	melanogenys *	(Boulenger)	1898	sand
	nigrolabiata *	Poll	1951	sand

506

ochrogenys bathyphilus *	Poll	1956	sand
ochrogenys ochrogenys *	(Boulenger)	1914	sand
ornatipinnis *	Boulenger	1901	sand
sima *	Boulenger	1899	sand
spilopterus *	Poll & Stewart	1975	sand
tenuidentata *	Poll	1951	floor/sand

* denotes an endemic genus or species.

LIST OF SPECIES **as per 30 June, 1985***

2. *NON-CICHLIDS:*

N Genus and Species	Orig. Descr.	Date	Habitat
LEPIDOSIRENIDAE			
1 Protopterus aethiopicus	Heckel	1851	swamps
POLYPTERIDAE			
2 Polypterus endlicheri congicus	Boulenger	1898	swamps
ornatipinnis	Boulenger	1902	swamps
CLUPEIDAE			
1 Limnothrissa * miodon *	(Boulenger)	1906	pelagic
1 Stolothrissa * tanganicae *	Regan	1917	pelagic
MORMYRIDAE			
1 Mormyrops deliciosus	(Leach)	1818	rivers
1 Pollimyrus nigricans	(Boulenger)	1906	swamp/river
1 Hippopotamyrus discorhynchus	(Peters)	1852	swamp/rock
1 Gnathonemus longibarbis	(Hilgendorf)	1888	swamp/river
1 Marcusenius stanleyanus	(Boulenger)	1897	swamp/river
1 Mormyrus longirostris	Peters	1852	swamp/river
KNERIIDAE			
1 Kneria wittei	Poll	1944	torrents
ALESTIDAE			
2 Hydrocynus vittatus	(Castelnau)	1861	pelagic
goliath	(Boulenger)	1898	pelagic
3 Alestes imberi	Peters	1852	coastal
macrophthalmus	Günther	1867	coastal
rhodopleura	Boulenger	1906	coastal
1 Bryconaethiops boulengeri	Pellegrin	1900	rivers
1 Micralestes stormsi *	Boulenger	1902	Lukuga
CITHARINIDAE			
3 Distichodus fasciolatus	Boulenger	1898	Lukuga
maculatus	Boulenger	1898	Malagarazi
sexfasciatus	Boulenger	1897	coastal
1 Citharinus gibbosus	Boulenger	1899	swamps
CYPRINIDAE			
18 Barbus altianalis	Boulenger	1900	rivers
apleurogramma	Boulenger	1911	rivers
caudovittatus	Boulenger	1902	typical Congo River spec.
eutaenia	Boulenger	1904	rivers
kerstenii	Peters	1868	rivers
lineomaculatus	Boulenger	1903	rivers
lufukiensis *	Boulenger	1917	Lufuku River
minchini	Boulenger	1906	rivers
miolepis	Boulenger	1902	rivers

* denotes an endemic genus or species.

	nicholsi	Vinciguerra	1928	Lukuga
	oligogrammus *	David	1937	rivers
	paludinosus	Peters	1852	rivers
	pellegrini	Poll	1939	rivers
	serrifer	Boulenger	1900	rivers
	taeniopleura *	Boulenger	1917	rivers
	tropidolepis *	Boulenger	1900	rivers
	urostigma *	Boulenger	1917	rivers
	urundensis *	David	1937	rivers
4	*Varicorhinus leleupanus* *	Matthès	1962	coastal/river
	ruandae	Pappenheim & Boul.	1914	rivers
	stappersi *	Boulenger	1917	coastal/river
	tanganicae *	Boulenger	1905	coastal/river
6	*Labeo cylindricus*	Peters	1852	coastal/river
	dhonti *	Boulenger	1919	Lukuga
	fuelleborni	Hilg. & Papp.	1903	rivers
	kibimbi *	Poll	1949	rivers
	lineatus	Boulenger	1898	rivers
	velifer	Boulenger	1898	Lukuga
5	*Barilus moorii*	Boulenger	1900	river estuary
	neavii *	Boulenger	1907	coastal
	salmolucius	Nich. & Grisc.	1917	river estuary
	tanganicae *	Boulenger	1900	coastal
	ubangensis	Pellegrin	1901	river estuary
2	*Engraulicypris congicus*	Nich. & Gris	1917	Lukuga
	minutus *	(Boulenger)	1906	pelagic

BAGRIDAE:

1	*Bagrus docmak*	(Forsskal)	1775	estuaries
1	*Bathybagrus tetranema* *	Bailey & Stewart	1984	?
6	*Chrysichthys grandis* *	Boulenger	1917	ubiquitous
	graueri *	Steindachner	1911	ubiquitous
	brachynema *	Boulenger	1900	ubiquitous
	platycephalus *	Worthington & Ricardo	1936	ubiquitous
	sianenna *	Boulenger	1906	deep bottoms
	stappersii *	Boulenger	1917	sand/mud
1	*Lophiobagrus* * *aquilus* *	Bailey & Stewart	1984	rock
	asperispinis *	Bailey & Stewart	1984	rock
	brevispinis *	Bailey & Stewart	1984	rock
	cyclurus *	(Worth. & Rich.)	1937	rock
2	*Phyllonemus* * *filinemus* *	Worth. & Rich.	1937	rock
	typus *	Boulenger	1906	rock
1	*Auchenoglanis occidentalis*	(Valenciennes)	1840	ubiquitous
1	*Leptoglanis brevis*	Boulenger	1915	torrents

MOCHOKIDAE

7	*Synodontis dhonti* *	Boulenger	1917	rock
	polli *	Gosse	1982	rock
	granulosus *	Boulenger	1900	rock
	lacustricolus *	Poll	1953	rock
	multipunctatus *	Boulenger	1898	rock
	nigromaculatus	Boulenger	1905	swamps
	petricola *	Matthès	1959	rock
1	*Chiloglanis lukugae*	Poll	1944	torrents
	pojeri	Poll	1944	torrents

AMPHILIIDAE

2	*Amphilius platychir*	(Günther)	1864	torrents
	kivuensis	Pellegrin	1933	torrents

CLARIIDAE

1	*Heterobranchus longifilis*	(Valenciennes)	1840	ubiquitous
1	*Dinotopterus cunningtoni*	Boulenger	1906	rock
4	*Clarias liocephalus*	Boulenger	1898	swamps
	mossambicus	Peters	1852	swamps
	ornatus *	Poll	1943	swamps
	theodorae	Weber	1897	mud/coast
1	*Tanganikallabes* * *mortiauxi* *	Poll	1943	rock

MALAPTERURIDAE

1	*Malapterurus electricus*	(Gmelin)	1789	ubiquitous

CYPRINODONTIDAE

1	*Aplocheilichthys pumilus*	(Boulenger)	1906	swamps
1	*Lamprichthys* * *tanganicanus* *	(Boulenger)	1898	open/coast

CENTROPOMIDAE

4	*Lates (Luciolates) angustifrons* *	Boulenger	1906	open
	mariae *	Steindachner	1909	open
	microlepis *	Boulenger	1898	open
	stappersi *	Boulenger	1914	open

ANABANTIDAE

1	*Ctenopoma muriei*	(Boulenger)	1906	swamps

MASTACEMBELIDAE

12	*Mastacembelus albomaculatus* *	Poll	1953	rock
	cunningtoni *	Boulenger	1906	sand
	ellipsifer *	Boulenger	1899	rock
	flavidus *	Matthès	1962	rock
	frenatus *	Boulenger	1901	rock/rivers
	micropectus *	Matthès	1962	rock
	moorii *	Boulenger	1898	rock
	ophidium *	Günther	1893	sand/rock
	plagiostomus *	Matthès	1962	rock
	platysoma *	Poll & Matthès	1962	rock

tanganicae *	Günther	1893	rock
zebratus *	Matthès	1962	rock
TETRADONTIDAE			
1 *Tetraodon mbu*	Boulenger	1899	estuary (Malagarazi)

Chart of Geological, Climatic, and Ecological Features Leading to a Differentiation Between the Southern and Northern Biotopes of the Lake.

Features	Northern biotopes	Southern biotopes
Shores and floors	mostly magmatic or volcanic rock	mostly sedimentary or magmatic rock
Average depth	200m and silted	400m
Rainy season length	nearly 8 months	between 4.5 and 5 months
Distribution of rains	nearly 7 months	concentrated in 3.5 months
Hydrographic basin	2 main rivers, about 100 brooks and torrents	1 main river, about 30 affluents
Plankton bloom start	from mid-October	from end of December
Plankton density	very high	average
Dry season drop in temperature of water (surface layer)	1-2°C at most	2-3°C
Lateral visibility	down to 1m	not less than 4m
Maximum vertical visibility	at best 10 m	up to 20m
Dominant winds	SSE	SSE (rainy season); E (dry season)
Biocover	down to approx. 10m; rather thin	down to minimum of 15m; very lush
Sponges	dense and varied	scarce, 1 or 2 species common
Jellyfish	year-round and often in dense clouds	more seasonal and scattered
Leeches	very rare	very common
Endoparasites on rock-grazing cichlids	unheard of	50% of *Tropheus* infected

CHAPTER XII

A FEW HINTS TO THE HOBBYIST

As Dr. Herbert R. Axelrod discovered when we first met in 1956, I am not a very good hobbyist. I was much too busy starting my ornamental fish export business and much too preoccupied by research in the field to devote much time to the fun of watching the fishes behave in my tanks. At one time I was collecting, acclimatizing, and shipping 10,000 fish a week.

But if I lack finesse in the setting up of a tank, there is one point about which I can be of some help—how to get the Lake Tanganyika cichlids to feel more at home in captivity. I know what they do in the lake and what they need in order to be happy. This is especially important with Rift Lake cichlids because more than any other fish I know of, they have reached an exceptional level of sophisticated behavior. Most are certainly not dumb, and a few have already been recognized as being capable of unexpected individual initiatives. What the Tanganyika cichlids often lack in gaudy colors (compared to the Malawi species) they more than make up for with an astounding array of specific behaviors.

Back in 1972, after one year spent collecting and observing the lake fishes, my daughter and I agreed that, after so many years spent collecting in the Congo River basin, the lake fauna was rather boring. "Always cichlids," we used to complain. Life as a fish collector on the lake was rather dull. Now, years later, after thousands of hours spent underwater, our precocious boredom has given way to awe. Day after day *our* cichlids provide us with new clues to their incredibly rich diversity.

Here and there in this book I have tried to share with you some of our most fascinating discoveries, like the inflatable egg-dummies of *Simochromis,* the various nest-building activities of *Callochromis macrops melanostigma,* the topographical memory of several sedentary rock-dwellers, the sequential spawns of *Julidochromis,* and so many other unusual behaviors. None of these are useless, for each of them help the fish to fit into the immensely complex panorama of life on a submerged slope.

In several cases one gets the feeling that some fishes do not display only behavior inherited through their genes, but that we are at a boundary, for us still very hazy, between what is pure instinct and the first manifestations of what might be called intelligent behavior. This is the capacity for an individual fish to change its ways when confronted with a situation that the species hadn't met before.

With fishes like these you must understand that the setup of your tank will have an immediate impact on their behavior. They may remain more or less what they were in the wild, or they may become distorted, exacerbated, or muted. Cichlids are *pliable* animals; they will adapt, they will survive, but even they can stand only so much.

So I am not going to talk much about the chemistry of your water—important as it might be. Many publications have already dealt with this problem. Taking it more or less for granted, I will instead deal more at leisure with the social relationships that will develop according to the tank layout you decide on for your tank.

Water chemistry

With a pH between 8.0 and 9.0, DH between 10 and 14°, conductivity between 500 and 600 µS, and temperature between 24 and 26°C, your Tanganyika cichlids should be happy and in good condition. You must avoid the buildup of ammonia at all costs, so your filtration should be very good, and an occasional change of 1⁄3 to 1⁄4 of the water approximately twice a month in a large tank, more often in a small one, will help. Avoid also the buildup of gases in the gravel. Your cichlids will be able to tolerate any deficiency better in one of the above-mentioned water parameters if the other qualities of the water are good. This means that it is a combination of these deficiencies that is the main cause of losses in your tank.

We have kept *Tropheus* in very murky waters with a high level of ammonia without trouble in a 100 m³ pond, but the simple addition of a small bunch of *Elodea canadensis* killed them all overnight. With proper pH and DH I have also kept *Tropheus* in water that dropped below 15°C at night.

To fill your tank initially you can use tap water and try to adjust the pH and DH with chemicals. Or you can start with rain water or distilled water and use the so-called Tanganyika salts available in the trade or simply marine mixes (without the sodium chloride of course). Anyhow, you can not duplicate exactly the lake water chemistry, especially with regard to the magnesium carbonate and trace elements, which might be insoluble. Chlorine and chloramines are deadly for the lake fishes and should be eliminated from your water before you start putting fishes into your tank.

Using a proper type of gravel will help stabilize the pH and DH of your tank. Coral sand has been used but dolomitic gravel is probably the best medium both in the tank and in the filter. Toxic wastes should be taken care of by an oversized filtration system, strong oxygenation, and periodic changes of water, as I said above. If you follow these rules your pets should be able to settle down to life in a confined space and be reasonably happy.

Now then, why do aquarists have some trouble with their Rift Lake cichlids? I see two main reasons why their fishes do not adjust: 1. The type of food that is given to the fishes is not sufficient for them to live on. 2. The socialization with other fishes leads to unbearable stresses. These two reasons can be compounded of course when the water in your tank is not exactly suited to their requirements.

Feeding

The basic food of most species in the lake consists of a variety of tiny shellfish (daphnia, cyclops, and shrimp) moving about in the plankton clouds or on the surface of the substrate whether it is rock or sand. There are practically no worms or insect larvae in the lake. Some fishes are vegetarians, but most of them do not feed on aquatic plants, but on algae belonging to the phytoplankton or the biocover. Many are more or less predatory, especially on eggs and fry, and some will dispatch the remains of dead fishes.

From this you should be able to figure out what is wrong with the diet you impose on your pets. Probably you are giving them too much starch and fats. Lake cichlids will gorge themselves on tubifex worms, bloodworms, beef heart, etc. but this diet is definitely not the best for their ultimate health. *Tropheus* feed on fatty meats, develop liver troubles and eventually die.

Most cichlids will take to flake food, but this diet should be complemented by some live or frozen foods, including shrimp and crab meat, fish eggs, etc. Vegetarian fishes such as the rock-grazers should not be fed only plant flakes, as in nature the biocover they feed upon hides many invertebrates that are swallowed along with the algae.

Feeding should be often, as the fishes in the lake spend most of their active hours eating. More, of course, should be given in the early morning and at sunset. Feeding captive fish only once a day makes them hungry, and when they are hungry they get into a bad temper and become aggressive. Feeding them often (three or four times a day) helps to keep the tank in good harmony.

Cohabitation

Very few hobbyists have a sound approach to the problem of which species they are going to put into a tank. Some of them will try to stuff the tank with as many species as they can lay their hands on and that the tank can possibly hold. Some aquarists on the other hand specialize in one type of cichlids, like the rock-grazers, and fill their tank with several *Tropheus* varieties, a few *Petrochromis,* and perhaps a *Simochromis.* Then they start to wonder why the fishes kill each other or get emaciated and die.

Many fishes live together on a slope, and their density can reach incredible levels. Dozens of species and tens of thousands of fishes might crowd a single slope. This doesn't mean that you can put much more than two dozen fishes in a 500 liter tank and still hope for the best. Try as you might your tank will never have the combination of features that makes life tolerable for a crowd such as the one you find on a slope. These features are:

1. A volume of water that buffers changes in chemistry and allows the fish to move about and do what they wish with a minimum of disruption by other species.
2. A very wide variety of shelters from which the fish can pick their own territory. Even if there is such a territory every 30cm, two territories of individuals belonging to the same species can still be far apart.
3. A wide choice of foods available more or less at all times from which each fish can pick what it wants most often without being in competition with another fish. With none of them starved or hungry, feeding doesn't develop into a general commotion ending in a savage rush in which some fish are always the losers.

4. A very wide array of ecological niches, so that each species can live without too many contacts with competing or antagonistic species.

None of these features can be duplicated in a tank due to lack of space for appropriate shelters, lack of food, and because the fishes must live permanently in contact with each other when they would perhaps have avoided contact (and could do so) in the lake. In a tank promiscuity is permanent, which is why it is so important to pick your pets properly, instead of according to your preferences. Too many aquarists force their selection of pets to live together and then wonder why they fight.

As you have gone through this book you know already that there are mistakes one should avoid. If you put together fishes that have the same habits and the same way of feeding, you know that you are heading for trouble. *Petrochromis* and *Tropheus* compete in the lake and sometimes fight each other. In a tank this aggressiveness will be exacerbated, and as *Petrochromis* are bulkier the *Tropheus* will be the ones to go. Two *Tropheus* males will also fight when there is a female around. If you are not interested in breeding *Tropheus* you can keep several together and have them behave if you pick only males. As soon as you put a female into the tank, there you go. You will wind up with only one male left. So it is better to have rock-grazers with nestbreeders or plankton feeders in a medium-sized tank. They do not compete with each other in the lake and never quarrel. They stand

excellent chances to adapt to each other in captivity. Similarly, don't put two pairs of *Julidochromis* into a tank if you can't give them well-separated territories.

If you wish to have your pets duplicate in your tank what they do in the wild, know their behavior. If a fish is mobile it needs room; if sedentary a shelter; if polygamous provide a male with several females, otherwise the bride will not be heard if she claims "to have a headache." If a species is gregarious to the point of *Lamprologus brichardi,* why secure only one or a single pair that will not display the rich social life typical of the species?

The aquarist will also make mistakes in the setup of the landscape in his tank. It is much better to distribute the rock piles around the tank, separating them by bare sand patches, than to build one single huge pile, so that the fish living in the rocks can more or less be isolated from each other. The size of the shelters, caves, and openings should be such that they can accommodate fishes of various sizes. The smaller species should be able to retreat into recesses where they are free from intrusions and live in peace.

Once the fishes get acquainted, they will adjust and hierarchies will develop and be accepted by all. In the wild these hierarchies do not develop so often and don't reach the same intensity, because the fishes can always avoid a bully. In a tank they cannot, and hierarchies are the safety valves preventing death. But they put a heavy stress on some of the fishes at the lower rung of the ladder. If you add a new specimen to the roster, by the time he has adjusted to his quarters he has been bullied so much that he often cannot stand the stress and dies. If he survives, which means that he fights back, the entire hierarchy might be upset and some other fish might not adapt to the new situation.

The thoughtful aquarist should remember this problem when he wants to start a pair breeding. In some pairs the male is the boss, in others the female has heard about women's lib. With many nestbreeding cichlids cooperation by the two partners is hard to come by. Usually one is ready to spawn and the other is just not in the mood yet. It will be bullied around and might even get killed. When one has ascertained which is the boss, take the weaker one out first, put it in another tank, get it acquainted with its new quarters and settled down for a few days, then bring in its mate. The newcomer in strange surroundings will have lost much of its combativity leaving time for its partner to get ready for sex.

Compatibility between fishes in the narrow confines of a tank is not easy to come by. The hobbyist should use common sense and draw from what he knows about the behavior of the fishes in the lake and their biotope.

I have told about the various coastal habitats of the lake, especially the sandy bottoms and the rock habitats with their respective specialized faunas. The aquarist will of course wish to have fishes from the two faunas in his tank, let's say sand-dwelling *Xenotilapia, Callochromis,* or *Reganochromis,* along with

some of the typical rock-dwellers like *Tropheus,* or the many nestbreeders, such as *Julidochromis, Lamprologus brichardi, L. compressiceps, L. leleupi,* etc.

This is not an easy task, as sand-dwelling cichlids need room to move about, which in nature means expansive sandy bottoms. One cannot wish to have members of the two faunas happy in cramped quarters. Sand-dwellers are gregarious, so one needs to have several of a kind if, again, one wants to enjoy their antics. If you wish to have a tank harboring two dozen rock-dwellers and as many sand-dwellers, plan a 500 liter tank as a minimum, with a bottom surface of about one meter square, at least half of it covered with sand.

On the other hand, species living in midwater, like *Cyathopharynx, Ophthalmotilapia,* or *Cyprichromis,* have lately become more common. These fishes need water free from any obstacle in which to move. Thus if you want to have such fishes you will need a tank providing headroom, at least 60 or 70cm above the sand bottom.

A high tank will provide the added profit of good landscaping, as you can plan small cliffs and ledges duplicating a typical rocky slope. The rocks you will use for landscaping and providing the fishes with shelters should be rounded and devoid of any sharp edges on which they might get scratched or bruised. It is hopeless to try to duplicate the appearance of a rock-strewn slope in a tank, but stones in shades of beige, brown, and green should help. Quartz, which is the basic

material of many coasts, also is excellent and takes a nice patina. The sand should be coarse, not too fine, and buff colored. If you wish to add plants, anchor them securely and choose them rather coarse. Although most of the Lake Tanganyika cichlids are not plant-eaters, some of them, like *Petrochromis* and *Tropheus* (to a lesser degree), will make up for the shortage of biocover by nibbling the leaves.

When you plan the layout of your tank think about the need for periodic cleanups. Sedentary cichlids are very sensitive to any change in their surroundings and the arrangement of their shelters. It has happened that a pair of *Julidochromis* living together in peace for several months started to fight and tried to kill each other after a change in the appearance of their shelter. Plan the landscape in your tank in such a way that the tank can be cleaned without removing the setup, and when you clean it do it cautiously.

If you have succeeded in choosing compatible fishes and the layout of your tank is proper you may see your pets start to spawn. Their population will start increasing slowly because many eggs and fry will fall prey to the other fishes in the tank. This is a normal situation that can be alleviated by providing excess shelters for the young fish. If you didn't crowd your tank from the start, you will now be able to reap the profits of your initial restraint. Your fishes, old and young, will live in harmony.

Lighting

I have very seldom seen a well-lit tank of Rift Lake cichlids. Most hobbyists follow the normal pattern—one or two fluorescent plant tubes providing a uniform flood of light over the whole tank. Little does this layout duplicate the atmosphere of a lake slope. Close to the surface the rocks are bathed in strong sunlight; sunbeams play their crisscross patterns on the bottom while caves and anfractuosities remain in darkness. Deeper down the sun shafts move about in the twilight of dull gray-green water, pervading the whole landscape with an aura of mystery. If you want to enhance your display of Rift Lake cichlids and make it really spectacular, cut out the fluorescent fixtures and put a few narrow spot–lights directly overhead.

Breeding

Wild-caught Lake Tanganyika cichlids are often expensive for reasons explained elsewhere. However, most species breed readily in a tank, providing the amateur with an opportunity to sell young fish and recover his initial investment. This is very good, because a lot of people who cannot afford the prices of adult imported fishes can get acquainted with the African cichlids and become more involved with these fascinating pets.

A very important bonus is the wealth of information amateur breeders gather about the breeding behavior and other features of fishes that had not yet attracted the attention of scientific circles. In this respect the amateurs have really done pioneering work on the ethology of many little-known species. Of course it is easy to stretch a good thing too far. Some amateur breeders have increased the scope of their activities to the point where they mass produce second-rate fishes or try to hybridize them and market artificial varieties.

Let's take *Tropheus* as an example. In nature *Tropheus* will probably breed not more than 5-6 times a year, with an average of 12 to 15 eggs laid at each spawn. With predators lurking about the female will delay the time when she will release and abandon her fry in the rock rubble so that they can survive. The full breeding cycle will then last about two months. In a tank with no predation to worry about, she most probably will cut buccal incubation by one or two weeks and release her fry when they are much smaller. She is thus ready for another spawn more often than in the wild and willingly or not she cannot escape the attentions of the males. Let us not forget that she doesn't live in a natural biotope and is thus not as healthy as she would be in the lake. She certainly suffers from some deficiencies although her sexual activity is increased. No wonder that if her offspring are often weak and of poor esthetic value.

This problem can get worse if, with a view to increasing productivity, an amateur removes the fertilized eggs from a female's mouth to incubate them without her help. She will then be ready for another spawn a few days afterward and might spawn 10 or 12 times a year. But the eggs will be smaller, less loaded with vital proteins, and the resulting fry will be hard put to survive in good health. The problem of degeneration is also compounded and the strain weakened if the small number of breeders used brings about inbreeding.

Hybridization of Tanganyika cichlids is something that should be avoided at all costs. Some amateurs want to improve on Nature's ways and try to start new strains, especially of *Tropheus,* by cross-breeding valuable races. They thus hope to bring about varieties with very gaudy colors that would yield a very nice profit. They don't realize the harm they could do.

If some of the lake cichlids used by scientists to study genetic problems are hybrids and not the pure natural genetic stock they think they are working on, how will they be able to research the genealogy of the lake fishes? Remember that Lake Tanganyika is probably the only lake on earth where the fishes, starting with a few ancestors, have evolved into so many lines, born at various times but still living together. Lake Tanganyika is to freshwater biology what the Grand Canyon of Arizona is to geology.

To disrupt the study of the genetic background of the lake fishes by hybridization, when so many species and local races remain to be discovered and studied, is totally irresponsible. Even if attempts at hybridization are not made with a view toward financial rewards but for a personal feeling of achievement, they are harmful. It will be difficult for an amateur to resist spreading a "successful" hybrid among his friends. So,

516

for God's sake don't hybridize! There are enough fish species and varieties in the lake to satisfy your wildest dreams and keep you busy with their behaviors.

Prices of Lake Tanganyika Cichlids

Hobbyists often complain about the prices they have to pay for many Rift Lake cichlids, but there are several good reasons why they cannot be cheap.

Collecting is very expensive in the middle of Africa. In countries where petrol has to be imported by trucks, collecting trips on a lake 450 miles long are made mostly by means of outboard-powered boats. Longer trips last several days, at the end of which you may bring back perhaps a thousand fishes at most, often much less. Moreover, most fishes are individually hand-picked during a dive. There can be only a few of these each day. Such a major trip can cost as much as $5,000-6,000! Proper acclimation is a slow and costly process that often doubles if not triples the cost of a fish. But this is money well spent as it guarantees that the fish will travel to the importer abroad in top condition. From the company I founded in Burundi consignments arrive in the hands of their clients with an average loss of less than 1%. This is why my customers agree that I am one of the very best fish suppliers in the whole world. Airline freight costs have quadrupled in 10 years time, although the airlines haven't improved that much on the service they offer.

The main reason for high prices is the much slower turnover experienced by all people concerned with the trade of African Rift Lake cichlids, compared with the one experienced with community tank fishes that sell briskly. Only 2 or 3 percent of all aquarists are interested in cichlids, but they are very keen and knowledgeable (in my opinion the best in the hobby), thus they are choosy. In short it means that they don't enter a petshop and just buy a "bunch" of fishes. They want pairs or trios and probably know more about the behavior of the fishes they buy than the dealer does. Any fish with the smallest defect will be left over. This situation is passed back down the line to the collector on the lake, who has to comply with very specific instructions. Now, imagine that you are a collector fishing 100 feet down for *Cyphotilapia,* trailing a huge net 50m long by 6m high, that you see a small group of the banded fish and try to get them into the net. Then you sort them out and from, let's say, perhaps 50 fish you release half of them back into the lake. After two days of decompression you now bring back only 20 or 25 fish!... How much do these remaining *Cyphotilapia* cost the collector after they are acclimatized? Suppose, also, that if by misfortune there are more males than females, you are then left with a few unsaleable fish on your hands. *Tropheus, Petrochromis, Julidochromis, Lamprologus*

compressiceps, L. calvus, L. leleupi, L. furcifer, etc....you name them...are all caught one by one. Only a few species are caught in small groups. But every single fish coming from the lake has been captured selectively with a method adapted to their behavior when hunted down. It is not by sheer luck that a fish is caught in a big haul with a large seine. It or the school had been sighted and forced into the net. Hence the cost.

Because prices are high every one in the trade feels a strong temptation to take things into his own hands. For the wholesaler this means starting his own collecting stations on the lakes and for a retailer to get into direct contact with the exporter thus getting rid of the importer-wholesaler. For the amateur this means going to the lake by himself during an African safari and collecting a few fishes while enjoying a splendid vacation with a taste of adventure *a la Stanley* and making perhaps a few dollars.

Such ventures seldom succeed. One doesn't improvise fish collecting nowadays without a solid background knowledge of Africa, nor import large quantities of fishes without a lot of money and experience. Wading in African waters can be dangerous...and I am not talking so much about the obvious snakes, insects, crocs, malaria, sleeping sickness, or bilharzia. Local people spending their days scratching out a minimal living from a hostile nature have of course some problem understanding that you are bent not on catching fishes you can eat,

517

but on the smaller ones that for them are useless. So you more often than not look suspicious. A few months ago an amateur was looking around for a few fishes to bring back home. He was poaching without any authorization, was mistaken for a spy or a mercenary, and eventually spent a couple of months in jail in a small remote city in the middle of Africa. Believe me, there are better ways to spend a vacation. African people are often very friendly and might go to great lengths to make you feel welcome...once they have learned to appreciate your qualities. If not, even a handful of dollar bills will not keep you out of trouble.

Hints to neophytes

Many amateurs ponder whether or not they are going to switch from their decorative, but boring, "community" to a tankful of Rift Lake cichlids. They have heard so many bad stories about cichlids—usually from retailers who don't handle them—that it's little wonder if they are very reticent. Let's leaf through the bad press.

1) The setup is expensive:

Cichlids need giant tanks and costly filtration. If you wish to entertain a school of *Cyphotilapia* and *Boulengerochromis* it's true you will need something in the range of the Steinhart Aquarium in San Francisco. The size of your tank will depend on the number of fishes and their size. But many species just don't need so much room, and keeping a pair of small-sized cichlids from Lake Tanganyika, such as the dwarf shell-dwelling *Lamprologus*, or *L. leleupi*, *Julidochromis transcriptus*,

Telmatochromis bifrenatus or *T. vittatus*, or *L. brichardi*, would not call for more than a 60-liter tank (20-gallon). You can probably keep two or three pairs happy and breeding in a 40-gallon tank.

It all depends on whether you pick sedentary and territorial fishes or nomads, like *Tropheus*, which need more elbow room. You cannot hope to keep a dozen medium-sized species in less than 250 liters (60 gallons) if you wish to avoid promiscuity. As for fishes that grow big, such as *Cyphotilapia* (35cm), *Lobochilotes* (45cm), and *Boulengerochromis* (in excess of 50cm), you will need a 1000-liter tank. But you can still have the cichlids and not go broke.

2) Cichlids are aggressive

I have already explained why cichlids can be aggressive in captivity. Most often it's because they are not happy and you are a bad host.

Lake Tanganyika cichlids are not especially aggressive, and what aggressiveness they display is quite acceptable. Seldom do they kill each other when antagonists are kept well apart in different shelters. The hierarchy that will be established after a short while will ensure that major fights will not occur. The dominated fishes will seek shelter, the dominant ones will be happy with a show of power and leave it at that. Seldom will you see a serious bite, except between partners starting prenuptial antics and one is not ready. You can avoid fights, as I said, by refraining from putting concurrent species in narrow confines. On the other hand, several small cichlids and even some large

ones (*Cyathopharynx* guild, *Cyprichromis*, and *Xenotilapia*, to give but a few examples) have never been seen in a fight and are quite peaceful.

I know of many other families of fishes that are quite intolerant of each other and of other fishes and that are kept in community tanks by the mainstream aquarist. Of course there are some species that should best be avoided, the typical examples of which may be found in the scale-rippers (*Perissodus*). Most fishes in the lake feed on microorganisms and are thus not predators by trade. Of course any cichlid will jump on fish-eggs or fry. But if you know of a single fish that wouldn't, tell me.

If you wish to keep aggressiveness in check, don't put fishes belonging to closely-related ecological niches together, such as *Tropheus* and *Petrochromis* or *Simochromis*.

Outright predators, such as *Lamprologus elongatus*, cruising in the open and capable of swallowing a smaller fish, are not necessarily trouble in a tank. They will not bother any fish they cannot eat whole. They are much less of a nuisance than is commonly thought.

3) Cichlids ruin plants:

This generalization is stupid. Some cichlids, being vegetarians, of course eat plants, but there are very few of them in the lake and, as I stressed, most Tanganyika cichlids live on zooplankton or zoobiocover. I have never seen any *Lamprologus*, *Julidochromis*, *Chalinochromis*, or *Telmatochromis* nibble at plants in the lake. Most mouthbrooder's don't either, especially all those that pick up their food in midwater or dig in the sand.

Even in areas where there are broad-leafed *Vallisneria* growing between the rubble you don't find their stems shortened by grazing. Systematically browsing fishes go after the vegetal biocover and leave aquatic plants alone. Only *Tilapia nilotica* and *Limnotilapia dardennei* are plant eaters.

Cichlids can uproot plants when they dig a crater nest, which is of course a nuisance. Now don't tell me that you can't find a way to anchor your plants securely so that they cannot be uprooted. One way is to plant them between two rocks in a place fishes won't pick for a nest site. You might expect that if you keep some rock-grazers where there is a shortage of edible biocover they might attack your plants. Soft plants like *Cabomba, Myriophyllum,* amazons, etc. should be left out and preference given to plants having harder leaves.

4) Rift Lakes cichlids are very expensive:

We have seen why they can indeed be very costly, especially the rare or newly found varieties. But these fishes are not automatically the most appealing, if one forgets about outdoing the Joneses next door.

Many Lake Tanganyika cichlids are not expensive. Even some very good varieties of *Tropheus* are rather medium-priced if one sticks to one rule—don't buy fully grown specimens of the largest species. Buy half-grown or juvenile wild-caught specimens. They offer several good points. First they

acclimatize better because they are more adaptable than adult, perhaps old, fishes having spent years in the lake ecosystem. Second, their fertility is probably better and will last longer. Third, they will be much cheaper to start with. Not that the exporter sells large specimens at a higher price, but simply because of . killing freight costs. One example will illustrate this point. *Cyphotilapia* sent from Bujumbura, Burundi to New York:

Size	Number per box	Freight per fish
30cm	2	US $20.00
20cm	6	7.00
15cm	10	4.00
12cm	15	2.50
6cm	40	1.00

If you still find the price of Tanganyika cichlids very high, remember that basically they are hardy fishes. Once established in a well-managed tank they will live for years. Most cichlids in the lake appear to have an excellent longevity, at least 5 years, and there are some specimens in our tanks that have been with us for 12 years.

If you buy young fishes you know that you will enjoy them for many years to come. But don't buy them too young. I would advise against buying barely free-swimming fry, as they are still very delicate. Rock-grazers should be between 2.5 and 3cm long (1"-1¼"), not smaller; nestbreeders should be between 3 and 4cm. By that time a prospective buyer should be able to check on

eventual defects that might still be hidden in smaller fry. Tanganyika cichlids, provided they have been well treated against diseases and parasites by exporters or raised by good breeders selling reliable products, are not very sensitive to infections and recuperate very well from wounds. Special care should be given to infection by Ich. An outbreak can be serious in fishes that have not built up any immunity against a disease that doesn't exist in their lacustrine biotope.

I will end this chapter with a list of the species with which a neophyte amateur could start his initiation with the Lake Tanganyika cichlid. The list is divided according to the biotopes the fishes come from, warm water river affluents, cold water mountain torrents, or the lake itself.

1. Warm rivers (more than 20°C, pH around 7.0, DH between 3 and 10°): *Astatoreochromis straeleni, Astatotilapia bloyeti, A. burtoni; Aplocheilus pumilus* (an excellent and hardy cyprinodont).
2. Cold mountain torrents (less than 20°C, pH around 7.0, DH between 3° and (6°): *Orthochromis malagaraziensis, Barbus* sp., kneriids, *Amphilius.*

3. Lacustrine biotopes (temperature between 24 and 27°C, pH 8.0-9.0, DH 12°-14°): *Ectodus, Eretmodus, Chalinochromis, Julidochromis, Lamprologus* spp. (*brichardi, brevis, buescheri, caudopunctatus, wauthioni, fasciatus, leleupi, moorii, niger, modestus, mondabu, multifasciatus, "magarae", ocellatus, savoryi, sexfasciatus, tretocephalus, tetracanthus, toae*), *Telmatochromis* (all), *Triglachromis, Chrysichthys sianenna, Synodontis* (all), *Phyllonemus, Lophiobagrus*.

-Needing more room: the large *Lamprologus, Lepidiolamprologus, Tilapia, Limnotilapia, Cyphotilapia*, and *Lobochilotes*.

-Should be considered as rather delicate and left to experienced aquarists: *Cyprichromis, Cyathopharynx, Ophthalmotilapia, Cunningtonia, Asprotilapia, Aulonocranus, Reganochromis, Trematocara*, most *Xenotilapia* (except for *flavipinnis*), as well as a few *Lamprologus* species (among them *compressiceps, calvus*, and *prochilus*), and among non-cichlids *Lamprichthys* and *Mastacembelus*.

Pelagic species of cichlids and non-cichlids alike appear poorly adapted to surviving in captivity, at least as adults and under the conditions existing in amateur's tanks. The same can be said of the many large or very large non-cichlid fishes roaming over the lake bottom. Anyhow, there are more than enough species coming from the lake, with many more still to be discovered, to keep you happy for a very long time.

Ovophagous or Egg-eating Guild

Lake Victoria harbors at least one haplochromine ovophagous (egg-eating) predator. We at first started looking among the mouthbrooding cichlids of Lake Tanganyika to see if we could detect the lock-jaw technique first (I think) described by Greenwood — but found none. Several specialized egg-eaters, however, did exist under our very noses and it was only by piecing together many accidental observations and concentrating on this phenomenon that we could determine that several species of *Telmatochromis* had developed as ovophagous predators. So far *T. temporalis, T. dhonti*, and *T. bifrenatus* have been implicated. As far as the other two species of *Telmatochromis* are concerned, the case is unclear due to lack of consistent observations, but very probably they behave similarly in their home waters (around Kalemie on the west coast for *T. burgeoni* and on the southern coast for *T. vittatus*). The three species for which data are available apparently do not prey on the spawns of mouthbrooders but on spawns of their fellow nestbreeders.

The behavior of *Telmatochromis temporalis* and *T. dhonti* strongly suggest that they specialize in annihilating the spawns of *Lamprichthys tanganicanus*, the sole cyprinodont in the lake which, itself, preys on juvenile fishes living above the bottom or for one reason or another swim in midwater.

Lamprichthys is a giant of a killie, reaching a length of 15 cm (males), and spawns in pairs on rock ledges or corners of the rubble. The female lays what appears to be a considerable number of eggs. How many is difficult to determine as they are crystal clear. Only by the closest scrutiny with a magnifying glass on the effect they have on the rock underneath allows the observer to make them out. The place to look is determined only by observing where the pair were seen vibrating side by side on a rock. The eggs are probably just as difficult to detect by any predator that is not on the spot when spawning occurs.

It is in this respect that one can say that *T. temporalis* and perhaps less so *T. dhonti* are specialized. Any time a pair of *Lamprichthys* descends from midwater where the male has been courting the female and chasing other males away and heads directly for a spot in the rubble that he had previously selected, all *T. temporalis* and *T. dhonti* within a radius of two to three meters rush to the spawning site and start their vigil. They are not alone. Soon juvenile *Mastacembelus moorii* join the assemblage. The wait is not long. With all fins erect the spawning partners descend, the female laying the eggs while the male fertilizes them. The rush by *T. dhonti*, and especially by *T. temporalis*, is spectacular, some of them managing to sneak under the mating pair and gobble up the eggs even as they come from the female's vent. They become so obnoxious that they often bump into the spawners, forcing them to cut their spawning short or to even abandon any attempt at beginning.

As the male *Lamprichthys* is so eager and thus oblivious to the "baby-snatchers" around him it is always the female who takes the initiative — either to descend to rest among the crowd, or to stop the session in mid course and ascend again to the open waters. But since she is bursting with eggs she often again follows the male to the same spot where the "fiends" meanwhile have been polishing off the eggs that had already been laid.

T. temporalis and *T. dhonti* circle around the rocks, darting in between them and into the holes into which the eggs might have rolled. The mastacembelid eels are able to poke their pointed snouts into the tiniest slits in search of eggs.

The predators are now ready for the next round. Their wait will not be long. Ten, twelve, perhaps even more times the pair of cyprinodonts will repeat their suicidal behavior with the same zeal and, unfortunately, the same results.

The male will sometimes attempt to find an alternate spawning place but it is usually too close to the previous one and, in the emergency situation he is in, probably less carefully chosen. As it is, the predators (*Telmatochromis* and mastacembelid eels) are watching the male's moves and as soon as the female joins him on the new site, they surround the pair and dash in again. The spawning pair may then try to fend off their enemies by alternating the sessions between the two sites — but to no avail. Every spawn is destroyed.

Very seldom are other cichlids living in the neighborhood attracted by the commotion. It appears that the two *Telmatochromis* species, and often only *T. temporalis*, the most common of the two, are the only ones that are systematically aware of and on the lookout for a pair of *Lamprichthys* getting ready to spawn. One might wonder how *Lamprichthys* manages to survive as a species with their suicidal spawning behavior. The answer is quite simple. They are perennial wanderers along a coast. When a male and a ripe female pair together they spawn on the ground below them, wherever they happen to be. If the bottom consists of dense rubble among which populations of *Telmatochromis* and mastacembelid eels are dense, the spawn is destroyed.

If the pair selects an isolated rock in the midst of a sandy area, odds are high that it is not occupied by these specialized predators and the spawn will remain undetected and hatch.

Telmatochromis bifrenatus
This is one of the really small cichlids in the lake. It lives on rocky bottoms seldom, if ever, venturing far from the rocks that mark their territory and in which they find shelter and a breeding place. It is usually solitary, but also aggregates in small groups mostly seen feeding on tiny invertebrates picked from the bottom. One might dismiss them as a harmless microvore were it not for the keen appetite they display when they happen to detect a spawning session occurring nearby, especially if it happens to be one by a pair of *Boulengerochromis microlepis*, the goliath among all cichlids. This species most often spawns over sand in a crater nest, often also near boulders. They have to beware of the sand-dwelling predators like *Lamprologus cunningtoni*, *L. tetracanthus*, *L. callipterus* (individually or in small packs), *L. modestus*, *L. mondabu*, and the like. These predators are most usually kept at bay by the watchful vigil of the spawning pair and seldom manage to close in on the nest site. Whenever the nest is close to rocks the danger from predators can come as well from fishes that are not specialized in dining on fish eggs but will never pass them by. Of these fishes there are many. Then also the rocks shelter countless other predatory species like mastacembelid eels and catfishes. But the most dangerous spot to select as a nestsite is obviously the top of a low-lying boulder. Very few fishes select an open flat surface close to the ground on which to lay their eggs in full view of their neighbors. Such fishes are usually mouthbrooders and the spawn does not remain more than a few seconds on the bare rock.

Among nestbreeding cichlids only *Boulengerochromis* occasionally selects such a

521

site. If it does, it is probably because it knows that with its mate it will be more than a match for any predator around. As a matter of fact they tolerate only one, *Telmatochromis bifrenatus*. This is a mistake. Does the huge cichlid dismiss the dwarfs as harmless? Anyway, *Boulengerochromis* is not short-sighted. When it discovers the *bifrenatus* among its eggs it simply chases them away by moving forward a few inches in their direction. They scatter, but not far and they soon return. Does the male *Boulengerochromis* weigh the relatively small harm half a dozen *bifrenatus* can do to the 10,000 or so eggs deposited on the rock against the wholesale destruction of the eggs by larger predators were it to abandon the nest to chase the *bifrenatus* more seriously? Who knows?

It is obvious that the pygmy predators know their host well. They creep under his belly as soon as he turns his back on them, gobble up half a dozen eggs each before he turns around and confronts them, and then flee. They never approach the spawn in front of the guarding male, they always approach from the rear. And they return many times.

Thus one can say that *T. bifrenatus* is well acquainted with the behavior of its giant opponent. How large is the damage done? One might guess that probably a few hundred eggs are taken over the two or three days it takes the *Boulengerochromis* eggs to hatch. The fry, after hatching, are moved in the parents mouth from the rock to a small depression in the sand bottom. There at long last they

will be free from the pygmy pests and will cope more successfully with what we will call normal predators.

Tanganyika Cichlids: The Mouth as a Tool

Many cichlids commonly use their mouth not only to pick up food or fight, but also as a tool with which to shovel a mouthful of sand. This behavior is used either to grab the invertebrates (essentially shrimps) buried within the sand or to dig their nest. At first glance the anatomical mechanisms (mouth and gullet) put to use for these activities appear to work in an identical fashion: the sand is shoveled or picked up in the mouth, then expelled through the gill openings as well as spewed out by the fish through its mouth. In fact, the muscles of the mouth and gullet can reverse their action when the fish is sifting the sand for food or simply carrying the sand away.

Let us take as an example the case of *Lamprologus modestus*, an ubiquitus cichlid from Lake Tanganyika. It is an omnivorous feeder and scavenger that also commonly feeds by sifting sand in search of invertebrates. It also digs a tunnel before spawning.

1. When feeding, this fish scoops a mouthful of sand from the bottom and, in small batches, processes the material through its throat and gill rakers. The larger invertebrates are retained by the gill raker seive and ultimately are swallowed while the fine sand grains are expelled through the gill openings. When the whole mouthful has been processed

in this way a few grains and large particles, such as flakes of micaschist, remain in the mouth and throat. These are brought together and then spewed out. In this process the main part of the sand is expelled through the gills and only very little through the mouth.

2. On the other hand, when the fish is building its nest (deep and rather tunnel-shaped, buttressed by a stone or big *Iridina* mussel shell), it will carry many mouthfuls of sand from the site to a place where it will not fall back into the hole and there spew it out in one powerful blow. Those particles in the back of the mouth that have slipped into the throat and could not be expelled with the bulk of the sand are then disposed of through the gills. The second process is thus the exact opposite of the first, a proof of the singular flexibility of a cichlid mouth and the efficiency of this tool.

The fish can also use this tool as a crane to lift and carry much bulkier material. This was demonstrated during one of our dives by the sight of a *Lamprologus brichardi* clenching between its jaws a twig or a root longer than its body and 2-3 mm across and carrying it out of its nest-site. A male *Callochromis macrops melanostigma* a bit more than 10 cm long can carry a pebble 3 cm in diameter and 1 cm thick.

A much higher level of behavioral sophistication linked with the use of the mouth as a tool perhaps exists among several species of *Tropheus* of the lake. In our fish farm in Burundi we raise about two dozen geographical races of this fascinating fish. It so happened that some of the ponds stocked with *Tropheus* became colonized by an aquatic snail after we had put some plants in the ponds. We soon discovered that the snails were systematically attacked and eaten by the fishes (especially by *T. brichardi* and *T. moorii moorii*). We found most of the empty snail shells concentrated in several piles in each pond with only very few shells scattered here and there. The piles were from 30 to 50 cm across and involved up to about a hundred shells. We were astounded to find most of the surviving snails following the shallowest rims of the ponds, more or less out of reach of the fishes. Very few snails "dared" to venture deeper, where they were at risk.

The problem is: why do the *Tropheus* in our ponds carry the snails to a picnic area instead of eating them on the spot? Other cichlids in the lake also leave piles of snails or mussels on a big stone in small shards.

Tropheus were often seen with a snail clenched between their teeth as they were nibbling at the flesh, an additional proof that they are also carnivorous. In addition, all of the *Tropheus* caught by Dr. Max Poll in 1946-47 (eleven specimens altogether) were caught on hook and line baited with worms, and none were caught by any other means.

We were much more startled to discover a male *Tropheus* in one of our tanks displaying an old empty snail shell in front of a female. He would leave it on the ground, then come back to pick it up again and again and

carry it about like a trophy or a plaything. Whether this shell could have been used, because of its egg-like shape, as an artifact and a preliminary display inducing the female to think about mating cannot be told at this stage of our observations, but was thought worth recording. It was, anyhow, a very strange sight to behold. Given the degree of elaborate behaviors already recorded among the Lake Tanganyika cichlids, one can expect a wealth of similar observations to be accumulated in the years to come.

PREDATOR—PREY RELATIONSHIPS

A few examples of the amount of predation from which rock-dwelling fishes suffer will illustrate the problems they meet when looking after their youngsters. In these dramas the predators will be "amateurs", which means that they do not belong to the specialized killers and that their diet does not consist normally of other fishes but mainly of small invertebrates — they are micro-carnivorous feeders. Given a chance (and there are many on the crowded slopes) they will go after the eggs and babies of other species (and perhaps even their own species) as long as their prey can be gobbled up whole. The victims of their attacks will be the spawns of "professional" predators whose diet, when adult, is composed of fishes swallowed whole or ripped apart. Later on, when adult, the surviving fry of the predatory species (ex. *Boulengerochromis, Lepidiolamprologus elongatus,*

and *Perissodus microlepis*) will attack the adults or youngsters of the very species that brought them so much trouble in the past, *Lamprologus brichardi* and *L. modestus*. Both are nestbreeders, but if the first species is known for its sociability, gregarious behavior, and overall peacefulness, the second fish is a loner and picks up invertebrates from the sand as well as from zooplankton.

Boulengerochromis vs. *Lamprologus brichardi*:

One would think that the giant *Boulengerochromis*, whose pairs most often are made up of fish about 40 to 50 cm long, could defend their eggs and young fry very well against a fish as small as an adult *L. brichardi*. In fact, the eggs are less often attacked by *brichardi* because more often than not they are located on sand and *brichardi* schools prefer to hover over rocks where they raise their offspring in communal nurseries. At this early stage the eggs and newly hatched fry of the giant cichlid are preyed upon by the tiny *Telmatochromis bifrenatus* and occasional intruders passing by.

If the pressure from predation grows too strong the *Boulengerochromis* parents will take their eggs in the mouth and leave, choosing another location either nearby or up to 20 meters away. They dig another small cavity in the sand and put their eggs in it. This is common practice and can be repeated very often. We have witnessed a pair moving the eggs twice within a 30-minute time span. A pair of *Boulengerochromis* can in this way move the eggs a dozen times in the three to four days it takes the eggs to hatch and the fry to become free-swimming. One might imagine that this behavior might have been one of the first steps leading to mouthbrooding.

The very compact fry swarm, swirling on itself in spirals and rising from the bottom, moves about very much. It is at this time that the *Lamprologus brichardi* become dangerous as the swarm can thus move from a relatively safe area into one, near rubble, that teems with other cichlids, foremost among them *L. brichardi*. If this happens, the *brichardi* will rush to attack the fry and gobble them up from all sides. If the pair of parents comes in to protect their brood, the intruders flee into the open water or take shelter in the rubble. They come in again as soon as the pair has turned away. Dozens, perhaps even hundreds, of the *Boulengerochromis* fry disappear in these repeated attacks. The swarm doesn't look any smaller, however, as the total number of fry can reach well over 10,000.

As in the case of *Telmatochromis bifrenatus* we have never seen one of the harassed parents really try to kill even a small *L. brichardi*, which they could swallow whole. They just repel them for a short time, and sometimes even do nothing at all to stop the slaughter going on, at least for a period of several seconds. Is it because their attention at that time has been devoted to other, more dangerous predators in the neighborhood? It is difficult to tell.

We have witnessed attempts by one of the pair to kill one lone predator, more often than

not either *Telmatochromis temporalis* or *Lamprologus modestus*, coming from its shelter under a rock. In this case the *Boulengerochromis* parent repeatedly tried to bite and seize the predator in its jaws, but couldn't because its big head couldn't squeeze inside of the crevice. Losing time trying to kill a single predator might mean that the swarm could fall prey to a pack of other "wolves." On the other hand, killing a single predator from such a pack could very well frighten the other fishes around and yield better results. Once I saw the entire swarm of fry sent tumbling pell-mell by one powerful thrust of the parent's tail, moving them well away from the danger area and over the open sand floor. The adult was at the time

trying to repel one intruder and I couldn't fathom if the beneficial swipe had been intentional or a chance happening. It sounds farfetched but still well in line with other behaviors of this fantastic fish.

Lepidiolamprologus elongatus vs *L. brichardi*:
Many occasional predators attack the fry of *L. elongatus*. Among them, given a chance —if the breeding pair gets involved in other attacks— the *brichardi* will jump the fry. The babies will try to escape detection by rushing to the bottom and remaining still so that they blend with the sand. But *brichardi* are sharp-sighted for they need to be able to catch tiny invertebrates drifting around, and the

Pierre Brichard working with African fishes at his field station. The fishes are held and treated for varying lengths of time before being shipped. Photo by Dr. Herbert R. Axelrod.

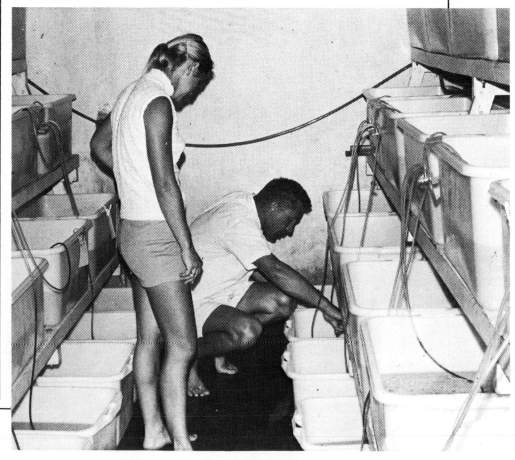

elongatus fry, even as they jump here and there, are no match for their killers. Eventually, one of the breeders returns, which is enough to send the *brichardi* scampering to their shelters. The surviving fry will then rise again from the bottom to meet their parents.

Perissodus microlepis vs. *L. brichardi*:

The scale-eating *Perissodus* is a mouthbrooder, but as we have seen one of the most prolific, with a hundred fry or more per spawn for a fish about ten to twelve centimeters in length. Thus the eggs have to be small. The larvae hatch quickly, and the fry soon absorb their yolk sac, making mouthbrooding difficult. Anyhow, the fry have to feed. The female soon releases them on the bottom, usually on top of a big stone, and takes them back into her mouth only when there is danger. The male lends a hand in protecting the fry to the point of taking care of the mouthbrooding while the female repels intruders.

At this stage the two mates often display a very strong sexual dimorphism, much more so than when they mated. The mouthbrooding partner still has the deep blue sheen from the mating session; the other, standing guard or resting (its activity is quite limited), has a very pale pattern. Essentially, and most significantly perhaps, the underparts of its body are silvery. This mate will not stand over the fry but on the side of the flock, and will not take them in its mouth. The more brightly colored one will hover above them and might

be the one most involved in repelling the predators, although it has some babies in its mouth. Repelling them is easy: just a short forward move is sufficient.

The fry, when in danger, flatten out on the rock, with which their mottled pattern blends, and remain quiet. When the mate in full nuptial garb comes back they rise toward its mouth and try to get in. They will not head for the mate lacking the nuptial garb. Although I have seen many such actions of the *Perissodus microlepis* fry, I never could detect what in the hovering adult pattern told the fry to rush for its mouth. Very often they try, but the haven remains closed.

One would think that a predator like *Perissodus microlepis*, which all fish around appear to know very well and fear, would be safe in protecting its fry, especially from such a timid fish as *brichardi. L brichardi* as an adult often has lost a few scales to the attacks of the scale-ripper. Far from it. I have seen a lone *brichardi* attack a swarm of babies in full view of the defending mates and repeatedly gobble up a few fry, whether or not they were resting on the bottom or swimming above ground. Again, the defending pair never brought its attack to its logical conclusion — the death of the intruder — the most efficient way to bring the other predators to more common sense and caution.

What can we conclude from these few examples of fry preyed upon when just free-swimming and still quite small (all of them about 6-7mm long)? First, that they are then

in danger even from fishes that otherwise are quite harmless and will not prey upon the fry when they grow a bit bigger (probably about 1 cm), or swim faster, or are difficult to gobble up by a fish like *L. brichardi*. The havoc brought about by this fish among the fry of other fishes is very significant because its schools can be so large (up to 100,000 fish over an area one hectare). With a density reaching up to 10 *brichardi* per square meter of rubble, any spawn of very small fry is in danger and their cumulative effects can be devastating.

Second, it shows the beneficial effects of having the fry released in the open at the largest possible size, exemplified by the long mouthbrooding care of *Tropheus*.

Third, the *brichardi* do not attack large fry, whose predation is then taken over mostly by the specialized carnivorous species. This shows that through the entire process of development the juveniles fall prey to different predators acting in successive waves. One doesn't need more to figure out the deadly drama of life and death of a young fish on a rocky slope.

Index of Geographical Places with Coordinates

Place	Latitude (South)	Longitude (East)
Albertville (now Kalemie)	5°56	29°12
Banza (Cape)	4°03	29°15
Bemba (or Munene)	3°40	29°09
Bujumbura	3°20	29°40
Bulu Island	6°00	29°43
Edith Bay	6°30	29°57
Ikola	6°15	30°25
Kabimba Bay	5°33	29°21
Kabogo (Cape)	5°27	29°46
Kala	8°07	30°59
Kalemie (Albertville)	5°56	29°12
Kapampa	7°40	30°10
Karema	6°49	30°27
Kasaba Bay	8°31	30°42
Kasanga	8°28	31°10
Kasimia (village & Bay)	4°31	29°10
Kavalla Island	5°40	29°25
Kibwe Bay	5°22	29°47
Kigoma	4°52	29°38
Kinyamkolo (Nyamkole-Mpulungu)	8°48	31°06
Kirando	7°25	30°35
Lagosa	5°57	29°53
Lueba	4°	29°06
Lugufu (or Rugufu) River delta	5°21	29°47
Luhanga	3°31	29°09
Lukuga River and outlet	5°55	29°12
Magara (Cape)	3°43	29°20
Malagarazi river delta	5°12	29°47
Mboko (or Kifumbwe Islands)	3°47	29°06
Minago	3°47	29°21
Moliro	8°13	30°34
Mpala	6°45	29°31
Mpimbwe (Cape)	7°06	30°30
Mpulungu	8°48	31°06
Mtoto	6°59	29°44
Mvua (village & Bay)	8°05	30°34
Mwerasi (village & River)	7°20	30°16
Nyanza Lac (town and Bay)	4°21	29°36
Rumonge	3°59	29°26
Ruzzizi delta	3°16	29°14
Tembwe (Cape)	6°30	29°29
Tumpa	7°04	29°48
Utinta Bay	7°08	30°31
Uvira	3°24	29°08
Yungu (village and Bay)	4°50	29°07
Zongwe (Cape and River)	7°18	30°13

GLOSSARY

Air bladder (swim bladder): A usually gas-filled sac in the upper body cavity that helps provide buoyancy for the fish.

Alluvion: Fine sediment.

Anfractuosity: A narrow, dark, winding cave or crevice.

Axil: The region between the point of insertion of the pectoral or pelvic fin and the body.

Canine: A sharply pointed, usually large, tooth.

Carnivore: An animal that feeds largely on other animals.

Caudal fin: The tail fin.

Caudal peduncle: The part of the trunk of a fish from the base of the caudal fin to a line between the last rays of the dorsal and anal fins; the "tail stem."

Compressed: Flattened from side to side.

Depressed: Flattened from top to bottom.

Dorsal: Top. The dorsal fin is the fin on the topline of the body.

Fin formula: Method of expressing the number of spines and rays in the fins of a fish. Roman numerals stand for spines, Arabic numerals for soft rays. If the spines and rays form a continuous fin, the numbers are separated by a comma; if the spines are in a fin separate from the rays, a dash is used. Thus: X,12 is a continuous fin with ten spines and 12 soft rays; III-6 is a divided fin, the front part with three spines, the back with six soft rays.

Gill-raker: Usually small and stout projections from the inner side of the gill arch (the red gill filaments are on the outer side).

Herbivore: An animal that feeds largely on plants.

Infeudated: Tied ecologically to a certain niche or area; restricted.

Lateral line: Sensory organ along the sides of a fish that consists of tubulated or pored scales that receive low-frequency vibrations such as water movement.

Molariform: Resembling a molar tooth, thus usually low and rounded; often associated with diets that consist of hard foods such as mollusks.

Ocellus: A central spot of color surrounded by a ring of another color and thus resembling to some extent an eye.

Omnivore: An animal that feeds to about the same extent on both plant and animal matter.

Opercle: The large bone forming the posterior part of the operculum. The preopercle is the usually boomerang-shaped bone below and behind the eye.

Operculum: The gill-cover; the bones and skin that shield the gills from the outside.

Pharyngeal teeth: Patches of teeth on the upper and/or lower gill arches behind the mouth cavity. Usually the upper and lower pharyngeals oppose each other or are opposed by a thickened pad of tissue used as a grinding surface.

Ray: One of the external supporting elements of a fin. A spine is a hardened, one-piece ray; a soft ray is flexible and usually has visible segments.

Standard length: The length of a fish from the tip of the upper lip to the base of the caudal fin.

Synonym: A scientific name for a species, genus, or subspecies proposed after the currently accepted name and not in use. Synonyms are validly proposed names and may eventually be found to apply to a distinct taxon.

Taxon: A systematic category such as the subspecies, species, and genus.

Unpaired fins: The dorsal, anal, and caudal fins; the paired fins are the pectorals and pelvics.

Ventral: Bottom; the ventral fins precede the anal fin and are more formally called the pelvics.

Measurement Conversion Factors

When you know—	Multiply by—	To find—
Length:		
Millimeters (mm)	0.04	inches (in)
Centimeters (cm)	0.4	inches (in)
Meters (m)	3.3	feet (ft)
Meters (m)	1.1	yards (yd)
Kilometers (km)	0.6	miles (mi)
Inches (in)	2.54	centimeters (cm)
Feet (ft)	30	centimeters (cm)
Yards (yd)	0.9	meters (m)
Miles (mi)	1.6	kilometers (km)
Area:		
Square centimeters (cm^2)	0.16	square inches (sq in)
Square meters (m^2)	1.2	square yards (sq yd)
Square kilometers (km^2)	0.4	square miles (sq mi)
Hectares (ha)	2.5	acres
Square inches (sq in)	6.5	square centimeters (cm^2)
Square feet (sq ft)	0.09	square meters (m^2)
Square yards (sq yd)	0.8	square meters (m^2)
Square miles (sq mi)	1.2	square kilometers (km^2)
Acres	0.4	hectares (ha)
Mass (Weight):		
Grams (g)	0.035	ounces (oz)
Kilograms (kg)	2.2	pounds (lb)
Ounces (oz)	28	grams (g)
Pounds (lb)	0.45	kilograms (kg)
Volume:		
Milliliters (ml)	0.03	fluid ounces (fl oz)
Liters (L)	2.1	pints (pt)
Liters (L)	1.06	quarts (qt)
Liters (L)	0.26	U.S. gallons (gal)
Liters (L)	0.22	Imperial gallons (gal)
Cubic centimeters (cc)	16.387	cubic inches (cu in)
Cubic meters (cm^3)	35	cubic feet (cu ft)
Cubic meters (cm^3)	1.3	cubic yards (cu yd)
Teaspoons (tsp)	5	millimeters (ml)
Tablespoons (tbsp)	15	millimeters (ml)
Fluid ounces (fl oz)	30	millimeters (ml)
Cups (c)	0.24	liters (L)
Pints (pt)	0.47	liters (L)
Quarts (qt)	0.95	liters (L)
U.S. gallons (gal)	3.8	liters (L)
U.S. gallons (gal)	231	cubic inches (cu in)
Imperial gallons (gal)	4.5	liters (L)
Imperial gallons (gal)	277.42	cubic inches (cu in)
Cubic inches (cu in)	0.061	cubic centimeters (cc)
Cubic feet (cu ft)	0.028	cubic meters (m^3)
Cubic yards (cu yd)	0.76	cubic meters (m^3)
Temperature:		
Celsius (°C)	multiply by 1.8, add 32	Fahrenheit (°F)
Fahrenheit (°F)	subtract 32, multiply by 0.555	Celsius (°C)

BIBLIOGRAPHY

AXELROD, G. S., 1977. A new species of *Tropheus* (Pisces Cichlidae) from Lake Tanganyika. J.L.B. Smith Instit. of Ichthyology, Special Publ. nr. 17: fig. 1-4, pl. 1-4, map, 6 tables.

AXELROD, G. S. and HARRISON, J. A., 1978. *Simochromis margaretae,* a new species of cichlid fish from Lake Tanganyika. Ibid., Special Publ. nr. 19: 1-16, 7 fig.

AXELROD, H. R. and BURGESS, W. E. 1988. African Cichlids of Lakes Malawi and Tanganyika. Ed. 12. T.F.H. Publ., Neptune, NJ. 448 pages.

AXELROD, H. R., W. E. BURGESS, N. PRONEK, and J. G. WALLS. 1989. Dr. Axelrod's Atlas of Freshwater Aquarium Fishes. Ed. 3. T.F.H. Publ., Neptune, NJ. 782 pages.

BAILEY, R. G. 1968. Fishes of the genus *Tilapia* (Cichlidae) in Tanzania, with a key for their identification. East African Agricult. and Forestry J., XXXIV (2): 194-202, 5 fig.

BAILEY, R. and STEWART, D. J., 1977. Cichlid fishes from Lake Tanganyika. Additions to the Zambian Fauna including two new species. Occas. Papers Mus. Zool. Univers. Michigan, nr. 679: 3 fig., 3 pl., 5 tables.

BAILEY, R. M., ROBINS, C. R. and GREENWOOD, P. H., 1980. *Chromis* Cuvier in Desmarest 1814 (Osteichthyes, Perciformes, Pomacentridae): Proposal to place on official list of generic names in Zoology and that generic names ending in -*chromis* be ruled to be masculine. Z.N. (S)2329. Bull. Zool. Nomencl., XXXVII, part 4 : 247.

BORODIN, N. A., 1931. Some new cichlid fishes from Lakes Victoria and Tanganyika, Central Africa. Proc. New England Zool. Club., XII: 49-54.

BORODIN, N. A., 1936. On a collection of freshwater fishes from Lakes Nyasa, Tanganyika and Victoria in Central Africa. Zool. Jahrb. (System.), LXVIII(1): 1-34.

BOULENGER, G.A., 1898. Report on the collection of fishes made by Mr. J. E. S. Moore in Tanganyika during his Expedition 1895-1896. Trans. Zool. Soc. London, XV, pt. 1, nr. 1: 1-30, 2 fig., pl. I-VIII.

ID.,1899. Second contribution to the ichthyology of Lake Tanganyika. On the fishes obtained by the Congo Free State Expedition under Lieut. Lemaire in 1898. Trans. Zool. Soc. London, XV, pt. 9: 87-95, pl. XVIII-XX.

ID., 1899. Materiaux pour la faune du Congo. Poissons nouveaux du Congo (5 partie). Ann. Mus. Congo, Zool. ser. 1, I, 5: 97-127, pl. XI-XLVII.

ID., 1900. Materiaux pour la faune du Congo. Poissons nouveaux du Congo. Ann. Mus. Congo, Zool. ser. 1, I, 6: 131-164, pl. XLVII-LVI.

ID., 1901. Diagnoses of new fishes discovered by Mr. J. E. S. Moore in Lakes Tanganyika and Kivu. Ann. Mag. N.H., (7) VII: 1-6.

ID., 1901. Third contribution to the ichthyology of Lake Tanganyika. Report on the collection of fishes made by Mr. J. E. S. Moore in Lakes Tanganyika and Kivu during his second Expedition 1899-1900. Trans. Zool. Soc. London, XVI, pt. III: 137-178, pl. XII-XX.

ID., 1901. Les Poissons du bassin du Congo. Bruxelles, 1901. xii—532 pp., 25 pl., map., and fig.

ID., 1905. A list of freshwater fishes of Africa. Ann. Mag. Nat. Hist., (7) 16: 36-60.

ID., 1906. Fourth contribution to the ichthyology of Lake Tanganyika. Report on the collection of fishes made by Dr. W. A. Cunnington during the third Tanganyika Expedition, 1904-1905. Trans. Zool. Soc. London, XVII, pt. VI, no. 1: 537-576, pl. XXX-XLI.

ID., 1914. Diagnoses de poissons nouveaux. I. Acanthopterygiens, Opisthomes, Cyprinodontes (Mission Stappers au Tanganyika-Moero). Rev. Zool. Afr., III: 442-447.

ID., 1915. Catalogue of the Freshwater Fishes of Africa in the British Museum (Nat. Hist.). London. Vol. III: 526 pp., 351 fig.

ID., 1919. On a collection of fishes from the Lake Tanganyika with description of three new species. Proc. Zool. Soc. London, 1919 (I): 17-20. fig. 1-3.

ID., 1920. Poissons de la mission Stappers

1911-1913 pour l'exploration hydrographique et biologique des lacs Tanganyika et Moero. Rev. Zool. Afr., VIII (1): 1-57.

BRICHARD, P., 1978. Fishes of Lake Tanganyika. T.F.H. Publ., Neptune, NJ. 448 pp.

BURGESS, W. E. 1989. An Atlas of Freshwater and Marine Catfishes. A Preliminary Survey of the Siluriformes. T.F.H. Publ., Neptune, NJ. 784 pages.

COLOMBE, J. and ALLGAYER, R. 1985. Description de *Variabilichromis, Neolamprologus* et *Paleolamprologus,* genres nouveaux du lac Tanganyika, avec redescription des genres *Lamprologus* Schilthuis 1891 et *Lepidiolamprologus* Pellegrin 1904. Rev. Franc. des Cichlidophiles, no. 49: 9-28, fig.

COULTER, G. W., 1965-1966. The deep benthic fishes at the south of Lake Tanganyika with special reference to distribution and feeding in *Bathybates* species, *Hemibates stenosoma* and *Chrysichthys* sp. Fish. Rev. Bull. Zambia, 4: 31-38. 2 tab., graphs.

DAVID, L., 1936. Contribution a l'etude de la faune ichthyologique du Lac Tanganika. Rev. Zool. Bot. Afr., XXVIII (2): 149-160, 5 fig.

ID., 1937. Poissons de l'Urundi. Ibid., XXIX (4): 413-420. 2 fig.

DAVID, L. and POLL, M., 1937. Contribution a la faune ichthyologique du Congo Belge: collection du Dr. H. Schouteden (1924-1926) et d'autres recolteurs. Annales Mus. R. Congo Belge, ser. 4 (Zool.), III (5): 189-294, fig.

FRYER, G. and T. D. ILES. 1972. The Cichlid Fishes of the Great Lakes of Africa. Oliver & Boyd, Edinburgh. 642 pages.

GERY, J. 1977. Characoids of the World. T.F.H. Publ., Neptune, NJ. 672 pages.

GREENWOOD, P. H., 1954. On two species of cichlid fishes from the Malagarazi River (Tanganyika), with notes on the pharyngeal apophysis in species of the *Haplochromis* group. Ann. Mag. Nat. Hist., (12) 78: 401-414, 3 fig.

ID., 1978. A review of the pharyngeal apophysis and its significance in the classification of African cichlid fishes. Bull. Brit. Mus. (N.H.)., Zool. ser., 33(5): 297-323, 18 fig.

ID., 1979. Towards a phyletic classification of the genus *Haplochromis* (Pisces Cichlidae) and related taxa. Part I. Ibid., 35(4): 265-322, 20 fig.

ID., 1980. A new species of cichlid fish from the Malagarazi swamps and river, Tanzania, East Africa. Ibid., 38(3): 159-163.

ID., 1983. The *Ophthalmotilapia* assemblage of cichlid fishes reconsidered. Ibid., 44(4): 249-290. 24 fig.

GUENTHER, A., 1893. Descriptions of Reptiles and Fishes collected by Mr. E. Coode-Hore on Lake Tanganyika. Proc. Zool. Soc. London, 1893: 628-632, pl. LVIII and fig.

LADIGES, W. 1959. Fische aus dem Tanganyika-See. Die Aquar. und Terrar. Zeitschr. (DATZ), 12(5): 130-134; 12(6): 165-166 (8 photos).

LIEM, K. F., 1979. Modulatory multiplicity in the feeding mechanism in cichlid fishes, as exemplified by the invertebrate pickers of Lake Tanganyika. J. Zool. Soc. London, 189: 93-125, 13 fig.

ID., 1981. A phyletic study of the Lake Tanganyika cichlid genera *Asprotilapia, Ectodus, Lestradea, Cunningtonia, Ophthalmochromis* and *Ophthalmotilapia*. Bull. Mus. Comp. Zool., 149(3): 191-214, 9 fig.

LIEM, K. F. and GREENWOOD, P. H., 1981. A functional approach to the phylogeny of the pharyngognath teleosts. Amer. Zool., 21: 83-101, 10 fig.

LIEM, K. F. and STEWART, D. J., 1976. Evolution of the scale-eating cichlid fishes of Lake Tanganyika: a generic revision with a description of a new species. Bull. Mus. Comp. Zool., 147(7): 319-350, 31 fig.

LOWE, R. H. (McCONNEL), 1956. The breeding behaviour of *Tilapia* species (Pisces Cichlidae) in natural waters: observations on *T. karomo* Poll and *T. variabilis* Boul. Behaviour, IX(2-3): 140-163, 3 fig., pl. VII-IX.

MARLIER, G., 1959. Observations sur la biologie littorale du lac Tanganika. Rev. Zool. Bot. Afr., LIX(1-2): 164-183,3 fig., 2 pl.

MATTHES, H., 1958. Un Cichlidae nouveau du lac Tanganika: *Julidochromis transcriptus* n. sp. Folia Scient. Afr. Centr., IV(4): 85-86.

ID., 1959. Un Cichlidae nouveau du lac Tanganika: *Julidochromis transcriptus* n. sp. Rev. Zool. Bot. Afr., LX(1-2): 126-130, fig.

ID.,1959. Un Cichlidae nouveau du lac Tanganika: *Petrochromis orthognathus* n. sp. Folia Scient. Afr. Centr., V(1): 17-l8.

ID., 1959. Un Cichlidae nouveau du lac Tanganika: *Petrochromis orthognathus* n. sp. Rev. Zool. Bot. Afr., LX(3-4): 335-341, 3 photos.

ID., 1959. Une sous-espece nouvelle de *Lamprologus leleupi*: *L. leleupi melas.* Fol. Scient. Afr. Centr., V(1): 18.

ID., 1961. *Boulengerochromis microlepis*, a Lake Tanganyika fish of economical importance. Bull. Aquatic Biology (Amsterdam), III(24): 1-15, 7 fig.

ID., 1962. Poissons nouveaux ou interessants du lac Tanganika et du Ruanda. Ann. Mus. Roy. Afr. Centr. Tervuren, Ser. Zool., no. 111: 27-88, 5 fig., 4 pl.

MATTHES, H. and TREWAVAS, E., 1960. *Petrochromis famula* n. sp., a cichlid fish of Lake Tanganyika. Rev. Zool. Bot. Afr., LXI(3-4): 349-357, 2 fig.

MYERS, G. S., 1936. Report on the fishes collected by N. C. Raven in Lake Tanganyika in 1920. Proc. U.S. Nat. Mus., Washington, 84(2998): 1-15, pl. I.

NELISSEN, M., 1977. *Pseudosimochromis*, a new genus of the family Cichlidae (Pisces) from Lake Tanganyika. Rev. Zool. Afr., 91(3): 730-731.

ID., 1977. Description of *Tropheus moorii kasabae* n. ssp. (Pisces Cichlidae) from the south of Lake Tanganyika. Rev. Zool. Afr., 91(1): 237-242, 1 fig., 1 tab.

ID., 1978. Description of *Simochromis pleurospilus* sp. n., a sibling species of *S. babaulti* from Lake Tanganyika. Ibid., 92(3): 627-638, 3 fig.

ID., 1979. A taxonomic revision of the genera *Simochromis, Pseudosimochromis* and *Tropheus* (Pisces Cichl.). Ann. Mus. Roy. Afr. Centr., 229: 54 pp., fig. 1-13, tab. 1-5.

NELISSEN, M. and THYS VAN DEN AUDENAERDE, D., 1975. Description of *Tropheus brichardi* sp. n. from Lake Tanganyika (Pisces Cichlidae). Rev. Zool. Afr., 89(4): 974-980, 2 fig.

NICHOLS, J. T. and LA MONTE, F. R., 1931. A new *Lamprologus* from Lake Tanganyika. Amer. Mus. Novitates, no. 478: 1-2, fig. 1.

PELLEGRIN, J., 1903. Contribution a l'etude anatomique, biologique et taxinomique des poissons de la famille des Cichlidae. Mem. Soc. Zool. Fr., 16: 41-402, 41 fig., pl. IV-VII.

ID., 1927. Description de Cichlides et d'un Mugilide nouveaux du Congo Belge. Rev. Zool. Afr., 15(1): 52-57.

ID., 1927. Mission Guy Babault. Poissons du lac Tanganika. Bull. Mus. Nat. H.N. Paris, 33: 499-501.

POLL, M., 1932. Contribution a la faune des Cichlidae du lac Kivu (Congo Belge). Rev. Zool. Bot. Afr., XXIII(1): 29-35, 1 photo, 2 pl.

ID., 1942. Cichlidae nouveaux du lac Tanganika appartenant aux collections du Musee du Congo. Ibid., XXXVI(4): 343-360.

ID., 1943. Descriptions de Poissons nouveaux du lac Tanganika, appartenant aux familles des Clariide et Cichlidae. Ibid., XXVII(3-4): 305-318.

ID., 1944. Descriptions de Poissons nouveaux recueillis dans la region d'Albertville (Congo Belge) par le Dr. G. Pojer. Bull. Mus. Roy. Hist. Natur. Belgique, XX(3): 1-12, 10 fig.

ID., 1946. Revision de la Faune ichthyologique du lac Tanganika. Ann. Mus. Congo Belge, Zool., ser I, IV(3): 141-364, fig. 1-87, pl. I-III.

ID., 1948. Descriptions de Cichlidae nouveaux recueillis par le Dr. J. Schwetz dans la riviere Fwa (Congo Belge). Rev. Zool. Bot. Afr., XLI(1): 91-104, 6 fig., pl. XXI.

ID., 1948. Descriptions de Cichlidae nouveaux recueillis par la mission hydrobiologique belge au lac Tanganika (1946-1947). Bull. Mus. Roy. Hist. Natur. Belgique, XXIV(26): 31 pp., 22 fig.

ID., 1949. Deuxieme serie de Cichlidae nouveaux recueillis par la Mission hydrobiologique belge au lac Tanganika (1946-1947). Bull. Inst. Roy. Sci. Natur. Belgique, XXV(33): 55 pp., 33 fig.

ID., I950. Descriptions de deux Poissons petricoles du lac Tanganika. Rev. Zool. Bot. Afr., 43(4): 292-302, 6 fig.

ID., 1951. Troisieme serie de Cichlidae nouveaux recueillis par la mission hydrobiologique belge au lac Tanganika (1946-1947). Bull. Inst. Roy. Sci. Natur. Belgique, XXVII(29): 1-11, 2 fig.

ID., 1951. Troisieme serie de Cichlidae nouveaux recueillis par la mission hydrobiologique belge au lac Tanganika (1946-1947), suite 1. Bull. Inst. Roy. Sci. Natur. Belgique, XXVII(30): 1-11, 5 fig.

ID., 1951. Troisieme serie de Cichlidae nouveaux recueillis par la mission hydrobiologique belge au lac Tanganika (1946-1947), suite 2 et fin. Bull. Inst. Roy. Sci. Natur. Belgique, XXVII(31): 1-5, 2 tab.

ID., 1951. Histoire du peuplement et origine des especes de la Faune ichthyologique du lac Tanganika. Ann. Soc. Roy. Zool. Belg., LXXXI: 111-140, 3 pl.

ID., 1952. Quatrieme serie de Cichlidae nouveaux recueillis par la mission hydrobiologique belge au lac Tanganika (1946-1947). Bull. Inst. Roy. Sci. Natur. Belgique, XXVIII(49): 1-20, 2 fig.

ID., 1952. Resultats scientif. Explor. hydrob. belge au lac Tanganika (1946-1947). Introduction. Les Vertebres. I: 103-165, 17 pl., fig.

ID., 1956. Resultats scientif. Explor. hydrob. belge au lac Tanganika (1946-1947). Poissons Cichlidae. III(5 B): 619 pp., 131 fig., 10 pl., 1 map.

ID., 1957. Les Genres de Poissons d'eau douce de l'Afrique. Ann. Mus. Roy. Congo Belge, Zool., 54: 191 pp., 425 fig. (Also published in: Publ. Dir. de l'Agriculture, des Forets et de l'Elevage, Bruxelles, Ministre des Colonies, 1957.)

ID., 1974. Contribution a la faune ichthyologique du lac Tanganika d'apres les recoltes de P. Brichard. Rev. Zool. Afr., 88(1): 99-110, 3 fig.

ID., 1976. *Hemibates bellcrossi* sp. n. du lac Tanganika. Ibid., 90(4): 1017-1020, 1 fig.

ID., 1978. Contribution a la connaissance du genre *Lamprologus* Schth. Description de quatre especes nouvelles, rehabilitation de *L. mondabu* et synopsis remanie des especes du lac Tanganika. Bull. Cl. Sciences, Ac. Roy. Belg., LXIV: 725-758, 6 fig.

ID., 1979. Un *Haplochromis* rouge du lac Tanganika, femelle de *H. benthicola* Matthes 1962 (Pisces Cichlidae). Rev. Zool. Afr., 93(2): 467-475, fig.

ID., 1981. Contribution a la faune ichthyologique du lac Tanganika. Revision du genre *Limnochromis* Regan 1920. Description de trois genres et d'une espece nouvelle: *Cyprichromis brieni*. Ann. Soc. Roy. Zool. Belg., 3(14): 163-179, 3 pl., 1 fig.

ID., 1983. *Greenwoodochromis* nom. n. Cybium, VII(1).

ID., 1984. *Haplotaxodon melanoides* sp. n. du lac Tanganika. Rev. Zool. Afr., 98(3): 677-683, 1 fig.

ID., 1984. Un Cichlidae meconnu du lac Tan-

ganika: *Lamprologus finalimus* Nich. et La Monte 1931. Cybium, VIII(4): 88-91, 1 fig., 3 tab.

ID., 1985. Description de *Xenotilapia flavipinnis* sp. n. Rev. Zool. Afr., 99.

ID., 1987. Classification des Cichlidae du lac Tanganika: Tribus, genres et especes. Mem. Acad. Roy. Belgique, XLV(2): 1-156.

POLL, M. and MATTHES, H., l962. Trois poissons remarquables du lac Tanganika. Ann. Mus. Roy. Afr. Centr., Zool., 111: 1-26, 2 fig., 6 pl.

POLL, M. and STEWART, D., 1975. A new cichlid fish of the genus *Xenotilapia* from Lake Tangkanyika, Zambia (Pisces Cichlidae). Rev. Zool. Afr., 89(4): 920-924, 2 fig., 1 tab.

ID., 1977. Un nouveau *Lamprologus* du sud du lac Tanganika (Zambia). Ibid., 91: 1047-1056, 1 fig., 2 tab.

POLL, M. and ·THYS VAN DEN AUDENAERDE, D., 1974. Genre nouveau *Triglachromis* propose pour *Limnochromis otostigma* Regan 1920, Cichlidae du lac Tanganika. Ibid., 88(1): 127-130, 1 fig.

POLL, M. and TREWAVAS, E., 1952. Three new species and two new subspecies of the genus *Lamprologus*. Cichlid fishes of the Lake Tanganyika. Bull. Inst. Roy. Sci. Natur. Belgique, XXVIII(50): 16 pp., 5 fig.

REGAN, C.T., 1920. The classification of the fishes of the family Cichlidae. I. The Tanganyika genera. Ann. Mag. (N.H.), ser 9, X: 33-53.

ID., 1920. A new cichlid fish of the genus *Limnochromis* from Lake Tanganyika. Ibid., ser 9, V: 152.

ID., 1932. The cichlid fishes described by Borodin from lakes Tanganyika and Victoria. Proc. New Engl. Zool. Club., XIII: 27-29.

SCHEUERMANN, H., 1977. A partial revision of the genus *Limnochromis* Regan 1920. Cichlidae (British Cichlid Association), III(2): 69-73, 4 photos, 1 tab.

STAECK, W., 1975. A new cichlid fish from Lake Tanganyika: *Julidochromis dickfeldi* sp. n. Rev. Zool. Afr., 89(4): 981-986, 1 fig., 1 tab.

ID., 1978. Ein neuer Cichlidae aus dem sudlichen Tanganyikasee: *Lamprologus nkambae* n. sp. (Pisces Cichlidae). Ibid., 92(2): 436-441, 1 fig., 1 tab.

ID., 1980. Ein neuer Cichlide vom Ostufer des

Tanganyikasee: *Lamprologus leleupi longior* n. ssp. Ibid., 94(1): 11-14, fig. 1-2.

ID., 1982. Hanbuch der Cichlidenkunde. Buntbarsche Arten, Verhaltensbiologie, Pflege und Zucht. Kosmos Handbuch, Stuttgart. 200 pp., 153 fig., photos.

ID., 1983. *Lamprologus buescheri* sp. n. from the Zambian part of the Lake Tanganyika. Senckenbergiana Biol., 63(5-6): 325-328, 3 photos.

STEINDACHNER, F., 1909. Uber enige neue Fischarten aus dem Tanganyikasee gesammelt durch A. Horn und Marie Horn. Sitz. Math. Natuurwiss. kl. Ak. Wien, XXIV: 399-404.

ID., 1909. Uber zwei neue Cichlidenarten aus dem see Tanganyika. Stitz. Math. Naturwissen. kl. Ak. Wiss. Wien, XXV: 425-428.

ID., 1909. Uber eine neue *Tilapia* und *Lamprologus*-art dem Tanganyikasee und uber *Brachyplatysoma (Taemonema) platynema* Blgr. aus der Umgebung von Para. Sitz. Math. Naturwissen. kl. Ak. Wiss. Wien, XXVI: 443-446.

ID., 1911. Beitrage zur Kenntnis der Fischfauna des Tanganyikasees und des Kongogebietes. Anz. Kaiserl. Akad. Wiss. Wien, XXVIII(27): 528-530.

ID.,1911. Beitrage zum Kenntnis der Fischfauna des Tanganyikasees und des Kongogebietes. Sitzungsber. Kaiserl. Akad. Wiss. Wien, 120(10): 1171-1186, pl. I-III.

THYS VAN DEN AUDENAERDE, D., 1960. Note sur le statut et la position taxonomique de *Tilapia christyi* Boulenger 1915 (Pisces Cichlidae). Rev. Zool. Bot. Afr., LXI(3-4): 342-348, 2 tab.

ID., I963. La distribution geographique des *Tilapia* au Congo. Bull. Acad. Roy. Sci. Outre-mer, 3: 570-605, 2 tab., 6 fig.

ID., 1964. Revision systematique des especes congolaises du genre *Tilapia* (Pisces Cichlidae). Ann. Mus. Roy. Afr. Centr., Zool., 124: 1-155, 24 fig., 11 pl.

ID., 1968. An annotated bibliography of *Tilapia* (Pisces Cichlidae). Ann. Mus. Roy. Afr. Centr., Docum. Zool., 14: 405 pp.

TREWAVAS, E., 1946. The types of African cichlid fishes described by Borodin in 1931 and 1936 and on two species described by Boulenger in 1901. Proc. Zool. Soc. London, 116(II): 240-246.

ID., 1953. A new species of the cichlid genus *Limnochromis* from Lake Tanganyika. Bull. Int. Roy. Sci. Nat. Belgique, 29(6): 1-3, fig. I.

ID., 1973. On the cichlid fishes of the genus *Pelmatochromis*, with proposal of a new genus for *P. congicus*: on the relationship between *Pelmatochromis* and *Tilapia* and the recognition of *Sarotherodon* as a distinct genus. Bull. Br. Mus. (Nat. Hist.), Zool., 25(1): 3-26.

ID., 1981. Addendum to *Tilapia* and *Sarotherodon*? Buntbarsche Bull., 87.

ID., 1981. Nomenclature of the tilapias of southern Africa. J. Limnol. Soc. South Afr., 7(1): 42.

ID., 1983. Tilapiine Fishes of the Genera *Sarotherodon, Oreochromis* and *Danakilia*. British Mus. (Nat. Hist.), London. 583 pp., 188 fig., 119 tab.

WICKLER, W., 1962. Ei-Attrappen und Maulbruten bei Afrikanischen Cichliden. Zeitschr. fur Tierphysiologie, 19(2): 129-164, 17 fig.

ID., 1963. Naturliche Augen und Ei-Attrappen an Fischen : Innerartliche Mimikry als Sonderfunktion der Korperfarbung. Veroffentlichungen des Instit. fur Meeres Forschung in Bremerhaven. Sonderband 1963. Drittes meeresbiologisches Symposium: 222-227.

WHITLEY, G. P., 1929. Studies in Ichthyology, nr. 3. Rec. Austr. Mus., 17(3): 101-143.

VAILLANT, 1899. *Ectodus foae*. Bull. Mus. Paris: 221.

YAMAOKA, K., 1983. A revision of the cichlid fish *Petrochromis* from Lake Tanganyika, with description of a new species. Japanese J. Ichthyology, 30(2): 129-141, fig. 1-7.

APPENDIX

Diagnoses of new species related to *Lamprologus brichardi*

Our 1984 explorations of the western coastline of Lake Tanganyika between Kalemie and Moliro (6° to 8° south), although much too short, yielded about a dozen as yet unidentified and probably undescribed species. Of special interest is a group of several *Lamprologus* similar to the well-known *Lamprologus brichardi* Poll. They have in common a very crescentic caudal fin, a feature that is quite uncommon among the lacustrine species of the genus; the unpaired fins are filamentous; and they share several important proportions and counts.

Five of these new species are described here. All are included in the key to *Lamprologus*, and their proportions and counts are summarized at the end of each species discussion. Type specimens are currently in the author's possession and will be deposited shortly. Also included here are brief comparative discussions of *L. brichardi* and *L. pulcher*.

Lamprologus brichardi Poll, 1974

This species is widespread in the northern half of the Lake, exhibiting only minor differences among local populations. It is sympatric with *L. falcicula* on the Magara coast of Burundi. It is more local in the southern half of the Lake, where it is known from the Kungwe Mountains area on the eastern coast and from Moba to Cape Kapampa on the western coast. At Kapampa it is sympartric with *L. gracilis* n. sp., while at Cape Zongwe it is sympatric with *L. splendens* n. sp. This is a strongly gregarious species, as discussed in the main text. The maximum total length is about 8 cm.

The body is beige in color, the underparts off-white. Some populations have the scales of the sides rimmed with orange, especially on the caudal peduncle. The jet black opercular spot is connected to a metallic orange stripe extending from the eye through the opercular spot and then sloping downward. All the unpaired fins have strong filaments and have a wavy pattern of yellow-orange. The tail fin is crescentic, with very long creamy white filaments. All the unpaired fins are edged in white. The cheek is beige with wavy sky-blue lines.

The scales at the bases of the dorsal and anal fins form sheaths raised above the outline of the body. The gill-rakers vary from 7 to 14. There are 10.5 to 12.5 scales in a lower transverse row (counted diagonally up from the first anal spine).

Proportions and Meristics: Body height 29-33% SL (Standard Length); head length 31-34% SL; dorsal fin origin to mouth 30-34% SL; dorsal fin base 55-59% SL. Eye diameter 23-26% HL (Head Length); interorbital width 23-28% HL; lower jaw 38-43% HL. Scales in longitudinal line (counted in a zigzag) 34-36; scales in upper lateral line 18-22; scales in lower lateral line 8-11; scales in lower transverse row (from first anal spine diagonally upward) 10.5-12.5. Dorsal fin XVIII-XIX,8-9; anal fin V-VI,5-7. Gill-rakers 7-14. Scales absent from cheek, preopercle, and opercle; many scales on unpaired fins; basal sheaths strong.

Lamprologus pulcher Poll, 1949

This species appears to be restricted to the southernmost part of the Lake, where it is allopatric to all other related species. The maximum total length is about 8 cm.

The color is much like that of *Lamprologus brichardi*, but the orange scale edgings of the sides are more vivid and the cheek is a dark blue-green with a pattern of wavy light blue stripes. The postorbital orange stripe is broken, each of the two sections slanting downward, the first in front of the black opercular spot, the second behind it.

There are 8 to 10 gill-rakers and 9.5 scales in a lower transverse row, one to three scales fewer than in *L. brichardi*.

Proportions and Meristics: Body height 31-36% SL; head length 31-33% SL; dorsal fin origin to mouth 33% SL; dorsal fin base 54-60% SL. Eye diameter 24-27% HL; interorbital width 25-29% HL; lower jaw 40-48% HL. Scales in longitudinal line 32-34;

scales in upper lateral line 20; scales in lower lateral line 6; scales in lower transverse row 9.5. Dorsal fin XVIII-XIX,8-9; anal fin V-VIII,6. Gill-rakers 8-10. Scales absent from cheek, preopercle, and opercle; many scales on unpaired fins; strong basal sheaths.

Lamprologus OLIVACEOUS Brichard, n. sp.

This is the smallest species of the group, seldom exceeding 5.5 cm in overall length. It is similar to *L. pulcher* but differs from all the species of the group by having only 8.5 scales in a lower transverse row and only 7 gill-rakers. There are a few scales on the opercles, very few on the unpaired fins, and the sheaths at the bases of the dorsal and anal fins are not much raised. The scales on the sides appear very high and large for the size of the fish, an effect caused by the low number of scales in a lower transverse row.

The color is pale greenish beige, the scales of the sides outlined with faint brown rims. There is no black opercular spot, and the postorbital brown stripe is faint. The filaments on the unpaired fins are long and white in color.

L. olivaceous is found in and around the Bay of Luhanga on the western coast of the Lake, where it is sympatric with *L. crassus* n. sp. It is moderately gregarious, usually a few fish being found together. The maximum total length is about 5.5 to 6 cm.

Proportions and Meristics: Body height 29-31% SL; head length 33-37% SL; dorsal fin origin to mouth 33-34% SL; dorsal fin base 54-61% SL.

Eye diameter 25-28% HL; interorbital width 22-25% HL; lower jaw 37-45% HL. Scales in longitudinal line 32-35; scales in upper lateral line 15-19; scales in lower lateral line 3-8; scales in lower transverse row 8.5. Dorsal fin XVIII-XIX,8-9; anal fin V-VI,5-6. Gill-rakers 7. Scales absent from cheek and preopercle, present on opercle; unpaired fins with few scales; basal sheaths strong.

Lamprologus CRASSUS Brichard, n. sp.

This uniformly slate-blue species with a sky-blue eye is distinct from the other members of the group in the relatively deeper and more plump body, the lack of a color pattern on the head, the blue rims of the unpaired fins, and the rather short blue-white filaments on the unpaired fins. There are few scales on the unpaired fins, and the protective sheaths at the bases of the dorsal and anal fins are poorly developed. There are 9 or 10 gill-rakers.

L. crassus is found in and around the Bay of Luhanga on the western coast of the Lake, where it is sympatric with *L. olivaceous* n. sp. It usually lives in pairs and has a maximum size of about 7 cm total length.

Proportions and Meristics: Body height 30-35% SL; head length 33-34% SL; dorsal fin origin to mouth 32-33% SL; dorsal fin base 56-59% SL. Eye diameter 23-25% HL; interorbital width 26-29% HL; lower jaw 41-43% HL. Scales in longitudinal line 33-36; scales in upper lateral line 19-23; scales in lower lateral line 6-7; scales in lower transverse row 10.5-11.5. Dorsal fin XIX-XX,8; anal fin VI,6. Gill-rakers

9-10. Scales absent from cheek and preopercle, present on opercle; few scales on unpaired fins; basal sheaths weak.

Lamprologus SPLENDENS Brichard, n. sp.

In this species the length of the head is 36-39 percent of the standard length (SL), longer than in the other species (except some *L. falcicula* n. sp.). This measurement is also reflected in the longer distance from the first spine of the dorsal fin to the mouth. The number of tubulated scales in the upper lateral line, 18-26, is relatively high. The preopercle is scaly, the opercle very scaly, there are many scales on the fins, and the sheaths at the bases of the unpaired fins are strong.

The coloration of *L. splendens* is striking: The body and fins are dark purple or black, with a metallic crimson opercular spot. The unpaired fins have long filaments and are rimmed with sky-blue, including the filaments. The caudal fin is strongly concave, rimmed in the middle with a lemon-yellow crescent.

This is a cave-dweller found near Cape Zongwe. It is gregarious in small schools. Maximum total length is about 7-8 cm.

Proportions and Meristics: Body height 31-34% SL; head length 36-39% SL; dorsal fin origin to mouth 36-38% SL; dorsal fin base 55-60% SL. Eye diameter 20-25% HL; interorbital width 22-27% HL; lower jaw 41-47% HL. Scales in longitudinal line 33-36; scales in upper lateral line 18-26; scales in lower lateral line 6-10; scales in lower transverse row 10.5-11.5.

Dorsal fin XVIII-XIX,8-9; anal fin V-VI,6-7. Gill-rakers 8-9. Scales absent from cheek, many on preopercle and opercle; many scales on unpaired fins; basal sheaths strong.

Lamprologus GRACILIS Brichard, n. sp.

This relatively elongated and colorful species has a shallow body, only 25-28 percent of SL (29-36 percent in the other species). There are 21-26 tubulated scales in the upper lateral line and 11.5-13.5 scales in a lower transverse row. The gill-rakers are very numerous, 16 to 18, distinctly above the usual counts in related species. The unpaired fins are heavily scaled, and the sheaths at the bases of the dorsal and anal fins are strong and high.

The fish is beige, the underparts lighter, with a pattern of distinct white dots on the sides. The unpaired fins are checked with white dots extending onto the caudal peduncle. The filaments of the unpaired fins are long. There is a black opercular spot.

This distinctive species is found along with L. brichardi at Cape Kapampa. It lives singly or in small groups and is relatively large, reaching 8 to 9 cm in total length.

Proportions and Meristics: Body height 25-28% SL; head length 31-33% SL; dorsal fin origin to mouth 32-36% SL; dorsal fin base 55-61% SL. Eye diameter 22-25% HL; interorbital width 22-27% HL; lower jaw 38-47% HL. Scales in longitudinal line 33-36; scales in upper lateral line 21-26; scales in lower lateral line 7-13; scales in lower transverse row 11.5-13.5. Dorsal fin XIX-XX,7-8; anal fin VI,5-7. Gill-rakers 16-18. Scales absent from cheek and preopercle, present on opercle; unpaired fins with many scales; basal sheaths strong.

Lamprologus FALCICULA Brichard, n. sp.

This species (discussed in more detail in the main text) is similar to Lamprologus brichardi in shape and has an exaggerated crescentic caudal fin. The entire body is drab beige, including the unpaired fins, which lack brighter markings. There is no postorbital stripe. The filaments of the caudal fin are not well developed.

Sympatric with L. brichardi on the Magara coast of Burundi, it is a territorial, aggressive species that seldom leaves the floor and doesn't form schools.

Proportions and Meristics: Body height 29-33% SL; head length 33-38% SL; dorsal fin origin to mouth 31-34% SL; dorsal fin base 57-64% SL. Eye diameter 22-25% HL; interorbital width 23-30% HL; lower jaw 38-45% HL. Scales in longitudinal line 31-34; scales in upper lateral line 19-25; scales in lower lateral line 4-9; scales in lower transverse row 10.5. Dorsal fin XIX,8-9; anal fin V-VI,5-7. Gill-rakers 8-9. Scales absent from cheek, preopercle, and opercle. Many scales on unpaired fins; basal sheaths strong.

APPENDIX II: RECENT OBSERVATIONS

Page 482: *Synodontis multipunctatus*

The following notes and the discussion of the phenomenon deal with an unusual aspect of predation that perhaps may have been better inserted in the chapter dealing with predation and defense.

As I mentioned before, we have found at times young fry of S. multipunctatus in the mouth of the mouthbrooding cichlids. These occurrences were rare and didn't involve many fish. Perhaps 20 Synodontis fry were found over the years in the mouth of Tropheus, Simochromis, Cyphotilapia, and especially Ophthalmotilapia. Dr. Sato in turn started a systematic study of several cichlid species to discover how often these fish had Synodontis fry or eggs in their mouth and what was the reason for such behavior. He reached the conclusion that the catfish laid their eggs in the cichlid's mouth, where they developed, hatched, and started to eat the host's spawn. He called this behavior an obligatory behavior of the catfish, which means that they are unable to raise their spawn unless it is sheltered in the cichlid's mouth.

I must admit to having strong reservations regarding such an obligatory behavior on the part of the catfish for several reasons:

- Synodontis multipunctatus is an exceedingly common fish. On the 20 km of coastline between Kitasa and Magara, the main rocky coast in Burundi waters, there are probably more than 100,000 of the catfish. We had, over the years, about 20 of their fry found in cichlid mouths. Across the lake where Dr. Sato studied these fishes he had about 100 fry found in cichlids. Very few fry, indeed, for such a big population of adults.

Were all Synodontis born and raised in a cichlid mouth we would have found over the years thousands upon thousands of fry in the mouths of mouthbrooding cichlids, more so because the foster parents belong to species that are collected in large numbers and exported as ornamental fishes.

On the contrary, to find a Synodontis fry in the mouth of a cichlid mouthbrooder is quite accidental. Were the catfish to put their spawn into the cichlid's mouth systematically one would have to admit that the spawns suffer very high losses, as one finds so few fry being brooded by a cichlid, the opposite result from what would be expected with this behavior.

- Were in fact the cuckoo spawns highly efficient and systematic, several species of cichlids would be endangered if the fry of the catfish fed on the host eggs while in the mouth and the cichlid spawns were destroyed.

One has to admit that *Synodontis* might introduce their eggs into the foster parent's mouth, but it is very hard to believe that the survival of *Synodontis multipunctatus* depends on this technique.

- The main problem is to understand how this can be done with fishes like *Ophthalmotilapia* and *Cyphotilapia*, which live in midwater and not on the bottom, while the catfish live on the bottom and never go into midwater. One has to accept that the insertion of the catfish's eggs can happen either when a pair of cichlids comes to the bottom for a spawning session or
at night when they come down to rest among the rubble.

It has been said that in a tank in which cichlids lived with a pair of *S. multipunctatus*, the latter started to get excited and spawned after a pair of cichlids had a breeding session.

Although thousands of cichlids have been observed spawning in the lake, we have never seen a single catfish nearby and interested in what the cichlids did. Thus in the wild it doesn't appear that cichlid spawning behaviors have anything to do as a trigger for the catfish's spawns. But in a tank one can speculate that the pheromones emitted by a mating pair of cichlids can be sensed by the catfish and be a strong sexual stimulus for them as well.

- Catfish are awake at night and forage while the cichlids are in a deep slumber and rest on the bottom. Thus it might happen that a wandering *Synodontis multipunctatus* comes near a mouthbrooding cichlid and inserts some eggs into the sleeper's mouth, or leaves them nearby so that when the cichlid wakes up she thinks she has let some of her eggs go and picks them up. There is only one hitch to this nicely laid out screenplay. A female *Synodontis* has several hundred eggs to lay and she can only insert a few into each cichlid. She will thus need at least 100 cichlids to brood her entire spawn. A large, fully grown *S. multipunctatus* will carry more than 1,000 eggs, and, as I have said, there are thousands upon thousands of the catfish wanting to spawn.

If most eggs do not get into a cichlid's mouth and get lost, one might say that the obligatory spawning system adopted by the catfish is indeed a very bad one.

Remaining within the realm of possibility, we might envision that *Synodontis* lay some eggs (or all their eggs) near mouthbrooding cichlids, to which they are attracted by the smell of the eggs in the mouth of the sleeping mouthbrooder. The cichlid waking up in the morning mistakes the eggs on the ground for some of her own and picks them up. Perhaps some

of the catfish eggs hatch and the fry then develop inside the mouth and start nibbling at the host's own eggs. Perhaps some young fry foraging on their own are similarly attracted to the jaws of the mouthbrooder by the smell of the eggs and enter the mouth?

As the matter now stands the evidence appears to be weighed in favor of accidental brooding of *Synodontis multipunctatus* fry by cichlid mouthbrooders and not a systematic do-or-die cuckoo spawn by the catfish.

Page 50, line 35
A smaller species, barely 10mm long, parasitizes sand-dwelling fishes like *Reganochromis*.

Page 202, col. 1, par. 3
Lamprologus callipterus schools are exclusively males. Females, much smaller, remain separate from the males and wander much less.

Page 234, col. 3, par. 3
Ophthalmotilapia nasutus females enter the nest while the male hovers over the rim of the nest and circles above projecting sperm all over the nest. Then he stands still, turning his back to the female, his tassels lying on the ground. She lays a couple of eggs, takes them into her mouth, and pulls at the tassels, a signal for the unseeing male to deposit more sperm. She doesn't mistake the tassels for one of the eggs. *Cunningtonia* males do not have tassels at the fine tips of the pelvic fins that could be mistaken for eggs—but the female still pulls and mouths the tips of the male's pelvic fins.

Page 266, col. 1, par. 1
But altogether *Tropheus, Eretmodus, Simochromis,* and *Petrochromis* bear the brunt of the attacks.

Page 286: *Chalinochromis brichardi*
Species of *Chalinochromis* spawn in the rubble rather than in deep caves. Most often they just spawn in an anfractuosity. One day as I was checking on a spawn with fry about 15 mm long I discovered that some of them were pecking at one of the parent's sides. I didn't believe at first that the fry were feeding discus-style on mucus from the parent's skin, but I had to admit that this was the case as more and more fry came out from hiding and started to nibble in earnest from their parent's side. I cannot from this single sighting state that all *Chalinochromis* fry feed as discus fry normally do, or if they start early, or if they at first also complement this type of feeding by browsing on the biocover. One would certainly like to learn more about this unexpected behavior, which has as yet not been reported for any other African cichlids.

Page 289, col. 2, par. 1
A new subrace has recently been discovered in the north, *Cyathopharynx furcifer* "Ruziba." This race has the head striped with bright blue and the unpaired fins tipped with a broad yellow patch.

Page 293, col. 3, par. 1
A strong solution of iodine is the best disinfecting drug to clear wounds that otherwise might fester.

Page 298, col. 1, par. 4
Two *Cyprichromis* species, as yet unidentified, have been found in the Moba area of Zaire. One is entirely metallic orange, the second, much smaller, has two black stripes extending the full length of the dorsal fin on a pure white background.

Page 320, col. 3, par. 1
Besides the two populations found respectively in the south (Mpulungu) and the north of the Lake, a huge habitat extending from Luhangwa Bay and Mtoto on the West Coast was found by our team in 1984. *Julidochromis ornatus* there display several distinct pattern features.

Page 333, col. 1, par. 1, line 4
But this is exceptional, as *Lamprologus brichardi* feed ravenously on freshly hatched fry of many fishes including those of *Boulengerochromis* and even of the scale-ripper *Perissodus microlepis*.

Page 337, col. 1, line 5
The females are barely half the size of the males.

Page 340, col. 1, par. 3
Our exploration in 1984 of the Zaire coast discovered many populations of *L. caudopunctatus* that progressively led to *L. leloupi* Poll. *L. caudopunctatus* is thus simply a local race of *L. leloupi* and not a valid species.

Page 348, col. 1, par. 3
More races have been found at Luhangwa Bay and Moba on the southwestern coast. *L. leleupi* is also common around the fifth parallel south on the West Coast.

Page 371, col. 2, par. 3
Lamprologus tretocephalus on the West Coast reaches Mtoto, just north of Moba.

Page 388, col. 1, bottom
Kachese southern race: metallic orange all over.

Page 412, col. 3, line 2
They are pedophilic predators and thus attack eggs and young fry.

Page 413, col. 1, par. 2
As aquarium specimens they have one shortcoming–they are ravenous, specialized egg-snatchers.

Page 422, col. 1, par. 2
We managed to catch a few late in 1988 and got them to breed in a pond, but then the fish died.

Index

Page numbers referring to color photographs are printed in **bold**.

539

543